P9-ELH-699

Macmillan Encyclopedia of Physics

MACMILLAN ENCYCLOPEDIA OF PHYSICS

John S. Rigden

Editor in Chief

Volume 1

MACMILLAN REFERENCE USA
Simon & Schuster Macmillan
NEW YORK

Simon & Schuster and Prentice Hall International
LONDON   MEXICO CITY   NEW DELHI   SINGAPORE   SYDNEY   TORONTO

Simon & Schuster Macmillan
1633 Broadway
New York, NY 10019

Library of Congress Catalog Card Number: 96-30977

PRINTED IN THE UNITED STATES OF AMERICA

Printing Number
1   2   3   4   5   6   7   8   9   10

LIBRARY OF CONGRESS CATALOGING-IN-PUBLICATION DATA

Macmillan Encyclopedia of Physics / John S. Rigden, editor in chief.
      p.   cm.
   Includes bibliographical references and index.
   ISBN 0-02-897359-3 (set).
   1. Physics—Encyclopedias.   I. Rigden, John S.
   QC5.M15   1996
   530′.03—dc20   96-30977
                                                         CIP

This paper meets the requirements of ANSI-NISO Z39.48-1992
(Permanence of Paper).

CONTENTS

Editorial and Production Staff

Brian Kinsey
Project Editor

David K. Ekroth
Executive Editor

Debra H. Alpern
Dafna Gorfinkle
Editorial Assistants

Jonathan G. Aretakis   Christopher Curioli   Cynthia Klingensmith
Mary Ellen Mormile   Robert J. Reilly   Todd C. Reiss
Ingrid Sterner   Beth Wilson
Copy Editors

Bill Drennan
Suzanne Martinucci
Greg Teague
Helen Wallace
Proofreaders

AEIOU, Inc.
Indexer

G & H Soho, Inc.
Art Studio

Graphic Sciences Corp.
Compositor

Rose Cappozelli
Production Manager

MACMILLAN REFERENCE
Elly Dickason, *Publisher*
Paul Bernabeo, *Editor in Chief*

PREFACE

We build monuments to our national heros: former presidents, military veterans, and athletes. An encyclopedia is a monument to the largely invisible community of scholars who diligently and privately work to expand the boundaries of our knowledge. The *Macmillan Encyclopedia of Physics* is a monument to a particular group of scholars called physicists.

I like to think of the natural sciences as a dialogue between a scientist and Nature. The character of the dialogue depends on the maturity of the science. The science of physics is a very mature science. Over the past three centuries, powerful and sweeping theories have been developed in terms of which very specific and probing questions, by means of carefully designed instruments, can be put to Nature. Hence, the dialogue between physicists and Nature is highly formalized. Through this dialogue, theories are tested; the answers Nature provides bring new insights into how the physical world works.

The twentieth century, now coming to an end, has been a remarkable testimony to the efficacy of the physicist's dialogue with Nature: the ability of the human intellect to conceive probing questions and to interpret subtle answers. These answers have pushed the boundaries of understanding. During this century, our knowledge of the universe we inhabit has reconciled the stars shining in the night sky with the fusion of unseen atomic nuclei; we now understand where and how the elements that make up the material world, including the bodies of living organisms, originate. The physics of the twentieth century has brought structure to the atom and the molecules that come together to adorn our natural environment.

The view is often expressed that science advances by discounting or disproving what previously had been accepted as true. This is largely a mistaken view. As physical knowledge increases, it is not that old ideas are found to be false; rather, it is that old ideas are found to be limited in their applicability. This typically happens as new instruments permit physicists to surpass previous limitations: going faster, seeing further and with greater resolution, and observing things smaller. As new domains are explored, old ideas must sometimes be adapted for the new domain; in the old domain, the old ideas are typically just as valid as before.

Physics has its roots in antiquity; however, physics, as we know it, began in the seventeenth century with the work of Johannes Kepler, Galileo Galilei, and Issac Newton. The eighteenth and nineteenth centuries built on this beginning and prepared the way for the outburst of the accomplishments in basic science and technology during the twentieth century. The *Macmillan Encyclopedia of Physics* is about the subject of physics from the insights of the past to the present.

Physics captivates physicists, but there are things about the subject that fascinate many nonphysicists as well. The laws of physics, paradigms of certainty, allow the precise prediction of future events and bring the comforting sense that all is right with the world. The basic concepts of physics, few in number, are provocative and powerful. In the direction of small things, these concepts extend from our living rooms to the inner workings of the atom. In the other direction, these same basic concepts extend from our front yards to the edge of the observable universe, where atoms congeal into the stars and galaxies. The theories of physics apply to the stars with lifetimes of billions of years just as they apply to particles with lifetimes much less than a trillionth of a second. Quantum mechanics, the language of atoms, fires the imagination.

While the language of physics is mathematics, many entries herein have been written without the use of mathematical symbols. However, some ideas of physics, by their vary nature, are so dependent on the logic of mathematics for their expression that symbols and mathematics must enter into their presentation. A table of symbols and abbreviations is included at the beginning of each volume to assist readers unfamiliar with any mathematical or scientific notation that might arise.

A diverse range of topics is included in this encyclopedia. The principal reason for this is that the *Macmillan Encyclopedia of Physics* is intended for a broad range of readers. The various subdisciplines and branches of physics that provide career paths are included to give students some knowledge about the substance of the various areas of basic and applied physical research. There are entries about the history of physics focusing on particular concepts and theories. And there are biographical entries that feature a select group of physicists who have made seminal contributions to our understanding of the physical world. The number of physicists whose work clearly deserves recognition is large. Therefore, no living physicist whose work is still in progress has been included. For students and teachers there are topics that, by their nature, can stimulate animated discussion. Of course, these same topics can activate the imagination of any reader. There are also entries dealing with the physics of everyday phenomena, such as white clouds, tornadoes, and the yearly seasons.

The entries in the *Macmillan Encyclopedia of Physics* are arranged in alphabetical order. To provide additional guidance to users of this reference work, especially for teaching purposes, I have organized a reader's guide, which follows this preface, to identify some of the key entries in the encyclopedia's plan. Cross-references appearing at the end of each encyclopedia entry are designed to guide users further in finding related discussions. Some concepts, important in themselves, appear in a more succinct form in the Glossary, located at the end of volume 4. Finally, a general index, also at the end of volume 4, provides thorough guidance to subjects covered herein.

Many individuals have contributed to this encyclopedia. The editors, Jim McGuire, Helen Quinn, David Schramm, Roger Stuewer, and Carl Tomizuka, carried a major responsibility, and they did so with all the professionalism one would expect. There are 516 authors who have contributed to this work; without these authors, the encyclopedia could not have taken form. I extend my personal word of thanks to the editors and authors. Finally, the staff at Macmillan Reference have been outstanding. Elly Dickason, as Publisher at Macmillan, helped guide the project, and Brian Kinsey, also at Macmillan, managed the logistics of the project on a day-to-day basis. I thank them both.

I end on a personal note. My wife, Diana Wyllie Rigden, spent many evenings alone while I was in my study engaged in the various activities that fell to me. I thank her for her patience.

JOHN S. RIGDEN

READER'S GUIDE

Conductor
Current, Alternating
Current, Direct
Electrical Conductivity
Electrical Resistance
Electrical Resistivity
Electromotive Force
Inductance
Inductance, Mutual
Inductor
Insulator
Resistor
Self-Inductance
Semiconductor
Short Circuit

Fundamental Constants
Avogadro Number
Boltzmann Constant
Charge, Electronic
Cosmological Constant
Dielectric Constant
Fine-Structure Constant
Gas Constant
Gravitational Constant
Hubble Constant
Light, Speed of
Mathematical Constants
Particle Mass
Planck Constant
Rydberg Constant

Fundamental Theories
Big Bang Theory
Electromagnetism
Grand Unified Theory
Newtonian Mechanics
Quantum Mechanics and Quantum Field Theory
Relativity, General Theory of
Relativity, Special Theory of
Thermodynamics

Historical Topics
Accelerator, History of
Atomic Theory, Origins of
Copernican Revolution
Cosmic Microwave Background Radiation,
    Discovery of
Electromagnetism, Discovery of
Electron, Discovery of
Laser, Discovery of
Neutrino, History of

Neutron, Discovery of
Newtonian Synthesis
Physics, History of
Quantum Mechanics, Creation of
Quantum Theory, Origins of
Radioactivity, Discovery of
Relativity, General Theory of, Origins of
Relativity, Special Theory of, Origins of
Scientific Revolution
Telescope, History of
Thermodynamics, History of
Transistor, Discovery of
Universe, Expansion of, Discovery of
Wave–Particle Duality, History of
X Ray, Discovery of

Laws
Ampère's Law
Biot–Savart Law
Boyle's Law
Bragg's Law
Brewster's Law
Charles's Law
Conservation Laws
Coulomb's Law
Electromagnetic Induction, Faraday's Law of
Gauss's Law
Gravitational Force Law
Hooke's Law
Ideal Gas Law
Inverse Square Law
Kepler's Laws
Kirchhoff's Laws
Lenz's Law
Newton's Laws
Ohm's Law
Snell's Law
Stefan–Boltzmann Law

Observed Phenomena
Aurora
Avalanche
Cloud, Gray
Cloud, White
Color
Doppler Effect
Earthquake
Eclipse
Friction
Hurricane
Lightning
Moon, Phases of

Rainbow
Reflection
Refraction
Seasons
Sky, Color of
Sound
Thunder
Tides
Tornado

Particles
Atom
Boson, Gauge
Electron
Elementary Particles
Excitation, Collective
Fermions and Bosons
Hadron
Lepton
Magnon
Molecule
Neutron
Nucleon
Phonon
Photon
Plasmon
Polaron
Proton
Quark
Soliton

Principles
Archimedes' Principle
Bernoulli's Principle
Complementarity Principle
Correspondence Principle
Cosmological Principle
Equivalence Principle
Franck–Condon Principle
Least-Action Principle
Mach's Principle
Pascal's Principle
Pauli's Exclusion Principle
Uncertainty Principle

Space
Advanced X-Ray Astrophysics Facility
Astrophysics
Astrophysics, X-Ray
Big Bang Theory
Black Hole
Black Hole, Kerr

Black Hole, Schwarzschild
Cosmic Background Explorer Satellite
Cosmic Microwave Background Radiation
Cosmic Ray
Cosmic String
Cosmological Constant
Cosmological Principle
Cosmology
Cosmology, Inflationary
Einstein Observatory
Galaxies and Galactic Structure
Gamma Ray Bursters
Gamma Ray Observatory
Interstellar and Intergalactic Medium
Magnetosphere
Neutrino, Solar
Neutron Star
Nova
Nucleosynthesis
Planetary Magnetism
Planetary Systems
Planet Formation
Pulsar
Pulsar, Binary
Quasar
Red Giant
Redshift
Solar System
Solar Wind
Stars and Stellar Structure
Supernova
Uhuru Satellite
Universe
Universe, Expansion of
X-Ray Binary

States of Matter
Gas
Liquid
Plasma
Solid

Subdisciplines
Acoustics
Astrophysics
Atmospheric Physics
Atomic Physics
Basic, Applied, and Industrial Physics
Biophysics
Celestial Mechanics
Chemical Physics
Computational Physics

LIST OF ARTICLES

Free Body Diagram
P. L. Altick

Free Fall
Thomas A. Moore

Freezing Point
Robert P. Bauman

Frequency, Natural
George Rosensteel

Fresnel, Augustin-Jean
Robert H. Silliman

Friction
Robert P. Bauman

Fusion
E. Michael Campbell

Fusion Bomb
Stanley Goldberg

Fusion Power
Stewart J. Zweben

G

Galaxies and Galactic Structure
Marianne Takamiya

Galilean Transformation
David Halliday

Galilei, Galileo
William R. Shea

Galvanometer
Dennis Barnaal

Gamma Ray Bursters
Angela V. Olinto

Gamma Ray Observatory
Jean M. Quashnock

Gas
Stefan Machlup

Gas Constant
Richard H. Dittman

Gauge Invariance
Stephen L. Adler

Gauge Theories
Stephen L. Adler

Gauss's Law
James L. Monroe

Geiger Counter
William R. Wharton

Geode
William G. Melson

Geophysics
G. N. Rassam

g-factor
Sankoorikal L. Varghese

Gibbs, Josiah Willard
Elizabeth Garber

Grand Unified Theory
Howard Georgi

Grating, Diffraction
Sumner P. Davis

Grating, Transmission
Jefferson L. Shinpaugh

Gravitational Assist
Roger E. Diehl

Gravitational Attraction
Thomas A. Moore

Gravitational Constant
Andrew Strominger

Gravitational Force Law
Sung Kyu Kim

Gravitational Lensing
J. N. Hewitt

Gravitational Wave
Richard H. Price

Rutherford, Ernest
Lawrence Badash

Rydberg Constant
Stanley Bashkin

S

Sakharov, Andrei Dmitrievich
Andrew M. Sessler

Scalar
Ian R. Gatland

Scale and Structure
Helen R. Quinn

Scanning Probe Microscopies
Christine Orme
Bradford G. Orr

Scanning Tunneling Microscope
Barbara H. Cooper

Scattering
John Reading

Scattering, Light
Corinne Vinches

Scattering, Raman
Alvin D. Compaan

Scattering, Rayleigh
John T. Park

Scattering, Rutherford
Robert L. Stearns

Schrödinger, Erwin
Walter Moore

Schrödinger Equation
Ronald J. W. Henry

Scientific Method
Allan Franklin

Scientific Notation
Robert C. Hilborn

Scientific Revolution
Richard S. Westfall

Scintillation Counter
Howard Nicholson

Seasons
Charles J. Peterson

Seismic Wave
T. C. Wallace

Seismology
T. C. Wallace

Selection Rule
Linda Young

Self-Inductance
David A. Dobson

Semiconductor
J. Spector

Servomechanism
Carol Zwick Rosen

Shock Wave
Steven N. Shore

Short Circuit
Giulio Venezian

Signal
William Bickel

Signal-to-Noise Ratio
William Bickel

Significant Figures
Thomas J. Loredo

SI Units
Mario Iona

Skin Effect
F. R. Yeatts

Sky, Color of
Gary Waldman

LIST OF CONTRIBUTORS

Isaac D. Abella
University of Chicago
Radar
Radio Wave

S. C. Abrahams
Southern Oregon State College
Ferroelasticity
Pyroelectricity

Robert K. Adair
Yale University
Barrier Penetration
Motion, Brownian

Gregory S. Adkins
Franklin and Marshall College
Mathematical Constants

Carl G. Adler
East Carolina University
Event
Light, Speed of

Stephen L. Adler
Institute for Advanced Study, Princeton, NJ
Gauge Invariance
Gauge Theories

David C. Ailion
University of Utah
Nuclear Magnetic Resonance

L. A. Akers
Arizona State University
Electronics

Ralph W. Alexander Jr.
University of Missouri, Rolla
Birefringence
Refraction

Marcelo Alonso
Florida Institute of Technology (retired)
Electricity
Electrostatic Attraction and Repulsion

P. L. Altick
University of Nevada, Reno
Free Body Diagram

Jeeva Anandan
Max Planck Institute for Physics, Munich, Germany,
and *University of Oxford, England*
Locality
Quantum Theory of Measurement

David L. Anderson
Oberlin College
Cathode Ray
Electron
Electron, Discovery of

William Anderson
University of Virginia
Spectroscopy, Laser

Allen Apblett
Tulane University
Polymer

Thomas Archibald
Acadia University, Canada
Light, Electromagnetic Theory of

John Askill
Millikin University
Infrasonics
Ultrasonics

Gordon J. Aubrecht II
Ohio State University
Cyclotron
Electrolytic Cell
Lawrence, Ernest Orlando
Nucleon
Wave Motion

Kevin Aylesworth
Aylesworth Technical Services, Cambridge, MA
Electroluminescence
Electron, Conduction
Ferrimagnetism
Holes in Solids
Photoconductivity

James F. Babb
Harvard-Smithsonian Center for Astrophysics
Einstein–Podolsky–Rosen Experiment

Lawrence Badash
University of California, Santa Barbara
Radioactivity, Discovery of
Rutherford, Ernest

Marie Baehr
Elmhurst College
Mass

A. D. Baer
China Lake, CA
Emission, Thermionic
Work Function

Ralph Baierlein
Wesleyan University, Middletown, CT
Negative Absolute Temperature

Yehuda B. Band
Ben-Gurion University, Beer Sheva, Israel
Elements, Rare Earth

Grace A. Banks
Chestnut Hill College
Coaxial Cable
Transformer

Anders Bárány
University of Stockholm, Sweden
Spectroscopy, Atomic

William A. Barletta
Lawrence Berkeley National Laboratory, Berkeley, CA
Accelerator

Dennis Barnaal
Luther College
Ammeter
Galvanometer
Ohmmeter
Potentiometer
Voltmeter

R. Michael Barnett
Lawrence Berkeley National Laboratory, Berkeley, CA
Hadron
Quark

H. H. Barschall
University of Wisconsin, Madison
Nuclear Physics

D. F. Bartlett
University of Colorado, Boulder
Archimedes' Principle

Stanley Bashkin
University of Arizona (emeritus)
Rydberg Constant

Kenneth Batchelor
Brookhaven National Laboratory, Upton, NY
Radiation, Synchrotron
Synchrotron

Robert P. Bauman
University of Alabama, Birmingham
Convection
Evaporation
Freezing Point
Friction
Heat
Heat Engine
Phase, Change of
Temperature Scale, Fahrenheit
Temperature Scale, Kelvin
Water

William E. Baylis
University of Windsor, Canada
Atomic Mass Unit
Atomic Number
Clifford Algebra
Polarized Light

Benjamin F. Bayman
University of Minnesota
Liquid-Drop Model
Nuclear Moment
Nuclear Size
Nuclear Structure
Twin Paradox

Mitchell C. Begelman
University of Colorado, Boulder
Astrophysics

Robert J. Beichner
North Carolina State University
Education

Carl M. Bender
Washington University
Theoretical Physics

Birger Bergersen
University of British Colombia, Canada
Maxwell Speed Distribution
Maxwell–Boltzmann Statistics

Steven D. Bernstein
Raytheon (Research Division), Lexington, MA
Piezoelectric Effect

Robert T. Beyer
Brown University
Acoustics

Chander P. Bhalla
Kansas State University
Auger Effect

William Bickel
University of Arizona
Fourier Series and Fourier Transform
Noise
Signal
Signal-to-Noise Ratio

Nicholas P. Bigelow
University of Rochester
Atom Trap

George Bissinger
East Carolina University
Centrifugal Force
Centripetal Force
Kerr Effect

Roger D. Blandford
California Institute of Technology
Active Galactic Nucleus
Quasar

John P. Blewett
Brookhaven National Laboratory, Upton, NY
(retired)
Accelerator, History of

Mary L. Boas
DePaul University (emeritus)
Coordinate System, Cartesian
Coordinate System, Cylindrical
Coordinate System, Polar
Coordinate System, Spherical

James J. Boyle
Harvard-Smithsonian Center for Astrophysics
Davisson–Germer Experiment

Robert W. Brehme
Wake Forest University
Lorentz Force
Lorentz Transformation

B. Alex Brown
National Superconducting Cyclotron Laboratory,
Michigan State University
Decay, Alpha
Decay, Beta
Decay, Gamma
Decay, Nuclear
Nuclear Binding Energy
Nuclear Shell Model
Proton

H. R. Brown
University of Oxford, England
Quantum Theory of Measurement

Laurie M. Brown
Northwestern University
Neutrino, History of
Pauli, Wolfgang
Yukawa, Hideki

Dana A. Browne
Louisiana State University
Broglie Wavelength, de

Reinhard Bruch
University of Nevada, Reno
Ionization Chamber

Mark S. Bruno
Gateway Community Technical College
Planck, Max Karl Ernst Ludwig
Radiation, Blackbody

H. C. Bryant
University of New Mexico
Relaxation Time

S. G. Buccino
Tulane University
Precession

Manfred Bucher
California State University, Fresno
Ice
State, Equation of
Triple Point

Jed Z. Buchwald
Massachusetts Institute of Technology and *Dibner*
Institute for the History of Science and
Technology
Light, Wave Theory of

Tomasz Bulik
University of Chicago
Pulsar, Binary

Armin Bunde
Intitut für Theoretische Physik III, Giessen, Germany
Diffusion

Kieron Burke
Rutgers University, Camden Campus
Ionic Bond
Molecule

A. F. Burr
New Mexico State University
Magnet
Magnetic Flux
Magnetic Pole
Magnetization
Solenoid

C. Denise Caldwell
University of Central Florida
Faraday Effect
Interferometry
Polarized Light, Circularly

John R. Cameron
University of Wisconsin, Madison
Biophysics
Body, Physics of
Thermoluminescence

David K. Campbell
University of Illinois, Urban-Champaign
Chaos

E. Michael Campbell
Lawrence Livermore National Laboratory,
Livermore, CA
Fusion

Claude R. Canizares
Massachusetts Institute of Technology
Advanced X-Ray Astrophysics Facility

Geoffrey Cantor
University of Leeds, United Kingdom
Young, Thomas

Donald S. L. Cardwell
UMIST, Manchester, England
Thermodynamics, History of

David C. Cassidy
Hofstra University
Heisenberg, Werner Karl

Michael J. Cavagnero
University of Kentucky
Precession, Larmor

Tu-Nan Chang
University of Southern California
Bohr Magneton
Harmonics, Spherical
Motion, Harmonic
Oscillator, Harmonic

Marvin Chester
University of California, Los Angeles
Quantum Statistics
Wave Function

M. C. Chidichimo
University of Waterloo, Canada
Phase Velocity

Soshin Chikazumi
University of Tokyo, Japan (emeritus)
Magnetic Permeability

Timothy E. Chupp
University of Michigan, Ann Arbor
Photoelectric Effect

Sam J. Cipolla
Creighton University
Diffraction, Fresnel
Thin Film

Carmen Cisneros
Universidad Nacional Autonoma de Mexico, Cuernavaca
Luminescence
X Ray

C. Lewis Cocke
Kansas State University
Ion

Bernard L. Cohen
University of Pittsburgh
Energy, Nuclear
Mass Defect

Samuel A. Cohen
Princeton University
Plasma

Lawrence A. Coleman
University of Arkansas, Little Rock
Exponential Growth and Decay
Heat Pump
Order and Disorder
Pascal's Principle
Right-Hand Rule

Lee A. Collins
Los Alamos National Laboratory, Los Alamos, NM
Resonance

Alvin D. Compaan
University of Toledo
Infrared
Scattering, Raman
Solar Cell

Robert N. Compton
Oak Ridge National Laboratory, Oak Ridge, TN, and University of Tennessee
Diffraction
Fraunhofer Lines
Mass Spectrometer
Oscillation
Oscilloscope

Gayle Cook
California Polytechnic State University, San Luis Obispo
Distribution Function
Probability
Specific Heat

Barbara H. Cooper
Cornell University
Scanning Tunneling Microscope

Donald Correll
Lawrence Livermore National Laboratory, Livermore, CA
Basic, Applied, and Industrial Physics

Eric J. Cotts
State University of New York, Binghamton
Calorimetry
Heat Capacity
Thermal Expansion

John M. Cowley
Arizona State University
Bragg's Law

Elisabeth Crawford
Université Louis Pasteur, Strasbourg, France
Nobel Prize Winners

Michael D. Crisp
U.S. Department of Energy
Elements, Transuranium

Kyle Cudworth
University of Chicago
Parallax

Basil Curnutte
Kansas State University
Lens, Compound
Spectroscopy, Mass

Lorenzo J. Curtis
University of Toledo
Double-Slit Experiment
Oscillator, Forced
Reflection

James T. Cushing
University of Notre Dame
Physics, Philosophy of

J. W. Darewych
York University, Canada
Impulse

Cary N. Davids
Argonne National Laboratory, Argonne, IL
Chain Reaction
Elements
Mass Number
Nuclear Force
Nucleus
Nucleus, Isomeric

Ronald C. Davidson
Princeton Plasma Physics Laboratory, Princeton, NJ
Tokamak

Richard S. Davis
International Bureau of Weights and Measures, Cedex, France
Metrology

Sumner P. Davis
University of California, Berkeley
Grating, Diffraction

David S. P. Dearborn
Lawrence Livermore National Laboratory, Livermore, CA
Photosphere
Red Giant
Stars and Stellar Structure
Stellar Evolution

Robert K. DeKosky
University of Kansas
Copernican Revolution
Copernicus, Nicolaus

Robert H. Dickerson
California Polytechnic State University, San Luis Obispo
Carnot Cycle
Distribution Function
Probability
Specific Heat

Ulrike Diebold
Tulane University
Electron Microscope

Roger E. Diehl
California Institute of Technology
Gravitational Assist

L. Thomas Dillman
Ohio Wesleyan University
Radiation Physics

Joseph F. Dillon Jr.
Yale University
Magnetic Domain
Magneto-Optical Effects

Michael Dine
University of California, Santa Cruz
Antimatter
Charge Conjugation
CPT Theorem

Maria Cristina Di Stefano
Truman State University
Conduction
Electrical Conductivity
Electrical Resistivity

Richard H. Dittman
University of Wisconsin, Milwaukee
    Boyle's Law
    Charles's Law
    Gas Constant
    Hooke's Law

Deborah J. Dixon
Lawrence Berkeley National Laboratory,
    Berkeley, CA
    Velocity, Angular
    Velocity, Terminal

David A. Dobson
Beloit College
    Inductance
    Inductance, Mutual
    Self-Inductance

Scott Dodelson
Fermi National Accelerator Laboratory, Batavia, IL
    Dark Matter
    Great Attractor
    Great Wall

Fred E. Domann
University of Wisconsin, Platteville
    Energy, Mechanical

Denis P. Donnelly
Siena College
    Field Lines

T. M. Donovan
Palo Alto, CA
    Emission, Thermionic
    Work Function

Persis S. Drell
Cornell University
    Quark, Bottom

Max Dresden
Stanford Linear Accelerator Center
    Particle
    Superconductivity, High-Temperature
    Wave–Particle Duality

Mildred S. Dresselhaus
Massachusetts Institute of Technology
    Condensed Matter Physics
    Materials Science

Robert Drullinger
National Institute of Standards and Technology,
    Boulder, CO
    Atomic Clock

John E. Drumheller
Montana State University
    Curie Temperature
    Hysteresis
    Magnetic Moment

Russell J. Dubisch
Siena College
    Space Travel

Douglas Duncan
University of Chicago and *Adler Planetarium*
    Hubble Space Telescope

Robert V. Duncan
University of New Mexico
    Superconducting Quantum Interference
    Device

David W. Duquette
Complete Light Solutions, Minneapolis, MN
    Nicol Prism

Dennis Ebbets
Ball Aerospace Systems Division, Boulder, CO
    Accuracy and Precision
    Pressure, Atmospheric

M. E. Eberhart
Colorado School of Mines
    Covalent Bond
    van der Waals Force

David L. Ederer
Tulane University
    Optics, Geometrical
    Spectroscopy, X-Ray

Alan K. Edwards
University of Georgia
    Energy
    Energy, Kinetic

Leonard Eisenbud
State University of New York, Stony Brook
    (emeritus)
    Newton's Laws

Michael Elbaum
Weizmann Institute of Science, Rehovot, Israel
Phase Rule

John Ellis
*European Organization for Nuclear Research
(CERN), Geneva, Switzerland*
Supergravity

Roy M. Emrick
University of Arizona
Creep
Crystal Defect
Whiskers

Rolf C. Enger
United States Air Force Academy
Normal Force

Roger A. Erickson
Stanford Linear Accelerator Center
Positron
Positronium

David J. Ernst
Vanderbilt University
Oscillator
Oscillator, Anharmonic

Matthias Ernzerhof
Tulane University
Hydrogen Bond

William E. Evenson
Brigham Young University
Center of Mass
Condensation, Bose–Einstein
Momentum
Momentum, Conservation of

C. W. F. Everitt
Stanford University
Maxwell, James Clerk

Isobel Falconer
Open University, Scotland
Thomson, Joseph John

James M. Feagin
California State University, Fullerton
Beats
Wave Speed

Heidi Fearn
California State University, Fullerton
Current, Eddy
Levitation, Electromagnetic
Pauli's Exclusion Principle
Surface Tension

Alfred Feitisch
Spectra-Physics Lasers, Inc., Mountain View, CA
Laser, Ion

Joseph L. Feldman
Naval Research Laboratory and *Carnegie Institution
of Washington, Washington, DC*
Specific Heat, Einstein Theory of

Edward J. Finn
Georgetown University
Heat, Mechanical Equivalent of

Douglas K. Finnemore
Ames Laboratory, Iowa State University
Cryogenics
Superconductivity

Ephraim Fischbach
Purdue University
Fifth Force

Charlotte Froese Fischer
Vanderbilt University
Ionization Potential

C. L. Foiles
Michigan State University
States, Density of
Thermocouple
Thermoelectric Effect

Peter Fong
Emory University
Entropy
Thermodynamics

Kenneth W. Ford
American Institute of Physics (retired)
Conservation Laws
Energy, Conservation of

Robert Fox
University of Oxford, England
Carnot, Nicolas-Léonard-Sadi
Heat, Caloric Theory of

Ronald F. Fox
Georgia Institute of Technology
H-Theorem
Phase Space
Statistical Mechanics

Allan Franklin
University of Colorado, Boulder
Scientific Method

Melissa Franklin
Harvard University
Quark, Top

Bretislav Friedrich
Harvard University
Stern–Gerlach Experiment

Joshua Frieman
Fermi National Accelerator Laboratory, Batavia, IL,
and *University of Chicago*
Dark Matter, Baryonic
Dark Matter, Cold
Dark Matter, Hot

Wolfgang Fritsch
Hahn-Meitner Institute, Berlin, Germany
Phase

Edward S. Fry
Texas A & M University
Bell's Theorem

Robert Q. Fugate
USAF Phillips Laboratory, Kirtland AFB, NM
Optics, Adaptive

Richard M. Fuller
Gustavus Adolphus College
Aerodynamic Lift
Aerodynamics
Battery
Electric Generator
Motor

Thomas Gallagher
University of Virginia
Spectroscopy, Laser

Cynthia Galovich
University of Northern Colorado
Energy, Activation

Kinetic Theory
Mean Free Path

Elizabeth Garber
State University of New York, Stony Brook
Clausius, Rudolf Julius Emmanuel
Gibbs, Josiah Willard

Ian R. Gatland
Georgia Institute of Technology
Scalar
Tensor
Vector

Timothy Gay
Behlen Laboratory, University of Nebraska, Lincoln
Polarizability
Resolving Power

Mark Gealy
Concordia College
Energy Levels

Clayton A. Gearhart
St. John's University, Collegeville, MN
Boltzmann, Ludwig

Howard Georgi
Harvard University
Conservation Laws and Symmetry
Grand Unified Theory

Edward Gerjuoy
University of Pittsburgh (emeritus)
Interference

F. A. Gianturco
University of Rome, Italy
Molecular Physics

Sarah L. Gilbert
National Institute of Standards and Technology,
Boulder, CO
Laser Cooling

C. Stewart Gillmor
Wesleyan University, Middletown, CT
Coulomb, Charles Augustin

J. D. Goddard
University of California, San Diego
Rheology

Stanley Goldberg
University of Maryland Baltimore County
Fission Bomb
Fusion Bomb
Nuclear Bomb, Building of
Relativity, Special Theory of, Origins of

Alfred S. Goldhaber
State University of New York, Stony Brook
Magnetic Monopole

Edwin Goldin
American Institute of Physics, College Park, MD
Spectral Series

Mitchell M. Goodsitt
University of Michigan, Ann Arbor
CAT Scan

Robert H. Gowdy
Virginia Commonwealth University
Relativity, General Theory of

William G. Graham
Queen's University of Belfast, Northern Ireland
Fiber Optics

Geoffrey Greene
Los Alamos National Laboratory, Los Alamos, NM
Neutron

Edwin W. Greeneich
Arizona State University
Transistor

Thomas Greenlee
Bethel College, St. Paul, MN
Compton Effect

David Griffiths
Reed College
Conductor
Diode
Electromagnetic Radiation
Inductor
Photon

David John Griffiths
Oregon State University
Lightning
Thunder

Jonathan E. Grindlay
Harvard University
Astrophysics, X-Ray
Einstein Observatory
Uhuru Satellite
X-Ray Binary

Ø. Grøn
Oslo College of Engineering, Norway
Tides

Howard E. Haber
University of California, Santa Cruz
Boson, Higgs

Edwin E. Hach III
University of Arkansas, Fayetteville
Electric Susceptibility

David Hafemeister
California Polytechnic State University, San Luis Obispo
Society, Physics and

Richard F. Haglund Jr.
Vanderbilt University
Coherence
Optics, Nonlinear

Kenneth D. Hahn
Truman State University
Fluid Dynamics
Navier–Stokes Equation
Reynolds Number

Karl Hall
Harvard University
Landau, Lev Davidovich

Hans D. Hallen
North Carolina State University
Lens

David Halliday
University of Pittsburgh
Galilean Transformation

P. W. "Bo" Hammer
American Institute of Physics, College Park, MD
Center of Gravity
Ether Hypothesis

Edward R. Harrison
University of Massachusetts, Amherst
Olbers's Paradox

Michael J. Harrison
Michigan State University
Magnon
Phonon

Shlomo Havlin
Boston University and *Bar-Ilan University,*
Ramat-Gan, Israel
Diffusion

Robert L. Hawkes
Mount Allison University, Canada
Meteor Shower

Charles E. Head
University of New Orleans
Pendulum
Refraction, Index of

Mark A. Heald
Swarthmore College
Electromagnetic Induction, Faraday's
Law of
Lenz's Law

Peter Heering
Carl von Ossietzky University, Oldenburg,
Germany
Coulomb's Law
Inverse Square Law

Paul A. Heiney
University of Pennsylvania
Color Center
Crystal
Liquid Crystal

William R. Hendee
Medical College of Wisconsin
Vision

Ronald J. W. Henry
Georgia State University
Atomic Weight
Schrödinger Equation

Andrzej Herczyński
State University of New York, Buffalo
Density
Doppler Effect
Frame of Reference, Rotating

Michael F. Herman
Tulane University
Elements, Transition

Richard B. Herr
University of Delaware
Adhesion
Cohesion
Isochoric Process

Dudley Herschbach
Harvard University
Franklin, Benjamin
Stern–Gerlach Experiment

Norriss S. Hetherington
University of California, Berkeley
Hubble, Edwin Powell
Universe, Expansion of, Discovery of

J. N. Hewitt
Massachusetts Institute of Technology
Gravitational Lensing

Robert C. Hilborn
Amherst College
Dimensional Analysis
Nonlinear Physics
Scientific Notation

Ian Hinchliffe
Lawrence Berkeley National Laboratory,
Berkeley, CA
Boson, Gauge
Boson, *W*
Boson, *Z*

David Hitlin
California Institute of Technology
Quark, Charm

Paul Hodge
University of Washington
Milky Way

Dieter Hoffmann
Berlin, Germany
Mach, Ernst

James R. Hofmann
California State University, Fullerton
Ampère, André-Marie
Laplace, Pierre-Simon

Barry R. Holstein
University of Massachusetts, Amherst
Quantum Mechanical Behavior of
Matter

Ulrich Hoyer
*Westfälischen Wilhelms Universität, Münster,
Germany*
Atom, Rutherford–Bohr

James R. Huddle
United States Naval Academy
Eigenfunction and Eigenvalue
Ground State
Van de Graaff Accelerator

Jeff Hughes
Manchester University, England
Chadwick, James
Neutron, Discovery of

Lutz Hüwel
Wesleyan University, Middletown, CT
Aberration, Spherical

Mario Iona
University of Denver (emeritus)
Glossary
SI Units

Steven Iona
Horizon High School, Thornton, CO
Glossary

George E. Ioup
University of New Orleans
Pendulum, Ballistic
Pendulum, Foucault

Juliette W. Ioup
University of New Orleans
Brewster's Law

M. Frank Watt Ireton
American Geophysical Union, Washington, DC
Volcano

L. Donald Isenhower
Abilene Christian University
Bubble Chamber
Cloud Chamber

Elizabeth S. Ivey
University of Hartford
Temperature Scale, Celsius
Viscosity

James Jadrich
Calvin College
Newtonian Mechanics

Kannan Jagannathan
Amherst College
Electromagnetism

Philip B. James
University of Toledo
Apogee and Perigee
Escape Velocity
Kepler's Laws
Kepler, Johannes
Planetary Magnetism
Space
Space and Time
Van Allen Belts

Oleg D. Jefimenko
West Virginia University
Maxwell's Equations

Tung Hon Jeong
Lake Forest College
Holography

John W. Jewett Jr.
California State Polytechnic University, Pomona
Force

Brant M. Johnson
Brookhaven National Laboratory, Upton, NY
Dispersion

Karen E. Johnson
St. Lawrence University
Mayer, Maria Goeppert

E. Roger Jones
University of Kentucky
Causality
Determinism
Models and Theories

Frederick J. Keller
Clemson University
Kirchhoff's Laws
Ohm's Law

Kenneth F. Kelton
Washington University
Phase Transition

R. A. Kenefick
Texas A & M University
Inertia, Moment of
Optics, Physical

Murtadha A. Khakoo
California State University, Fullerton
Dichroism, Circular

Jeffrey T. Kiehl
National Center for Atmospheric Research,
Denver, CO
Atmospheric Physics

Timothy L. Killeen
Space Physics Research Laboratory, University of
Michigan, Ann Arbor
Ionosphere

Sung Kyu Kim
Macalester College
Field, Gravitational
Gravitational Force Law
Proper Length
Proper Time
Time Dilation

M. Kimura
Yamaguchi University, Japan
Wave Packet
Wavelength

Dale Kinsey
New York, NY
Born, Max

Roger D. Kirby
Behlen Laboratory, University of Nebraska, Lincoln
Electron, Drift Speed of
Paramagnetism

Joseph Klafter
Tel-Aviv University, Israel
Diffusion

Peter M. Koch
State University of New York, Stony Brook
Lamb Shift

Edward W. Kolb
Fermi National Accelerator Laboratory,
Batavia, IL
Big Bang Theory

Donald J. Kouri
University of Houston
Chemical Physics
Franck–Condon Principle

Laurie Kovalenko
Eckerd College
Optics

A. J. Kox
University of Amsterdam and *The Collected Papers*
of Albert Einstein, Boston University
Lorentz, Hendrik Antoon

Michael N. Kozicki
Arizona State University
Electrical Resistance
Resistor

Helge Kragh
University of Oslo, Norway
Dirac, Paul Adrien Maurice
Physics, History of
Quantum Theory, Origins of
Wave–Particle Duality, History of

Kenneth S. Krane
Oregon State University
Field, Electric
Field, Magnetic

Wilfred Krause
Munich, Germany
Bohr's Atomic Theory

Lawrence M. Krauss
Case Western Reserve University
Cosmological Constant
Michelson, Albert Abraham
Michelson–Morley Experiment

Andreas S. Kronfeld
Fermi National Laboratory, Batavia, IL
Quantum Chromodynamics

Walter Kutschera
Vienna, Austria
Radioactive Dating

G. L. Lamb Jr.
University of Arizona
Soliton

Paul Langacker
University of Pennsylvania
Baryon Number

W. Lauterborn
Universität Göttingen, Germany
Cavitation

David Layzer
Harvard University
Arrow of Time

David Lazarus
University of Illinois, Urbana-Champaign
Elastic Moduli and Constants
Elasticity

T. D. Lee
Columbia University
Parity

Lawrence S. Lerner
California State University, Long Beach
Acceptor
Donor

Michael Lieber
University of Arkansas, Fayetteville
Aberration
Equilibrium
Parallel Axis Theorem

Chii-Dong Lin
Kansas State University
Polarization

Chun C. Lin
University of Wisconsin, Madison
Wave Mechanics

Andrei Linde
Stanford University
Cosmology, Inflationary

S. M. Lindsay
Arizona State University
Electrochemistry

Sanford Lipsky
University of Minnesota
Fluorescence

Eugene C. Loh
University of Utah
Fly's Eye Experiment

Thomas J. Loredo
Cornell University
Data Analysis
Error, Experimental
Error, Random
Error, Systematic
Significant Figures

Jack Lotsof
Calspan Advance Technology Center, Snyder, NY
Industry, Physicists in

Michael S. Lubell
City College of the City University of New York
Electromagnetic Wave
Group Velocity

Robert R. Ludeman
Andrews University
Capacitance
Capacitor
Circuit, DC
Circuit, Integrated

Jane Luu
Harvard University
Solar System

K. B. MacAdam
University of Kentucky
Degeneracy

Stefan Machlup
Case Western Reserve University
Gas
Pressure

J. M. MacLaren
Tulane University
Dichroism
Diffraction, Electron
Diffraction, Fraunhofer

Mordecai-Mark Mac Low
Max Planck Institute for Astronomy, Heidelberg, Germany
Coriolis Force
Planet Formation
Planetary Systems

Martin M. Maltempo
University of Colorado, Denver
Propulsion

Paul C. Mangelsdorf Jr.
Swarthmore College (emeritus)
Chemical Potential
Degree of Freedom

Alfred K. Mann
University of Pennsylvania
Neutrino
Neutrino, Solar

Steven T. Manson
Georgia State University
Image, Optical
K Capture

M. Brian Maple
University of California, San Diego
Cooper Pair

Robert H. March
University of Wisconsin, Madison
Elementary Particles
Lepton
Muon

Paul V. Marrone
Calspan Advance Technology Center, Snyder, NY
Industry, Physicists in

Sergio Mascarenhas
University of São Paulo, San Carlos, Brazil
Biophysics
Electret

Grant J. Mathews
University of Notre Dame
Deuteron
Fallout
Isotopes
Nuclear Reaction
Radioactivity

John E. Mathis
Oak Ridge National Laboratory, Oak Ridge, TN, and *University of Tennessee*
Diffraction
Fraunhofer Lines
Oscillation

Richard A. Matzner
University of Texas, Austin
Black Hole
Black Hole, Kerr
Black Hole, Schwarzschild
Isothermal Process
Spinor
Time

P. V. E. McClintock
Lancaster University, United Kingdom
Quantum Fluid

Eugene J. McGuire
Sandia National Laboratory, Albuquerque, NM
Electromagnetic Spectrum

Matthew G. McHarg
United States Air Force Academy
Aurora

William G. Melson
Smithsonian Institution, Washington, DC
Avalanche
Geode

Eugen Merzbacher
University of North Carolina, Chapel Hill
Matrix Mechanics

R. A. Mewaldt
California Institute of Technology
Cosmic Ray

David D. Meyerhofer
University of Rochester
Laser
Laser, Discovery of
Maser
Quantum Optics

David A. Micha
University of Florida
Correspondence Principle
Kinematics

M. Coleman Miller
University of Chicago
Pulsar

Peter W. Milonni
Los Alamos National Laboratory, Los Alamos, NM
Pauli's Exclusion Principle

Raymond F. Missert
Calspan Advance Technology Center, Snyder, NY
Industry, Physicists in

Earl N. Mitchell
University of North Carolina, Chapel Hill
Color

Marvin H. Mittleman
City College of the City University of New York
Hyperfine Structure
Stark Effect

James L. Monroe
Pennsylvania State University, Beaver Campus
Gauss's Law

Eduardo C. Montenegro
Pontifícia Universidade Católica do Rio de Janeiro, Brazil
Atom

Delo E. Mook II
Dartmouth College
Circuit, AC
Irrotational Flow

Thomas A. Moore
Pomona College
Collision
Free Fall
Gravitational Attraction

Least-Action Principle
Mass-Energy
Spacetime

Walter Moore
Indiana University and University of Sydney (emertius)
Schrödinger, Erwin

Frederick J. Morgan
York University, Canada
Energy, Potential

Douglas R. O. Morrison
European Organization for Nuclear Research (CERN), Geneva, Switzerland
Error and Fraud

Robert Morriss
Tulane University
Image, Virtual
Mirror, Plane
Ray Diagram

Frank Moss
University of Missouri, St. Louis
Stochastic Processes

John Mottmann
California Polytechnic State University, San Luis Obispo
Carnot Cycle
Enthalpy
Molecular Speed

Pawel Mrozek
University of Arizona
Complementarity Principle

Alfred Z. Msezane
Clark Atlanta University
Trajectory

Dwight E. Neuenschwander
Southern Nazarene University
Ampère's Law
Bernoulli's Principle
Biot–Savart Law
Engine, Efficiency of
Joule Heating
Thermometer
Thermometry

Alice L. Newman
California State University, Dominguez Hills
Critical Points
Ideal Gas Law
Magnetic Cooling

Howard Nicholson
Mount Holyoke College
Scintillation Counter

Julian C. Niles
Clark Atlanta University
Trajectory

Monika Nitsche
University of Minnesota
Vortex

J. Rayford Nix
Los Alamos National Laboratory,
Los Alamos, NM
Elements, Superheavy

Kenneth Nordtvedt
Montana State University
Equivalence Principle

John D. Norton
University of Pittsburgh
Relativity, General Theory of,
Origins of

David Norwood
Tulane University
Photometry

Mary Jo Nye
Oregon State University
N Ray

Michael C. Ogilvie
Washington University
Particle Mass

Hans C. Ohanian
Rensselaer Polytechnic Institute, Troy, NY
Perturbation Theory

Angela V. Olinto
University of Chicago
Gamma Ray Bursters
Neutron Star

William F. Oliver III
University of Arkansas, Fayetteville
Brillouin Zone
Cyclotron Resonance
Fermi Surface

Christine Orme
H. M. Randall Laboratory, University of Michigan,
Ann Arbor
Scanning Probe Microscopies

Bradford G. Orr
H. M. Randall Laboratory, University of Michigan,
Ann Arbor
Scanning Probe Microscopies

Donald E. Osterbrock
Lick Observatory, University of California,
Santa Cruz
Telescope, History of

Richard E. Packard
University of California, Berkeley
Liquid Helium

Michael A. Paesler
North Carolina State University
Lens

Abraham Pais
Rockefeller University
Bohr, Niels Henrik David

John T. Park
University of Missouri, Rolla
Scattering, Rayleigh

E. N. Parker
University of Chicago (emeritus)
Alfven Wave
Hydrodynamics
Magnetic Behavior
Magnetohydrodynamics
Magnetosphere
Solar Wind

J. D. Patterson
Florida Institute of Technology
Boltzmann Constant

Marvin Paule
Colorado State University
Centrifuge

J. L. Peacher
University of Missouri, Rolla
Wave Packet
Wavelength

Roberto D. Peccei
University of California, Los Angeles
Boson, Nambu–Goldstone
Field, Higgs
Symmetry Breaking, Spontaneous
Vacuum State

Henry C. Perkins Jr.
University of Arizona
Adiabatic Process
Isobaric Process
Joule–Thomson Effect

Charles J. Peterson
University of Missouri, Columbia
Eclipse
Moon, Phases of
Seasons

Panos J. Photinos
Southern Oregon State College
Quantum

David Pines
University of Illinois, Urbana-Champaign
Bardeen, John
Excitation, Collective
Plasmon

Mark W. Plano Clark
University of Nebraska, Lincoln
Weight

P. M. Platzman
AT&T Bell Laboratories, Murray Hill, NJ
Polaron

Edward Pollack
University of Connecticut
Interferometer, Fabry-Pérot

Alan M. Portis
University of California, Berkeley
Jahn–Teller Effect

John Preskill
California Institute of Technology
Charge

Richard H. Price
University of Utah
Gravitational Wave
Mach's Principle
Tachyon
Weightlessness

James A. Purdy
Washington University
Medical Physics

Jean M. Quashnock
University of Chicago
Gamma Ray Observatory

Helen R. Quinn
Stanford Linear Accelerator Center
Approximation and Idealization
Bremsstrahlung
Color Charge
Feynman Diagram
Flavor
Interaction
Interaction, Electromagnetic
Interaction, Electroweak
Interaction, Fundamental
Interaction, Weak
Mass, Conservation of
Parameters
Particle Physics, Detectors for
Scale and Structure
Systems

Alastair I. M. Rae
University of Birmingham, United Kingdom
Uncertainty Principle

Shafiqur Rahman
Allegheny College
Quantum Number

Talat Rahman
Kansas State University
Snell's Law

Pierre Ramond
University of Florida
Supersymmetry

Patricia Rankin
University of Colorado, Boulder
Stefan–Boltzmann Law

G. N. Rassam
Optical Society of America
Geophysics

A. R. P. Rau
Louisiana State University
Zeeman Effect

John Reading
Texas A & M University
Scattering

Helmut R. Rechenberg
Max Planck Institute for Physics, Munich, Germany
Quantum Mechanics, Creation of

P. A. Redhead
National Research Council of Canada, Ottawa
Vacuum
Vacuum Technology

Andrew F. Rex
University of Puget Sound
Equipartition Theorem
Maxwell's Demon

Patrick Richard
Kansas State University
Electron, Auger

C. B. Richardson
University of Arkansas, Fayetteville
Energy, Free
Energy, Internal

John S. Rigden
American Institute of Physics, College Park, MD
Oppenheimer, J. Robert
Rabi, Isidor Isaac
Spectroscopy, Microwave
Waveguide

John C. Riley
University of South Carolina, Spartanburg
Energy, Radiant
Heat Transfer
Radiation, Thermal

Wolfgang Rindler
University of Texas, Dallas
Frame of Reference, Inertial
Inertial Mass

Michael Riordan
Stanford Linear Accelerator Center
Quarks, Discovery of
Transistor, Discovery of

Alan J. Rocke
Case Western Reserve University
Atomic Theory, Origins of

William B. Rolnick
Wayne State University
Acceleration
Acceleration, Angular

Alfred Romer
St. Lawrence University
Disintegration, Artificial
Röntgen, Wilhelm Conrad
X Ray, Discovery of

L. David Roper
Virginia Polytechnic Institute and State University
Rest Mass

Carol Zwick Rosen
Thermometrics, Inc., Edison, NJ
Pixel
Servomechanism
Transducer

George Rosensteel
Tulane University
Frequency, Natural

Robert Rosner
University of Chicago
Sun

Thomas D. Rossing
Northern Illinois University
Levitation, Magnetic
Microphone
Pressure, Sound
Sound
Sound, Musical
Sound Absorption

Laura M. Roth
State University of New York, Albany (emeritus)
Crystal Structure
Hall Effect
Solid

Steven J. Rothman
Journal of Applied Physics
Alloy
Metal
Metallurgy

Henry D. Royal
Washington University School of Medicine
Nuclear Medicine

M. Eugene Rudd
University of Nebraska, Lincoln
Ionization

Berkman Sahiner
University of Michigan, Ann Arbor
CAT Scan

M. B. Salamon
University of Illinois, Urbana-Champaign
Ferromagnetism

Mark A. Samuel
Oklahoma State University
Interaction, Weak
Quark, Down
Quark, Strange
Quark, Up

José M. Sánchez-Ron
Universidad Autónoma de Madrid, Spain
Action-at-a-Distance

Justin M. Sanders
University of South Alabama
Emission
Velocity

Jose M. Sasian
University of Arizona
Telescope

Michael R. Scheinfein
Arizona State University
Magnetostriction

Alfred S. Schlachter
Lawrence Berkeley National Laboratory, Berkeley, CA
Velocity, Angular
Velocity, Terminal

V. Hugo Schmidt
Montana State University
Ferroelectricity

Christopher H. Scholz
Lamont-Doherty Earth Observatory, Columbia University
Earthquake

Alec J. Schramm
Occidental College
Lepton, Tau

Alan Schriesheim
Argonne National Laboratory, Argonne, IL (Director Emeritus)
Reactor, Nuclear

Daniel V. Schroeder
Weber State University
Renormalization

Peter A. Schroeder
Michigan State University
Transport Properties

David R. Schultz
Oak Ridge National Laboratory, Oak Ridge, TN
Motion
Optical Fiber

John H. Schwarz
California Institute of Technology
Superstring

S. S. Schweber
Brandeis University
Feynman, Richard Phillips

Roy Schwitters
University of Texas, Austin
Particle Physics

Ivan Sellin
University of Tennessee
Line Spectrum

David J. Sellmyer
Behlen Laboratory, University of Nebraska, Lincoln
Diamagnetism
Magnetic Material
Magnetic Susceptibility

Mark D. Semon
Bates College
Potential Barrier
Quantum Electrodynamics

Andrew M. Sessler
Lawrence Berkeley National Laboratory, Berkeley, CA
Sakharov, Andrei Dmitrievich

R. Shankar
Yale University
Planck Constant
Quantization
Quantum Field Theory
Quantum Mechanics
Quantum Mechanics and Quantum Field Theory

Stuart L. Shapiro
Center for Astrophysics and Relativity, Ithaca, NY
Computational Physics

John P. Sharpe
California Polytechnic State University, San Luis Obispo
Center-of-Mass System
Specific Gravity
Statics

William R. Shea
Université Louis Pasteur de Strasbourg, France
Galilei, Galileo

Jefferson L. Shinpaugh
East Carolina University
Grating, Transmission

Steven N. Shore
Indiana University, South Bend
Shock Wave

Kristin Shrader-Frechette
University of South Florida
Ethics
Social Responsibility

Joseph Silk
University of California, Berkeley
Redshift
Universe
Universe, Expansion of

Robert H. Silliman
Emory University
Fresnel, Augustin-Jean

Isaac F. Silvera
Harvard University
Experimental Physics

Ruth Lewin Sime
Sacramento City College
Meitner, Lise

Paul R. Simony
Jacksonville University
Vector, Unit

Jules Six
Institut National de Physique Nucleaire et de Physique des Particules, Cedex, France
Curie, Marie Sklodowska
Radioactivity, Artificial

Crosbie Smith
University of Kent, Canterbury, United Kingdom
Kelvin, Lord

Winthrop W. Smith
University of Connecticut
Spectroscopy

Chris Sneden
University of Texas, Austin
Elements, Abundance of

John T. Snow
University of Oklahoma
Hurricane
Tornado

Paul G. Snyder
University of Nebraska, Lincoln
Doping

Michael I. Sobel
Brooklyn College of the City University of New York
Motion, Rotational
Rigid Body

Wlad T. Sobol
University of Alabama, Birmingham
Health Physics

J. Spector
Arizona State University
Semiconductor

John Stachel
Boston University
Einstein, Albert

Shane Stadler
Tulane University
Dipole Moment
Phosphorescence

George S. Stanford
Argonne National Laboratory, Argonne, IL
Fission
Reactor, Breeder
Reactor, Fast

H. Eugene Stanley
Boston University
Diffusion

Sumner Starrfield
Arizona State University
Nova

M. Stastna
University of Waterloo, Canada
Phase Velocity

David Statman
Allegheny College
Quantum Number

Brenton F. Stearns
Hobart and William Smith Colleges
Torricelli's Theorem

Robert L. Stearns
Vassar College (emeritus)
Scattering, Rutherford

Albert Stebbins
Fermi National Accelerator Laboratory, Batavia, IL
Cosmic String

Daniel L. Stein
University of Arizona
Spin Glass

Fredrick M. Stein
Colorado State University
Avogadro Number

Philip A. Sterne
University of California, Davis
Frame of Reference

Gay B. Stewart
University of Arkansas, Fayetteville
Electric Moment
Electric Potential
Electric Susceptibility

James H. Stith
Ohio State University
Motion, Circular

Thomas Howard Stix
Princeton University
Plasma Physics

Andrew Strominger
University of California, Santa Barbara
Gravitational Constant
Graviton
Planck Length
Planck Mass
Planck Time
Quantum Gravity

Roger H. Stuewer
University of Minnesota
Compton, Arthur Holly

Joseph Sucher
University of Maryland, College Park
Charge, Electronic

Woodruff T. Sullivan III
University of Washington
Cosmic Microwave Background Radiation, Discovery of

Leonard Susskind
Stanford University
Interaction, Strong
Quark Confinement

James C. Swihart
Indiana University, Bloomington
Absolute Zero
Josephson Effect

Laurence G. Taff
Johns Hopkins University
Celestial Mechanics

Randall Tagg
University of Colorado, Denver
Turbulence
Turbulent Flow

Marianne Takamiya
University of Chicago
Galaxies and Galactic Structure

Janet Tate
Oregon State University
Boltzmann Distribution
Condensation

Morton A. Tavel
Vassar College
Refrigeration

Charles E. Taylor
Antioch College
Hydrometer

Edwin F. Taylor
Boston University
Relativity, Special Theory of

Philip Taylor
Case Western Reserve University
Michelson, Albert Abraham

William J. Thompson
University of North Carolina, Chapel Hill
Stationary State

Carl T. Tomizuka
University of Arizona
Antiferromagnetism
Buoyant Force
Dielectric Constant
Electric Flux
Electromagnet
Electromagnetic Force
Electromotive Force
Ensemble
Ergodic Theory
Fluid Statics
Fractal
Liquid

Magnetic Resonance Imaging
Mössbauer Effect
Pressure, Vapor
PVT Relation
Specific Heat of Solids
Standard Temperature and Pressure
Temperature
Thermal Conductivity

James W. Truran
University of Chicago
Nucleosynthesis

Michael S. Turner
University of Chicago and *Fermi National Laboratory, Batavia, IL*
Cosmological Principle
Cosmology
Hubble Constant

R. Steven Turner
University of New Brunswick, Canada
Helmholtz, Hermann L. F. von

David A. Valone
California Institute of Technology
Millikan, Robert Andrews

James A. Van Allen
University of Iowa
Radiation Belt

Johannes van Ek
Tulane University
Motion, Periodic

Jean-François Van Huele
Brigham Young University
Fine-Structure Constant

Christiaan G. van Weert
University of Amsterdam, Netherlands
Critical Phenomena
Irreversible Process

Sankoorikal L. Varghese
University of South Alabama
Energy and Work
g-factor

James Veale
University of Virginia
Spectroscopy, Laser

Giulio Venezian
Southeast Missouri State University
Polarity
Short Circuit

Corinne Vinches
Tulane University
Scattering, Light

Howard G. Voss
Arizona State University
Current, Alternating
Current, Direct

J. J. Vuillemin
University of Arizona
Haas–van Alphen Effect, De

Gary Waldman
St. Louis Community College, Florissant Valley
(retired)
Cloud, Gray
Cloud, White
Light
Rainbow
Sky, Color of

Terry P. Walker
Ohio State University
Spallation

T. C. Wallace
University of Arizona
Seismic Wave
Seismology

Jianyi Wang
Tulane University
Atom, Rydberg

George H. Watson
University of Delaware
Insulator

Rand L. Watson
Texas A & M University
Spectrophotometry
Spectroscopy, Ultraviolet
Spectroscopy, Visible

Albert Wattenberg
University of Illinois, Urbana-Champaign
Fermi, Enrico

Larry Weaver
Kansas State University
Standing Wave

Charles A. Wert
University of Illinois, Urbana-Champaign
Strain
Stress

Richard S. Westfall
Indiana University, Bloomington
Newton, Isaac
Newtonian Synthesis
Scientific Revolution

William R. Wharton
Wheaton College
Geiger Counter

Bruce R. Wheaton
*Technology and Physical Science History Associates,
El Cerrito, CA*
Broglie, Louis-Victor-Pierre-Raymond de

J. Craig Wheeler
University of Texas, Austin
Supernova

Carl E. Wieman
University of Colorado
Laser Cooling

Frank Wilczek
Institute for Advanced Study, Princeton, NJ
Fermions and Bosons
Spin
Spin, Electron
Spin and Statistics

David Wilkinson
Princeton University
Cosmic Background Explorer Satellite
Cosmic Microwave Background Radiation

James F. Williams
University of Western Australia
Atomic Physics

L. Pearce Williams
Cornell University (emeritus)
Electromagnetic Induction
Electromagnetism, Discovery of
Faraday, Michael

Bruce Winstein
University of Chicago
Time Reversal Invariance

Roland Winston
University of Chicago
Energy, Solar

Thomas Winter
Pennsylvania State University, Wilkes-Barre Campus
Magnification

Lincoln Wolfenstein
Carnegie Mellon University
Field
Kaon
Symmetry

Yong Yan
University of Nevada, Reno
Ionization Chamber

F. R. Yeatts
Colorado School of Mines
Dielectric Properties
Reactance
Skin Effect

Orhan Yenen
University of Nebraska, Lincoln
Aberration, Chromatic

Joella G. Yoder
Renton, WA
Huygens, Christiaan

Donald G. York
University of Chicago
Interstellar and Intergalactic Medium

Linda Young
Argonne National Laboratory, Argonne, IL
Selection Rule

Harold S. Zapolsky
Rutgers University
Current, Displacement
Displacement

G. L. Zimmerman
Tulane University
Damping

Dean Zollman
Kansas State University
Circuit, Parallel
Circuit, Series

Stewart J. Zweben
Princeton Plasma Physics Laboratory, Princeton, NJ
Fusion Power
Tokamak

Bernard Zygelman
University of Nevada, Las Vegas
Excited State

Fredy R. Zypman
University of Puerto Rico, Humacao
Crystallography

COMMON ABBREVIATIONS AND MATHEMATICAL SYMBOLS

$=$	equals; double bond	$\|\ \|$	absolute value of
$\neq$	not equal to	$+$	plus
$\equiv$	identically equal to; equivalent to; triple bond	$-$	minus
		$/$	divided by
$\sim$	asymptotically equal to; of the order of magnitude of; approximately	$\times$	multiplied by
		$\oplus$	direct sum
		$\otimes$	direct product
$\approx, \cong$	approximately equal to	$\pm$	plus or minus
$\cong$	congruent to; approximately equal to	$\mp$	minus or plus
		$\sqrt{\ }$	radical
$\propto$	proportional to	$\int$	integral
$<$	less than	$\oint$	contour integral
$>$	greater than	Σ	summation
$\nless$	not less than	Π	product
$\ngtr$	not greater than	∂	partial derivative
$\ll$	much less than	$^{\circ}$	degree
$\gg$	much greater than	$^{\circ}\mathrm{B}$	degrees Baumé
$\leq$	less than or equal to	$^{\circ}\mathrm{C}$	degrees Celsius (centigrade)
$\geq$	greater than or equal to	$^{\circ}\mathrm{F}$	degrees Fahrenheit
$\nleq$	not less than or equal to	$!$	factorial
$\ngeq$	not greater than or equal to	$'$	minute
$\cup$	union of	$''$	second
$\cap$	intersection of	∇	curl
$\subset$	subset of; included in		
$\supset$	contains as a subset	ϵ_0	electric constant
$\in$	an element of	μ	micro-
$\ni$	contains as an element	μ_0	magnetic constant
$\rightarrow$	approaches, tends to; yeilds; is replaced by	$\mu\mathrm{A}$	microampere
		$\mu\mathrm{A\ h}$	microampere hour
$\Rightarrow$	implies; is replaced by	$\mu\mathrm{C}$	microcoulomb
$\Leftarrow$	is implied by	$\mu\mathrm{F}$	microfarad
$\downarrow$	mutually implies	$\mu\mathrm{g}$	microgram
$\Leftrightarrow$	if and only if	$\mu\mathrm{K}$	microkelvin
$\perp$	perpendicular to	$\mu\mathrm{m}$	micrometer
$\|$	parallel to	$\mu\mathrm{m}$	micron

μm Hg	microns of mercury
μmol	micromole
μs, μsec	microsecond
μu	microunit
$\mu\Omega$	microhm
σ	Stefan–Boltzmann constant
Ω	ohm
Ω cm	ohm centimeter
Ω cm/(cm/cm^3)	ohm centimeter per centimeter per cubic centimeter
A	ampere
Å	angstrom
a	atto-
A$_s$	atmosphere, standard
abbr.	abbreviate; abbreviation
abr.	abridged; abridgment
Ac	Actinium
ac	alternating-current
aF	attofarad
af	audio-frequency
Ag	silver
A h	ampere hour
AIP	American Institute of Physics
Al	aluminum
alt	altitude
Am	americium
AM	amplitude-modulation
A.M.	ante meridiem
amend.	amended; amendment
annot.	annotated; annotation
antilog	antilogarithm
app.	appendix
approx	approximate (in subscript)
Ar	argon
arccos	arccosine
arccot	arccotangent
arccsc	arccosecant
arc min	arc minute
arcsec	arcsecant
arcsin	arcsine
arg	argument
As	arsenic
At	astatine
At/m	ampere turns per meter
atm	atmosphere
at. ppm	atomic parts per million
at. %	atomic percent
atu	atomic time unit
AU	astronomical unit
a.u.	atomic unit
Au	gold

av	average (in subscript)
b	barn
b.	born
B	boron
Ba	barium
bcc	body-centered-cubic
B.C.E.	before the common era
Be	beryllium
Bi	biot
Bi	bismuth
Bk	berkelium
bp	boiling point
Bq	becquerel
Br	bromine
Btu, BTU	British thermal unit
C	carbon
c	centi-
c.	circa, about, approximately
C	coulomb
c	speed of light
Ca	calcium
cal	calorie
calc	calculated (in subscript)
c.c.	complex conjugate
CCD	charge-coupled device
Cd	cadmium
cd	candela
CD	compact disc
Ce	cerium
C.E.	common era
CERN	European Center for Nuclear Research
Cf	californium
cf.	confer, compare
cgs, CGS	centimeter-gram-second (system)
Ci	curie
Cl	chlorine
C.L.	confidence limits
c.m.	center of mass
cm	centimeter
Cm	curium
cm^3	cubic centimeter
Co	cobalt
Co.	Company
coeff	coefficient (in subscript)
colog	cologarithm
const	constant
Corp.	Corporation
cos	cosine
cosh	hyperbolic cosine
cot	cotangent
coth	hyperbolic cotangent

cp	candlepower	e.u.	electron unit
cP	centipoise	eu	entropy unit
cp	chemically pure	Eu	europium
cpd	contact potential difference	eV	electron volt
cpm	counts per minute	expt	experimental (in subscript)
cps	cycles per second	F	farad
Cr	chromium	*F*	Faraday constant
cS	centistoke	f	femto-
Cs	cesium	F	fermi
csc	cosecant	F	fluorine
csch	hyperbolic cosecant	fc	foot-candle
Cu	copper	fcc	face-centered-cubic
cu	cubic	Fe	iron
cw	continuous-wave	fF	femtofarad
D	Debye	Fig. (pl., Figs.)	figure
d	deci-	fL	foot-lambert
d.	died	fm	femtometer
da	deka-	Fm	fermium
dB, dBm	decibel	FM	frequency-modulation
dc	direct-current	f. (pl., ff.)	following
deg	degree	fpm	fissions per minute
det	determinant	Fr	francium
dev	deviation	Fr	franklin
diam	diameter	fs	femtosecond
dis/min	disintegrations per minute	ft	foot
dis/s	disintegrations per second	ft lb	foot-pound
div	divergence	ft lbf	foot-pound-force
DNA	deoxyribose nucleic acid	f.u.	formula units
Dy	dysprosium	*g*	acceleration of free fall
dyn	dyne	G	gauss
E	east	G	giga-
e	electronic charge	g	gram
E	exa-	G	gravitational constant
e, exp	exponential	Ga	gallium
e/at.	electrons per atom	Gal	gal (unit of gravitational force)
e b	electron barn	gal	gallon
e/cm3	electrons per cubic centimeter	g-at.	gram-atom
ed. (pl., eds.)	editor	g.at. wt	gram-atomic-weight
e.g.	exempli gratia, for example	Gc/s	gigacycles per second
el	elastic (in subscript)	Gd	gadolinium
emf, EMF	electromotive force	Ge	germanium
emu	electromagnetic unit	GeV	giga-electron-volt
Eng.	England	GHz	gigahertz
Eq. (pl., Eqs.)	equation	Gi	gilbert
Er	erbium	grad	gradient
erf	error function	GV	gigavolt
erfc	error function (complement of)	Gy	gray
Es	einsteinium	h	hecto-
e.s.d.	estimated standard deviation	H	henry
esu	electrostatic unit	h	hour
et al.	et alii, and others	H	hydrogen
etc.	et cetera, and so forth	*h*	Planck constant

H.c.	Hermitian conjugate	ks, ksec	kilosecond
hcp	hexagonal-close-packed	kt	kiloton
He	helium	kV	kilovolt
Hf	hafnium	kV A	kilovolt ampere
hf	high-frequency	kW	kilowatt
hfs	hyperfine structure	kW h	kilowatt hour
hg	hectogram	kΩ	kilohm
Hg	mercury	L	lambert
Ho.	holmium	L	langmuir
hp	horsepower	l, L	liter
IIz	hcrtz	La	lanthanum
I	iodine	LA	longitudinal-acoustic
ICT	International Critical Tables	lab	laboratory (in subscript)
i.d.	inside diameter	lat	latitude
i.e.	id est, that is	lb	pound
IEEE	Institute of Electrical and	lbf	pound-force
	Electronics Engineers	lbm	pound-mass
if	intermediate frequency	LED	light emitting diode
Im	imaginary part	Li	lithium
in.	inch	lim	limit
In	indium	lm	lumen
Inc.	Incorporated	lm/W	lumens per watt
inel	inelastic (in subscript)	ln	natural logarithm (base e)
ir, IR	infrared	LO	longitudinal-optic
Ir	iridium	log	logarithm
J	joule	Lr	lawrencium
Jy	jansky	LU	Lorentz unit
k, k_B	Boltzmann's constant	Lu	lutetium
K	degrees Kelvin	lx	lux
K	kayser	ly, lyr	light-year
k	kilo-	M	Mach
K	potassium	M	mega-
kA	kiloamperes	m	meter
kbar	kilobar	m	milli-
kbyte	kilobyte	m	molal (concentration)
kcal	kilocalorie	M	molar (concentration)
kc/s	kilocycles per second	m_e	electronic rest mass
kdyn	kilodyne	m_n	neutron rest mass
keV	kilo-electron-volt	m_p	proton rest mass
kG	kilogauss	M$_\odot$	solar mass (2×10^{33} g)
kg	kilogram	MA	megaamperes
kgf	kilogram force	mA	milliampere
kg m	kilogram meter	ma	maximum
kHz	kilohertz	mb	millibarn
kJ	kilojoule	mCi	millicurie
kK	kilodegrees Kelvin	Mc/s	megacycles per second
km	kilometer	Md	mendlelvium
kMc/s	kilomegacycles per second	MeV	mega-electron-volt; million
kn	knot		electron volt
kOe	kilo-oersted	Mg	magnesium
kpc	kiloparsec	mg	milligram
Kr	krypton	mH	millihenry

mho	reciprocal ohm	No.	number
MHz	megahertz	Np	neper
min	minimum	Np	neptunium
min	minute	ns, nsec	nanosecond
mK	millidegrees Kelvin; millikelvin	n/s	neutrons per second
mks, MKS	meter-kilogram-second (system)	n/s cm^2	neutrons per second per square centimeter
mksa	meter-kilogram-second ampere		
mksc	meter-kilogram-second coulomb	ns/m	nanoseconds per meter
ml	milliliter	O	oxygen
mm	millimeter	$o()$	of order less than
mmf	magnetomotive force	$O()$	of the order of
mm Hg	millimeters of mercury	obs	observed (in subscript)
Mn	manganese	o.d.	outside diameter
MO	molecular orbital	Oe	oersted
Mo	molybdenum	ohm^{-1}	mho
MOE	magneto-optic effect	Os	osmium
mol	mole	oz	ounce
mol %, mole %	mole percent	P	peta-
mp	melting point	P	phosphorus
Mpc	megaparsec	p	pico-
mph	miles per hour	P	poise
MPM	mole percent metal	Pa	pascal
Mrad	megarad	Pa	protactinium
ms, msec	millisecond	Pb	lead
mu	milliunit	pc	parsec
MV	megavolt; million volt	Pd	palladium
mV	millivolt	PD	potential difference
MW	megawatt	pe	probable error
mwe, m (w.e.)	meter of water equivalent	pF	picofarad
Mx	maxwell	pl.	plural
mμm	millimicron	P.M.	post meridiem
MΩ	megaohm	Pm	promethium
n	nano-	Po	polonium
N	newton	ppb	parts per billion
N	nitrogen	p. (pl., pp.)	page
N	normal (concentration)	ppm	parts per million
N	north	Pr	praseodymium
N, N_A	Avogadro constant	psi	pounds per square inch
Na	sodium	psi (absolute)	pounds per square inch absolute
NASA	National Aeronautics and Space Administration	psi (gauge)	pounds per square inch gauge
		Pt	platinum
nb	nanobarn	Pu	plutonium
Nb	niobium	R (ital)	gas constant
Nd	neodymium	R	roentgen
N.D.	not determined	Ra	radium
NDT	nondestructive testing	rad	radian
Ne	neon	Rb	rubidium
n/f	neutrons per fission	Re	real part
Ni	nickel	Re	rhenium
N_L	Loschmidt's constant	rev.	revised
nm	nanometer	rf	radio frequency
No	nobelium	Rh	rhodium

r.l.	radiation length	tanh	hyperbolic tangent
rms	root-mean-square	Tb	terbium
Rn	radon	Tc	technetium
RNA	ribonucleic acid	Td	townsend
RPA	random-phase approximation	Te	tellurium
rpm	revolutions per minute	TE	transverse-electric
rps, rev/s	revolutions per second	TEM	transverse-electromagnetic
Ru	ruthenium	TeV	tera-electron-volt
Ry	rydberg	Th	thorium
s, sec	second	theor	theory, theoretical (in subscript)
S	siemens	THz	tetrahertz
S	south	Ti	titanium
S	stoke	Tl	thallium
S	sulfur	Tm	thulium
Sb	antimony	TM	transverse-magnetic
Sc	scandium	TO	transverse-optic
sccm	standard cubic centimeter per minute	tot	total (in subscript)
		TP	temperature-pressure
Se	selenium	tr, Tr	trace
sec	secant	trans.	translator, translators; translated by; translation
sech	hyperbolic secant		
sgn	signum function	u	atomic mass unit
Si	silicon	U	uranium
SI	*Système International* (International System of Measurement)	uhf	ultrahigh-frequency
		uv, UV	ultraviolet
sin	sine	V	vanadium
sinh	hyperbolic sine	V	volt
SLAC	Stanford Linear Accelerator Center	VB	valence band
		vol. (pl., vols.)	volume
Sm	samarium	vol %	volume percent
Sn	tin	vs.	versus
sq	square	W	tungsten
sr	steradian	W	watt
Sr	strontium	W	West
STP	standard temperature and pressure	Wb	weber
		Wb/m^2	webers per square meter
Suppl.	Supplement	wt %	weight percent
Sv	sievert	W.u.	Weisskopf unit
T	tera-	Xe	xenon
T	tesla	Y	yttrium
t	tonne	Yb	ytterbium
Ta	tantalum	yr	year
TA	transverse-acoustic	Zn	zinc
tan	tangent	Zr	zirconium

JOURNAL ABBREVIATIONS

Acc. Chem. Res.
Accounts of Chemical Research
Acta Chem. Scand.
Acta Chemica Scandinavica
Acta Crystallogr.
Acta Crystallographica
Acta Crystallogr. Sec. A
Acta Crystallographica, Section A: Crystal
Physics, Diffraction, Theoretical, and General Crystallography
Acta Crystallogr. Sec. B
Acta Crystallographica, Section B: Structural
Crystallography and Crystal Chemistry
Acta Math. Acad. Sci. Hung.
Acta Mathematica Academiae Scientiarum
Hungaricae
Acta Metall.
Acta Metallurgica
Acta Oto-Laryngol.
Acta Oto-Laryngologica
Acta Phys.
Acta Physica
Acta Phys. Austriaca
Acta Physica Austriaca
Acta Phys. Pol.
Acta Physica Polonica
Adv. Appl. Mech.
Advances in Applied Mechanics
Adv. At. Mol. Opt. Phys.
Advances in Atomic, Molecular, and Optical
Physics
Adv. Chem. Phys.
Advances in Chemical Physics
Adv. Magn. Reson.
Advances in Magnetic Resonance
Adv. Phys.
Advances in Physics

Adv. Quantum Chem.
Advances in Quantum Chemistry
AIAA J.
AIAA Journal
AIChE J.
AIChE Journal
AIP Conf. Pro.
AIP Conference Proceedings
Am. J. Phys.
American Journal of Physics
Am. J. Sci.
American Journal of Science
Am. Sci.
American Scientist
Anal. Chem.
Analytical Chemistry
Ann. Chim. Phys.
Annales de Chimie et de Physique
Ann. Fluid Dyn.
Annals of Fluid Dynamics
Ann. Geophys.
Annales de Geophysique
Ann. Inst. Henri Poincaré
Annales de l'Institut Henri Poincaré
Ann. Inst. Henri Poincaré, A
Annales de l'Institut Henri Poincaré,
Section A: Physique Theorique
Ann. Inst. Henri Poincaré, B
Annales de l'Institut Henri Poincaré,
Section B: Calcul des Probabilites et
Statistique
Ann. Math.
Annals of Mathematics
Ann. Otol. Rhinol. Laryngol.
Annals of Otology, Rhinology, & Laryngology
Ann. Phys. (Leipzig)
Annalen der Physik (Leipzig)

Ann. Phys. (N.Y.)
Annals of Physics (New York)
Ann. Phys. (Paris)
Annales de Physique (Paris)
Ann. Rev. Mat. Sci.
Annual Reviews of Materials Science
Ann. Rev. Nucl. Part. Sci.
Annual Review of Nuclear and Particle
Science
Ann. Sci.
Annals of Science
Annu. Rev. Astron. Astrophys.
Annual Reviews of Astronomy and Astrophysics
Annu. Rev. Nucl. Part. Sci.
Annual Reviews of Nuclear and Particle
Science
Annu. Rev. Nucl. Sci.
Annual Review of Nuclear Science
Appl. Opt.
Applied Optics
Appl. Phys. Lett.
Applied Physics Letters
Appl. Spectrosc.
Applied Spectroscopy
Ark. Fys.
Arkiv foer Fysik
Astron. Astrophys.
Astronomy and Astrophysics
Astron. J.
Astronomical Journal
Astron. Nachr.
Astronomische Nachrichten
Astrophys. J.
Astrophysical Journal
Astrophys. J. Lett.
Astrophysical Journal, Letters to the Editor
Astrophys. J. Suppl. Ser.
Astrophysical Journal, Supplement Series
Astrophys. Lett.
Astrophysical Letters
Aust. J. Phys.
Australian Journal of Physics
Bell Syst. Tech. J.
Bell System Technical Journal
Ber. Bunsenges. Phys. Chem.
Berichte der Bunsengesellschaft für
Physikalische Chemie
Br. J. Appl. Phys.
British Journal of Applied Physics
Bull. Acad. Sci. USSR, Phys. Ser.
Bulletin of the Academy of Sciences of the
USSR, Physical Series

Bull. Am. Astron. Soc.
Bulletin of the American Astronomical Society
Bull. Am. Phys. Soc.
Bulletin of the American Physical Society
Bull. Astron. Instit. Neth.
Bulletin of the Astronomical Institutes of the
Netherlands
Bull. Chem. Soc. Jpn.
Bulletin of the Chemical Society of Japan
Bull. Seismol. Soc. Am.
Bulletin of the Seismological Society of
America
C. R. Acad. Sci.
Comptes Rendus Hebdomadaires des Seances
de l'Academie des Sciences
C. R. Acad. Ser. A
Comptes Rendus Hebdomadaires des Seances
de l'Academie des Sciences, Serie A:
Sciences Mathematiques
C. R. Acad. Ser. B
Comptes Rendus Hebdomadaires des Seances
de l'Academie des Sciences, Serie B: Sciences
Physiques
Can. J. Chem.
Canadian Journal of Chemistry
Can. J. Phys.
Canadian Journal of Physics
Can. J. Res.
Canadian Journal of Research
Chem. Phys.
Chemical Physics
Chem. Phys. Lett.
Chemical Physics Letters
Chem. Rev.
Chemical Reviews
Chin. J. Phys.
Chinese Journal of Physics
Class. Quantum Grav.
Classical and Quantum Gravity
Comments Nucl. Part. Phys.
Comments on Nuclear and Particle Physics
Commun. Math. Phys.
Communications in Mathematical Physics
Commun. Pure Appl. Math.
Communications on Pure and Applied
Mathematics
Comput. Phys.
Computers in Physics
Czech. J. Phys.
Czechoslovak Journal of Physics
Discuss. Faraday Soc.
Discussions of the Faraday Society

Earth Planet. Sci. Lett.
  Earth and Planetary Science Letters
Electron. Lett.
  Electronics Letters
Fields Quanta
  Fields and Quanta
Fortschr. Phys.
  Fortschritte der Physik
Found. Phys.
  Foundations of Physics
Gen. Relativ. Gravit.
  General Relativity and Gravitation
Geochim. Cosmochim. Acta
  Geochimica et Cosmochimica Acta
Geophys. Res. Lett.
  Geophysical Research Letters
Handb. Phys.
  Handbuch der Physik
Helv. Chim. Acta
  Helvetica Chimica Acta
Helv. Phys. Acta
  Helvetica Physica Acta
High Temp. (USSR)
  High Temperature (USSR)
IBM J. Res. Dev.
  IBM Journal of Research and Development
Icarus.
  Icarus. International Journal of the Solar System
IEEE J. Quantum Electron.
  IEEE Journal of Quantum Electronics
IEEE Trans. Antennas Propag.
  IEEE Transactions on Antennas and
    Propagation
IEEE Trans. Electron Devices
  IEEE Transactions on Electron Devices
IEEE Trans. Inf. Meas.
  IEEE Transactions on Instrumentation and
    Measurement
IEEE Trans. Inf. Theory
  IEEE Transactions on Information Theory
IEEE Trans. Magn.
  IEEE Transactions on Magnetics
IEEE Trans. Microwave Theory Tech.
  IEEE Transactions on Microwave Theory and
    Techniques
IEEE Trans. Nucl. Sci.
  IEEE Transactions on Nuclear Science
IEEE Trans. Sonics Ultrason. Ind. Eng. Chem.
  IEEE Transactions on Sonics Ultrasonics
    Industrial and Engineering Chemistry
Infrared Phys.
  Infrared Physics

Inorg. Chem.
  Inorganic Chemistry
Inorg. Mater. (USSR)
  Inorganic Materials (USSR)
Instrum. Exp. Tech. (USSR)
  Instruments and Experimental Techniques
    (USSR)
Int. J. Magn.
  International Journal of Magnetism
Int. J. Mod. Phys. A
  International Journal of Modern Physics A
Int. J. Quantum Chem.
  International Journal of Quantum Chemistry
Int. J. Quantum Chem. 1
  International Journal of Quantum Chemistry,
    Part 1
Int. J. Quantum Chem. 2
  International Journal of Quantum Chemistry,
    Part 2
Int. J. Theor. Phys.
  International Journal of Theoretical Physics
Izv. Acad. Sci. USSR, Atmos. Oceanic Phys.
  Izvestiya, Academy of Sciences, USSR,
    Atmospheric and Oceanic Physics
Izv. Acad. Sci. USSR, Phys. Solid Earth
  Izvestiya, Academy of Sciences, USSR, Physics
    of the Solid Earth
J. Acoust. Soc. Am.
  Journal of the Acoustical Society of America
J. Am. Ceram. Soc.
  Journal of the American Ceramic Society
J. Am. Chem. Soc.
  Journal of the American Chemical Society
J. Am. Inst. Electr. Eng.
  Journal of the American Institute of Electrical
    Engineers
J. Appl. Crystallogr.
  Journal of Applied Crystallography
J. Appl. Phys.
  Journal of Applied Physics
J. Appl. Spectrosc. (USSR)
  Journal of Applied Spectroscopy (USSR)
J. Atmos. Sci.
  Journal of Atmospheric Sciences
J. Atmos. Terr. Phys.
  Journal of Atmospheric and Terrestrial Physics
J. Audio Engin. Soc.
  Journal of the Audio Engineering Society
J. Chem. Phys.
  Journal of Chemical Physics
J. Chem. Soc.
  Journal of the Chemical Society

J. Chim. Phys.
Journal de Chemie Physique

J. Comput. Phys.
Journal of Computational Physics

J. Cryst. Growth
Journal of Crystal Growth

J. Electrochem. Soc.
Journal of Electrochemical Society

J. Fluid Mech.
Journal of Fluid Mechanics

J. Gen. Rel. Grav.
Journal of General Relativity and Gravitation

J. Geophys. Res.
Journal of Geophysical Research

J. Inorg. Nucl. Chem.
Journal of Inorganic and Nuclear Chemistry

J. Lightwave Technol.
Journal of Lightwave Technology

J. Low Temp. Phys.
Journal of Low-Temperature Physics

J. Lumin.
Journal of Luminescence

J. Macromol. Sci. Phys.
Journal of Macromolecular Science, [Part B] Physics

J. Mater. Res.
Journal of Materials Research

J. Math. Phys. (Cambridge, Mass.)
Journal of Mathematics and Physics (Cambridge, Mass.)

J. Math. Phys. (N.Y.)
Journal of Mathematical Physics (New York)

J. Mech. Phys. Solids
Journal of the Mechanics and Physics of Solids

J. Mol. Spectrosc.
Journal of Molecular Spectroscopy

J. Non-Cryst. Solids
Journal of Non-Crystalline Solids

J. Nucl. Energy
Journal of Nuclear Energy

J. Nucl. Energy, Part C.
Journal of Nuclear Energy, Part C: Plasma Physics, Accelerators, Themonuclear Research

J. Nucl. Mater.
Journal of Nuclear Materials

J. Opt. Soc. Am.
Journal of the Optical Society of America

J. Opt. Soc. Am. A
Journal of the Optical Society of America A

J. Opt. Soc. Am. B
Journal of the Optical Society of America B

J. Phys. (Moscow)
Journal of Physics (Moscow)

J. Phys. (Paris)
Journal de Physique (Paris)

J. Phys. A
Journal of Physics A: Mathematical and General

J. Phys. B
Journal of Physics B: Atomic, Molecular, and Optical Physics

J. Phys. C
Journal of Physics C: Solid State Physics

J. Phys. D
Journal of Physics D: Applied Physics

J. Phys. E
Journal of Physics E: Scientific Instruments

J. Phys. F
Journal of Physics F: Metal Physics

J. Phys. G
Journal of Physics G: Nuclear and Particle Physics

J. Phys. Chem.
Journal of Physical Chemistry

J. Phys. Chem. Ref. Data
Journal of Physical and Chemical Reference Data

J. Phys. Chem. Solids
Journal of Physics and Chemistry of Solids

J. Phys. Radium
Journal de Physique et le Radium

J. Phys. Soc. Jpn.
Journal of the Physical Society of Japan

J. Plasma Phys.
Journal of Plasma Physics

J. Polym. Sci.
Journal of Polymer Science

J. Polym. Sci., Polym. Lett. Ed.
Journal of Polymer Science, Polymer Letters Edition

J. Polym. Sci., Polym. Phys. Ed.
Journal of Polymer Science, Polymer Physics Edition

J. Quant. Spectros. Radiat. Transfer
Journal of Quantitative Spectroscopy & Radiative Transfer

J. Res. Natl. Bur. Stand.
Journal of Research of the National Bureau of Standards

J. Res. Natl. Bur. Stand. Sec. A
Journal of Research of the National Bureau of Standards, Section A: Physics and Chemistry

J. Res. Natl. Bur. Stand. Sec. B
    Journal of Research of the National Bureau
        of Standards, Section B: Mathematical
        Sciences
J. Res. Natl. Bur. Stand. Sec. C
    Journal of Research of the National Bureau
        of Standards, Section C: Engineering and
        Instrumentation
J. Rheol.
    Journal of Rheology
J. Sound Vib.
    Journal of Sound and Vibration
J. Speech Hear. Disord.
    Journal of Speech and Hearing Disorders
J. Speech Hear. Res.
    Journal of Speech and Hearing Research
J. Stat. Phys.
    Journal of Statistical Physics
J. Vac. Sci. Technol.
    Journal of Vacuum Science and Technology
J. Vac. Sci. Technol. A
    Journal of Vacuum Science and Technology A
J. Vac. Sci. Technol. B
    Journal of Vacuum Science and Technology B
JETP Lett.
    JETP Letters
Jpn. J. Appl. Phys.
    Japanese Journal of Applied Physics
Jpn. J. Phys.
    Japanese Journal of Physics
K. Dan. Vidensk. Selsk. Mat. Fys. Medd.
    Kongelig Danske Videnskabernes Selskab,
        Matematsik-Fysiske Meddelelser
Kolloid Z. Z. Polym.
    Kolloid Zeitschrift & Zeitschrift für Polymere
Lett. Nuovo Cimento
    Lettere al Nuovo Cimento
Lick Obs. Bull.
    Lick Observatory Bulletin
Mater. Res. Bull.
    Materials Research Bulletin
Med. Phys.
    Medical Physics
Mem. R. Astron. Soc.
    Memoirs of the Royal Astronomical Society
Mol. Cryst. Liq. Cryst.
    Molecular Crystals and Liquid Crystals
Mol. Phys.
    Molecular Physics
Mon. Not. R. Astron. Soc.
    Monthly Notices of the Royal Astronomical
        Society

Natl. Bur. Stand. (U.S.), Circ.
    National Bureau of Standards (U.S.),
        Circular
Natl. Bur. Stand. (U.S.), Misc. Publ.
    National Bureau of Standards (U.S.),
        Miscellaneous Publications
Natl. Bur. Stand. (U.S.), Spec. Publ.
    National Bureau of Standards (U.S.),
        Special Publications
Nucl. Data, Sect. A
    Nuclear Data, Section A
Nucl. Fusion
    Nuclear Fusion
Nucl. Instrum.
    Nuclear Instruments
Nucl. Instrum. Methods
    Nuclear Instruments & Methods
Nucl. Phys.
    Nuclear Physics
Nucl. Phys. A
    Nuclear Physics A
Nucl. Phys. B
    Nuclear Physics B
Nucl. Sci. Eng.
    Nuclear Science and Engineering
Opt. Acta
    Optica Acta
Opt. Commun.
    Optics Communications
Opt. Lett.
    Optics Letters
Opt. News
    Optics News
Opt. Photon. News
    Optics and Photonics News
Opt. Spectrosc. (USSR)
    Optics and Spectroscopy (USSR)
Percept. Psychophys.
    Perception and Psychophysics
Philips Res. Rep.
    Philips Research Reports
Philos. Mag.
    Philosophical Magazine
Philos. Trans. R. Soc. London
    Philosophical Transactions of the Royal Society
        of London
Philos. Trans. R. Soc. London, Ser. A
    Philosophical Transactions of the Royal Society
        of London, Series A: Mathematical and
        Physical Sciences
Phys. (N.Y.)
    Physics (New York)

Phys. Fluids
Physics of Fluids
Phys. Fluids A
Physics of Fluids A
Phys. Fluids B
Physics of Fluids B
Phys. Konden. Mater.
Physik der Kondensierten Materie
Phys. Lett.
Physics Letters
Phys. Lett. A
Physics Letters A
Phys. Lett. B
Physics Letters B
Phys. Med. Bio.
Physics in Medicine and Biology
Phys. Met. Metallogr. (USSR)
Physics of Metals and Metallography
(USSR)
Phys. Rev.
Physical Review
Phys. Rev. A
Physical Review A
Phys. Rev. B
Physical Review B: Condensed Matter
Phys. Rev. C
Physical Review C: Nuclear Physics
Phys. Rev. D
Physical Review D: Particles and Fields
Phys. Rev. Lett.
Physical Review Letters
Phys. Status Solidi
Physica Status Solidi
Phys. Status Solidi A
Physica Status Solidi A: Applied Research
Phys. Status Solidi B
Physica Status Solidi B: Basic Research
Phys. Teach.
Physics Teacher
Phys. Today
Physics Today
Phys. Z.
Physikalische Zeitschrift
Phys. Z. Sowjetunion
Physikalische Zeitschrift der Sowjetunion
Planet. Space Sci.
Planetary and Space Science
Plasma Phys.
Plasma Physics
Proc. Cambridge Philos. Soc.
Proceedings of the Cambridge Philosophical
Society

Proc. IEEE
Proceedings of the IEEE
Proc. IRE
Proceedings of the IRE
Proc. Natl. Acad. Sci. U.S.A.
Proceedings of the National Academy of
Sciences of the United States of America
Proc. Phys. Soc. London
Proceedings of the Physical Society, London
Proc. Phys. Soc. London, Sect. A
Proceedings of the Physical Society, London,
Section A
Proc. Phys. Soc. London, Sect. B
Proceedings of the Physical Society, London,
Section B
Proc. R. Soc. London
Proceedings of the Royal Society of London
Proc. R. Soc. London, Ser. A
Proceedings of the Royal Society of London,
Series A: Mathematical and Physical Sciences
Prog. Theor. Phys.
Progress of Theoretical Physics
Publ. Astron. Soc. Pac.
Publications of the Astronomical Society of the
Pacific
Radiat. Eff.
Radiation Effects
Radio Eng. Electron. (USSR)
Radio Engineering and Electronics (USSR)
Radio Eng. Electron. Phys. (USSR)
Radio Engineering and Electronic Physics
(USSR)
Radio Sci.
Radio Science
RCA Rev.
RCA Review
Rep. Prog. Phys.
Reports on Progress in Physics
Rev. Geophys.
Reviews of Geophysics
Rev. Mod. Phys.
Reviews of Modern Physics
Rev. Opt. Theor. Instrum.
Revue d'Optique, Theorique et Instrumentale
Rev. Sci. Instrum.
Review of the Scientific Instruments
Russ. J. Phys. Chem.
Russian Journal of Physical Chemistry
Sci. Am.
Scientific American
Sol. Phys.
Solar Physics

Solid State Commun.
Solid State Communications
Solid State Electron.
Solid State Electronics
Solid State Phys.
Solid State Physics
Sov. Astron.
Soviet Astronomy
Sov. Astron. Lett.
Soviet Astronomy Letters
Sov. J. At. Energy
Soviet Journal of Atomic Energy
Sov. J. Low-Temp. Phys.
Soviet Journal of Low-Temperature
Physics
Sov. J. Nucl. Phys.
Soviet Journal of Nuclear Physics
Sov. J. Opt. Technol.
Soviet Journal of Optical Technology
Sov. J. Part. Nucl.
Soviet Journal of Particles and Nuclei
Sov. J. Plasma Phys.
Soviet Journal of Plasma Physics
Sov. J. Quantum Electron.
Soviet Journal of Quantum Electronics
Sov. Phys. Acoust.
Soviet Physics: Acoustics
Sov. Phys. Crystallogr.
Soviet Physics: Crystallography
Sov. Phys. Dokl.
Soviet Physics: Doklady
Sov. Phys. J.
Soviet Physics Journal
Sov. Phys. JETP
Soviet Physics: JETP
Sov. Phys. Semicond.
Soviet Physics: Semiconductors
Sov. Phys. Solid State
Soviet Physics: Solid State
Sov. Phys. Tech. Phys.
Soviet Physics: Technical Physics
Sov. Phys. Usp.
Soviet Physics: Uspekhi
Sov. Radiophys.
Soviet Radiophysics
Sov. Tech. Phys. Lett.
Soviet Technical Physics Letters
Spectrochim. Acta
Spectrochimica Acta
Spectrochim. Acta, Part A
Spectrochimica Acta, Part A: Molecular
Spectroscopy

Spectrochim. Acta, Part B
Spectrochimica Acta, Part B: Atomic
Spectroscopy
Supercon. Sci. Technol.
Superconductor Science and Technology
Surf. Sci.
Surface Science
Theor. Chim. Acta
Theoretica Chimica Acta
Trans. Am. Cryst. Soc.
Transactions of the American Crystallographic
Society
Trans. Am. Geophys. Union
Transactions of the American Geophysical
Union
Trans. Am. Inst. Min. Metall. Pet. Eng.
Transactions of the Amercian Institute of
Mining, Metallurgical and Petroleum
Engineers
Trans. Am. Nucl. Soc.
Transactions of the American Nuclear Society
Trans. Am. Soc. Mech. Eng.
Transactions of the American Society of
Mechanical Engineers
Trans. Am. Soc. Met.
Transactions of the American Society for
Metals
Trans. Br. Ceramic Society
Transactions of the British Ceramic Society
Trans. Faraday Society
Transactions of the Faraday Society
Trans. Metall. Soc. AIME
Transactions of the Metallurgical Society of
AIME
Trans. Soc. Rheol.
Transactions of the Society of Rheology
Ukr. Phys. J.
Ukrainian Physics Journal
Z. Anal. Chem.
Zeitschrift für Analytische Chemie
Z. Angew. Phys.
Zeitschrift für Angewandte Physik
Z. Anorg. Allg. Chem.
Zeitschrift für Anorganische und Allgemeine
Chemie
Z. Astrophys.
Zeitschrift für Astrophysik
Z. Elektrochem.
Zeitschrift für Elektrochemie
Z. Kristallogr. Kristallgeom. Krystallphys. Kristallchem.
Zeitschrift für Kristallographis, Kristallgeome-
trie, Krystallphysik, Kristallchemie

Z. Metallk.
　　Zeitschrift für Metallkunde
Z. Naturforsch.
　　Zeitschrift für Naturforschung
Z. Naturforsch. Teil A
　　Zeitschrift für Naturforschung, Teil A Physik,
　　Physikalische Chemie, Kosmophysik

Z. Phys.
　　Zeitschrift für Physik
Z. Phys. Chem. (Frankfurt am Main)
　　Zeitschrift für Physikalische Chemie (Frankfurt
　　am Main)
Z. Phys. Chem. (Leipzig)
　　Zeitschrift für Physikalische Chemie (Leipzig)

ABERRATION

An ideal optical instrument (microscope, telescope, camera, etc.) should form a real or virtual image of any object whose geometrical shape is identical to that of the object but possibly of different size and orientation. The deviations from the perfect imaging are called aberrations and can be classified as chromatic and monochromatic.

Chromatic aberrations are due to the variation of the refractive index of materials through which the light must pass, according to the wavelength of the light. The net effect is to give the instrument a different focal length for different colors.

Monochromatic aberrations are formed by optical instruments that have an axis of symmetry, for example, are cylindrical in their configuration of lenses and mirrors. The standard theory of image formation in such an instrument was developed by Karl Friedrich Gauss in 1843. It assumes that the rays of light coming from the object are paraxial, that is, close to and making a sufficiently small angle u with the axis that one may approximate $\sin(u)$ by simply u (measured in radians). This "first order" or Gaussian theory predicts that a planar object will form a planar image, that is, every point in the object plane comes to a sharp focus at a point in the image plane, assuming both planes are perpendicular to the axis. However, the trigonometric approximation is of limited accuracy. Although sufficient for, say astron-omy, where objects are of small angular size, the rise of photography in the mid-nineteenth century demanded better image quality. A more accurate theory, based upon the better approximation $\sin(u) = u - u^3/6$, was developed by Ludwig Seidel in 1856. He showed that the inclusion of the cubic term in the approximation led to five new terms in the equations for image formation. The goal of instrument designers is to reduce these third-order terms to the point where their presence will not lead to objectionable deterioration of the image, and depends, in part on what the user is willing to pay. Each of the five "Seidel aberrations" has a traditional name: spherical aberration, coma, astigmatism, curvature of field, and distortion.

Spherical aberration arises when light rays from a point on (or near) the optical axis but passing near the outer edges of the system relative to the axis are brought to a different focal point from those passing nearer to the axis. This causes the image to be blurred. It is the only aberration for an on-axis object. Coma is a blurring of the image caused by the variation in image position according to whether the rays pass through the outer or central regions of the system. It occurs only for off-axis objects. In astigmatism, radial lines are brought to a different focus from perpendicular (i.e., tangential) lines. This blurring of the image may be caused by a small cylindrical distortion in the lens or mirror system or by improper centering of elements. Curvature of field is closely related to astigmatism. It implies that a planar object perpendicular to the axis is brought to an image that is a curved surface. In distortion, a sharp image of a square object may not be square; it may have sides that bow inward (pin cushion distortion) or outward (barrel distortion).

Today's computers can trace exactly hundreds of thousands of rays through an optical system with sufficient speed to allow for adjustment of the number, structure, position, and composition of optical elements to overcome the aberrations to the desired degree without invoking the mathematical formulas of Seidel and his successors.

See also: ABERRATION, CHROMATIC; ABERRATION, SPHERICAL; IMAGE, OPTICAL; OPTICS; OPTICS, GEOMETRICAL; VISION

Bibliography

BORN, M., and WOLF, E. *Principles of Optics*, 6th ed. (Pergamon, Oxford, Eng., 1980).

WELFORD, W. T. *Aberrations of the Symmetrical Optical System* (Academic Press, New York, 1974).

MICHAEL LIEBER

ABERRATION, CHROMATIC

The index of refraction depends upon the wavelength. According to Snell's law, when rays of different wavelengths (different colors) impinge on the same point of an interface, they have different deflections. When polychromatic light passes through a simple lens, the two consecutive refractions on either side of the lens result in a differential deflection for each wavelength. The lens has a different focal point for each wavelength, and the image is

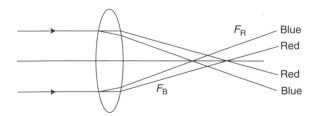

Figure 1 Axial chromatic aberrations.

blurred. The term "chromatic aberration" refers to the imperfections of an image formed by a lens due to variations in the position of the focal point of the lens as a function of the wavelengths present in the initial rays. Figure 1 illustrates axial chromatic aberrations for a collimated beam made of a mixture of blue and red light. In Fig. 1, F_B and F_R refer to the focal points of red and blue colored lights, respectively. Since the index of refraction for blue light is slightly greater than the index of refraction for the red light, the focal distance for the blue light is smaller than that of the red light. For a lens of nominal focal length of 40 in., the variation of the focal length over the visible spectrum may amount to about one in., the precise number depending on the index of refraction of the lens. In addition to the axial chromatic aberrations, the wavelength dependence in the transverse direction causes the lateral chromatic aberration.

With a thick simple converging lens, it is easy to observe chromatic aberrations. The real image of a point-source that produces polychromatic light is surrounded by a halo. As one moves the observation screen away from the lens, the halo changes its color from orange-red to blue-violet. The best image appears between these extremes.

The word "achromatic" refers to a lens system that minimizes chromatic aberration. It is possible to build achromatic lenses by combining a converging lens with a divergent one. This combination is called a doublet. The chromatic aberration in one of the lenses is partially canceled by the chromatic aberration in the second element. Such an arrangement corrects only for two specific wavelengths. For more demanding applications, the achromatic lens is made of many elements often cemented together. Modern designs of achromatic lenses are aided by ray-tracing computer programs. Although it is impossible to eliminate all aberrations, it may be possible to design lens systems that minimize the most troublesome aberration for the application.

Historically, the chromatic aberration was a very serious problem for the makers of optical instruments in the seventeenth century. Isaac Newton gave up on eliminating the chromatic aberrations and shifted his efforts from the refracting to the reflecting telescope since chromatic aberrations do not occur in mirrors. The invention of achromat by Chester Moor Hall between 1729 and 1733 and further developments by John Dollond in 1758 had significant impact on the design of optical instruments for navigation, astronomy, and surveying.

See also: ABERRATION; ABERRATION, SPHERICAL; LENS; LENS, COMPOUND; OPTICS; REFRACTION; SNELL'S LAW; WAVELENGTH

Bibliography

BORN, M., and WOLF, E. *Principles of Optics,* 2nd ed. (Macmillan, New York, 1964).

HECHT, E., and ZAJAC, A. *Optics* (Addison-Wesley, Reading, MA, 1979).

JAECKS, D. H. "Developments in 18th-Century Optics and Early Instrumentation" in *The History and Preservation of Chemical Instrumentation,* edited by J. T. Stock and M. V. Orna (Reidel, New York, 1986).

ORHAN YENEN

ABERRATION, SPHERICAL

Optical systems containing mirrors and/or lenses exhibit spherical aberration if they do not transform object points on the optical axis into corresponding image points. Perfect point-to-point imaging—ignoring diffraction-imposed limitations—requires special conditions, which are typically absent in single element optics. Consider, for example, light rays falling parallel to the optical axis onto a spherical mirror. Rays with increasing distance from the optical axis are reflected to cross the optical axis ever closer to the mirror. The envelope of all reflected rays is called a caustic and contains all the reflected light intensity. At a point somewhere between the foci of paraxial and marginal rays, the caustic has a minimum cross section, which is called the circle of least confusion and represents the blurred image of the infinitely distant object point (see Fig. 1). At the cost of diminished image intensity, the radius of this circle can be decreased by reducing the aperture size and thereby the distance of the marginal rays from the optical axis. Paraboloidal mirrors or spherical mirrors augmented by appropriately shaped, refracting entrance plates avoid spherical aberration altogether (the latter solution is used in the so-called Schmidt telescope).

Completely analogous to spherical aberration of mirrors, a single spherical, refracting surface as well as a spherical lens with two refracting surfaces produce blurred images of axial object points. In this case, a formal theory involves, in its simplest form, ray tracing using only the first two leading terms in an expansion of the sine function in Snell's law. Within this third-order theory (third order in the angle of incidence), spherical aberration is one of five basic types of aberrations and the only one affecting images of the on-axis points. The theory predicts a cubic variation of the spherical aberration with focal length and a dependence on the lens shape factor

$$q = \frac{r_1 + r_2}{r_1 - r_2},$$

where r_1 and r_2 are the radii of curvature of the two lens surfaces, respectively. From all the lenses with the same focal length, a lens with a shape factor

$$q_{best} = \frac{2(n^2 - 1)}{n + 2},$$

where n is the index of refraction of the lens glass, focuses parallel rays into the smallest circle of least confusion. With a typical value of $n \approx 1.5$ for glass, the best form lens should have $q_{best} \approx 0.7$. Since $q = 1$ corresponds to a plano-convex lens with the curved side toward the incoming parallel light, such a configuration is often used as an economical, although not complete correction for spherical aberration.

Spherical aberration can be entirely eliminated by different methods. Because it varies as the cube of the focal length, one can find a combination of one positive and one negative lens that has zero spherical aberration. Such a doublet has the added advantage of sufficient design flexibility so that one can simultaneously achieve minimal chromatic aber-

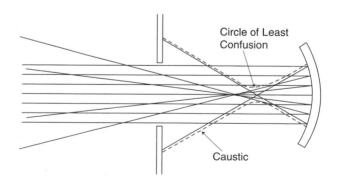

Figure 1   Reflected rays.

3

ration. Improvements in lens manufacturing also have made it possible to overcome spherical aberration for specific wavelengths by either giving lenses the appropriate, nonspherical shape or, alternatively, providing spherically shaped lenses with a graded refractive-index profile, varying axially in the material of the lens.

See also: ABERRATION; ABERRATION, CHROMATIC; IMAGE, OPTICAL; LENS; SNELL'S LAW

Bibliography

JENKINS, F. A., and WHITE, H. E. *Fundamentals of Optics,* 3rd ed. (McGraw-Hill, New York, 1957).
O'SHEA, D. C. *Elements of Modern Optical Design,* (Wiley, New York, 1985).

LUTZ HÜWEL

ABSOLUTE ZERO

Is there a limit as to how cold matter can be? The answer is yes. In contrast, there is no limit as to how hot it can be. This can be understood as follows. In classical physics, the temperature is a measure of the kinetic energy or energy of motion of the atoms in the matter. There is a lower limit on the kinetic energy that occurs at zero motion of the atoms, or when all motion stops. According to quantum mechanics, the minimum temperature occurs when the matter is in the lowest energy or ground state. On the other hand, there is no upper limit to the kinetic energy that the atoms can have.

The lowest possible temperature is $-273.15°C$ on the Celsius temperature scale. In low-temperature physics, the Kelvin temperature scale is used. Zero Kelvin represents absolute zero.

Because of quantum effects, there is still motion in the ground state or at absolute zero. One case where this readily shows up is in ^{4}He. At absolute zero in any solid, there is motion of quantum mechanical origin, which is called zero-point motion. For a ^{4}He crystal, this motion leads to a velocity of the He atoms of approximately 500 m/s. Then, because of the small attraction between the He atoms, the solid will tear apart and form a liquid no matter how low the temperature unless a pressure of more than 25 atm is applied. We also have motion of the electrons in a metal in which some of the electrons are moving with velocities of approximately 10^6 m/s. This is more than 2 million mph, and it occurs even at absolute zero.

According to the third law of thermodynamics, it is impossible to go all the way to absolute zero, although it is possible to get as close as we like. The closer we get to zero temperature, the harder it is to make further progress. It is approximately true that it is as difficult to cool from one degree to 0.1 K as it is to cool from 10 K to 1 K. So, by making additional effort, we can keep going down by factors of 10, but we can never get to zero.

There is more than one method of cooling, and generally several are used one after the other to reach the lowest temperatures. The initial refrigeration is with a machine similar to the one in the home refrigerator or air conditioner, and with it one can cool ^{4}He gas down to where it condenses into a liquid at 4.2 K.

We can go to still lower temperatures by allowing part of the He to boil away. This is the same effect as the cooling of our wet skin by the evaporation of water. By pumping away the He gas, we increase the evaporation rate and lower the temperature to about 0.9 K.

For the next stage, one can use the rare isotope of helium, ^{3}He, which boils at 3.2 K at atmospheric pressure. ^{4}He can be pumped on to cool below 3.2 K and then used to condense ^{3}He gas to the liquid. Finally, the ^{3}He liquid can be pumped on to cool it down to approximately 0.3 K.

The next step in going to lower temperatures is to use a helium-3/helium-4 dilution refrigerator that can operate routinely down to 0.015 K, or 15 mK. Below 0.8 K, a He liquid containing both ^{3}He and ^{4}He separates into two immiscible liquids (as a consequence of the third law of thermodynamics), the denser one being rich in ^{4}He, which goes to the bottom, and the other being rich in ^{3}He, which goes to the top. At these low temperatures, the ^{4}He acts like an inert background, and we can forget about it. The ^{3}He rich liquid behaves like liquid ^{3}He, while the ^{4}He rich liquid behaves like gaseous ^{3}He. The cooling is accomplished by "evaporating" the liquid ^{3}He into the "gaseous" ^{3}He.

One can go still lower than 15 mK by using adiabatic demagnetization. The spin of a nucleus behaves like a very small bar magnet or compass needle. If material with these spins is placed in a large magnetic field, the tiny magnets or spins tend to rotate to line up with the magnetic field. On the other hand, thermal motion tends to randomize the directions of the spins. At high temperatures, we need a very strong magnetic field to overcome the thermal disorder, while at low temperatures the field does not need to be so strong. We can turn this around and say that if we see the spins lined up in a small magnetic field, the spins must be at a low temperature.

Using these ideas and starting at 20 mK, we apply a large external magnetic field, thereby lining up the spins as much as possible. The sample must be in contact with other material (a thermal reservoir) at 20 mK to eliminate the heat generated while the spins are lining up. Then the thermal contact with the reservoir is broken and the magnetic field is turned down. The spins are still lined up but now in a low field. So they are at a lower temperature than 20 mK.

For some materials, such as silver (Ag), the interaction between the nuclear spins and the electrons and phonons (lattice vibrations) is quite weak so that the nuclear spins can maintain a much lower temperature for the order of three hours. For other materials, such as indium, the interaction of the nuclear spins with the rest of the material is much stronger so that the electron and lattice temperature remains nearly equal to the nuclear spin temperature at all times. This means that for indium (In), when the spins are cooled down by the adiabatic demagnetization, they are immediately warmed up by the rest of the material, which in turn is cooled. Thus one cannot reach as low a temperature with In as with Ag, but the temperature one does reach is the temperature of the entire material and not just of the spins.

One might well ask, why cool things down to such low temperatures? Let us examine what has been discovered. In 1908 Kamerlingh Onnes in Leiden succeeded in liquefying ^{4}He, and in 1911 he found that Hg lost all electrical resistance at about 4 K. He had discovered superconductivity. Years later it was discovered that ^{4}He becomes a superfluid at 2.2 K. (It has zero resistance to flow under certain conditions.) In 1962, when physicists could finally get to the sub-milliKelvin temperature range, Douglas D. Osheroff, Robert C. Richardson, and David M. Lee at Cornell discovered the long sought after superfluidity in liquid ^{3}He with a transition temperature 2.5 mK. Since a ^{3}He atom is a Fermion, the same as an electron, while ^{4}He is a boson, the superfluidity in ^{3}He has more in common with superconductivity in metals than it does with superfluidity in ^{4}He. Thus,

many exciting things have been found at these very low temperatures.

The record low temperature for a sample as a whole is 7 μK (0.000007 K). One can go to the much lower temperature of 1 nK (0.000000001 K) for nuclear spins alone, in which they are one thousand times colder than the rest of the sample. An intense area of research at these extremely low temperatures is a study of the magnetism that occurs with the nuclear magnetic moments interacting among themselves—so-called nuclear magnetism.

A recent development in ultralow-temperature physics is the discovery of a Bose–Einstein condensation of a dilute gas of ^{87}Rb atoms near a temperature of 170 nK. This is the first observation of the Bose–Einstein condensation in essentially noninteracting atoms. The work was carried out by Eric Cornell, Carl Wieman, and their group in 1995 at the National Institute of Standards and Technology and at the University of Colorado, both in Boulder. The atoms were cooled first by laser cooling to approximately 10 μK. Then these atoms were trapped by a special arrangement of magnetic fields and cooled to the final temperature by evaporation.

In the Bose–Einstein condensation, a large fraction of the atoms condense into the same extended quantum state. This is a quantum mechanical phenomenon that can occur only with bosons, particles with an integral spin. It was first predicted for noninteracting bosons by Albert Einstein in 1924, building on work by Satyendranath Bose in the same year. A necessary condition for this to happen is for the quantum-mechanical wave functions of the individual atoms to overlap each other to a large extent, forming a coherent state. Very low temperatures are necessary for the atomic wave functions to spread out sufficiently for this to occur.

See also: CONDENSATION, BOSE–EINSTEIN; JOULE–THOMSON EFFECT; LIQUID HELIUM; NEGATIVE ABSOLUTE TEMPERATURE; REFRIGERATION; TEMPERATURE; TEMPERATURE SCALE, KELVIN

Bibliography

LOUNASMA, O. V. "New Methods for Approaching Absolute Zero." *Sci. Am.* **221** (Dec.), 26–35 (1969).

LOUNASMA, O. V. "Nuclear Magnetic Ordering at Nano-Kelvin Temperatures." *Phys. Today* **41** (Oct.), 26–33 (1989).

MENDELSSOHN, K. *The Quest for Absolute Zero: The Meaning of Low Temperature Physics* (McGraw-Hill, New York, 1966).

POBELL, F. "Solid-State Physics at Microkelvin Temperatures: Is Anything Left to Learn?" *Phys. Today* **45** (Jan.), 34–40 (1993).

JAMES C. SWIHART

ABSORPTION

See SOUND ABSORPTION

ACCELERATION

The acceleration of an object is the rate at which its velocity changes with time. Velocity is a vector quantity, possessing both size (called speed) and direction; consequently, acceleration is also a vector quantity, as we shall see explicitly below. Vector quantities are usually written in boldface ($\mathbf{v}$) or with an arrow over the symbol ($\vec{v}$). The change of speed, direction, or both is referred to as an acceleration of the object. When the speed of an object is decreasing with time the word "deceleration" is sometimes used to describe this change.

Average Acceleration

A precise definition of average acceleration ($\mathbf{a}$), during a particular time interval, is the ratio of the change of velocity ($\Delta \mathbf{v}$) to the size of the time interval:

$$\mathbf{a} = \frac{\Delta \mathbf{v}}{\Delta t}.$$

More explicitly, for a time interval that runs from t_1 to t_2, $\Delta \mathbf{v}$ is the vector difference between the velocity at t_2 ($\mathbf{v}_2$) and the velocity at t_1 ($\mathbf{v}_1$), $\Delta \mathbf{v} = \mathbf{v}_2 - \mathbf{v}_1$, and $\Delta t = t_2 - t_1$.

Instantaneous Acceleration

To obtain the acceleration of an object at a particular instant, we must consider smaller and smaller time intervals containing that instant. The limit of the average acceleration as the time interval approaches zero is the instantaneous acceleration. This limit is defined in differential calculus as the derivative of the velocity with respect to time,

$$\mathbf{a} = \lim_{\Delta t \to 0} \frac{\Delta \mathbf{v}}{\Delta t},$$

and is written

$$\mathbf{a} = \frac{d\mathbf{v}}{dt}.$$

Units

In Standard International (SI) units, the unit of $\mathbf{v}$ is meters per second (m/s) and the unit of acceleration is $(\text{m/s}) \cdot \text{s}^{-1}$, which follows from the definition of $\mathbf{a}$. The unit $(\text{m/s}) \cdot \text{s}^{-1}$ is usually abbreviated m/s^2.

Linear Motion

For motion along a straight line, we dispense with the boldface and arrow for vectors; their directions are indicated by the use of plus or minus signs. We can plot the velocity (v) versus t and calculate the average acceleration as the ratio of Δv to Δt, which are shown in Fig. 1.

The instantaneous acceleration at any given time t_p is equal to the slope of the line tangent to the v–t curve at t_p. If we know v as a function of t, then the instantaneous acceleration can be obtained by taking the derivative of v with respect to t.

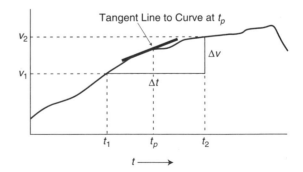

Figure 1 Plot of velocity versus time. The average acceleration is the ratio of Δv to Δt.

Motion in Two or Three Dimensions

The equations (above) that define **a** in terms of $\Delta\mathbf{v}$ are more complicated in two or three (or more) dimensions. The vector difference, $\Delta\mathbf{v} = \mathbf{v}_2 - \mathbf{v}_1$, can be obtained graphically by the usual method of subtracting vectors or can be obtained by treating the Cartesian (i.e., *x, y,* and *z*) components separately. In the component method, each component is treated as a one-dimensional problem and the components of **a**, so obtained, are then combined to form the vector **a**.

Acceleration Associated with Changes of Direction—Uniform Circular Motion

An object in uniform circular motion is moving in a circle at constant *speed.* Nevertheless its velocity is changing with time. The vector difference between $\mathbf{v}_2$ and $\mathbf{v}_1$ as $t_2 \to t_1$ points toward the center of the circle; consequently, this type of acceleration is called centripetal acceleration. The size of the centripetal acceleration is v^2/r, where v is the speed and r is the radius of the circle. For an object to move in a circle, it must be experiencing a physical centripetal force to cause the required centripetal acceleration.

Historical Background and Connections to Other Concepts

The first major contributor to our modern understanding of acceleration was Galileo Galilei, who spent many years studying the behavior of moving objects. He used logical and mathematical reasoning as well as ingenious experiments. For example, in the study of falling objects, he slowed down the motion by having the objects roll down very smooth inclined planes; consequently, using the crude methods of measuring time available to him, he found the connection between distance traveled and time (d varies as t^2). Galileo's studies overturned the long-held (Aristotelian) belief that heavier objects fall faster than lighter objects, in proportion to the ratio of their weights. He showed that all objects, when made in the same shape, fall at the same rate (i.e., with the same constant acceleration); the shape dependence of the acceleration downward is due to air resistance. His study of uniform (unaccelerated) motion led to his discovery of the law of inertia. He enunciated the law of inertia for moving objects by imagining an object moving on an ideally smooth infinite horizontal plane. He proposed that it would continue its uniform motion indefinitely. He also proposed that were an object to fall from a great height, it would eventually reach a speed at which the air resistance would cancel the pull of gravity. It would thereafter descend with uniform motion (constant velocity). The fact that uniform motion is the natural motion of objects when no net outside force is present is the essence of the law of inertia. However, this would not be true to an accelerated observer (i.e., in an accelerated frame of reference. (In modern language, we refer to observers in relative motion with the words "frame of reference.") Frames of reference in which the law of inertia holds are called inertial frames and Galileo's discovery of the law of inertia on Earth shows that it is very nearly an inertial frame. He also showed that if an object is dropped on a moving ship, it will fall straight down to those on the ship, while to those on shore it will have just the right horizontal velocity to keep up with the ship's motion; thus, to those on shore it moves in a parabola. In the frame of reference of those on the ship, the motion of an object dropped on the ship is the same as the motion of an object dropped on shore is to those on shore. This leads to the notion that in a reference frame in uniform motion (with respect to an inertial frame) the laws of nature must be mathematically identical to their form in an inertial frame. (In other words, the frame in uniform motion is also an inertial frame.) This concept is referred to as Galilean relativity.

The foremost contributor to our modern understanding of motion, force, and the formulation of classical (pre-twentieth-century) physics was Isaac Newton. He created the modern form of the law of inertia, which is sometimes called Newton's first law: In an inertial frame, an object at rest tends to remain at rest and an object in uniform motion tends to remain in uniform motion, unless acted upon by an outside force. Thus, in the absence of outside forces, the acceleration is zero. Newton's second law provided the precise connection between the motion of an object and the forces exerted on it. It showed that the acceleration of an object is proportional to the net force **F** (a vector sum of all the external forces on it):

$$\mathbf{F} = m\mathbf{a} = m\frac{d\mathbf{v}}{dt},$$

where m is the mass of the object. There is another formulation of this law:

$$\mathbf{F} = \frac{d\mathbf{p}}{dt},$$

where $\mathbf{p} = m\mathbf{v}$ is the momentum of the object. (This form of Newton's second law, in terms of momentum, is incorporated into Einstein's generalization of the laws of motion.) Newton discovered the mathematical form of the force of gravity between objects, which he used with his second law to account for the astronomical observations of the motion of planets and moons, as well as the motion of ordinary objects on the surface of Earth. (He invented calculus in order to accomplish this feat.) This deterministic, precise mathematical formulation of the laws of motion became the prototype of the formulations of physical laws. This deterministic understanding of the world also had great impact on philosophical concepts. A new worldview, with strict mathematical laws determining the future, was widely disseminated even in popular writings.

Accelerated Frames and "Fictitious" or Apparent Forces

A frame of reference that is accelerated with respect to an inertial frame is not an inertial frame; Newton's laws of motion do not hold in the accelerated frame. Objects at rest in the inertial frame (experiencing no net force) are measured to be accelerating (in the opposite direction) in the accelerated frame. "Fictitious" (or apparent) forces are often postulated to take account of the noninertial behavior of objects in accelerated frames. Then Newton's laws can be used in those frames. Common examples are the centrifugal forces in rotating frames (e.g., on molecules in a centrifuge and on patrons in a rotating carnival ride). These outward (apparent) forces are felt because the rotating frame is noninertial; they are not being exerted by any physical interaction. Other examples include the forces we feel in accelerating or decelerating vehicles.

Relativity and Gravity

Albert Einstein added a new dimension to our understanding of the motion of objects when he incorporated electromagnetism into the principle of relativity. In order for the laws of electromagnetism

to be the same in frames moving uniformly with respect to a given inertial frame, the transformations of position and time from one frame to the other have to be modified. This leads to time playing the role of a fourth dimension and to the mass in the momentum increasing as the speed of the particle approaches the speed of light. Einstein also showed that no massive object can be accelerated to the speed of light (c), that is, its speed must be less than c. His reasoning also led to the possible existence of massless particles, that are always traveling at the speed of light. Photons (quanta of light) are examples of massless particles. This expansion of the principle of relativity is called the special theory of relativity. Einstein also enunciated the principle of equivalence, which states that when an observer is in free fall (e.g., in a closed elevator), the effects of gravity are not observed. Everything in the elevator is falling at the same rate so that no relative motion will be observed. This is referred to as "weightlessness" in orbiting spaceships. In general, in an accelerated frame (like the falling elevator) we experience effects similar to the pull of gravity. (In the falling elevator the apparent forces oppose the pull of gravity.) Einstein generalized the principle of relativity to include accelerated frames of reference; he showed that gravity can be thought of as resulting from a distortion or curving of space time due to the presence of masses. This enlarged formulation of relativity, which includes the theory of gravitation, is called general relativity.

Accelerometers

Meters have been constructed to measure acceleration; they are called accelerometers. Most exploit the existence of an apparent force in an accelerated frame. Usually the apparent force causes a part of the accelerometer to move, and that change in position is detected electrically or electronically. Accelerometers are commonly used in vehicle design work; they are the heart of the triggering devices of the air bags in our cars.

Accelerators

The study of matter often involves bombarding a target (or a beam of particles) with a beam of high energy particles, to observe the resulting interactions. In order to accelerate the beam particles to high energies, many very sophisticated accelerators have been constructed. Some have oval beam paths that are larger than cities. Some of the information obtained from these huge accelerator probes the fundamental structure of matter, as well as conditions in the very early universe.

See also: ACCELERATION, ANGULAR; EINSTEIN, ALBERT; FRAME OF REFERENCE, INERTIAL; GALILEI, GALILEO; MOTION; MOTION, CIRCULAR; NEWTON, ISAAC; NEWTONIAN MECHANICS; NEWTON'S LAWS; RELATIVITY, SPECIAL THEORY OF; VECTOR; VELOCITY

Bibliography

COHEN, I. B. *The Newtonian Revolution, with Illustrations of the Transformation of Scientific Ideas* (Cambridge University Press, Cambridge, Eng., 1980).

COHEN, I. B. *The Birth of a New Physics,* revised and updated (W. W. Norton, New York, 1985).

GIANCOLI, D. C. *The Ideas of Physics* (Harcourt Brace Jovanovich, New York, 1974).

WILLIAM B. ROLNICK

ACCELERATION, ANGULAR

Angular acceleration is a concept that arises in the study of the orientation of a rigid object (body). It is usually used to describe the motion of the object when it is rotating about a fixed axis. A reference axis in space is drawn from the axis of rotation, in a plane perpendicular to the rotation axis. The angle (θ) between a similar line in the object (body axis), and that space axis denotes the angular position of the object, as shown in Fig. 1.

The angle θ is considered positive (negative) when the body axis has moved counterclockwise (clockwise) with respect to the reference axis. The rate at which the angle θ changes with time is the angular velocity (ω) of the object. The rate at which the angular velocity ω changes with time is the angular acceleration (α). These relationships among angular position, angular velocity, and angular acceleration, are identical to the relationships among the position, velocity, and acceleration of linear motion.

Average Angular Acceleration

The definition of average angular acceleration (α), during a particular time interval, is the ratio of the change of angular velocity ($\Delta\omega$) to the size of the time interval:

$$\alpha = \frac{\Delta\omega}{\Delta t}.$$

More explicitly, for a time interval that runs from t_1 to t_2, $\Delta\omega$ is the difference between the angular velocity ω_2 at t_2 and the angular velocity ω_1 at t_1 (i.e., $\Delta\omega = \omega_2 - \omega_1$, and $\Delta t = t_2 - t_1$).

Instantaneous Angular Acceleration

To obtain the angular acceleration of an object at a particular instant, we must consider smaller and smaller time intervals containing that instant. The limit of the average angular acceleration as the time interval approaches zero is the instantaneous angular acceleration. This limit is defined in differential calculus as the derivative of the angular velocity with respect to time,

$$\alpha = \lim_{\Delta t \to 0} \frac{\Delta\omega}{\Delta t}$$

and is written as

$$\alpha = \frac{d\omega}{dt}.$$

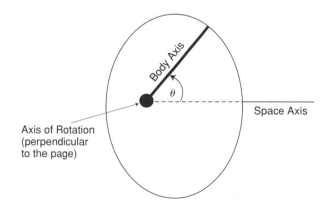

Figure 1   Angular positioning of an object.

Usual Units

The unit of angle (θ) can be either degrees (360° corresponds to a full revolution), radians (2π radians corresponds to a full revolution), or revolutions; the unit of ω is either degrees per second (°/s), radians per second (rad/s), revolutions per second (rev/s), or revolutions per minute (rpm); and the unit of acceleration is either (degrees per second) per second ((°/s) /s), usually abbreviated as degrees per second squared (°/s²), radians per second squared (rad/s²), or revolutions per second squared (rev/s²).

Graphical Representation

We can plot the angular velocity (ω) versus t and calculate the average angular acceleration as the ratio of $\Delta\omega$ to Δt, which are shown in Fig. 2. The instantaneous angular acceleration at any given time t_p is equal to the slope of the line tangent to the ω-t curve at t_p.

Relationship to Tangential Acceleration

Each point on the rigid object moves in a circle around the axis of rotation as the object rotates. A point on the object, which is a perpendicular distance r from that axis, moves through an arc of length $r\theta$, with θ expressed in radians, when the object rotates through an angle θ. The point has an instantaneous velocity and instantaneous acceleration that are tangent to that arc at the instantaneous location of that point, and are related to the instantaneous angular velocity (ω) and the instantaneous angular acceleration (α) by the equations

$$v_{\text{tangential}} = r\,\omega$$

and

$$a_{\text{tangential}} = r\,\alpha,$$

with ω expressed in radians per second and α expressed in radians per second squared.

Relationship of Angular Acceleration to Torque and Moment of Inertia

Consider a rigid body constrained to rotate about an axis. When a force is exerted on the rigid body, it causes a twist (torque) about that axis, which depends on the direction and location of the force. The torque (τ) may be clockwise or counterclockwise about the axis. When there are many forces, the net torque on the object will be the difference between (the sum of the counterclockwise torques) and (the sum of the clockwise torques). A nonzero net torque causes a corresponding angular acceleration of the object. The size of that angular acceleration depends on the mass distribution of the object about the rotation axis. The relevant mass-distribution quantity is called the moment of inertia of the object (about that axis) and is denoted by the letter I. The relationship between the torque and angular acceleration is

$$\tau_{\text{net}} = I\,\alpha,$$

for fixed I, which is analogous to the equation $F = m\,a$ for linear motion.

Effect of a Change of the Moment of Inertia

Even in the absence of a torque, the angular velocity of an object may change with time. The equation relating the net torque to the angular motion of the object, allowing for a changing I, is

$$\tau_{\text{net}} = \frac{dL}{dt},$$

where L is the angular momentum, defined as $L = I\,\omega$.

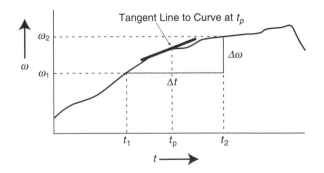

Figure 2 Plot of the angular velocity versus time to calculate the average angular acceleration.

Thus, the net torque equals the time rate of change of the angular momentum, analogous to the relationship of force to the change of momentum for linear motion. In the absence of a net torque, the angular momentum remains constant. However, if the moment of inertia changes with time, then the angular velocity must change as well so as to keep the angular momentum unchanged. This is the basis of the technique used by ice skaters to increase their angular velocities by pulling in their arms (which results in a decrease of their moments of inertia).

Three-Dimensional Rotations

When an object is not constrained to rotate about a fixed axis, three angles are required to describe its orientation. The orientation of three body axes with respect to three space axes is employed. The usual description employs two sets of orthogonal axes (one set fixed in the body and one space set) and angles called "Euler angles." The corresponding changes of angular velocities are not usually referred to as "angular accelerations." The behavior of these angles with respect to time is described by equations known as "the Euler equations."

See also: ACCELERATION; INERTIA, MOMENT OF; MOTION; VELOCITY, ANGULAR

Bibliography

SERWAY, R. A., and FAUGHN, F. S. *College Physics,* 3rd ed. (Saunders, Fort Worth, TX, 1992).

WILLIAM B. ROLNICK

ACCELERATOR

Particle accelerators denote the general class of devices that use electric and magnetic fields to raise the energy of electrons, protons, and ions to energies far above the values found in typical thermally produced beams of particles. In this process the particle velocities are increased (particles are accelerated), often to levels extremely close to the speed of light. The technologies used in modern accelerators are diverse. They include radio-frequency power generators, fast electronics, complex diagnostic instrumentation and controls, ultrahigh vacuum systems, superconducting materials, cryogenics, and powerful, high-precision magnets.

Purpose of Accelerators

Throughout the twentieth century, physicists have used many forms of ionizing radiation as microprobes (generalizing the idea of a microscope) to study the structure of matter on an ever smaller scale. Particle accelerators are the largest and most complex of these microprobes. For the study of nuclear and subnuclear processes, the probe particles (and often the targets) are protons, electrons, and ions. The first sources of radiation for the new "microscopy" were naturally radioactive materials such as radium. Radioactive materials can provide both particle probes, such as α particles (helium nuclei), and photon probes, such as x rays. Accelerators now produce all these probes at high energy. If the desired probes are x rays or gamma rays, the starting particles in the accelerator are electrons that produce photons in the desired energy range (x or gamma rays) via the bremsstrahlung or synchrotron radiation process.

As the resolution of any radiation probe is limited by the wavelength of the radiation, and as quantum mechanics tells us that the wavelength of the probe radiation is directly proportional to its energy, the advance of microscopy has demanded ever higher energy probes. Whether to improve the signal-to-noise ratio or to uncover rare processes in acceptably short periods of time, the advance of microscopy also demands ever brighter sources of radiation. These two characteristics, probe energy and probe brightness (or intensity), define the technical frontiers that continue to challenge accelerator builders.

General Characteristics and Principles of Operation

Every accelerator begins with an injector (a directional source of low-energy, charged particles such as electrons, protons, or other ions) plus associated high-voltage electrodes that extract the beam. The source generates the particle beam, which is characterized by an initial energy, current, cross-sectional size, and average angular divergence. The product of the beam size and divergence determines the beam quality, or emittance (smaller is better), and, ultimately, the brightness (current divided by emittance) of the final high-energy beam of particles. In its simplest form, the accelerator increases the particle energy by subjecting the beam to a high-voltage electric field.

The earliest particle accelerators used electrostatic fields to accelerate electrons, protons, and ions. However, the energies were limited by the ability of insulators (and the imperfect vacuum) in the accelerator to sustain high electric fields without breakdown. Electrostatic acceleration of electrons and ions to energies of 30 to 100 keV (kilo-electron-volt) is still used in the injectors of many modern accelerators.

If the accelerating voltage is applied for a very brief period (less than 1 μs), or if the field oscillates at radio frequencies (rf), the accelerator can sustain higher electric fields than in the case of static voltages. The use of rf fields presents a potential difficulty: The sign of the field rapidly alternates between accelerating and decelerating. Hence the direction of the force on a particle may also oscillate. Two principal approaches to overcoming this difficulty have been in use for decades.

Linear Accelerators

In 1928 Rolf Wideröe first exploited rf fields to accelerate particles in a short, linear accelerator ("linac") structure composed of coupled, resonant, radio-frequency cavities. Wideröe's approach (and that of the many subsequent variations of the rf linac) exploits the fact that the electromagnetic fields vary sinusoidally both in space and in time. If the phase velocity of the field is properly matched to the velocity of the particles, then as the particles run down the accelerator, they continually leave the regions where the accelerating field is decreasing (or even reversing) and enter regions where the field is increasing. Thus the beam experiences continual acceleration like a surfer riding a wave.

For protons or ions, the particle velocities can change markedly during the acceleration process. Consequently, the phase velocity of the wave v_{ph} in the accelerating structure must likewise increase along the structure. (The electron has a much smaller mass and hence reaches relativistic velocity at much lower energy than a proton or any ion.) For electrons, which can be injected at relativistic velocities (i.e., very close to the speed of light), the accelerator structure is typically designed to have a constant phase velocity equal to the speed of light.

The spatial variation of the electric field also demands that the length of the "bunch" of particles should be short with respect to the rf wavelength, in order to ensure that all particles in the bunch experience roughly equal acceleration. Nonetheless, for any bunch of finite length, the actual accelerating field varies within the bunch. This variation can lead to focusing the bunch in the axial (i.e., the travel) direction. Especially for nonrelativistic ions, both the spread in initial beam energies and the longitudinal electrical space charge forces (i.e., the Coulomb forces between particles in the bunch) tend to spread the beam. In particular, space charge forces accelerate the head of the beam and decelerate the tail. If the center of the bunch is located at the proper phase with respect to the rf field, then the tail can be accelerated more strongly than the head. Thus, the tendencies of space charge and energy spread can be counteracted. As a particle tends to slip back relative to the phase of the accelerating field, the larger rf field will drive the particle toward the central phase. For appropriate values of the central phase, the particles will oscillate in energy and in phase with respect to a central value, which may also be time varying. This technique is referred to as phase stability; the oscillations are called synchrotron oscillations. Phase stability is important to linear ion accelerators; it is essential to modern circular accelerators.

Circular Accelerators

The invention of the cyclotron in 1930 by Ernest O. Lawrence and M. Stanley Livingston demonstrated a second approach to exploiting rf acceleration, which involves removing the beam from the region containing the electric fields when the fields are of the wrong phase. In the cyclotron, protons (or heavier ions) circulate along a spiral path; on each turn they cross a gap with time-varying electric fields. During the times in which the fields have the wrong phase, the beam is not in the gap. Rather it is contained in a metallic structure that the rf fields cannot penetrate. This principle of using spatially localized rf fields is employed in almost all circular accelerators as well as in certain types of linacs.

Magnetic Bending and Focusing

Lawrence's cyclotron demonstrated a second important concept, namely, the use of powerful dipole magnets to bend the beam in spiral or circular orbits that pass repeatedly through the same accelerating cells. This procedure raises the effective energy delivered per unit length of accelerator: In the cyclotron, the dipole field is constant. As the particles gain energy they spiral outward. An alternative approach used in synchrotrons is to keep the particles on a constant circular orbit by increasing the magnetic field during the acceleration cycle. In the synchrotron, the accelerating fields are confined to one or more resonant rf cavities, placed at intervals around the circle. As the particles in the beam have a spread in energy, and, therefore, slightly different orbits, the use of phase stability is especially crucial to maintaining the bunch structure in synchrotrons.

In all types of accelerators, magnets provide another crucial function: focusing. All real beams have a tendency to expand in radius due to mutual electrostatic repulsion of the particles in the beam bunches, and to transverse thermal velocities. If the beam is surrounded by magnetic fields with a direction perpendicular to the beam axis and with a strength proportional to the distance outward from the desired beam axis, then the particles can be confined near the beam axis by harmonic potential. Quadrupole magnets provide such fields. However, while the field of the quadrupole focuses the particles in one plane transverse to the average motion, it defocuses in the other transverse plane. The "strong focusing" principle notes that the combination of two or three quadrupoles arranged with alternating focusing and defocusing planes leads to net focusing in all directions perpendicular to the beam motion. Typically the functions of bending and focusing beams are performed by different magnets, although "combined function magnets" are useful in some designs. The higher the energy of the beam, the stronger the bending and the focusing fields needed. Proton supercolliders at multi-trillion-electron-volt energies (for "microscopy" on the subnuclear scale) require tens of kilometers of powerful (several tesla), superconducting, bending and focusing magnets with tiny apertures and exquisite field precision and temporal stability to keep the submicrometer-sized beams in collision.

Storage Rings

Circular accelerators can serve not only to increase the energy of particle beams but also, in the case of storage rings, to maintain large circulating currents of particles at a constant high energy. The motivations for storage rings are twofold: "light sources" for use in atomic physics, material science, chemistry, and medicine; and colliding beams for nuclear and elementary particle research.

Light sources. When charged particles of energy E and mass m travel in a circular orbit of radius R, they radiate electromagnetic energy (synchrotron radiation) with a power proportional to $(E/m)^4 R^{-1}$. The average energy of the emitted photons scales is $(E/m)^3 R^{-1}$. This scaling implies that at giga-electron-volt particle energies, electrons emit copious synchrotron radiation with energies in the ultraviolet and x-ray range. In modern storage rings, the electron beam can have a small transverse size (tens of micrometers); hence, the x-ray beam can be extremely bright. This radiation can serve as a powerful tool for probing the structure of materials such as biological macromolecules, for microfabrication, and for the identification of minute levels of impurities and contaminants in a surface.

Colliders. For both ions and electrons, counter-rotating beams that collide at a few "interaction points" provide probes of the fundamental forces of nature at the highest energy. An advantage of accelerators using colliding beams over those limited to using beams incident on fixed targets is the significantly improved scaling of available center-of-mass energy (discovery potential, or "physics reach") with accelerator energy (and, therefore, accelerator size). The physics reach in a collider scales linearly with the beam energy E, in contrast with scaling as $E^{1/2}$ for beams directed onto fixed targets.

The principal figure of merit for colliders is the luminosity, which measures the number of collisions per second expected for a process with a given interaction probability. Under typical operating conditions, the luminosity scales linearly with beam energy, and with beam current, and inversely with the beam radius. The energy of the beam in the collider is set by the energy scale of the physics to be investigated; to achieve sufficient luminosity the circulating current may be required to exceed 1 A.

The synchrotron radiation power lost by the electron beam also scales linearly with the average current of the circulating beam. This power must be replenished by rf power supplied through resonant cavities in the ring. As the beam energy becomes tens of gigaelectronvolts, the rf power levels in colliders exceed tens of megawatts. The Large Electron Positron (LEP) collider, which is 28 km in circumference and located at the European laboratory CERN, will reach 100 GeV per beam. Because of practical limits in operating and construction costs, this energy is generally regarded as a practical limit for electron synchrotrons. To produce electron collisions at trillion-electron-volt energy at practical costs becomes the new challenge.

Linear Colliders

The aforementioned challenge can be met with linear colliders. As the name suggests, the linear collider consists of two high-energy linacs pointed at each other. The beams, one of electrons and the other of positrons, collide in a single pass and travel on to a beam dump. Presently the largest linac is the two-mile electron linac built at Stanford in 1970. The Stanford linac was upgraded to an energy of 50 GeV and converted to the first linear colliding beam facility, the Stanford Linear Collider (SLC) in the late 1980s. Trillion-electron-volt linear colliders, which are limited by cost considerations to 20 or 30 km in length, will require economical approaches to achieve accelerating fields an order of magnitude beyond those commonly found in existing accelerators. Progress in the design of linacs has been made through the design of new structures to contain the accelerating fields and new, high-power and high-frequency sources of radio-frequency power. In order for the luminosity to be large, the beam densities must also be very high. In the SLC the beam radius is roughly 2 μm. In trillion-electron-volt linear colliders, the beams may have a size of 10 nm. Producing such minute beams and keeping them in collision will require powerful, very stable magnets and sophisticated feedback electronics to keep the focusing elements in precise alignment.

Proton Accelerators

Another means of exploring the energy frontier of particle physics is to collide protons on protons. As protons are composite particles, the energy of a proton supercollider must be roughly five times higher than that of an electron-positron linear collider to achieve equivalent discovery potential. The great advantage of the proton supercolliders over electron colliders is that the power lost to synchrotron radiation is dramatically reduced because the larger mass of the proton diminishes this effect at a given beam energy. For this reason, even for en-

ergies as high as 100 TeV per beam, proton colliders require only conventional synchrotron technology.

The major limitation to proton synchrotrons is the cost and complexity associated with their very large size, which is dictated by the technology of superconducting magnets that bend the particle orbits into a circle. Accelerator magnets typically employ a ductile superconductor, niobium-titanium. At 4.2 K, NbTi will not support fields above 12 T (the critical field value H_c); practical NbTi magnets are limited to less than 10 T. Eventually, limits on collider size and complexity may demand magnets with fields in the range of 15 to 20 T. As no known ductile materials have H_c in this range, very-high-field dipole magnets must use brittle superconductors, such as Nb_3Sn with H_c above 20 T. As the magnetic pressures associated with such fields are hundreds of atmospheres, the design challenges are great.

Medical Applications of Accelerators

Accelerators are an essential, practical tool for some areas of therapeutic and diagnostic medicine. Many hospitals use small electron linacs to produce energetic x rays for radiation therapy. A much smaller number of hospitals use proton cyclotrons or synchrotrons to generate beams used to treat a wide variety of solid tumors. The protons have the advantage of depositing their energy more locally than x rays. Thus the tumor can be irradiated with less damage to the surrounding healthy tissue.

One of the most exciting diagnostic techniques associated with accelerators is positron emission tomography (PET). Positron emission tomography relies on the use of radioisotopes that emit a low-energy positron when they decay. The subsequent positron annihilation produces a pair of gamma rays that are recorded by two detectors. One of the most important radioisotopes for PET is O^{17}, which has a half-life of 2 min and which is readily incorporated into fructose. The imaging of the metabolizing of fructose in the brain has been an important indication of the loss of brain functionality associated with diseases such as Alzheimer's. Because of the short half-lifes of most PET radioisotopes, the technique is only currently becoming widespread, due to the commercial availability of small, easy-to-operate accelerators suitable for isotope production in hospitals.

Multi-kilo-electron-volt x-ray beams from synchrotron light sources have considerable potential to reveal the structures of viruses and proteins. By identifying the details of a substance known to play a key role in a disease, scientists will be able to custom-design pharmaceuticals that are more potent against their disease-causing targets and less harmful to other structures in the body. Hence, the adverse side effects of such drugs will be minimized.

See also: ACCELERATOR, HISTORY OF; ATOMIC PHYSICS; CYCLOTRON; LAWRENCE, ERNEST ORLANDO; PARTICLE PHYSICS; SYNCHROTRON

Bibliography

MONTH, M., and DIENES, M. *AIP Conference Proceedings,* No. 153: *Physics of Particle Accelerators* (American Institute of Physics, New York, 1987).

MONTH, M., and DIENES, M. *AIP Conference Proceedings,* No. 249: *Physics of Particle Accelerators* (American Institute of Physics, New York, 1992).

WILLIAM A. BARLETTA

ACCELERATOR, HISTORY OF

The particle accelerator is a device for accelerating electrons or nuclear particles, usually protons (nuclei of hydrogen atoms), to energies high enough to provide significant information in nuclear or particle physics. Particle energy is measured in "electron-volts" (eV). Particle acceleration takes place in an electric field, the strength of which is measured in "volts per meter" (V/m). Thus a particle that travels 1 m in a field of, say, 100 V/m is said to have an energy of 100 eV. The particle energy achieved by accelerators has increased roughly exponentially from less than 1 million eV in 1930 to about 1,000 billion eV in 1994. The following abbreviations are generally accepted:

1,000 eV = 1 kilo-electron-volt (keV)

1 million eV = 1 mega-electron-volt (MeV)

1 billion eV = 1 giga-electron-volt (GeV) (formerly 1 BeV)

1 trillion eV = 1 TeV.

The First Artificial Disintegration

Until 1932 nuclear physics experiments were done with the ions emitted by radioactive nuclei. It was be-lieved that nuclear disintegration could not be observed with accelerated ions of energies less than 1 MeV, and accelerators of so high an energy had not yet been built. But quantum theory showed that the "nuclear barrier" could be penetrated with ions of somewhat lower energy, and the first "artificial disintegration" was performed in 1932 by John D. Cockcroft and E. T. S. Walton in Cambridge, England, using 400-keV protons to bombard a lithium target. The high voltage was provided by a conventional voltage-multiplying circuit and was applied to an evacuated pipe in which electrodes provided the electric fields necessary for acceleration of the protons.

The First Linear Accelerator

Four years before the appearance of Cockcroft and Walton's results an event occurred that marks the beginning of accelerator history. Rolf Wideröe, a Norwegian, published a paper describing a method for accelerating to twice the applied voltage. The particles to be accelerated passed through three coaxial tubes. The first and third tubes were connected to one terminal of a high-frequency alternating voltage and the other terminal was connected to the central tube. The particles were accelerated at the first gap between the first and second tubes, then the alternating field reversed so quickly that, by the time the particle reached the second gap, it was accelerated again. Wideröe built a model in which he accelerated sodium and potassium ions and demonstrated that ions were accelerated to twice the energy corresponding to the peak applied voltage. It was also evident that more gaps could be added and the process repeated many times. This is the procedure used in modern linear accelerators.

The Cyclotron and the Electrostatic Generator

During the 1930s two accelerators were the favorites of nuclear physicists: the cyclotron and the electrostatic generator.

In 1929 Ernest O. Lawrence at the University of California discovered Wideröe's paper and realized that if the protons could somehow be returned to the first gap again and again, multiple acceleration could take place. A magnetic field, which has the effect of deflecting a moving charged particle onto a curved path, could do this. The coaxial tubes in Wideröe's accelerator would be replaced by something that would look like a flat pillbox cut in half

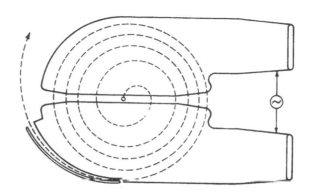

Figure 1 Particle orbits in the cyclotron.

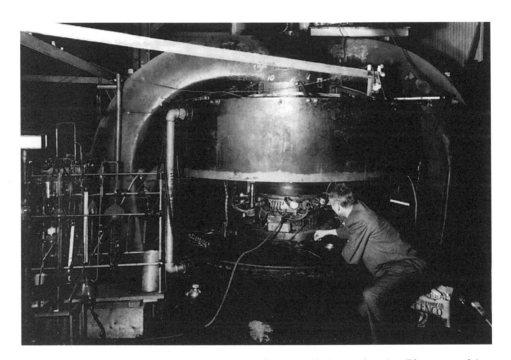

Figure 2 Ernest O. Lawrence and a cyclotron. (Science Service Photograph)

along its diameter—the half boxes would come to be known as D's because of their shape. The orbit in this structure would be as shown in Fig. 1. The orbits have the felicitous property that, as they grow longer from gap to gap, the ions travel faster and so they continue to take the same time—hence the electric field always has the same frequency. The D's are mounted between the poles of an electromagnet that supplies the magnetic field to curve the paths of the ions.

Lawrence suggested this idea to M. Stanley Livingston. Livingston constructed a model 4 in. in diameter. With 2,000 V between the D's, hydrogen

ions were accelerated to 80,000 V and the cyclotron principle was conclusively verified.

Lawrence and Livingston then built a series of cyclotrons for higher and higher energies. One MeV was achieved by 1932, 5 MeV in 1934, and 8 MeV in 1936. The "sixty inch" reached 16 MeV in 1939. Figure 2 shows Lawrence beside one of these machines.

Members of Lawrence's team were invited to many centers to supervise construction of local machines, and commercial firms began to build cyclotrons on contract.

Eventually a limit appeared. Because of relativistic effects, as ions approached the velocity of light they fell out of step with the alternating field and ceased to be accelerated. This began to happen at some tens of megaelectronvolts. The solution of this problem came after the end of World War II.

The idea of charging an insulated metal sphere to high voltage by spraying charge onto an insulating belt and removing the charge when the belt entered a hole in the sphere dates back to 1890, but it was not exploited until 1929 when Robert J. Van de Graaff at Princeton mounted two spherical electrodes (each 2 ft in diameter) on 7-ft glass rods, each with a motor-driven silk belt to transport the charge. Between the spheres he achieved a potential difference of 1.5 million volts limited only by electrical breakdown.

Van de Graaff moved to MIT and set about constructing a huge installation in an airplane hangar in south Massachusetts. Between two 12-ft diameter spheres he developed over 5 million volts, but difficulty in mounting an accelerating tube between the spheres proved insurmountable and the apparatus was rebuilt and moved to MIT.

In the meantime a less ambitious electrostatic machine was built at the Carnegie Institution by Merle A. Tuve, L. R. Hafstad, and Odd Dahl. It involved a single sphere charged to about 1.3 MV; Fig. 3 is a photo of this installation. It was operating reliably by 1935. Many important nuclear physics discoveries were made with this machine while Van de Graaff was still struggling with his huge spheres.

The electrostatic machine had an important advantage: The particle energy was sharply defined; cyclotron beams had some energy spread so nuclear resonances with well-defined energies were more easily measured with electrostatic machines. Their useful energy was raised up to and above 10 MeV when they were encased in pressure tanks. They are operated in many centers. Postwar, Van de Graaff and his associates formed the High Voltage Engineering Corporation to build electrostatic machines.

The Betatron

The two machines just described were used to accelerate positive ions. In 1940 an electron accelerator began a brief career—the betatron, in which electrons were accelerated by the electric field induced by a changing magnet flux linking their orbit. In effect it was a transformer, with the magnet winding as the primary and an electron beam as the secondary. The electrons traveled in circular paths guided by part of the field from the same magnet

Figure 3 Electrostatic machine at the Carnegie Institution. (Carnegie Institution of Washington)

structure. This idea was described by Wideröe in 1928 but he was unable to make the device work. In 1940 Donald W. Kerst and Robert Serber at the University of Illinois realized that the magnetic field must be shaped to provide focusing and prevent the electrons from escaping. Kerst built a 2.3 MeV model that worked immediately. He then moved to the General Electric Research Laboratory, where he built a 20 MeV machine. Finally, he returned to Illinois, where he built an 80 MeV machine and then a 300 MeV machine. This was the largest betatron ever built.

Back in Schenectady two General Electric engineers, Ernest Charlton and Willi Westendorp, interested in producing intense x-ray beams, built a 100 MeV betatron; 100 MeV was enough energy to make detectable the relativistic radiation emitted by electrons traveling on curved paths. This radiation, now known as "synchrotron radiation," is emitted in a small cone directly ahead of the electron. Its intensity increases extremely rapidly with electron energy and its spectrum is visible at 100 MeV. At higher energies it moves into the ultraviolet and then the x-ray range. It could not be seen in the opaque vacuum chamber of the 100 MeV machine, but its effect on the electron orbits were measurable, with the results in good agreement with the predictions of electromagnetic theory.

Synchrotron radiation has become important for studies of physics of condensed matter and surfaces, for chemistry, for biology, and in industry, and a large number of synchrotron light sources have been built. The National Synchrotron Light Source at the Brookhaven National Laboratory, for example, has more than 2,000 users.

Since World War II the accelerator art has been radically advanced by the discovery of two principles: phase stability and alternating gradient (or strong) focusing. These have made accessible particle energy ranges extending to the trillion-electron-volt range and limited only by construction costs.

Phase Stability and the Synchrotron

The principle of phase stability says that particles arriving at an incorrect phase at a gap between electrodes excited by a high frequency electric field will experience an automatic correction toward the correct phase. The consequences of this principle make it possible to pass the cyclotron energy limit mentioned above. Appropriate changes in the cyclotron structure result in the "synchrocyclotron." Also, it suggests a new machine, the "synchrotron," a roughly circular machine in which magnetic fields deflect the particles into their orbit, and acceleration is by radio-frequency electric fields across one or more gaps between electrodes.

This principle was discovered by Vladimir I. Veksler in the Soviet Union in 1944. Early in 1945, before Veksler's paper reached the United States, the same principle was published by Edwin McMillan of the University of California. McMillan immediately started designing an electron synchrotron for 300 MeV. Before he could complete his machine, however, two smaller synchrotrons were brought into operation. First was an 8 MeV synchrotron using an old betatron magnet built by Frank K. Goward and D. E. Barnes in England. The second was a 75 MeV machine assembled in the G. E. Research Laboratory. This machine had a transparent vacuum chamber that made possible the first viewing of synchrotron radiation. About a dozen large synchrocyclotrons were built in the United States and elsewhere. The peak energy achievable

appeared to be about 700 MeV. Throughout the world a similar number of large electron synchrotrons operated during the 1950s and 1960s at energies as high as 1200 MeV (1.2 GeV).

The first proton synchrotron was built at the newly formed Brookhaven National Laboratory. Opened in 1947, the laboratory had, as one of its goals, an accelerator to provide protons with energies over 1 GeV. M. Stanley Livingston of cyclotron fame was to be the project head. It was clear that the machine must be a proton synchrotron, which presented many difficult problems not encountered in the electron synchrotron. The many problems were finally solved and the machine came into operation in 1952 at an energy of 3 GeV. Figure 4 is a photograph of the Cosmotron, as the machine was named by its builders. Its ring of magnets was 70 ft across and 8 ft high. Its injector was a 4 MeV electrostatic machine built by High Voltage Engineering.

At the same time as the Cosmotron was approved, the University of California was approved for construction of a 6 GeV machine—enough energy to produce the antiproton if it existed. The Berkeley machine, called the Bevatron, was finished in 1954, and, shortly thereafter, the existence of the antiproton was demonstrated.

Alternating Gradient Focusing

In 1952 a dozen European nations were in the process of organizing a European international laboratory to be located in Geneva, Switzerland, and to be called CERN. One of its first projects was to be a proton synchrotron, which was to be a Cosmotron scaled up to 10 GeV. The head of the project would be Odd Dahl of Norway, one of the team that built the Carnegie Institution electrostatic machine. His deputy would be Frank Goward of England, who had built the first synchrotron. Together with their consultant Rolf Wideröe they made a trip to Brookhaven to ask advice from the builders of the Cosmotron.

In preparation for their visit, Livingston organized a study that, by a series of fortunate speculations, arrived at alternating gradient (strong) focusing. It is well-known that, if a charged particle beam is strongly magnetically focused in one plane, it will be strongly defocused in the other plane. What was realized by the team of Ernest D. Courant, Livingston, and Hartland S. Snyder was that the sequence of focusing and defocusing lenses is relatively strongly focusing no matter which lens comes first. The

Figure 4   The Cosmotron. (Brookhaven National Laboratories)

focusing thus achieved can be much stronger than the "weak focusing" used by Kerst in the Betatron, in the Cosmotron, and in the Bevatron.

Brookhaven immediately made a design for a 30 GeV alternating gradient synchrotron—the AGS—and CERN abandoned their 10 GeV Cosmotron in favor of a 28 GeV alternating gradient synchrotron. These machines came into operation in 1960 and 1959, respectively.

At the time of the discovery of alternating gradient focusing, it was pointed out that it could also be used in proton linear accelerators. This was incorporated in a 50 MeV linear accelerator that served as the injector for the AGS.

During the 1960s it was agreed that the next goal in the United States should be 200 GeV. A site in the suburbs of Chicago was chosen and Robert R. Wilson of Cornell was chosen as director. The machine, about 4 miles in circumference, operated at 200 GeV in 1972 and was pushed to 400 GeV in 1974. The laboratory is known as Fermilab (Fermi National Accelerator Laboratory); it is now the major center in the United States for high-energy physics.

Stanford Linear Accelerator Center

A second important center for high-energy physics is the Stanford Linear Accelerator Center (SLAC), where a 2-mile-long electron linear accelerator provides electrons and positrons (antielectrons) at energies up to 50 GeV. Work at Stanford began in the 1930s under William W. Hansen, who began the development of the 3,000 MHz power sources needed for such an accelerator. Postwar, under the direction of Wolfgang ("Pief") K. H. Panofsky, the development culminated in the 2-mile machine that has been used in many important studies in particle physics.

Colliding Beams

When a high-energy particle strikes a target at rest, only a fraction of its energy is available for nuclear disintegration and particle production. The rest goes into giving momentum to the struck particle. But if particles with equal and opposite velocities collide, all of the energy is available. At one time it was considered impractical to achieve sufficient numbers of collisions between two particle beams. But a pioneering group at CERN under Kjell Johnsen undertook in the 1960s to build intersecting storage rings to store 28 GeV protons from the CERN proton synchrotron and make them collide at several points. The experiment was a success and was the forerunner of a number of colliding beam systems. In 1981 CERN was first to accelerate a beam of antiprotons in the CERN SPS (super proton synchrotron) to energies eventually to reach 900 GeV and to make them collide with a proton beam.

At Fermilab the magnets in the synchrotron were replaced with superconducting magnets (operating at more than 400°F below zero). This made much higher magnetic fields possible; the energy of the accelerator was raised to 900 GeV (almost 1 TeV) and the machine was renamed the Tevatron. In 1987 an antiproton beam was produced and, since then, colliding beam experiments with almost 2 TeV in the center of mass can be performed.

A daring operation was completed at CERN when an electron-positron colliding beam ring known as LEP was brought into operation at about 20 GeV per ring. To decrease energy losses due to synchrotron radiation, the radius of this ring must be very large. The final ring was 17 miles in circumference, and housed in a tunnel partly under the mountains in France.

The Future

In the late 1980s physicists agreed that the outstanding problems in particle physics should be attacked with energies of about 20 TeV in each of proton-antiproton colliding beams. This would call for rings about 60 miles in circumference and was estimated to cost about $4 billion. However, the cost estimates steadily increased, and, when they reached about $10 billion, work on the Superconducting Supercollider (SSC) was canceled by the U.S. Congress.

In the meantime, at CERN, a project for building a proton-antiproton collider in the LEP tunnel was receiving serious consideration. With high-field superconducting magnets, 8.5 TeV per beam could be achieved for 17 TeV in the center of mass. This collider is called the LHC (large hadron collider).

With the demise of the SSC, much debate is in progress in the United States about the future program in high-energy physics.

Conclusion

Thus far, in the interests of brevity we have neglected many important high-energy accelerators— some in the United States and many elsewhere in the world. For example, in Germany there are major machines in Hamburg and Darmstadt. In Russia for a time the highest energy accelerator in the world was operating at 70 GeV at Serpukhov; also many important contributions have come from Novosibirsk from the laboratory of the late G. I. Budker. In France valuable work has been done at Saclay, Orsay, and Caen. In Italy pioneering work on particle storage went on at Frascati. The TRIUMF accelerator in Vancouver is Canada's center for high-energy physics. And in Japan the KEK laboratory near Tokyo is a major accelerator and high-energy physics center.

Heavy ion accelerator facilities have been developed at Berkeley and Darmstadt and the new RHIC (relativistic heavy ion collider) has been developed at Brookhaven.

See also: ACCELERATOR; ATOMIC PHYSICS; CYCLOTRON; LAWRENCE, ERNEST ORLANDO; PARTICLE PHYSICS; SYNCHROTRON; VAN DE GRAAFF ACCELERATOR

Bibliography

HEILBRON, J. L., and SEIDEL, R. W. *Lawrence and His Laboratory* (University of California Press, Berkeley, 1989).

LEDERMAN, L., and TERESI, D. *The God Particle* (Dell, New York, 1993).

LIVINGSTON, M. S., ed. *The Development of High-Energy Accelerators* (Dover, New York, 1966).

LIVINGSTON, M. S., and BLEWETT, J. P. *Particle Accelerators* (McGraw-Hill, New York, 1962).

JOHN P. BLEWETT

ACCEPTOR

The room-temperature electrical conductivity and other electrical properties of semiconductors are profoundly affected by the presence of very small amounts of certain impurities or crystal imperfections. The deliberate addition of small amounts of impurities that decrease the electron concentration is called acceptor doping, and the impurities are called acceptors. Acceptor doping increases the number of holes available for charge transport in the valence band of the semiconductor, and it is best understood as complementary to doping by donors, which increases the number of electrons available for charge transport in the conduction band.

The simplest and most important type of acceptor is substitutional. Semiconductors from group IV of the periodic table (Si or Ge) are normally acceptor doped using the group III impurities B, Al, Ga, and In. Because these impurities have one valence electron fewer than the atoms of the host lattice, they tend to remove an electron from the valence band, binding it in a localized state. The hydrogen-like model of doping applies to acceptors as well as donors; for In-doped Si, the experimentally measured necessary energy is $E_a = 1.6 \times 10^{-2}$ eV, a quantity readily available from thermal processes at room temperature ($k_B T = 2.6 \times 10^{-2}$ eV). However, this process is best regarded as releasing a hole to the valence band rather than binding an electron from that band.

It must be remembered that ionization of a hole (increase of its energy) involves its displacement to a lower state in the band-structure (energy-level) diagram. The unionized donor states (the ground states for the holes) thus lie just above the valence band edge, in the forbidden gap of the semiconductor.

A large proportion of the holes are ionized at room temperature, in contrast to a very small proportion of intrinsic holes, whose ionization energy is 0.55 eV. The crystal can readily be doped so that the acceptor holes dominate the electrical properties of the crystal.

In real semiconductor crystals, small concentrations of donor impurities are inevitably present. A sufficient concentration of acceptor impurities serves to empty electrons from the donor states so that the crystal exhibits the properties of a purely acceptor-doped semiconductor. A semiconductor that satisfies these conditions is called p type.

Like n-type semiconductors, p-type semiconductors can exhibit impurity conduction at low temperatures.

See also: DONOR; DOPING; ELECTRICAL CONDUCTIVITY; ENERGY LEVELS; IONIZATION; SEMICONDUCTOR

Bibliography

ASHCROFT, N. W., and MERMIN, N. D. *Solid-State Physics* (Saunders, Orlando, FL, 1976).

KITTEL, C. *Introduction to Solid-State Physics,* 7th ed. (Wiley, New York, 1996).

LAWRENCE S. LERNER

ACCURACY AND PRECISION

The goal of most scientific investigations is to obtain an estimate for the value of some quantity of interest. This may be as familiar as the temperature at which a liquid boils, or as esoteric as the distance to a remote galaxy. The answers we get are never perfect. There is always some uncertainty in the numerical value. In addition to simply reporting the results, scientists assess their validity, analyze the errors, and estimate the accuracy and precision of the new information.

In common usage the word "error" implies a mistake or something that is simply "wrong." In discussions of scientific data, errors are a recognition of uncertainty. There are two broad categories of errors, which we call illegitimate and legitimate. The first includes mistakes that can be eliminated by performing the work with care. Malfunctioning equipment can be repaired or replaced. Typographical errors in computer programs, digits transposed in a number, or the use of inappropriate data or algorithms all lead to erroneous results. These types of problems cannot be tolerated in quality work, and must be avoided or corrected. The category of legitimate measurement errors is further divided into systematic and random errors. Systematic errors tend to bias results in a consistent manner, while random errors create fluctuating uncertainties with a range of magnitudes. For a result to be of scientific value, it should be free of blunders, have small systematic errors, and allow quantitative estimates of the random errors.

Accuracy means that systematic errors are small, and the result is close to the "true" value. Two ways that accuracy can be established are by calibrating the technique and by obtaining consistent results with several independent measurements. Calibrating entails using the equipment, data collection, and analysis methods to measure a quantity that is well-known or accepted as a standard. An experiment to measure temperatures could be used to determine the freezing and boiling points of pure water. If the results are consistent with the accepted values, we develop confidence that the technique is reliable, and we believe that measurements of other temperatures should be accurate. The second approach is to use several different techniques, expecting that a flaw in one will be revealed by a result that is inconsistent with the others. This approach is often used when an accepted standard is not available. For example, a current issue in astronomy is the measurement of the size of the universe. Because distances to remote galaxies are not well-known, scientists use many different indicators, such as the pulsation periods of variable stars and the peak brightness of supernovae, in an attempt to avoid systematic errors in any one method. Systematic errors often can be identified and characterized. Once they are understood the techniques may be refined to reduce them or a correction may be applied to the raw measurement.

Precision is related to the random errors. A single measurement does not provide much insight into either accuracy or precision. If it is repeated several times the individual values will disagree with each other at some level. Their spread can indicate the range within which the true value is likely to be, and their average may be a useful estimate of that value. A carefully designed experiment will produce a larger sample to which statistical analysis can be applied. Parameters such as the "mean," "median," or "mode" are estimates of the value that are justified by the data. The precision is quantified by the spread of the individual values about the estimated true value. Parameters such as the "standard deviation," "variance," and "probable error" are all derived from the distribution. The smaller these parameters the smaller the uncertainty and the better the precision. The statistical item called the "standard deviation of the mean" is often regarded as the best measure of precision. Numerically, it is the standard deviation of the sample divided by the square root of the number of samples. If the errors are truly random the precision can be improved by increasing the number of measurements from which the mean is derived.

Often an experiment produces data showing relationships between two or more variables rather than multiple estimates of a single item. Such data are usually described by some kind of curve fitting, such as a least-squares fit. The concept of a mean is replaced by optimized values of the fitting parameters, and the uncertainty of a single value is replaced by entities such as "residuals," "chi-squared parameters," and "confidence intervals." The power of statistical analysis is that the values of derived parameters can be significantly more reliable than any of the individual measurements from which they are derived. This is especially true if the result has a known functional form, such as a single scalar value, or the slope of a linear relationship between two variables.

All possible sources of error should be examined, and the reported uncertainties must account for both systematic and random effects. The uncertainty may be implied by the number of significant digits quoted. An explicit statement should include the numerical value, its units (percent of mean or same absolute units as the mean), and indicate the method of calculation (sample standard deviation, standard deviation of the mean, and so on). When plotting results the individual data points should be accompanied by "error bars" whose lengths convey the magnitude of the uncertainties. The numerical meaning of the error bars should be explicitly stated.

Historical records indicate that by the thirteenth century B.C.E. Chinese astronomers had accurately determined the length of the year to be $365\frac{1}{4}$ days. By 1276 C.E. they had refined their estimates to be within 27 seconds of the currently accepted value. This represents a precision of approximately one part in a million. By contrast, the best modern estimates of the age of the universe range from 8 to 16 billion years. This large uncertainty indicates that this quantity has not yet been measured with either accuracy or precision.

See also: ERROR, EXPERIMENTAL; ERROR, RANDOM; ERROR, SYSTEMATIC; ERROR AND FRAUD; ETHICS

Bibliography

BEVINGTON, P. R. *Data Reduction and Error Analysis for the Physical Sciences* (McGraw-Hill, New York, 1969).

MEYER, S. L. *Data Analysis for Scientists and Engineers* (Wiley, New York, 1975).

TAYLOR, J. R. *An Introduction to Error Analysis: The Study of Uncertainties in Physical Measurements* (University Science Books, Mill Valley, CA, 1982).

DENNIS EBBETS

ACOUSTICS

The word "acoustics" derives from the Greek word *akouein*, to hear, and the origins of the subject certainly stem from the human voice making sounds and the human ear hearing them. Acoustics there-

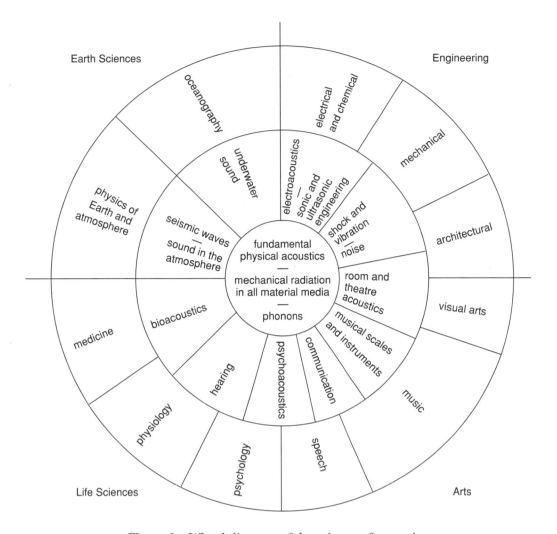

Figure 1   Wheel diagram of the science of acoustics.

fore begins with speech and hearing. Over the years, the field has expanded by inclusion, adding first the sounds of music, and through them, the field of vibrations. Since then have come many topics including electroacoustics, room acoustics and ultrasonics, and numerous others, until the ramifications of acoustics reach out into many branches of science.

A way of viewing this was provided by Bruce Lindsay in what has come to be known as Lindsay's wheel (Fig. 1). From this viewpoint, the core of the subject is physical acoustics, while the first ring represents the various subfields of acoustics. The second ring enumerates the major disciplines with which acoustics is associated, while the outermost portions are the basic areas of intellectual activity to which these disciplines belong.

Physical acoustics involves, primarily, artificial sources of sound, such as the vibrations of strings, air columns, membranes and solids, the manner in which sound propagates through continuous media in the form of waves (with velocity specific to the particular medium), the modes of attenuation of the sound as it passes, and, finally, the reception of the sound by various methods.

In the air, sound propagates under adiabatic conditions, and at a speed equal to $(\gamma kT/m)^{1/2}$, where γ is the ratio of specific heats, k is Boltzmann's constant, equal to 1.3×10^{-23} J/K, and m is the mass of a molecule of the gas in kilograms. The sound wave travels in straight lines; if the source is small, the sound wave will travel in all directions. Such a wave is called a spherical wave. Under certain other conditions (e.g., a distant source or special focusing), the sound wavefront can be considered to be plane (so-called plane waves). In practice, however, most sound sources yield more complex patterns. The source usually yields a diffraction pattern with a central lobe of highest amplitude, and many side lobes of lesser strength.

The intensity of a sound wave varies over an enormous range. It is usually specified in terms of a sound pressure level equal to $10 \log I/I_0$, where I is the actual intensity in watts per square meter, and I_0 is a reference level, approximately equal to the minimum audible sound intensity. Usually, I_0 is chosen to be 10^{-12} W/m^2. The level is measured in decibels (dB). The number of decibels can range from 0 to 160, which means a range in actual intensity of 16 powers of 10. On this scale, 30 dB corresponds to a quiet room, 60 to 70 dB corresponds to noisy traffic, and levels of 90 dB and higher are levels that can be injurious to one's hearing.

In propagating through matter, sound is attenuated in different ways. For a spherical wave, the sound intensity decays geometrically; that is, the intensity is inversely proportional to the square of the distance. In addition, the sound intensity is reduced by effects of shear viscosity and heat conduction, as well as by the exchange of energy between the sound wave and internal degrees of freedom of the substance.

Physical acoustics also can be involved in the theory of condensed matter, where the vibrations of the material are expressed in terms of elastic waves (phonons), and in the study of matter at extemely low temperatures, including second, third, and fourth sound in liquid helium. In an unbounded solid, the sound wave can be either longitudinal or transverse in nature (in fluids, only the longitudinal mode exists). There are also surface waves, both longitudinal and transverse, that propagate along the surface of solids. They are widely used in engineering for control purposes.

Traditionally, acoustic signals have always been assumed to be of very small amplitude. The particle velocities involved in very loud sounds in air are still very small, only of the order of a few centimeters per second, as compared with the speed of sound of about 340 m/s. This has been advantageous to those studying the subject, since it enabled them to drop some terms in the wave equation, making it a linear equation, and thus more amenable to mathematical solution. However, studies since the 1940s have looked more and more into the behavior of sound when such approximations are not made. This is the

field of nonlinear acoustics. The shock wave set up when the speed of an airplane exceeds the local sound velocity is an extreme example of nonlinearity in acoustics.

Most recently, the study of physical acoustics has led to research in the creation and behavior of acoustic solitons, and to the study of chaos.

If we follow the spokes of the wheel (Fig. 1) in a clockwise direction, we find that, in engineering, acoustics includes various electric and electronic devices that produce and amplify sound, such as piezoelectric crystals and ceramics, loudspeakers and speaker systems, acoustic amplifiers, microphones, and recorders and recording equipment, including records, magnetic tapes, and compact discs.

Another important application of engineering acoustics has been in the field of nondestructive testing. The use of ultrasound (high-frequency sound waves) in detecting flaws, cracks, and other weak points in rails, bridges, and buildings has been an important safeguard for humanity. On the engineering side there are also structural shock and vibration, and techniques involved to reduce their effect. This brings us to noise, which is omnipresent in our society. Combating noise involves the use of absorbent materials, mounting, and, more recently, active noise suppression (i.e., playing back the noise sounds with appropriate phase so as to achieve substantial cancellation), as well as the various means employed to reduce and control noise. Also part of engineering acoustics are the widespread studies of acoustic signal processing and computer modeling of acoustic systems.

On the border between engineering and the arts are room and theater acoustics, where acoustics contributes to architecture and the visual arts. Here acoustics is concerned with delivering sound to the audience in a manner most satisfactory for the listener. The acoustician assists in the proper design of auditoriums and other listening spaces. The work includes the choice of appropriate absorbing materials for surfaces of the room, and the use of properly located and phased loudspeakers to assist in the sound delivery.

Next is music, the understanding of how musical instruments produce sound, and how to improve these instruments. Here acousticians must bear the burden of having developed amplification of such instruments as the guitar, not to mention the "boom box." Advanced techniques of optical and electronic scanning of the vibrating parts of musical instruments have been widely employed in attempts to improve acoustics of musical instruments. Mention also should be made of electronic music.

In recent years, there has been an enormous growth in research in the fields of speech and hearing. Use of the latest electronic equipment and mathematical tools in studying the physical, physiological, and psychological aspects of these fields has become standard. The study of speech and its characteristics is a major research area. The detailed study of the cochlea has become a field of its own. Great attention has been paid to deafness and the ways of combating its effects.

In the field of medicine, there has been widespread application of acoustical techniques, largely ultrasonic, to the study of various soft tissues in the human body, such as fetuses and kidneys. Specialized fields, such as the study of the effect on humans of noise and various mechanical vibrations, also have developed. Ultrasound also has been used in therapy, and shock waves have found application in lithotripsy.

In the field of earth sciences, acoustics has entered into the study of seismic waves and also into the study of the atmosphere. Acoustic study of the ocean began with underwater sound as a device to detect underwater objects such as submarines. The history of this research since World War II is one of innovation in equipment, theoretical studies of long-range propagation, and signal analysis. At the same time, improved knowledge of the ocean and the life in it were useful by-products. Now scientists have turned more to the study of the ocean itself, as well as of its sea life, through acoustic oceanography, where underwater sound has become an important research tool. Such studies have ranged from long-range transmissions over half the globe for the detection of possible global warming, to the pursuit of schools of fish for industrial fishing.

Throughout all its branches, acoustics remains a discipline with strong theoretical and experimental components. Although many of the fields have drifted far apart in their subject matter, the mathematics used in them has been central and common, so that advances in one field very quickly find application in another. The same is true for much of the electronic equipment that has been developed. Thus, acoustics continues to make significant contributions to the store of human knowledge and to the betterment of the quality of life.

See also: SEISMOLOGY; SOUND; SOUND, MUSICAL; SOUND ABSORPTION; ULTRASONICS

Bibliography

LINDSAY, R. B., ed. *Acoustics: Historical and Philosophical Development* (Dowden, Hutchinson & Ross, Stroudsburg, PA, 1973).

ROBERT T. BEYER

ACTION-AT-A-DISTANCE

To describe the interaction among particles, physicists have coined three concepts, namely action in a continuous medium (fields), action-at-a-distance, and the more familiar and unproblematic action by impact. The problem is, in James Clerk Maxwell's words (1876), "that of the transmission of force. We see that two bodies at a distance from each other exert a mutual influence on each other's motion. Does this mutual action depend on the existence of some third thing, some medium of communication, occupying the space between the bodies, or do the bodies act on each other immediately, without the intervention of anything else?" In this last case, in which the bodies act on each other, modifying their trajectories without the intervention of anything else, we say that the interaction takes place by means of actions-at-a-distance, or direct interactions.

Actions-at-a-distance reached a position of prominence in physics owing to their central role in Isaac Newton's mechanics and theory of gravitation, as developed in his *Principia* (1687). One of the main parts of this work, Book III, which deals with gravitation, is based on the famous inverse square law, of standard action-at-a-distance structure, since it depends only on the positions of the interacting bodies, taken at the same time. In this sense, it may be thought that the action-at-a-distance component of Newton's system is constrained to his theory of gravitation. As it turns out this is not the case; Newton's assumption that when bodies at a distance are moving relative to one another the third law still holds implies in principle that the interaction between them takes place instantaneously, for if the interaction between them takes time, then the action of, let us say, A on B may not be simultaneous with that of B on A, and therefore not equal to it at all times. But the only instantaneous interaction that makes sense is instantaneous action-at-a-distance.

The first edition of the *Principia,* aside from an oblique reference in Newton's preface, contains neither a suggested cause of gravitation nor a justification for the omission of such mechanism. We know now that such omission was more the demonstration of Newton's inability to solve this problem, as well as his pragmatic approximation to physics, than his implicit support of actions-at-a-distance, which he considered philosophically absurd. Indeed, action-at-a-distance theory had to fight against René Descartes's more intuitive vortex theory of planetary motion, in which all natural phenomena ought to be explained in principle on the basis of a continuous medium. Eventually, the Cartesian system collapsed in favor of Newtonianism, but even so some looked for other alternatives. For example, Christiaan Huygens, who although convinced by Newton that the hypothesis of Cartesian vortices was untenable, sought a different kind of continuous medium that might produce the effects attributed to central forces.

The success of Newton's dynamics throughout the eighteenth century in the explanation and prediction of physical phenomena gave finally, although not permanently, the action-at-a-distance concept a status that intuition had denied it. Initially, electric and magnetic phenomena seemed to favour actions-at-a-distance. Thus, in the 1820s André-Marie Ampère developed, on the basis of Hans Christian Oersted and Jean-Baptiste Biot and Félix Savart's previous works, an action-at-a-distance force for the interaction between line elements of two currents. This was the first significant contribution toward the development of electrodynamics. Indeed, throughout the nineteenth century there were, especially in Germany, continuous efforts to establish satisfactory action-at-a-distance electromagnetic dynamics. Carl Friedrich Gauss, Wilhelm Weber, Hermann von Helmholtz, Bernhard Riemann, Rudolf Clausius and Walter Ritz figure among those who made important contributions in that direction.

However, actions-at-a-distance suffered a serious blow with the development of Maxwell's successful field electrodynamics. Since then the field concept has dominated physical interactions. The formulation in 1915 of Albert Einstein's gravitational general theory of relativity, also a field theory, as well as the firm position of fields in the quantum realm, which has led to quantum theories of the remaining basic physical forces (strong, weak, and electromagnetic), explain this situation.

It is important to point out, however, that actions-at-a-distance have proved to be more general than it was thought in Newton's time. Special relativity does not, for instance, imply that only field theories are permissible, although the fact that the already available Maxwell's field electrodynamics was a special relativistic theory tended to obscure this point. Particularly important in this connection was the work of Karl Schwarzschild, who in 1903 showed that Maxwell's equations could also be derived from an action-at-a-distance theory (he used a variational principle where only particle variables appear). Later on, Hugo Tetrode (1922) and Adriaan Fokker (1929) developed that approach, emphasizing the possible advantages that not resorting to fields might have in the quantum domain.

Two features of the Schwarzschild–Tetrode–Fokker formulation of electrodynamics must be noted. The first is that it relies on *non-instantaneous,* Lorentz-invariant, actions-at-a-distance. In this way, it is consistent with special relativistic requirements. Newton's old concept was therefore considerably generalized. The second is that to derive Maxwell's equations Schwarzschild needed retarded as well as advanced interactions between the charged particles, a situation in principle rather paradoxical, as we seem not to be affected by the future. Perhaps more important was that neither of these scientists provided mechanisms to understand the source of the force experienced by the charge itself as a result of its motion (radiation-reaction).

These problems were finally solved by John Wheeler and Richard Feynman in two extraordinary papers published in 1945 and 1949 in the *Reviews of Modern Physics.* What they did was to develop a somewhat complex mechanism in which advanced signals combine in such a way as to make it appear—except for the phenomenon of radiative reaction—that each particle generates only the usual and well-verified retarded forces.

As in Tetrode and Fokker's cases, Wheeler and Feynman were attracted to this approach by problems they encountered in quantum theory: If many of the difficulties that appear in relativistic quantum theory are recognized as due to the infinite degrees of freedom introduced by fields, why not dispense with them in the most straightforward manner; that is, resorting to actions-at-a-distance? Soon, however, mechanisms to cope with these difficulties within the domain of field theories were developed (renormalization), while nobody has been able so far to do something similarly successful with the action-at-a-distance approach. And so, despite a few attempts at revival within the framework of steady-state cosmological theories (J. E. Hogart, Fred Hoyle, and Jayant Narlikar), actions-at-a-distance remain far from the main center of activity in today's physics.

See also: AMPÈRE, ANDRÉ-MARIE; CLAUSIUS, RUDOLF JULIUS EMMANUEL; HELMHOLTZ, HERMANN L. F. VON; HUYGENS, CHRISTIAAN; INTERACTION, ELECTROMAGNETIC; INTERACTION, STRONG; INTERACTION, WEAK; MAXWELL, JAMES CLERK; NEWTONIAN MECHANICS

Bibliography

HESSE, M. B. *Forces and Fields* (Littlefield Adams, Lanham, MD, 1965).

HOYLE, F., and NARLIKAR, J. V. *Action at a Distance in Physics and Cosmology* (W. H. Freeman, San Francisco, 1974).

NIVEN, W. D., ed. *The Scientific Papers of James Clerk Maxwell,* Vol. II (Dover, New York, 1965).

WHEELER, J. A., and FEYNMAN, R. P. "Interaction with the Absorber as the Mechanism of Radiation." *Rev. Mod. Phys.* **17,** 157–181 (1945).

WHEELER, J. A., and FEYNMAN, R. P. "Classical Electrodynamics in Terms of Direct Interparticle Action." *Rev. Mod. Phys.* **21,** 425–433 (1949).

JOSÉ M. SÁNCHEZ-RON

ACTIVATION ENERGY

See ENERGY, ACTIVATION

ACTIVE GALACTIC NUCLEUS

An active galactic nucleus (AGN) is the hyperactive nucleus of a distant galaxy. The most spectacular type of AGN is a quasar, although it appears that most galaxies, including our own, exhibit some form of activity in their nuclei. (The phrase AGN is sometimes restricted to sources where a surrounding

galaxy can be seen clearly. However, here it is supposed that quasars are located in galactic nuclei and so all of these objects are AGN.) Other types of AGN include "radio galaxies," "Seyfert galaxies," and "BL Lac objects." Astronomers are beginning to understand the relationship between the different types of AGN.

It is generally believed that the most luminous AGN are powered by gas falling onto a black hole with a mass in the range of a million to a billion times the mass of the Sun. A black hole is formed when a large amount of gas accumulates in a very small region and the inward pull of gravity is so large that the gas collapses in upon itself and eventually prevents light from escaping. The size of a million solar mass black hole will be of order three million kilometers; that of a billion solar mass object, three billion kilometers.

After they have formed, black holes can continue to attract and ingest gas from their surroundings. The infalling gas should have some motion around the hole and eventually it will settle into an "accretion disk" that lies in its equatorial plane. The orbital period of the gas in this disk should decrease with increasing radius, just as is the case for planets in the solar system. This, in turn, implies that there will be a frictional force that causes the gas to move inwards, liberating its gravitational binding energy in the process. Most of this energy should be liberated close to the black hole and the overall efficiency should approach 10^{16} J for every kilogram of gas consumed. To put this into context, it should be noted that explosives like TNT release only about 10^7 J/kg, which is a billion times less efficient. Even hydrogen bombs have roughly a hundred times smaller efficiency.

Black holes have many interesting properties. We think of them as being surrounded by a surface called an "event horizon." When gas crosses this surface, it is lost from view and cannot be retrieved. However, the mass of the black hole, as measured by its gravitational attraction, will increase. As the gas is likely to be orbiting the black hole, it will also tend to spin it up. Stars that approach a million solar mass black hole will be pulled apart by the strong tidal forces; those that are swallowed by more massive, billion solar mass black holes will cross the event horizon before this happens.

There is a characteristic, maximum luminosity that can be radiated by the gas that accretes onto a black hole. This is known as the "Eddington limit." At this limit the outwardly directed pressure of the radiation balances the inward pull of gravity, preventing further accretion of gas onto the black hole. The Eddington limit for a million solar mass black hole associated with a low luminosity AGN is comparable with the luminosity of a normal bright galaxy like our own, while the Eddington luminosity of a billion solar mass black hole is a thousand times greater and matched to that from an ultraluminous quasar. When there is sufficient gas around the black hole, it will radiate like a black body. For a power given by the Eddington limit and a size of a few black hole radii the characteristic temperature of the gas is in the ultraviolet part of the spectrum. This roughly matches the observations. However, some of the gas will be at lower density and much higher temperature so that it can emit x rays.

In addition to the release of gravitational energy by accreting gas, there is a second possible way that black holes can fuel nuclear activity. It turns out that the rotational energy can also be extracted by surrounding gas when there is a strong magnetic field linking this gas to the black hole. Under these conditions, it is expected that large numbers of relativistic electrons and gamma rays will be produced. Processes like this are suspected to be responsible for the relativistic jets that power radio sources.

Although most of the power associated with bright AGN may be generated quite close to a massive black hole, much of the escaping radiation is reprocessed by surrounding gas into optical and ultraviolet emission lines and infrared radiation. As a consequence our view of an AGN and our classification of it may depend strongly upon our orientation with respect to the black hole and its gas supply.

Several pieces of observational evidence can be cited in favor of this model of AGN, although none of them constitute a definitive proof that black holes are present. Firstly, it is known that, at least on the average, the efficiency of energy release must be very high, in excess of that associated with nuclear reactions. The only reasonable possibility appears to be the release of gravitational energy from a potential well as deep as that formed by a black hole. Secondly, rapid variability in the emission suggests that the sources are very compact, in extreme cases as small as a black hole. Thirdly, those sources that contain radio jets also require the existence of a good gyroscope in the nucleus of the associated galaxy. A good candidate for this gyroscope is a spinning black hole. Finally, the central masses in

nearby galaxies that may once have harbored active nuclei have been estimated using the measured velocities of orbiting gas and stars and these are typically in the million to billion solar mass range, as anticipated.

Although the black hole model is widely adopted to explain luminous AGN like quasars, the lower luminosity sources like "Liners" and "Starburst galaxies" may be largely powered by a dense central cluster of young stars even if there is a black hole present.

See also: ASTROPHYSICS; ASTROPHYSICS, X-RAY; BLACK HOLE; BLACK HOLE, KERR; BLACK HOLE, SCHWARZSCHILD; FIELD, GRAVITATIONAL; GALAXIES AND GALACTIC STRUCTURE; GAMMA RAY BURSTERS; GRAVITATIONAL ATTRACTION; GRAVITATIONAL CONSTANT; GRAVITATIONAL FORCE LAW; GRAVITATIONAL LENSING; GREAT ATTRACTOR; MASS-ENERGY; STARS AND STELLAR STRUCTURE; X RAY

Bibliography

FRANK, J.; KING, A.; and RAINE, D. *Accretion Power in Astrophysics,* 2nd ed. (Cambridge University Press, Cambridge, Eng., 1992).

THORNE, K. S. *Black Holes and Time Warps: Einstein's Outrageous Legacy* (W. W. Norton, New York, 1994).

ROGER D. BLANDFORD

ADHESION

Adhesion is the physical bonding together of dissimilar substances as a result of the intermolecular electrostatic forces that are manifest when the substances are brought into close contact. Bonding where a chemical change occurs at the interface is not, generally, termed "adhesion"; although, outside the study of physics, this may be an important consideration.

The intermolecular forces responsible for adhesion are often called van der Waals forces after the Dutch physicist Johannes D. van der Waals, who modified the equation of state for an ideal gas to allow for the size and interactions of the constituent molecules. Various electrical forces may come into play depending on how the charges are distributed in the interacting molecules. Some molecules have asymmetric charge distributions that result in a permanent electric dipole. Even molecules without a permanent electric dipole may have their charge distributions modified by proximity to another molecule, resulting in an induced dipole.

Adhesion and cohesion arise from the same electrostatic forces; the only difference is whether the interacting molecules are of the same species (cohesion) or are dissimilar (adhesion).

It is at the surface of a body that comparison is made between the inward directed cohesive forces and the adhesive forces of that material for a dissimilar substance. Actual surfaces are often too rough at the molecular scale to make good contact and are usually contaminated by the molecules of substances foreign to the desired comparison. Because of the short-range nature of these forces, even a single molecular layer of adsorbed gases will confound a measurement, which must, therefore, be made between freshly cleaned surfaces in an ultrahigh vacuum. Under such conditions cold welding of metals is possible, but even a minute exposure to air will oxidize the surfaces and prevent cohesion (or proper adhesion if the pieces are of different metals).

Pressure, of course, enhances the contact between surfaces. Common examples are graphite on paper, where the force applied to a pencil and the sharpness of its point increase the pressure and yield better adhesion. Similarly, chalk on slate is much more difficult to erase if the lecturer uses a heavy hand to apply it. By contrast, chalk on glass is a case where the cohesion of the chalk is stronger than its adhesion to the glass.

The best contact between surfaces is made when one substance is a liquid. Liquids typically have very low cohesion and are drawn by adhesion to the other material, thus wetting it. Compare ink on paper to graphite on paper.

See also: COHESION; ELECTROSTATIC ATTRACTION AND REPULSION; VAN DER WAALS FORCE

Bibliography

LEE, L. *Fundamentals of Adhesion* (Plenum, New York, 1991).

RICHARD B. HERR

ADIABATIC PROCESS

The term "adiabatic process" is used to denote any process in which no energy transfer as heat occurs across the boundaries of the system under study. An adiabatic wall is, therefore, one that stops all heat transfer. While it may be difficult to achieve zero heat transfer, the ideal of an adiabatic process is a very useful one. In practice, many hardware devices, such as valves, nozzles, and turbines, may be realistically idealized as adiabatic. To achieve this, excellent insulation is required to minimize conduction heat transfer and, when necessary, a highly reflecting surface is used to minimize radiation heat transfer. Since a vacuum provides no medium for conduction and a silvered surface is an excellent reflector, we can see why the vacuum bottle provides a means to achieve a near-adiabatic wall. The adiabatic model is the process for two of the four processes in the Carnot cycle. The other two processes in that cycle are modeled as isothermal.

A particularly important case occurs when an adiabatic process is also reversible. An adiabatic-reversible process is known as one of constant entropy, usually denoted as isentropic. One can idealize some mechanical devices as reversible if there is minimal friction. For example, a well-insulated, smooth nozzle can be treated as both reversible and adiabatic, and, hence, as isentropic. A well-designed, insulated turbine can also approach this idea. The isentropic process provides an example of an ideal case because, in the absence of friction and heat transfer, the maximum amount of work output that can be achieved (e.g., when using a turbine) is reached for this case. When work is required, as for a pump, the minimum input occurs for the isentropic case. The concept of a reversible process is important. A process beginning from an initial state is reversible if, during or after the process, both the system of interest and the environment can be restored to their beginning states. Thus, reversibility implies the ability to restore. One can say that a reversible process "leaves no footprints in the sands of time." There are many examples of possibly reversible processes, such as a resistanceless circuit, the frictionless and adiabatic nozzle, and the ideal pendulum. The frictionless spring-mass, which can in theory oscillate forever, is another such example. Processes that are idealized as reversible include restrained expansion or compression, frictionless motion, elastic stretching of a solid, and zero-resistance electric circuits. The reversible process is an idealization, but a very useful one, even if it cannot be achieved. Since friction is always present and since heat transfer, in general, requires a temperature difference, real processes are not reversible and are referred to as irreversible. The reversible adiabatic process is an unattainable ideal, but can be used as a benchmark for the effectiveness of a particular design.

An adiabatic process of historical interest involves the calculation of the speed of sound in air. Isaac Newton attempted to do this calculation assuming that these small pressure waves move through the air in an isothermal (constant-temperature) manner. In fact, the propagation of sound waves is an adiabatic process. The error in the speed of sound between the two modes is about 15 percent at room temperature. The correct sound velocity in air at 70°F is 1,125 ft/s.

When combustion takes place in a well-insulated combustor, the products leave at the adiabatic flame temperature. This temperature represents the highest value that can be reached in any particular combustion process. Its value depends on the fuel being used. The adiabatic flame temperature is an important parameter in both air pollution calculations and when selecting appropriate materials to withstand combustion conditions.

An adiabatic saturator is a device to measure humidity. Here, an air-water-vapor mixture of unknown humidity flows into the insulated device and is saturated with water when the mixture flows over a water surface. The mixture then leaves at saturated conditions. Measurement of the inlet and exit pressure and temperature is sufficient to determine the inlet humidity. An adiabatic demagnetization is a process in which a magnetic field is suddenly removed while maintaining a near-zero heat transfer condition. This process can be used with paramagnetic salts, such as cerium magnesium nitrate, to achieve temperatures below 1 K.

See also: CARNOT CYCLE; ENTROPY; HEAT TRANSFER; INSULATOR; ISOTHERMAL PROCESS; SOUND

Bibliography

REYNOLDS, W. C., and PERKINS, H. C. *Engineering Thermodynamics*, 2nd ed. (McGraw-Hill, New York, 1977).
WARK, K. *Advanced Thermodynamics for Engineering* (McGraw-Hill, New York, 1995).

HENRY C. PERKINS JR.

ADVANCED X-RAY ASTROPHYSICS FACILITY

The Advanced X-Ray Astrophysics Facility (AXAF) is a space observatory for studying x rays from celestial objects. Its objectives include the detailed study of energetic, x-ray emitting objects such as highly active stars, neutron stars, and black holes in binary systems, remnants of supernova explosions, the active nuclei of distant galaxies, the most distant quasars, and the hot, diffuse gas in galaxies and between galaxies grouped in giant clusters. The capabilities of AXAF permit accurate measurements of the temperature, chemical composition, and physical condition of the emitting material, giving astronomers unique information about the nature and evolution of objects in the universe. It can also address fundamental questions of cosmology such as the location and quantity of dark matter and the size and age of the universe.

AXAF is a next-generation version of the Einstein Observatory, and, like the Hubble Space Telescope, is one of NASA's great observatories for astronomical research across the electromagnetic spectrum. AXAF is sensitive to x-ray energies of approximately 100 to 10,000 eV, corresponding to wavelengths of roughly 1 to 100 Å. AXAF weighs 10,000 lb, is 40 ft long with solar panels spanning more than 60 ft, and will be placed in a highly elliptical orbit of 10,000 × 100,000 km, circling the earth once every 42 hr.

AXAF consists of four sets of highly polished mirrors that focus x rays from distant objects onto one of several electronic cameras or detectors. In order to reflect x rays with high efficiency, the mirrors are extremely smooth, with roughness measured in angstroms, and then coated with a thin layer of metallic iridium. The x-ray images are focused to ~0.5 arc second, many times sharper than those of any previous or planned x-ray mission.

AXAF contains two kinds of cameras, either of which can be placed at the focus of the telescope by a mechanical drive mechanism. One produces slightly sharper images over a larger field, while the other can record the energy of each x-ray photon with an accuracy of roughly 10 percent as it builds up an image, giving simultaneous spectral and spatial information. Even finer energy information, with accuracies of ~0.1 percent, can be obtained by using one of the two transmission grating spectrometers carried by AXAF. These can be mechanically inserted between the telescope mirror and the cameras to disperse the x rays in much the same way that a prism disperses light into its colors. Special versions of the two cameras are used to read out the transmission gratings.

AXAF is being designed and built by a team of NASA, industry, and university scientists and engineers and will be open to use by the international astronomical community. It has a design life of five years, but it could well last longer. Each year, AXAF will make thousands of individual observations, answering old questions and very likely raising new ones.

See also: ASTROPHYSICS, X-RAY; COSMIC BACKGROUND EXPLORER SATELLITE; EINSTEIN OBSERVATORY; HUBBLE SPACE TELESCOPE

Bibliography

BRADT, H.; OHASHI, T.; and POUNDS, K. "X-Ray Astronomy Missions." *Annu. Rev. Astron. Astrophys.* **30,** 391–427 (1992).
CANIZARES, C. R., and SAVAGE, B. D. "Space Astronomy and Astrophysics." *Phys. Today* **44** (4), 60–68 (1991).

CLAUDE R. CANIZARES

AERODYNAMIC LIFT

The aerodynamic lift on an object moving through air is defined as the upward force on the object that opposes the downward force of gravity on the object. This lift force results from the difference between the pressure acting upward on the bottom of the object and the pressure acting downward on the top of the object. The total lift force is equal to the product of this pressure difference times the effective area of the object that is perpendicular to the force of gravity. For streamline airflow around a body, the pressure difference and the resulting lift force, Eq. (2), can be derived from Bernoulli's equation, Eq. (1), as follows:

$$P_b - P_t = \tfrac{1}{2}\rho((V_b)^2 - (V_t)^2), \qquad (1)$$

$$\text{lift force} = \tfrac{1}{2}\rho((V_b)^2 - (V_t)^2)A, \qquad (2)$$

where P is pressure, ρ is density of air, V is the velocity of the air, the subscript b refers to the bottom side of the airfoil, the subscript t refers to the top side of the airfoil, and A refers to the effective area of the airfoil.

The lift-to-drag ratio is considered to be an important measure of flight efficiency for a body moving through air. The lift is the upward force as just defined and the drag is the total force opposing the forward motion of the airfoil through the air. The drag force due to fast flow (eddy flow) around an object is proportional to the square of the flow speed and the lift force is given by Eq. (2) above.

Wind-tunnel studies show that the lift-to-drag ratio acting on a body increases by putting a small degree of curvature on a thin flat plate moving through the air. Figure 1 shows this comparison between a flat plate and an airfoil.

Wind-tunnel research leads to an optimization of the lift on a body moving through air by designing the body in the form of the traditional airfoil shape of aircraft wings shown in Fig. 2.

The lift-to-drag ratio increases with a small increase of the attack angle of the airfoil. This is the angle between the plane of the airfoil and the direction of motion through air as shown in Fig. 2. Jet airplane pilots abruptly increase the attack angle during take-off after reaching a specified speed; this action produces the necessary lift force for lift-off. However, in normal flight there is a critical attack angle that results in the breakdown of smooth air flow over the airfoil. At this angle the lift-to-drag

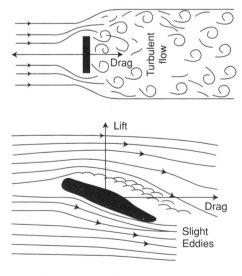

Figure 1 Lift-to-drag ratio.

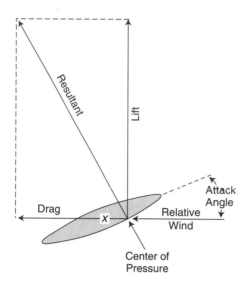

Figure 2 Traditional airfoil shape of aircraft wings.

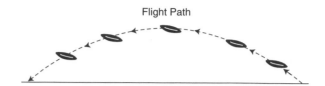

Figure 3 Typical flight path for a discus.

ratio begins to decrease rapidly and the airfoil stalls and is forced downward.

There are many examples of these aerodynamic principles applied to the projection of objects in sports. Consider the discus throw in track and field meets. At a low angle of attack, the lift-to-drag ratio on the discus is low, at a 10° attack angle the lift-to-drag ratio is significantly higher. Experiments indicate that the main effect of the spin of the discus is to stabilize the discus by maintaining a constant attack angle resulting in a flight as illustrated in Fig. 3.

The range of the discus throw is a function of the initial launch velocity, the projection angle, and the attack angle. Several sources report a maximum

range of about 155 ft is possible for an initial velocity of 70 ft/s, a projection angle of 35°, and an attack angle of 25°. Aerodynamic lift factors are the reason that the projection angle for maximum range is 35° rather than the 45° predicted for idealized projectile motion. Analysis of aerodynamic lift effects also leads to a conclusion that a discus thrower should be able to throw a discus several meters farther into a wind than would be possible in the absence of wind. This prediction is confirmed for wind velocities up to about 20 m/s.

The curve of a baseball in flight is a result of a sideways lift force on the ball produced by the rotation of the ball. The velocity of the air on the side of the ball where the spin velocity is parallel to the flight direction of the ball will be less than on the opposite side of the ball where the spin velocity is antiparallel to the air flow around the ball. Thus, according to the Bernoulli principle, the pressure on the high velocity side will be less than the pressure on the low velocity side. This force due to the rotation of the ball is called the Magnus effect. The magnitude of the Magnus force producing the curve of the baseball is the product of the pressure difference across the ball times the effective area of the ball (see Fig. 4).

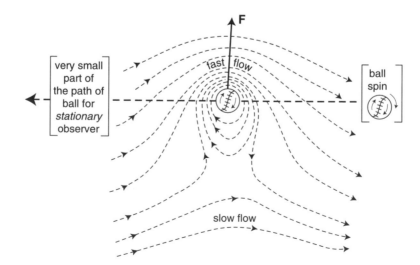

Figure 4 Forces acting on a baseball in flight.

32

You may demonstrate the Bernoulli lift effect by blowing on a piece of paper held in your hands so that the trailing edge sags under its own weight. As you blow just above the paper, you will see the lift effect on the paper.

We have shown that under the correct conditions the magnitude of the aerodynamic lift force acting on an aircraft can be greater than its total weight, thus making possible the flight of heavier-than-air aircraft. The application of the principles of aerodynamic lift plays an important part in the design of automobiles, trucks, and high speed trains as well as in the design of aircraft. Likewise, the analysis of the projection of any body through air involves an understanding of aerodynamic lift effects in order to predict the range of the projectile motion. This analysis can be applied to many examples in sports, such as the discus throw, the javelin throw, and the frisbee throw. Finally, we can apply the principles of aerodynamic lift to help us understand nature's own air fleet, which ranges from the seed pods of oak trees, bumble bees, and dragon flies to the largest of the known flying creatures, such as giant condors to prehistoric flying pterosaurs with wingspans as large as 15 m.

See also: AERODYNAMICS; BERNOULLI'S PRINCIPLE; FORCE; TURBULENCE; TURBULENT FLOW

Bibliography

MARION, J. B., and HORNYAK, W. F. *Physics for Science and Engineering* (Saunders, Philadelphia, 1982).

ROGERS, E. *Physics for the Inquiring Mind* (Princeton University Press, Princeton, NJ, 1960).

WALKER, J. *The Flying Circus of Physics with Answers* (Wiley, New York, 1977).

RICHARD M. FULLER

AERODYNAMICS

Aerodynamics is the physics that deals with the mechanics of objects moving through gases. In particular, it focuses on the forces acting on bodies moving through air.

Experiments show that the force on an object moving through air depends on the shape of the object and the speed of the object. There is an upward force on the object that is called the aerodynamic lift on the object. There is also a force in the opposite direction of the motion; this force is called the aerodynamic drag.

The field of aerodynamic engineering involves using the known aerodynamic principles to design and manufacture bodies that optimize desired aerodynamic effects of these bodies moving through air. For example, airplanes, cars and trucks, and projectiles of all kinds are designed based on aerodynamic principles. In each of these cases, it is desirable to minimize the aerodynamic drag on the body as it moves through air. Such designs are called "streamlining" because the design promotes streamline air flow around the body that is characteristic of minimum drag on the body.

The aerodynamic drag or air resistance on a standard car traveling at 40 mph (64 km/h) uses energy at a rate of 72 kW. Approximately 4.6 kW of this energy goes into overcoming the air resistance. Experimental studies show that the power needed for a car to overcome air resistance increases approximately as the cube of the speed of the car. Thus, a car traveling at 80 mph (129 km/h) requires about eight times the power required at 40 mph to overcome air resistance. During the energy crisis of the 1970s, streamlining of cargo trucks reduced air-resistance effects between 10 and 20 percent, and reducing the speed limit from 75 to 55 mph reduced the air-resistance effects by about 220 percent. These are two practical examples of aerodynamic analysis applied to real problems.

One of the most useful aerodynamic analysis is that of a sphere moving through air. The case of a sphere falling freely in air can be treated by using the drag relation known as Stokes's law given as: drag $= 6\pi\eta rv$, where η is viscosity of air, r is the radius of the sphere, and v is the speed of the sphere. This drag force opposes the force due to gravity to give the following equation that can be used to find the "terminal velocity" that produces the drag force that balances the gravity force and thus results in the constant terminal velocity for the rest of the fall, $mg = 6\pi\eta rv$, and $v = mg/6\pi\eta r$. This same type of analysis can be applied to any shape of a body freely falling through air, but the shape factor and velocity dependence must be determined by experiment.

The projected flight of a sphere through air is the basic situation of the flight of a baseball, a tennis ball, a golf ball, a basketball, and a shot put. In each case the retarding drag force on the moving ball is proportional to the square of the speed of the sphere through the air.

Likewise, the drag force on a runner with speed V running against a wind, with velocity v, will experience a drag force proportional to $(V - v)^2$.

In the running broad jump, the drag force affects the jumper by slowing down the runner before take-off and also reducing the speed of the jumper in the air. In this analysis the density of air also comes into play. For example, the Mexico City Olympics produced record jumps, which can be accounted for in part by the reduction of the air density at the altitude of the games.

It can be shown that the shielding effect of running directly behind another runner can reduce the drag force to between one-fifth to one-half of the drag on the front runner. This is the physics behind "drafting" strategies of experienced distance runners. This same principle also applies to cycling with even greater consequences since cycling speeds are more than twice as fast as running speeds. The shielding effect for cyclists shows that the lead cyclist has eight times the drag as that of the lead runner. The trailing cyclist will have the drag reduced by at least by one-third.

See also: AERODYNAMIC LIFT; CURRENT, EDDY; FORCE; TURBULENCE; TURBULENT FLOW

Bibliography

MARION, J. B., and HORNYAK, W. F. *Physics for Science and Engineering* (Saunders, Philadelphia, 1982).

ROGERS, E. *Physics for the Inquiring Mind* (Princeton University Press, Princeton, NJ, 1960).

WALKER, J. *The Flying Circus of Physics with Answers* (Wiley, New York, 1977).

RICHARD M. FULLER

ALFVEN WAVE

Alfven waves are transverse waves in an electrically conducting fluid permeated by a magnetic field. A highly conducting fluid (e.g., a liquid metal or an ionized gas), is tied to the field for a time t over any scale ℓ that is large compared to $(4\eta t)^{1/2}$, where η is the resistive diffusion coefficient of the fluid. Therefore, any sufficiently large-scale disturbance of the field and fluid involves mutual motion. A magnetic field **B** possesses a tension $\mathcal{T} = B^2/4\pi$ dyn/cm^2

along the field, so any local displacement in a direction perpendicular to the field is pulled back to its original position by the tension $\mathcal{T}$, sketched in Fig. 1. In precise analogy to a plucked guitar string, a displacement propagates with velocity $(\mathcal{T}/\rho)^{1/2}$ along the field, where ρ is the density of the string or fluid in gm/cm^3. Thus, the wave velocity in a field **B** is called the Alfven speed $B/(4\pi\rho)^{1/2}$. There is equipartition of energy in the Alfven wave, so that

$$\frac{1}{2}\rho < v^2 > = \frac{\Delta \mathbf{B}^2}{8\pi},$$

where $< v^2 >$ is the mean square fluid velocity and $\Delta \mathbf{B}$ is the magnetic field component of the wave. The waves are called Alfven waves after Hannes Alfven, who first pointed out their existence.

If the fluid is compressible, combinations of longitudinal and transverse displacements provide more complicated magnetohydrodynamic (MHD)

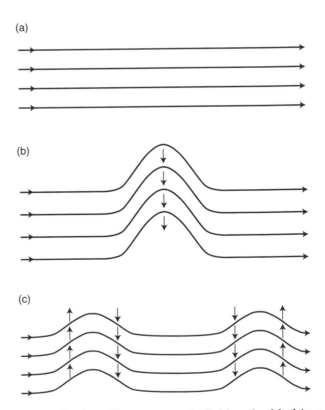

Figure 1 A uniform magnetic field embedded in a highly conducting fluid (a) is locally displaced and released from the configuration shown in (b). The tension in the field sets the fluid in motion, indicated by the short arrows, providing the two propagating Alfven waves shown in (c).

wave motions, propagating with speeds depending on both the Alfven speed and on the speed of sound in the fluid. The Alfven wave velocity is typically 600 km/s in the outer magnetosphere of the earth, where $B \cong 3 \times 10^{-4}$ G with $N = 1$ hydrogen atom/cm^3. In the solar wind outside, where $B \cong 10^{-4}$ G and $N = 5$, it is 100 km/s and small compared to the wind velocity. The Alfven speed is of the order of 2,000 km/s in the x-ray corona of the Sun, where $B \sim 10^2$ G, $N = 10^{10}$/cm^3; and in the coronal holes, where $B \sim 10$ G, $N = 10^8$/cm^3, which are the origin of the fast streams in the solar wind. The Alfven speed under typical interstellar conditions of 3×10^{-6} G and $N = 1$/cm^3 is 6 km/s. In the liquid metal core of the earth where $\rho \sim 10$ gm/cm^3 and the azimuthal magnetic field may be 20 G, the Alfven speed is about 2 cm/s and very much larger than the 1 mm/s convective motions. In a magnetic field of 10^2 G in a laboratory plasma of 10^{12} atoms/cm^3, it is 200 km/s.

See also: MAGNETOHYDRODYNAMICS

Bibliography

PARKER, E. N. *Cosmical Magnetic Fields* (Clarendon, Oxford, Eng., 1979).

ROBERTS, P. H. *An Introduction to Magnetohydrodynamics* (American Elsevier, New York, 1967).

E. N. PARKER

ALLOY

An alloy is a mixture of two or more chemical elements, at least one of which is a metal, and having metallic properties. Two examples of alloys are steel (a mixture of iron and carbon) and bronze (a mixture of copper and tin).

The history of metals and alloys goes back many thousands of years. The first uses of these materials were for decorative purposes (desirable properties: appearance and malleability), or for tools and weapons (desirable properties: hardness and strength). The malleable and aesthetically pleasing metals gold and copper, which occurred naturally, were too soft to be useful as tools or weapons. The first alloys were probably copper alloyed with impurities (arsenic?), doubtless reduced from an im-

pure oxidized ore in a campfire. The alloys were harder and stronger than native gold and copper, and therefore could be made into better tools and weapons. The accidental smelting of alloys developed into the purposeful alloying of copper with tin, around 3000 B.C.E., to make bronze. The Bronze Age ended around 1100 B.C.E. when people learned how to smelt iron, which is in turn stronger than bronze. The Iron Age lasted until methods were developed for the mass production of steel in the mid-nineteenth century.

At the present time, alloys are used because their properties can be tailored to a particular function. The main properties of interest are mechanical properties, such as strength, ductility, wear resistance, stiffness. However, alloying can also improve nonmechanical properties such as corrosion resistance, electrical and magnetic properties, formability, or machinability. In addition to the properties of an alloy, its cost is also important. The development of economical alloys with particular properties is an important part of metallurgy and materials science.

One of the factors determining the properties of an alloy is the way the atoms are arranged in the alloy, that is, the structure of the alloy. An alloy can be liquid or solid, but we shall consider mostly the latter. The atoms in a solid alloy are arranged on the points of a crystal lattice, that is, the alloy is crystalline, although the crystallinity is not reflected in the macroscopic structure, as it is in a mineral. That is because metallic alloys for practical use are made up of small individual crystallites called grains, usually 1 to 100 μm (10^{-6} to 10^{-4} m) in cross section.

Another important aspect of the structure is the distribution of the different components in the alloy. The atoms of the different components can be arranged more or less randomly, in a solid solution that is a solid analogue of an aqueous solution. Or the alloy can be heterogeneous, with some grains composed of one phase (a given composition and crystal structure) and other grains made up of another phase, with a different composition and crystal structure. For example, if the solubility of the second component is exceeded, a second phase will precipitate. A two-phase alloy can also be obtained by cooling certain compositions from the melt. The phases existing as a function of composition and temperature are shown on a phase diagram. Which phases appear and their structures depend on the electronic structure of the alloy; some interesting rules connecting these have been given by William Hume-Rothery.

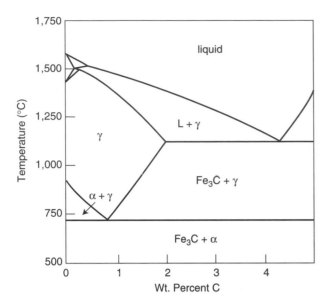

Figure 1 Simple phase diagram for the iron-carbon system.

A simplified phase diagram for the iron-carbon system is shown in Fig. 1. Below 738°C, the alloys are made up of two phases. One is α-iron, which has iron atoms located on the corners and in the center of the cubic lattice and carbon atoms (solubility 0.038% by weight), located inside the cubic lattice. The other is Fe_3C, a hard compound. The fraction of the latter depends on the carbon content of the steel; the carbon content also determines the tensile strength, the stress needed to break a bar by pulling. This property changes from about 30 MPa for 0 percent C to about 100 MPa for 1 percent C (1 Pa = 1 kg/m^2). However, increasing the carbon content lowers the ductility (elongation of the bar on breaking) from 42 percent at 0 percent C to 10 percent at 1 percent C.

The structure of the alloy can be changed not only by varying the composition, but also by heat treatment. If Fe-C alloys are heated into the γ phase (Fe atoms on the corners and face centers of the lattice cubes, carbon atoms at the body centers) and then quenched (cooled very rapidly), the cubic lattice is distorted, and the strength is raised greatly, to 200 MPa for 1 percent C, with a corresponding decrease in ductility. Better ductility can be obtained by heating to a low temperature (tempering). The original structure and the ductility can be restored, but with a corresponding loss in strength, by heating to a high temperature and cooling slowly (annealing). The mechanical properties in the paragraph above refer to the annealed state. Any distortion of the crystal lattice, caused by mechanical working,

precipitation, or quenching, for example, tends to increase the strength, lower the ductility, and increase the electrical resistivity.

A vast variety of alloys is available today. The following are a few examples that emphasize the relation between properties and use.

Structural steels used in buildings, bridges, and ship plates are made of low carbon (<0.3% C) steels because these alloys combine fair strength with ductility, easy formability by rolling or drawing, and very low cost per unit weight. They are used in the as-rolled or annealed condition.

Machine components such as shafts, gears, axles, and rails need great strength and toughness. This is obtained by quenching and tempering medium carbon (0.3 to 0.7% C) steel. Alloying elements such as Cr, Mn, or Ni are added to increase the hardenability (ease of quenching through a thick section). A typical ultrahigh strength aircraft steel is alloyed with 0.4 percent C, 0.7 percent Mn, 0.8 percent Cr, 1.8 percent Ni, and 0.25 percent Mo.

Tools and dies require great hardness and are made of high carbon (0.7 to 1.7% C) steels, quenched but lightly tempered. High-speed tools are alloyed with the refractory metals W or Mo to give them strength at high temperature.

Cutlery and kitchen utensils are made of stainless steels (18% Cr, 8% Ni, low carbon). The Cr contributes the corrosion resistance by oxidizing to a thin protective film of Cr_2O_3, the Ni stabilizes the γ phase to room temperature for better formability and improves the mechanical properties. Alloys based on this composition are also used as cladding for nuclear fuels because of their corrosion resistance in coolant, good chemical compatibility with the fuel material, and low neutron capture cross section.

Electrical conductors' most important property is low electrical resistance, so Cu is used. Alloying is kept to the minimum consistent with acceptable strength, because impurity atoms scatter electrons, thus increasing the resistivity.

Aircraft frames and skin are made of Al alloys because of their excellent combination of high strength and low density. The alloys are precipitation hardened, and clad with pure Al to form native oxide (Al_2O_3) skin for corrosion resistance.

Solders must have a low melting point and good wetting properties. Pb-Sn alloys are the least expensive, but environmental regulations against lead because of its toxicity are leading to a search for inexpensive lead-free solders.

Brass screws must machine easily, so the 70 percent Cu-30 percent Zn alloys are further alloyed with lead, which causes the chips to break off easily during machining, but at some cost of strength.

See also: CRYSTAL STRUCTURE; METAL; METALLURGY

Bibliography

BRAY, J. L. *Non-ferrous Production Metallurgy* (Wiley, New York, 1941).
COTTRELL, A. H. *An Introduction to Metallurgy* (Edward Arnold, London, 1967).
HUME-ROTHERY, W. *Atomic Theory for Students of Metallurgy* (The Institute of Metals, London, 1955).

STEVEN J. ROTHMAN

ALPHA DECAY

See DECAY, ALPHA

ALTERNATING CURRENT

See CURRENT, ALTERNATING

AMMETER

An ammeter is a device used to measure the presence of electric current (moving electric charge) in a circuit. If the current flowing in an element of the circuit is to be measured, the circuit must be opened and the ammeter placed in series with the circuit element so that all of the current passing through the element must pass through the ammeter. The reading given by an ammeter is in ampères (Cs/s). Ideally, the current passing in the circuit after the ammeter is placed in it should be the same as without the ammeter; therefore, the resistance to current flow of an ideal ammeter is zero. Actual ammeters are designed to have low resistance, typically a few ohms to a fraction of an ohm. An amme-

ter should *never* be connected directly between the terminals of a battery or other source of electromotive force (emf), for its low resistance permits a large current to flow, which may damage the ammeter components.

The common type of ammeter, which gives a reading by means of a needle moving over a dial, is based on a D'Arsonval galvanometer. The galvanometer is in fact a sensitive ammeter. Usually we wish to measure a current that is larger than the full-scale capability of the galvanometer I_G, which might be 50 μA. Then a shunt resistance R_s is connected in parallel with the galvanometer G, so that much of the current flowing through the ammeter is shunted around the galvanometer itself (Fig. 1). To make a multirange ammeter, capable of measuring different full-scale currents, several different shunt resistors are arranged so that a certain one may be switched in to obtain a certain full-scale capability.

Since a galvanometer is commonly made with permanent magnets, it is a dc, or direct current, instrument. Most analog (moving needle) ammeters are dc ammeters. Alternating current (ac) ammeters exist that are constructed with electromagnets or other special techniques, but they are now quite rare. Digital ammeters typically use a digital voltmeter to measure the voltage V_r that occurs across a small sensing resistor r through which passes the current to be measured. Then the current is displayed from $I = V_r/r$. Using this technique, an ac digital ammeter can be obtained by use of an ac digital voltmeter.

Referring to Fig. 1, suppose that we want to design the shunt resistance R_s for an ammeter that reads I amperes full-scale, and the galvanometer coil has sensitivity I_G and resistance R_m. (In practice R_m may include additional resistance connected in series with the galvanometer coil.) When the current I is flowing through the ammeter, current I_G flows through the galvanometer so that a voltage difference $I_G R_m$ appears across the coil. A current $(I - I_G)$

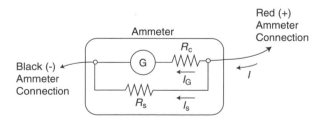

Figure 1 Ammeter constructed from a galvanometer and shunt resistor R_s.

flows through the shunt resistor R_s. Since the voltage difference $(I - I_G)R_s$ that appears across the shunt resistor must be the same as the voltage $I_G R_m$ across the coil, we have $R_s = I_G R_m / (I - I_G)$.

See also: ELECTRICAL RESISTANCE; GALVANOMETER; OHMMETER; VOLTMETER

Bibliography

GREGORY, B. *An Introduction to Electrical Instrumentation* (Macmillan, New York, 1973).

JONES, L. *Electrical and Electronic Measuring Instruments* (Wiley, New York, 1983).

STOUT, M. B. *Basic Electrical Measurements,* 2nd ed. (Prentice Hall, Englewood Cliffs, NJ, 1960).

DENNIS BARNAAL

AMPÈRE, ANDRÉ-MARIE

b. Lyons, France, January 20, 1775; *d.* Marseilles, France, June 10, 1836; *electrodynamics.*

Ampère was the only son of Jean-Jacques Ampère and Jeanne-Antoinette Desutieres-Sarcey, descendants of prosperous silk-merchant families in Lyons. One of his two sisters died quite young and the other later served him as a housekeeper in Paris. In 1793, when Ampère was eighteen, his father was unjustly executed by guillotine during the Reign of Terror, the most violent period of the French Revolution. In 1799 Ampère married Catherine-Antoinette Carron, and in 1800 she gave birth to their son, Jean-Jacques. Her health failed shortly thereafter, however, and she died in 1803. Ampère's personality took on a permanently melancholy cast following these tragic events of his youth.

Ampère received no formal education whatsoever. His father encouraged him to read extensively, and the young Ampère committed to memory entire articles of the famous *Encyclopedia* edited by Jean Le Rond D'Alembert and Denis Diderot. His parents were devout Catholics; Ampère often was tormented by tensions between his intellectual and spiritual inclinations. One of his earliest scientific interests was botanical classification, and he also developed an early talent for mathematics. Following some teaching and tutoring in Bourg and Lyons, in

1804 Ampère was appointed to teach mathematics at the École Polytechnique, the prestigious institute for scientific education in Paris. He wrote several mathematics memoirs on partial differential equations, and in 1814 he won election to the French Academy of Sciences as a mathematician. During this period, Ampère also kept abreast of developments in chemistry. He proposed a highly geometric conception of molecular structure, developed a classification scheme for the chemical elements, and gave an independent presentation of what is generally referred to as Avogadro's hypothesis.

Ampère did not devote detailed attention to physics until 1820, when he was forty-five years of age. In that year Hans Christian Oersted discovered that an electric current can alter the orientation of a suspended magnet. Ampère immediately followed up Oersted's work by discovering that two linear electric currents either repel or attract each other depending upon their mutual orientation. He then argued that all magnetic phenomena are due to similar forces between tiny molecular circuits within magnets. He discovered a mathematical formula for the force between any two infinitesimally short circuit elements, and he applied this formula to a wide variety of experimental arrangements of electric circuits and magnets, a new branch of physics he called "electrodynamics." In 1822 he came very close to discovering electromagnetic induction, subsequently discovered by Michael Faraday in 1831. Ampère's most creative contributions came during the transition from the electrostatics and magnetic theories of the early nineteenth century to the electrodynamic field theory that subsequently became a major component of classical physics. His experimental and theoretical ingenuity earned him recognition from the famous English physicist James Clerk Maxwell as the "Newton of electricity." In 1881 his name was commemorated as the unit of electric current.

In addition to his contributions to mathematics, chemistry, and physics, Ampère also maintained an interest in philosophy, particularly scientific methodology and classification schemes.

See also: AMPÈRE'S LAW; ELECTROMAGNETIC INDUCTION, FARADAY'S LAW OF; ELECTROMAGNETISM

Bibliography

CANEVA, K. "Ampère, the Etherians, and the Oersted Connection." *The British Journal for the History of Science* **13,** 121–138 (1980).

HOFMANN, J. "Ampère's Invention of Equilibrium Apparatus: A Response to Experimental Anomaly." *The British
Journal for the History of Science* **20**, 309–341 (1987).

HOFMANN, J. "Ampère, Electrodynamics, and Experimental Evidence." *Osiris* **3**, 45–76 (1987).

HOFMANN, J. *André-Marie Ampère, Enlightenment and Electrodynamics* (Cambridge University Press, Cambridge, Eng.,
1995).

WILLIAMS, L. P. "What Were Ampère's Earliest Discoveries
in Electrodynamics?" *Isis* **74**, 492–508 (1983).

WILLIAMS, L. P. "André-Marie Ampère." *Sci. Am.* **260**,
90–97 (1989).

JAMES R. HOFMANN

AMPÈRE'S LAW

Ampère's law relates the *static* magnetic field **B** to
the *steady* current I that produces it. First published
in 1827 by André Ampère, it is derived from the
Biot–Savart law, which says that the static magnetic
field **B**(**r**) (or "magnetic induction") produced by
the steady conventional current I is given by

$$\mathbf{B}(\mathbf{r}) = k_m I \int_C \frac{d\mathbf{r}' \times \mathbf{R}}{R^3} \qquad (1)$$

(see Fig. 1). In Eq. (1), **r** is the vector from the origin to the field point, **r**′ is the vector from the origin
to the source point (i.e., the point where the infinitesimal current element of $I d\mathbf{r}'$ resides; primed
coordinates refer to locations of sources, and un-
primed coordinates are the field points), $\mathbf{R} = \mathbf{r} - \mathbf{r}'$
is the vector from the source point to the field point,
C is the path followed by the current, and $k_m = 10^{-7}$
N/A^2 in SI units. We sketch the derivation of Ampère's law below. Conceptually, it is equivalent to
dotting Eq. (1) with the displacement vector $d\mathbf{r}$, integrating the result around an arbitrary closed path
Γ, then using some vector identities on the right-
hand side. We thereby construct Ampère's law in integral form,

$$\oint_\Gamma \mathbf{B} \cdot d\mathbf{r} = 4\pi k_m I_\Sigma. \qquad (2)$$

In Eq. (2), I_Σ is the net conventional current that
pierces *any* surface Σ that has the closed path Γ for

its boundary (see Fig. 2). The direction of **B** relative
to the direction of its source current finds expression in Ampère's law through a right-hand rule: If
you point the thumb of your right hand in the direction of the net conventional current as it crosses Σ,
then the fingers of your right hand will point in the
direction to go around the closed path Γ such that
$\mathbf{B} \cdot d\mathbf{r} > 0$. The coefficient in Ampère's law is often
written $\mu_0 \equiv 4\pi k_m$, the "permeability of free space."

The mathematically efficient derivation of Ampère's law begins with the Biot–Savart law as applied
to a steady current that occupies a finite volume V,

$$\mathbf{B}(\mathbf{r}) = k_m \iiint_V \frac{d^3\mathbf{r}' \, [\, \mathbf{j}(\mathbf{r}') \times \mathbf{R}]}{R^3}, \qquad (3)$$

where $\mathbf{j}(\mathbf{r}')$ is the density of conventional current,
and $d^3\mathbf{r}'$ is the volume occupied by an infinitesimal
current element. Evaluating the curl of Eq. (3) and
using some vector identities, we obtain Ampère's law
in differential form,

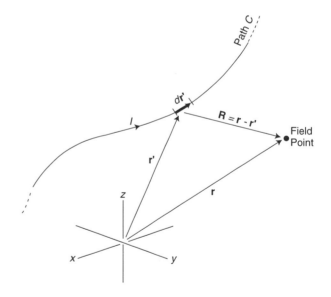

Figure 1 Relative to the coordinate system shown,
the vector **r** locates the field point (i.e., the point in
space where the field will be calculated). The vector **r**′ locates the source point (i.e., the point in
space where the infinitesimal source of the field
is located). For the magnetostatic field, that point
source is an infinitesimal segment of the current I
that flows along the path C. The vector $d\mathbf{r}'$ is tangent to the conventional current I at the source
point. The vector $\mathbf{R} = \mathbf{r} - \mathbf{r}'$ is the vector *from* the
source point *to* the field point.

$$\mathbf{V} \times \mathbf{B} = 4\pi k_m \mathbf{j}. \qquad (4)$$

Introducing any surface Σ bounded by the closed path Γ, we evaluate the flux through Σ of the quantities appearing on both sides of Eq. (4), then apply Stokes's theorem to the left-hand side of the result, to obtain Ampère's law in an integral form equivalent to Eq. (2),

$$\oint_\Gamma \mathbf{B} \cdot d\mathbf{r} = 4\pi k_m \iint_\Sigma \mathbf{j}(\mathbf{r}') \cdot \hat{\mathbf{n}} \, da'. \qquad (5)$$

In Eq. (5), da' is an infinitesimal patch of surface area on Σ, and $\hat{\mathbf{n}}$ is a unit vector normal to da'.

Since it was derived from the Biot–Savart law, the stipulations that must be met for Ampère's law to be valid are the same as those required of the Biot–Savart law; namely, that the current be *steady* and thus the magnetic field be *static*.

Since we already have the Biot–Savart law, why do we bother with Ampère's law? There are two impor-

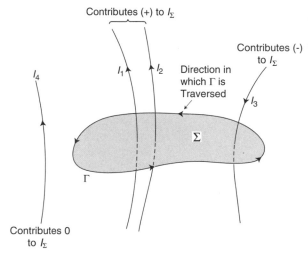

Figure 2 Currents I_1 and I_2 contribute positive values to the current I_Σ of Ampere's law; current I_3 contributes a negative value to I_Σ, and current I_4 contributes zero to I_Σ. Notice the sense of direction given to the closed path Γ forms the boundary of the surface Σ. Current I_4 does not pierce the surface Σ at all. Currents I_1 and I_2 pierce Σ in the direction opposite to I_3. Currents I_1 and I_2, and the sense of direction shown on Γ, are consistent with the right-hand rule, and thus these currents make a positive contribution to I_Σ.

tant reasons. First, in some highly symmetric cases, the integral form of Ampère's law provides a simpler tool for calculating $\mathbf{B}$ than does the cumbersome Biot–Savart law. Second, Ampère's law offers an elegant approach for generalizing the theory from the special case of magnetostatics to the general case of the coupled electric and magnetic fields that arise in time-dependent electrodynamics. Let us discuss each of these virtues in turn.

In cases where the symmetry of the current distribution ensures that (a) the direction of $\mathbf{B}$ is known everywhere by inspection, and (b) one can find a closed path such that the component of $\mathbf{B}$ tangent to $d\mathbf{r}$ is constant (or piecewise constant) everywhere along the path, then the line integral $\oint \mathbf{B} \cdot d\mathbf{r}$ reduces to the product of a length and the magnitude of $\mathbf{B}$. Under these circumstances Eq. 2 or Eq. (5) can be used as a powerful tool to find the magnitude of $\mathbf{B}$ everywhere along that path. Let us illustrate with an example.

Let a steady current I_0 of uniform density j flow along a straight cylindrical conducting rod of radius r_0. Let the rod be infinitely long and call its axis the z axis. From the right-hand rule applied to the cross product appearing in Eq. (3) and from the symmetry of the current distribution about the z axis, we know that the magnetic field lines will form concentric circles about the z axis, and that the magnitude of $\mathbf{B}$ can vary only with the distance ρ from the axis. It therefore remains to find this magnitude $B = B(\rho)$ (see Fig. 3). We have two regions to consider: $\rho > r_0$ and $\rho < r_0$. In either case if we choose for closed path Γ a circle of radius ρ whose axis is the z axis, then the left-hand side of Eq. (2) or Eq. (5) reduces to $(B)(2\pi\rho)$. For $\rho < r_0$, the right-hand side of Eq. (5) is $\mu_0 j\pi\rho^2 = \mu_0(I_0/\pi r_0^2)\pi\rho^2 = \mu_0 I_0(\rho^2/r_0^2)$. Thus interior to the wire Ampère's law gives $\mathbf{B}(\rho) = (\mu_0 I_0/2\pi r_0^2)\rho$. Exterior to the wire, where $\rho > r_0$, the current passing through Σ is simply I_0, so that Ampère's law gives $\mathbf{B}(\rho) = \mu_0 I_0/2\pi\rho$.

There is a striking analogy between the applications of Ampère's law and Gauss's law for the electric field $\mathbf{E}$: when expressed in integral form, these laws may provide effective tools for finding the magnitude of their respective fields. For this to happen it is necessary that the symmetry of the source distribution be sufficiently great that the field direction is known everywhere, and that surfaces (in Gauss's law for $\mathbf{E}$) or closed paths (in Ampère's law) can be found for which the flux or line integral reduces to a constant field magnitude times an area or length, respectively.

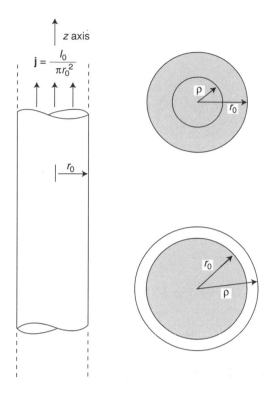

Figure 3 Showing the geometry of the calculation of the magnetostatic field inside and outside a wire of radius r_0 carrying a steady current density j.

When the sources of electric and magnetic fields are allowed to be time-dependent, then the requirements of local electric charge conservation, coupled with Gauss's law for **E,** suggest that Ampère's law for fields in vacuum be superseded by

$$\oint_\Gamma \mathbf{B} \cdot d\mathbf{r} = \iint_\Sigma \left[4\pi k_m \, \mathbf{j}(\mathbf{r}') + \left(\frac{1}{c^2}\right) \frac{\partial \mathbf{E}}{\partial t} \right] \cdot \hat{\mathbf{n}} \, da', \quad (6)$$

where c is the speed of light in vacuum. The differential form of Eq. (6) is

$$\mathbf{V} \times \mathbf{B} = 4\pi k_m \mathbf{j} + \left(\frac{1}{c^2}\right) \frac{\partial \mathbf{E}}{\partial t}. \quad (7)$$

Equation (6) and Eq. (7) may rightly be called the "Ampère–Maxwell law," since James Clerk Maxwell was the first person to mathematically model a time-dependent electric field as a "source" of a magnetic field, thereby completing the equations of electromagnetism known as "Maxwell's equations."

See also: BIOT–SAVART LAW; COULOMB'S LAW; ELECTROMAGNETISM; FIELD, ELECTRIC; FIELD, MAGNETIC; GAUSS'S LAW; MAXWELL'S EQUATIONS; VECTOR

Bibliography

FEYNMAN, R. P.; LEIGHTON, R. B.; and SANDS, M. *The Feynman Lectures on Physics,* Vol. 2 (Addison-Wesley, Reading, MA, 1963).

GRIFFITHS, D. J. *Introduction to Electrodynamics,* 2nd ed. (Prentice Hall, Englewood Cliffs, NJ, 1989).

HALLIDAY, D.; RESNICK, R.; and KRANE, K. S. *Physics,* 4th ed. (Wiley, New York, 1992).

JACKSON, J. D. *Classical Electrodynamics,* 2nd ed. (Wiley, New York, 1975).

LORRAIN, P.; CORSON, D. R.; and LORRAIN, F. *Electromagnetic Fields and Waves* (W. H. Freeman, New York, 1988).

PANOFSKY, W. K. H., and PHILLIPS, M. *Classical Electricity and Magnetism,* 2nd ed. (Addison-Wesley, Reading, MA, 1962).

ROJANSKY, V. *Electromagnetic Fields and Waves* (Prentice Hall, Englewood Cliffs, NJ, 1971).

SERWAY, R. A. *Physics for Scientists and Engineers with Modern Physics,* 3rd ed. (Saunders, Fort Worth, TX, 1990).

TIPLER, P. A. *Physics,* 2nd ed. (Worth, New York, 1982).

VANDERLINDE, J. *Classical Electromagnetic Theory* (Wiley, New York, 1993).

DWIGHT E. NEUENSCHWANDER

ANTIFERROMAGNETISM

Antiferromagnetism is a type of magnetic behavior observed in some metal oxides, such as MnO, and some sulfides, such as Cr_2O_3, in the presence of a magnetic field. This is a class of substance that can be characterized by an interaction favoring an antiparallel orientation of neighboring spins. This is in contrast to ferromagnetism, which favors parallel orientation of its spins. Antiferromagnetism was first proposed theoretically by Louis Néel in 1932 and later discovered experimentally in MnO. The feature that distinguishes antiferromagnetism is the manifestation of a peak when the magnetic susceptibility is plotted against temperature.

Antiferromagnetism can be described in terms of two interlocking sublattices, one made up of atoms of type A only and the other of type B. For instance, A atoms occupy corners of a cube in a unit cell and B

atoms can occupy the face-centers of the cube. At low temperature the negative exchange interaction between the spins dominate and the spins in the A sublattice and those in the B sublattice are aligned antiparallel to each other. Because of this interaction, the paramagnetic susceptibility in the direction of the magnetic field $\chi_\parallel$ is much less than that of ordinary paramagnetic solids. As the temperature is raised, antiferromagnetic ordering is disturbed and $\chi_\parallel$ actually increases. When an antiferromagnetic solid reaches a critical temperature known as the Néel temperature, T_N, it becomes a paramagnetic solid and obeys the Curie–Weiss law (Fig. 1). Unlike ferromagnetic substances beyond the Curie point, the linear extrapolation of $1/\chi$ versus temperature plot of an antiferromagnetic substance does not intersect the horizontal (temperature) axis at a point on the positive side. This is shown schematically in Fig. 2.

The relation between χ and T for para-, ferro- and antiferromagnetic cases can be expressed as

$$\chi = C/T \text{ (paramagnetism)},$$

$$\chi = C/(T - \theta) \text{ (ferromagnetism)},$$

and

$$\chi = C/(T + \theta) \text{ (antiferromagnetism)}.$$

A few additional comments are in order for the antiferromagnetism of MnO. For the strongest nega-

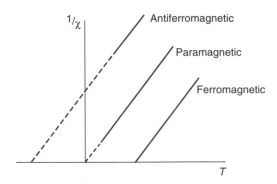

Figure 2 Schematic plotting of linear extrapolation versus temperature for an antiferromagnetic substance.

tive exchange interaction, magnetic moments of Mn ions do not interact directly with its neighboring Mn ions. Instead they interact indirectly through the neighboring O^{2-} ions. This type of interaction is known as the superexchange interaction.

The most direct experimental evidence for antiferromagnetism is in the results of neutron diffraction experiments. A neutron beam is electrically neutral, and it is scattered only by magnetic interactions including those between the neutron spin magnetic moment and the paramagnetic ions. The diffraction pattern analysis is similar to that of x ray and electron diffraction. The ordered magnetic sublattice gives rise to "superlattice" lines that are not present in magnetically disordered systems.

These experiments on neutron diffraction revealed that there are antiferromagnetic materials that are more complicated than those described above. For instance, rare earth metals terbium, dysprosium, holmium, erbium and thulium are ferromagnetic at low temperature, but they become antiferromagnetic with helical spin structure above their Néel temperature, and are paramagnetic.

See also: CURIE TEMPERATURE; FERROMAGNETISM; PARAMAGNETISM

Bibliography

FEYNMAN, R. P.; LEIGHTON, R. B.; and SANDS, M. *The Feynman Lectures on Physics,* 3 vols. (Addison-Wesley, Reading, MA, 1963–1965).

KITTEL, C. *Introduction to Solid-State Physics,* 6th ed. (Wiley, New York, 1986).

CARL T. TOMIZUKA

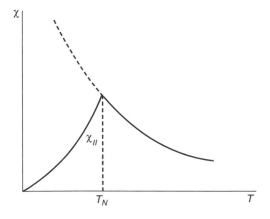

Figure 1 Diagram relating the change from antiferromagnetism to paramagnetism to the Néel temperature.

ANTIMATTER

The nature of antimatter is best illustrated by considering some examples. For every known subatomic particle, there is an antiparticle with the same mass and opposite electric charge. Associated with the electron, there is an antielectron, known as the positron. To a very high degree of accuracy, it is known that the electron has the same mass as the positron. Another important property, the magnetic moment, is also opposite for the electron and positron. The magnetic moment describes the magnetism of a particle. This magnetism is related to the spin of a particle. For an electron, the spin tends to line up along the magnetic field, while for a positron it tends to align opposite to the field. Theoretical arguments indicate that the masses of these particles should be exactly the same. Similarly, associated with the proton is the antiproton, with exactly the same mass and opposite charge of the proton. Electrically neutral particles can also have particles. The antineutron, for example, has no electric charge. Its magnetic moment, however, is exactly opposite to that of the neutron.

The existence of antiparticles was predicted in 1928 by the English theoretical physicist P. A. M. Dirac. Dirac played a major role in the development of quantum mechanics in the 1920s. In its earliest form, quantum mechanics did not incorporate the principles of Einstein's special relativity. Dirac attempted to develop a relativistic version of quantum mechanics. He postulated a relativistic generalization of the Schrödinger equation, known as the Dirac equation. He quickly recognized that this equation gave a correct description of the spin and magnetic properties of electrons. But it had the puzzling feature that for every state of positive energy, it predicted the existence of states with the opposite (and, hence, negative) energy. Since energies of particles can be arbitrarily large if their momentum is sufficiently large, this meant that there could be states of arbitrarily negative energy. Worse, since systems tend to fall into the state with the lowest available energy, electrons would continuously drop to ever lower energy states, emitting photons along the way. To resolve this problem, Dirac made a seemingly outlandish proposal. He suggested that in nature all of the infinite number of negative energy states were filled, in accord with the Pauli exclusion principle, which states that no two electrons can occupy the same quantum state. This banished the problem of stability. Moreover, Dirac reasoned that if a photon struck one of these negative energy electrons, it could excite it to a positive energy state. The remaining empty state, or "hole," would appear like a particle with electric charge opposite to that of the electron but with exactly the same mass. The minimal photon energy required to accomplish this would be twice the rest energy $(2m_ec^2)$ of the electron, where m_e is the electron mass and c is the speed of light. This object was the positron. It was discovered not long afterward by Carl Anderson in cosmic rays.

Dirac's theory also predicted that particles and antiparticles could annihilate, producing photons or other forms of energy. In Dirac's hole picture, an electron in a positive energy state could emit a pair of photons and fall into a negative energy state. The positive energy state (electron) and negative energy state (positron) would thus disappear, leaving two high-energy gamma rays. In modern quantum field theory, these phenomena can be described without reference to the negative energy states. In quantum electrodynamics, the theory of electrons interacting with photons, a photon passing through the electric field of an atomic nucleus, for example, can produce an electron-positron pair, provided that there is enough energy available. Similarly, electrons and positrons can annihilate, producing pairs of photons.

Antiparticles of virtually all other known particles have since been produced in particle accelerators and studied. The antiparticle of the proton, the antiproton, was produced in the Berkeley Bevatron in 1954. Not only have antiparticles of massive particles, such as the neutron and the mesons (described below) been produced, but also the antiparticles of the neutrinos.

In the framework of modern quantum field theory, this pairing of particles and antiparticles is automatic and holds for both fermions and bosons. The rates for processes involving the production and annihilation of particle-antiparticle pairs can be computed, often with high accuracy. Such processes are now a basic experimental tool. For example, there are several accelerators in Geneva, Palo Alto, Ithaca, and Beijing in which electrons and positrons are made to collide with each other at high energies. In these annihilations, other forms of matter are produced. Several new types of matter have been discovered and/or studied in this way ("charmed particles," and certain "beauty" particles, for example). At other accelerators, such as the SPS at CERN and the Tevatron at Fermilab, protons and antiprotons collide at high energies. In this case some of the quarks, which are the constitutents of the protons, and the antiquarks, which are the constitutents of the antiprotons, can annihilate, producing, again, new forms of matter such as the top quark. The W and Z particles, which are the carriers of the weak force, were discovered in this way.

It is possible for particles to be their own antiparticles if they do not carry electric charge. The most important example of this type is the photon. Another is the π^0 meson. Both of these particles are bosons, particles of integer spin that are subject to Bose statistics. In principle this phenomenon can also occur for fermions (particles of half integer spin like the electron and the neutrino), though no examples of this phenomena are presently known. The mesons offer other interesting examples of the pairing of particles and antiparticles. Like the π^0, the electrically neutral η is its own antiparticle. There are four that are electrically charged: the π^+ and its antiparticle, the π^-; and the K^+ and its antiparticle, the K^-. There are also two neutral mesons that are antiparticles of one another: the K^0 and $\overline{K}^0$. In the theory of strong interactions, the π^+ is made up of a u and a $\overline{d}$ quark bound together (the u and d quarks have electric charge $+\frac{2}{3}e$ and $-\frac{1}{3}e$, respectively, where e is the electron charge); the π^- is a bound state of their antiparticles, the $\overline{u}$ and d. Similar relations exist for the other quarks.

On quite large scales, there is evidence that our universe consists principally of matter, not antimatter. If there were large concentrations of antimatter in our galaxy, one would expect to see a significant amount of antimatter in cosmic rays, but in fact little, if any, antimatter appears in the primary cosmic rays (antimatter is produced when high energy cosmic rays collide with atoms in the atmosphere). Of course, the universe is much larger than our galaxy. In clusters of galaxies, there are large quantities of gas. If much of this gas consisted of antimatter, there would be a large flux of gamma rays from these clusters. The fact that there is no such flux from the Virgo Cluster means that there is not a large amount of antimatter there.

If matter and antimatter are very much alike, the question naturally arises as to why this is so. It could be that the universe was simply created this way. For example, if there were some net number of electrons per (comoving) volume at early times, this number (plus the number of neutrinos) would remain the same at all times. Yet it would be much more satisfying if such a fundamental aspect of nature had an answer within microphysical theory. At first sight, this might seem difficult, since it is not clear, as we have described them, in what way particles and antiparticles differ apart from their charge. Thus, whatever processes produced, say, protons, would be expected to produce equal numbers of antiprotons.

This problem of understanding the matter–antimatter asymmetry is usually formulated in a somewhat different way. Protons and neutrons are said to carry baryon number 1; their antiparticles carry baryon number -1. (In the quark model, quarks carry baryon number $\frac{1}{3}$; antiquarks carry baryon number $-\frac{1}{3}$. Protons and neutrons are each composed of three quarks.) Baryon number is known to be conserved to a very good approximation. If it were not, a proton would quickly decay, for example, to a positron, an electron neutrino, and a π^0 meson. The proton lifetime, however, is known to be larger than about 10^{31} years. In a universe symmetric between baryons and antibaryons (a universe with zero net baryon number) one would expect all of the protons and antiprotons to have annihilated one another in the early stages of the big bang. Similarly, electrons and positrons would also have disappeared. The current universe would consist exclusively of photons.

One might, of course, imagine that the universe was simply created with some net baryon and lepton number. However, there are in fact slight differences in the interactions of certain particles and antiparticles. The K^0 and $\overline{K}^0$ mesons, which are antiparticles of one another, are known to have slightly different properties. This slight asymmetry between particle and antiparticle is known as the phenomenon of *CP* violation, and was first observed in 1964. Shortly after this discovery, Andrei Sakharov pointed out that with this discovery one might hope to understand the presence of matter in the universe, provided that two additional conditions were satisfied: the microphysical laws must violate baryon number, and the universe, early in its history, must not have been in thermal equilibrium. If baryon number were not violated in the fundamental laws, then, whether or not *CP* is violated, if one starts with zero baryon number, one ends with zero baryon number. We know that baryon is conserved to a high degree (this is why protons appear stable). But we also know that it is violated, to a very slight degree, in the "standard model" of elementary particles. It is also violated in "grand unified theories" and string theories. It is possible that these violations are sufficient to produce the observed asymmetry between matter and antimatter.

Sakharov also noted that there is one further condition. According to the big bang theory, at very early times the universe is quite hot. When the universe was very young, about 10^{-5} s or less, matter and antimatter were present in almost equal amounts, comparable to the number of photons. From the success of the big bang theory in accounting for the abundances of the light elements, we know that the excess of particles and antiparticles was about one part in 10^{10} at that time. Sakharov noted that for this to occur, there must have been some departure from thermal equilibrium. Otherwise, general principles of statistical mechanics, coupled with the *CPT* theorem (which guarantees the equality of masses of particles and antiparticles), would have insured equal numbers of baryons and antibaryons. In the framework of modern particle physics, a number of sources for such departures are known. By now, there are several competing proposals for the formation of the asymmetry.

See also: BARYON NUMBER; *CPT* THEOREM; DIRAC, PAUL ADRIEN MAURICE; MAGNETIC MOMENT; PARTICLE PHYSICS; QUANTUM CHROMODYNAMICS; QUANTUM ELECTRODYNAMICS; QUANTUM FIELD THEORY; QUANTUM MECHANICS; SAKHAROV, ANDREI DMITRIEVICH

Bibliography

KANE, G. *The Particle Garden* (Addison-Wesley, Reading, MA, 1995).

PAIS, A. *Inward Bound, Of Matter and Forces in the Physical World* (Oxford, New York, 1986).

MICHAEL DINE

APOGEE AND PERIGEE

Kepler's first law of planetary motion states that the orbits of the planets are ellipses with the sun at one focus as shown in Fig. 1. The separation r of the planet and the sun is related to θ, the angle with respect to a fixed axis, by the equation for the orbit given by

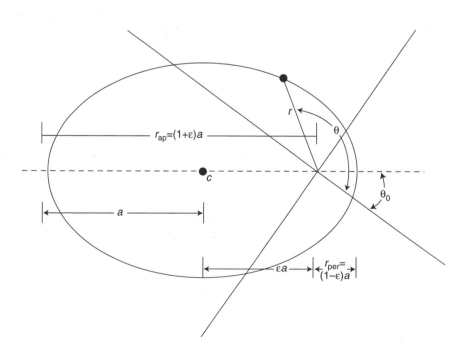

Figure 1 Elliptical planetary orbit.

$$r = \frac{a(1 - \epsilon^2)}{1 + \epsilon \cos (\theta - \theta_0)},$$

where θ_0 is an integration constant resulting from integration of Newton's second law, a is the semimajor axis of the planet's orbit, and ϵ is the eccentricity. The function $r(\theta)$ has two extrema, one maximum and one minimum, which can be found in terms of the astronomical parameters a and ϵ by differentiation. The minimum separation, $r = a(1 - \epsilon)$, occurs when $\theta = \theta_0$, and the maximum separation, $r = a(1 + \epsilon)$, occurs when $\theta = \theta_0 + \pi$. Thus the integration constant corresponds to the angle of minimum separation relative to the fixed axis and may be taken to be zero if the fixed axis is chosen along the direction corresponding to minimum separation. The point in the planet's orbit of minimum separation is called the perihelion of the orbit, and the point of maximum separation is called the aphelion.

Alternatively, the aphelion and perihelion may be described in terms of the total energy E and the angular momentum L, which are constants of the motion for the central force problem. Since $L = mr^2 d\theta/dt$, the angular dependence may be eliminated to provide a one-variable equation for conservation of energy:

$$\frac{1}{2}m\left(\frac{dr}{dt}\right)^2 + \frac{1}{2}m\left(\frac{L}{mr}\right)^2 - \frac{GMm}{r} = E,$$

where the total energy is less than zero for a bound orbit (that is, the kinetic energy is less than the magnitude of the negative gravitational potential energy). The extrema of $r(\theta)$ are, via the chain rule, also the points where the radial velocity vanishes, so the turning points or apsides of the orbit are simply the solutions of the algebraic equation

$$\frac{1}{2}m\left(\frac{L}{mr}\right)^2 - \frac{GMm}{r} = E;$$

since this is a quadratic equation, there are precisely two roots that lead to alternative expressions for the aphelion and perihelion in terms of the constants of the motion.

The terms for the minimum and maximum separation in an elliptical orbit are derived from the Greek prepositions *apo*, meaning away, and *peri*, meaning near. These are coupled with the Greek word for the sun, *helios*, to derive aphelion and perihelion for solar system objects. For a general orbit, the closest approach is sometimes called periapsis and the farthest is called apoapsis. In the case of a terrestrial satellite, apogee and perigee (derived from *gaia*, meaning earth) are the usual terms. For a binary star system, apastron and periastron are used. Other terms have been coined in special cases, for example, apjove and perijove for Jupiter.

See also: CELESTIAL MECHANICS; ENERGY, KINETIC; ENERGY, POTENTIAL; GRAVITATIONAL ASSIST; KEPLER, JOHANNES; KEPLER'S LAWS; PLANETARY SYSTEMS

Bibliography

MARION, J. B., and THORNTON, S. T. *Classical Dynamics of Particles and Systems,* 3rd ed. (Harcourt Brace Jovanovich, San Diego, CA, 1988).

PHILIP B. JAMES

APPLIED PHYSICS

See BASIC, APPLIED, AND INDUSTRIAL PHYSICS

APPROXIMATION AND IDEALIZATION

In seeking a mathematical description of the physical world physicists are faced with a huge variety of effects at many different scales. One of the most successful approaches to trying to understand a problem is to simplify it by dividing it up into several stages. Physicists neglect certain details of the behavior of a system and concentrate on the gross features first, with the idea that once we have understood the overall behavior we can later correct the theory to account for the details. Sometimes we do this without even noticing that we are ignoring an effect, simply because our early measurements are not refined enough to notice the small effects.

Limitations of Approximate Initial Conditions

The level of accuracy to which we can predict a system's future motion depends on the accuracy to which we know its current position and velocity (i.e., the initial conditions of the problem). In the real world situation, we never have exact knowledge of all the initial conditions that might influence the flight of a ball, for example, but we can often get very accurate predictions given good approximate knowledge.

To find equations that describe the behavior of a given system it is often helpful to begin with an ideal or simple system that is like the system under study in many ways, but simple enough to be described by a mathematics that is already well-understood. For example, no real billiard ball is a perfect sphere, but to treat the behavior of billiard balls by first understanding the behavior of colliding spheres is certainly a reasonable approach. Any actual billiard ball is approximately spherical, so the ideal ball is a good approximation of the real one and allows us to understand its behavior. Here the ideal for simple calculation is also ideal in the classical sense of being aesthetically perfect. This is why we use the term "ideal." However, physicists have generalized the word and use the term "idealization" when the simplification is just a practical matter for calculation rather than one with any aesthetic value.

Simplification by Ignoring Effects

As an example of the way approximation can simplify a problem, consider the trajectory of a baseball or tennis ball through the air. We can get a rough estimate of the expected path of the ball if we approximate this situation by the ideal situation of a ball traveling in a vacuum. The problem is simpler because in a vacuum we do not need to know how the wind is blowing or whether the ball is spinning, or anything about the roughness or other features of the surface of the ball. We have reduced this complicated real problem to the simple ideal one of an object moving under the influence of a single constant force, that of gravity. This is a problem that can be solved without calculus. Furthermore, the solution gives a good estimate of where the ball will land.

A more sophisticated treatment will include the friction from the air (air resistance). This causes a velocity-dependent force in a direction opposite to the direction of motion of the ball at any time. If the wind is blowing, that will change the direction of this friction force. In addition, as any tennis or baseball player knows, if the ball has some initial spin, then that alters the direction of the force on the ball due to the air and causes the flight path to curve. Clearly the calculation of the path becomes a little more complicated as each of these effects is included.

Another way a problem may be simplified or idealized is to treat the components of the system as simple objects described by a few parameters, thus ignoring all the details of their substructure at smaller scales. We do this, for example, whenever we describe a solid as a uniform medium with bulk properties such as elasticity or conductivity. We know that the atomic substructure of the material is of tiny, dense nuclei with intervening spaces populated only by electrons. However, as long as we do not measure anything that is sensitive to very short times or very short distances, it is a good approximation to treat the material in terms of its average properties. Separating the behavior of one scale of physics from another, either in distance scales or in time scales, is a powerful approximation technique to simplify problems.

Old Laws are Approximate Versions of New Ones

Two very famous changes in the history of physics occurred when new ideas and experiments revealed that equations originally believed to be exact treatments of problems were in fact approximations of a more precise theory. These were the introduction of special relativity and of quantum mechanics.

In special relativity the momentum **p** for a particle of mass m moving with velocity **v** (speed v) is

$$\mathbf{p} = \frac{m\mathbf{v}}{\sqrt{1 - (v/c)^2}},$$

where c is the speed of light. For v much less than c (as before) the denominator is close to 1 and can be ignored; $p = mv$ is a very good approximation for everyday speeds. However, as v/c approaches 1 the square root in the denominator approaches zero, and the relativistic form must be used. (This is why a massive object cannot be accelerated to the speed of light; the closer v comes to c the larger is the change in momentum that occurs for a small change in velocity. It would require an infinite force, or a finite force acting for an infinite time, to accelerate a massive object up to the speed of light.)

The changes in physics that occur because of quantum mechanics are even more fundamental. Physicists refer to the physics understood before the recognition of quantum mechanics as "classical" physics. It is now recognized that all physical phenomena are governed by the laws of quantum mechanics. However, there are many problems where the quantum effects are such tiny corrections to the classical answer that they can be ignored. Planck's constant is $h = 6.626 \times 10^{-34}$ J·s. Quantum effects are all proportional to this quantity, and, hence, tiny on everyday energy and time scales.

These two great revolutions in physics did not invalidate the earlier understanding. Instead, physicists learned that an answer previously thought to be exact was in fact approximate. In certain regimes the old (approximate) answer could be a very accurate one. Progress in the understanding of basic physical laws seems to follow this pattern. Today, whenever physicists state an exact physical law, we mean that *as far as is currently known* this is the correct law, but we know that future experiments may require the introduction of some corrections to this law that apply under special circumstances.

Initial Conditions

There is another way that approximation enters into the use of equations to predict the future behavior of a physical system. Even when physicists know the equations describing a system exactly, we cannot predict its future unless we know its current or starting state. To find where the ball will land, even in the simplified ideal problem, we must know its velocity at some definite position (e.g., at the point where it was thrown). The level of accuracy with which we can predict its future motion depends on the accuracy with which we know these quantities, called the initial conditions of the problem. We never have exact knowledge of all the factors that may influence the flight of the ball, but still we can often get very accurate predictions given good approximate knowledge.

In classical physics it was thought that whenever the initial conditions were determined accurately enough, then the future motion should be pre-

dictable. In quantum physics a new element enters here, the Heisenberg uncertainty principle, which tells us that it is impossible to determine precisely both the position and the momentum of an object at the same time. The result is that in quantum mechanics, calculations do not completely predict the future of a system but only allow us to calculate the relative probabilities of different possible outcomes for a given initial state.

Limitations of Approximations

Use of approximate knowledge to make approximate predictions does not work in systems that have a behavior that physicists call "chaotic." The behavior of a chaotic system depends very sensitively on the initial conditions. Two such systems that start out looking very similar evolve with time to very different conditions. Typically, it is not possible to obtain sufficiently accurate and detailed knowledge of the current state of such a system to make any long-term predictions about its future. Chaotic systems are often complex nonlinear systems with many components, such as weather systems, but there are also cases where relatively simple-looking systems exhibit chaotic behavior. Recent advances in the understanding of chaotic behavior do not invalidate the classical and quantum understandings of physics; where the systems are not chaotic these older methods still apply. Chaos theory is now beginning to give us some insights into the behavior of chaotic systems, behavior that we could not discuss at all with older methods.

In every physics problem one uses some approximations and idealizations. Even when the system is not in a chaotic regime, it is important to be aware of the approximations made, so that one can recognize the cases where they cease to be valid. One should always estimate the magnitude of corrections due to neglected effects or properties, to ensure that the approximation has not been used beyond the range of its validity.

See also: CHAOS; PARAMETERS; SCALE AND STRUCTURE; SYSTEMS

Bibliography

LORENTZ, E. N. *The Essence of Chaos* (University of Washington Press, Seattle, 1993).

HELEN R. QUINN

ARCHIMEDES' PRINCIPLE

Archimedes was a Sicilian mathematician who was killed during the battle of Syracuse (212 B.C.E.). His book on floating bodies has nine propositions, of which the following four are of particular importance:

2. The surface of any fluid at rest is the surface of a sphere whose center is the same as that of the earth.
3. Of solids, those that, size for size, are of equal weight with a fluid will, if let down into the fluid, be immersed so that they do not project above the surface, but do not sink lower.
5. Any solid lighter than a fluid will, if placed in the fluid, be so far immersed that the weight of the solid will be equal to the weight of the fluid displaced.
7. A solid heavier than a fluid will, if placed in it, descend to the bottom of the fluid, and the solid will, when weighed in the fluid, be lighter than its true weight by the weight of the fluid displaced.

Archimedes proved these propositions by assuming the contrary and showing that the fluid would flow until the conditions specified by his propositions are satisfied. A modern proof is similar. One imagines a parcel of the fluid to be surrounded by a bag of neglible weight. If the bag is not to rise or fall, it must be buoyed up by a force equal to the weight of the fluid within the bag. If the bag were replaced by a solid of the same shape and mass as that of the fluid displaced, it would neither rise nor fall. This condition of neutral buoyancy is expressed by proposition 3.

Alternatively, a solid of the same shape, but of less mass than the displaced fluid, will rise in accord with proposition 5; whereas one with more mass than the displaced fluid will sink. For example, wood, which is less dense than water, will float on the water's surface. Steel, which is more dense than water, sinks. The water does, however, buoy up the steel. The apparent weight of a steel sample in water is reduced from its true weight in agreement with proposition 7.

This reduction in weight was probably used by Archimedes to determine whether a crown given to King Hiero of Sicily was either pure gold or gold alloyed with the lighter element silver. In a modern derivation, T. L. Heath shows that the ratio w_1/w_2 of the weights of gold and silver in the crown is given by

$$\frac{w_1}{w_2} = \frac{F_2 - F}{F - F_1}.$$

Here F is the loss in weight of the crown, of true weight $W = w_1 + w_2$ when placed in water. F_1 and F_2 are the loss in weights of pure gold and pure silver samples, respectively, each of true weight W.

Archimedes' principle is often used today. It provides the buoyancy needed for ships to float and submarines to cruise underwater. The supporting fluid need not be water. Hot air balloons rise because the air they contain is less dense than the cool air they displace. A float buoyed by mercury has been used in a delicate measurement of the constant of Newtonian gravitation, G.

Density of Solids

The density of any substance is defined as the ratio of its mass to its volume, $\rho_s = M/V$. The ratio of the density of a particular material to that of water is called the specific gravity of the material. The specific gravity of any (insoluble) solid can be determined directly from Archimedes' principle. One compares the true weight of the solid with its weight when submerged in a container of water. The submerged weight is reduced from the true weight by an amount $\rho_{water}gV$. The specific gravity of the substance is just the ratio of the true weight to the loss in weight on submersion. The value of local acceleration of gravity g is immaterial; it cancels in this ratio.

To use the specific gravity to determine density, one must know the density of water. This density is not quite a constant, but depends on the temperature and purity of the water and its isotopic content. In the metric system the kilogram was specifically chosen as the unit of mass so that the density of water is approximately 1 kg/liter. At an earlier date, Isaac Newton recognized the density of spring water as 1,000 oz. (avdp.) per cubic foot.

Density of Liquids

The density of a liquid can be determined by both weighing and measuring the volume of a given sample. This determination can be reduced to a single step by using a hydrometer. A hydrometer is any device that determines the density of a liquid by the characteristics of a known solid that floats in the liquid. The most common contemporary example is used to find the density of antifreeze solutions. In this hydrometer, a series of colored balls of different densities are covered with the fluid. The density of the fluid is that of the ball that is just neutrally buoyant.

Another example of a hydrometer is the Archimedes float, an empty but weighted vessel that floats in the unknown liquid. The density of the liquid can be found by noting how much mass must be added to the top of the float to make it float with a predetermined point at the surface of the liquid.

See also: BUOYANT FORCE; DENSITY; SPECIFIC GRAVITY

Bibliography

BLACK, N. H., and LITTLE, E. P. *An Introductory Course in College Physics* (Macmillan, New York, 1956).

HALLIDAY, D.; RESNICK, R.; and WALKER, J. *Fundamentals of Physics,* 4th ed. (Wiley, New York, 1993).

HEATH, T. L. *The Works of Archimedes with the Method of Archimedes* (Cambridge University Press, Cambridge, Eng., 1897).

HODDESON, L. H. "How Did Archimedes Solve King Hiero's Crown Problem: An Unanswered Question." *Phys. Teach.* **8,** 14–19 (1972).

D. F. BARTLETT

ARROW OF TIME

Since Aristotle, time has been represented in mathematical physics by a line. Points on the line represent "nows," and each point divides the line into rays representing the past and the future. Everyday phenomena distinguish sharply between these rays. Friction transforms work into heat. Thermal conduction and convection smooth out temperature differences. Diffusion homogenizes inhomogeneous mixtures. These phenomena are said to be irreversible. They define a preferred direction in time, and so do the laws that govern them; the laws change form when the time variable t is replaced by $-t$.

During the second half of the nineteenth century, physicists discovered that all macroscopic processes are irreversible. They all generate a quantity to which Rudolf Clausius, its inventor, gave the name "entropy." The assertion that all physical pro-

cesses generate entropy is called the second law of thermodynamics, or the law of entropy growth. (The first law of thermodynamics, the law of energy conservation, states that no physical process creates or destroys energy.) Clausius showed that if the law of entropy growth were not valid, we would be able to construct a certain kind of perpetual-motion machine—not a machine that created energy, which would violate the first law, but one that converted heat extracted from a reservoir at a single temperature, such as a lake, entirely into useful work. Such machines would be able to reverse the effects of friction, gathering up heat from the environment and using it to lift weights, charge batteries, and run electric generators.

The law of entropy growth applies to processes, not to systems (unless they are isolated). A system may lose entropy, provided a more than compensating increase in entropy occurs elsewhere. For example, when heat flows from a warmer to a cooler body, the warmer body loses entropy while the cooler body gains entropy. The entropy gain exceeds the entropy loss, so there is a net increase of entropy. Similarly, living systems thrive at the entropic expense of their environment.

The law of entropy growth touches every branch of macroscopic physics and chemistry. It dictates the direction of every chemical reaction, including the reactions that take place in living cells. It explains why stars shine and why iron rusts. No physical law has wider ramifications or has been more rigorously tested.

Like energy and pressure, entropy is both measurable and calculable. One can calculate the entropy of a dilute gas from Clausius's definition and check the calculation against a table of measured entropies. Unlike energy and pressure, however, entropy is not reducible to the physical quantities that characterize individual molecules and their interactions. The energy of an ideal gas, for example, is just the sum of the energies of its molecules. The pressure that an ideal gas exerts on a wall is equal to the rate at which molecular impacts transfer momentum to a section of the wall of unit area. Entropy has no such molecular interpretation. It is not connected with any property of individual molecules, such as energy or momentum. Moreover, at the molecular level of description there is no arrow of time! The laws that govern the structure and interactions of atoms and molecules, as well as their interaction with light, make no distinction between the directions of the past and the future. They are completely time-reversible. Thus neither entropy nor the law of entropy growth figures in a microscopic description of the physical world. Both "emerge" at the macroscopic level of description.

In 1873 Ludwig Boltzmann explained how this happens. Earlier, James Clerk Maxwell had constructed a statistical description of a gas. The central element in Maxwell's theory is the distribution function of molecular positions and velocities. This function specifies the number of molecules whose coordinates and velocity components lie in given ranges. Maxwell proved that if gas in a box is isolated from external disturbances, collisions between its molecules cause the distribution function to approach a particular form. Once this form has been achieved, no further changes in the average properties of the gas take place. The relaxed or equilibrium distribution is spatially uniform; the average number of molecules per unit volume has the same value everywhere in the box. The equilibrium distribution is also isotropic; there are equal numbers of molecules traveling in every direction. Finally, Maxwell showed that the equilibrium distribution of molecular speeds has a universal form and that the average molecular speed is proportional to the square root of the temperature. Boltzmann carried Maxwell's work a step further. From the distribution function he constructed a quantity known as *H*. Boltzmann proved that molecular collisions cause *H* to increase until it reaches its largest possible value. The distribution function then takes the universal form discovered by Maxwell. Thus *H* has the most characteristic property of Clausius's entropy.

In fact, for dilute gases (the systems studied by both Maxwell and Boltzmann) *H* turns out to be a generalization of Clausius's entropy, in the following sense. Under conditions when Clausius's and Boltzmann's definitions both apply, Boltzmann's *H* satisfies the same mathematical relations as Clausius's entropy. But *H* is defined for a much broader class of states than Clausius's entropy. Thus Boltzmann's proof that *H* is a nondecreasing function of the time both justifies and generalizes the law of entropy growth, as applied to a dilute gas.

What does *H* mean? And how can molecular collisions, which do not define a preferred direction in time, cause a statistical property of the gas to exhibit a preferred direction?

The first question becomes easier to answer when *H* is considered more abstractly—not as a quantity constructed from a distribution function of mole-

cular positions and velocities but as a quantity constructed from *any* distribution function that assigns probabilities to a set of possible states. Viewed in this way, *H* is a measure of the randomness of the distribution it refers to; and, for a certain broad and well-defined range of applications, it is the only acceptable measure. If a distribution can be realized in *N* equally likely ways, then *H* is the logarithm of *N*. The more random a distribution, the greater the number of ways in which it can be realized. In this sense a good hand in bridge or poker is less random than a poor hand; it can be realized in fewer ways (and is therefore less common). Entropy, then, is a measure of molecular randomness or disorder, and Boltzmann's *H* theorem says that molecular collisions tend to make the distribution of molecular positions and velocities increasingly random.

This brings us to the second question. How can molecular collisions, which do not discriminate between the directions of the past and the future, cause the distribution of molecular positions and velocities to become increasingly random? To answer this question we first need to understand the role of initial conditions in scientific descriptions. Phenomena are *governed* but not *determined* by laws. Phenomena are determined jointly by laws and initial conditions. For example, the trajectory of a tennis ball is determined jointly by Newton's laws of motion and the ball's initial position, velocity, and spin. Now, statistical theories like Maxwell's and Boltzmann's have three components: laws governing molecules and their interactions, statistics, and initial conditions. The laws governing individual molecules and their interactions do not discriminate between the directions of the past and the future. Neither do statistical operations like classifying, binning, and averaging. The arrow of time must therefore arise from the initial conditions satisfied by systems that exhibit the arrow.

What are these initial conditions? Boltzmann showed that *H* would increase with time if correlations between two or more molecules were permanently absent. This condition means that the probability of finding a given molecule in a given place with a given velocity is independent of the positions and velocities of all the other molecules. Later work has shown that it is sufficient to assume that molecular correlations are absent *initially*.

As an illustration, consider the diffusion of perfume molecules from an opened bottle in a perfectly still room, perfectly isolated from its surroundings. Each evaporated molecule performs a random walk, traveling a short distance in a straight line, colliding with an air molecule, then setting off in a new direction. Eventually, the probability of finding a given molecule in a given region depends only on the volume of the region and not on its position; the perfume is now uniformly distributed. But suppose the velocity of every molecule in the room were suddenly reversed. Then the history of the entire collection of molecules would be reversed; the perfume molecules would zigzag their way back into the bottle. The initial condition that must be satisfied for this to happen (the state of the molecules just after all the velocities have been reversed) is highly correlated. The molecular correlations contain a record of the system's past history, and this record is what guides the perfume molecules back into the bottle. The independent motions of individual perfume molecules suffice to randomize their spatial distribution; an unimaginably intricate team effort is needed to reverse the process. The law of entropy growth rests on the premise that correlated initial states of the kind needed to generate, rather than degrade, order do not occur in nature.

Thus the work of Boltzmann and his successors implies that the law of entropy growth is not really a law at all—at least not in the same sense as the laws of conservation of energy and momentum. Rather, it is a disguised statement about the kinds of initial conditions that are regularly present in macroscopic systems. In effect, the law of entropy growth says that certain kinds of microscopic order, exemplified by molecular correlations in dilute gases, are regularly absent in the initial states of macroscopic systems.

Why are they regularly absent? One popular answer is that molecular correlations in a gas are very easily destroyed by external disturbances. There is no way of insulating a gas from such disturbances. Even if some way could be found to reverse the motions of all the molecules in the perfume experiment, gravitational disturbances (which cannot be screened out) would almost instantly corrupt the memory of the initial configuration stored in molecular correlations. Not all kinds of molecular correlation are equally vulnerable to external disturbances, however. This is shown by the so-called spin-echo experiment, in which an initially random-looking, but actually highly correlated, distribution of nuclear magnetic moments (permanent magnets) evolves into an almost perfectly aligned array.

But there is a more fundamental reason why an appeal to external disturbances cannot *fully* account for the absence of microscopic order in naturally occurring macroscopic systems: For such an explanation to work, the disturbances themselves must have a random component. Of course, the very fact that the law of entropy growth works tells us that external disturbances *do* have a random component. But that is precisely what needs to be explained. The laws governing molecules, atoms, and subatomic particles do not by themselves rule out a universe in which "anti-entropic" behavior like the spontaneous formation of ice cubes in glasses of tepid water is common.

Some people have sought the origin of the randomness needed to explain the law of entropy growth in the phenomenon of chaos. Some classical dynamical systems, including the simplest model of an ideal gas, are so sensitive to small changes in the initial conditions that their long-term behavior, though fully determined by laws and initial conditions, is unpredictable in practice. But while chaos resembles randomness in some ways, in other ways it is highly nonrandom. In any case, it remains unclear how chaos could provide the kind of randomness needed to explain the law of entropy growth.

There is a growing body of opinion that a complete explanation of the law of entropy growth must have a historical character. History links the initial conditions of macroscopic systems to the initial state of the universe, and it is there, perhaps, that the ultimate explanation of time's arrow is to be found.

See also: BOLTZMANN, LUDWIG; BOLTZMANN DISTRIBUTION; CHAOS; CLAUSIUS, RUDOLF JULIUS EMMANUEL; ENTROPY; MAXWELL, JAMES CLERK; MAXWELL–BOLTZMANN STATISTICS; MAXWELL SPEED DISTRIBUTION; THERMODYNAMICS; THERMODYNAMICS, HISTORY OF

Bibliography

LAYZER, D. *Cosmogenesis: The Growth of Order in the Universe* (Oxford University Press, New York, 1990).

DAVID LAYZER

ASTROPHYSICS

Astrophysics is a science dealing with the description and interpretation of astronomical systems according to the principles of physics. It differs from other branches of physics in that its practitioners cannot conduct controlled experiments on the systems they study but must determine the properties of distant astronomical objects through passive, remote observations. Progress in astrophysics depends largely on the comparison of observational data with the predictions of theoretical models. Topics studied by astrophysicists include the formation and evolution of stars; the properties of galaxies, which are large aggregates of stars and gases; the gases and dust that fill the spaces between stars and galaxies; and the origin, structure, and evolution of the universe as a whole.

Astrophysicists do not only study individual examples of interesting phenomena; they also strive to develop an overall picture of the contents of the universe, and an understanding of how the various components relate to each other. To accomplish this, they must take care to design observing programs that will collect statistically representative samples of data, so that inferences they draw about the cosmos will have general validity. Still, the inability of astrophysicists to perform direct experiments forces them to make two basic assumptions about the nature of the universe: First, that the physical laws that apply in earthbound laboratories apply everywhere in the universe; and second, that there is nothing special about the vantage point from which they observe. The second assertion, often called the Copernican principle, has been repeatedly challenged and vindicated, most recently during the 1920s when Edwin Hubble established the existence of galaxies outside the Milky Way.

Modern astrophysics traces its roots to the development of spectroscopy in the late nineteenth and early twentieth centuries. Armed with the newly formulated theoretical methods of quantum mechanics, statistical mechanics, and radiation physics, early twentieth-century astronomers found that they could determine the chemical compositions, temperatures, densities, and other properties of stars and gaseous nebulae by measuring the wavelengths and intensities of spectral lines. Velocities of objects toward or away from the observer could be measured by interpreting small shifts in the observed wavelengths of spectral lines in terms of the Doppler effect. Using this technique, Vesto Slipher found in 1912 that virtually all galaxies appear to be receding from us. About a decade later, Hubble's determination that the recession speed is proportional to the distance of the galaxy provided the crucial clue that the universe is expanding.

The theory of stellar structure and evolution is the prototype of a successful astrophysical theory. By combining concepts from fluid dynamics and nuclear physics with radiation, statistical, and quantum physics, astrophysicists in the 1920s through 1960s explained the observed relationships among the masses, sizes, and power outputs of stars, and showed how stars evolve in time as they use up their nuclear fuel. By following a complex network of nuclear reaction chains, the team of Margaret and Geoffrey Burbidge, William Fowler, and Fred Hoyle showed how the observed abundances of virtually all chemical elements could be explained as the consequence of nucleosynthesis inside stars.

The developments of quantum mechanics and relativity in the first three decades of the twentieth century led to predictions of the existence of exotic "compact objects": white dwarfs, neutron stars, and black holes. White dwarfs are remnants of burnt-out stars with masses comparable to that of the Sun but similar to Earth in size, supported against gravity by a strong repulsion among electrons ("degeneracy pressure") that results from the Pauli exclusion principle—an effect that results entirely from the quantum nature of matter. In 1930 Subrahmanyan Chandrasekhar showed that relativistic effects would cause this repulsion to weaken if the mass of the white dwarf exceeded about 1.4 times the mass of the Sun, resulting in the collapse of the star. While this prediction was not widely accepted at the time, in 1939 J. Robert Oppenheimer and George Volkoff predicted the existence of stars about a billion times denser than white dwarfs, supported by an analogous degeneracy pressure among neutrons. While nearby white dwarfs are readily observed with telescopes in visible light, neutron stars were only discovered in 1968 in the guise of pulsars, which are rapidly spinning neutron stars with large magnetic fields.

Even neutron stars are predicted to have a mass limit, which is about two to three times the mass of the Sun. Any more massive remnant should collapse to form a black hole, a gravitational singularity predicted by Karl Schwarzschild in 1916 using the then-new general theory of relativity. While black holes do not "shine" in the sense of ordinary stars, they can be detected through the effects of their intense gravitational fields on nearby matter and radiation. X-ray observations of close binary star systems have provided convincing evidence for the existence of black hole remnants of ordinary stars in the Milky Way Galaxy. Moreover, most astrophysicists now believe that "supermassive" black holes, with masses as large as that of 1 billion suns, lurk in the centers of many galaxies and are responsible for the tremendous outputs of energy associated with "active galactic nuclei," of which quasars are the most dramatic examples.

The collapse of a dying star or white dwarf to form a neutron star is anything but tranquil. The "bounce" of the star's outer layers as they crash into the core sends a shock wave racing outward, which blows off the outer part of the star in a violent explosion called a supernova. Walter Baade and Fritz Zwicky first suggested the connection between supernovae and the formation of neutron stars in 1934. In 1942 Jan Oort and his colleagues used historical records to connect the supernova of 1054 with the Crab Nebula, which exhibits a shell of gaseous filaments expanding from a common center at speeds of thousands of kilometers per second. The later discovery of a pulsar at the center of the Crab Nebula clinched this interpretation.

Supernovae are relatively rare events, with only one or two occurring per century in a typical galaxy. Most examples are observed in distant galaxies, with only the remnants of ancient explosions visible in our own. Therefore, the occurrence of Supernova 1987A (i.e., the first supernova discovered in 1987) in the nearby Large Magellanic Cloud (at a distance of some 180,000 light-years) was a stroke of luck, allowing astrophysicists to study the aftermath of the explosion in unprecedented detail. In a dramatic verification of the core collapse theory, underground detectors on Earth recorded the passage of the predicted burst of neutrinos (elusive subatomic particles) during the moments following the collapse, several hours before the explosion was visible at optical wavelengths. Astrophysicists have studied the subsequent evolution of the expanding remnant at wavelengths across the electromagnetic spectrum, yielding new insights into the nuclear, chemical, and dynamic processes associated with the explosion.

The difficulties encountered by astrophysicists trying to develop a theory of supernova explosions illustrates the level of complexity inherent in most astrophysical problems. While the basic ideas are relatively straightforward, producing a convincing "model" of a supernova explosion has proven to be very difficult. For example, the initial formation of the shock wave and its expansion through the star's envelope depend sensitively on turbulent mixing and chemical inhomogeneities deep within the star.

Astrophysicists now rely increasingly on supercomputers to simulate these kinds of complex effects. The wealth of new observational data on Supernova 1987A has predictably added to the sophistication demanded of acceptable supernova theories.

While the predictions of neutron stars and black holes are among the most dramatic successes of astrophysical theory, the majority of astronomical discoveries have taken astrophysicists by surprise. Maarten Schmidt's discovery of quasars in 1963 is a case in point. No one foresaw the existence of compact regions, no bigger than the solar system, which could generate energy at a rate exceeding that of an entire galaxy of 1 trillion stars. We now understand a quasar to be the nucleus of a galaxy containing a supermassive black hole that swallows gas from its surroundings. The energy is released as the gas swirls into the intense gravitational field surrounding the hole. To take another example, who would have predicted that the nuclei of galaxies would emit powerful, well-collimated streams of gas traveling at speeds close to the speed of light? Yet "cosmic jets" are not only a common manifestation of "active galactic nuclei," but analogous outflows are also observed to emerge from stars in the process of formation and from at least one binary star system (SS 433). The theoretical explanation for such jets is far from settled, although many researchers believe that electromagnetic forces both propel and focus the flows along the rotation axis of the central star or black hole.

Since the 1960s, technological advances have largely driven progress in astrophysics, especially by opening up new portions of the electromagnetic spectrum to observation. Since gamma-ray, x-ray, ultraviolet, and most infrared radiation cannot penetrate Earth's atmosphere, observations at these wavelengths require instruments operating in space. Observations in visible light, using traditional ground-based telescopes, have been revolutionized by sensitive photon counting devices such as charge-coupled devices (CCDs, also the basis of modern video cameras), and the active correction of telescope optics to compensate for the blurring caused by atmospheric turbulence, may soon be accomplished.

The advent of each technological advance or new observing technique has invariably led to important discoveries. Ground-based radio telescopes have revealed the existence of cosmic jets and pulsars, and found the background radiation left behind by the big bang. Telescopes observing at infrared and submillimeter wavelengths have shown us tremendous bursts of star formation occurring in the centers of distant galaxies, and telltale evidence of the earliest epochs of galaxy formation. CCDs on optical telescopes have detected faint arcs of light that prove to be the images of galaxies distorted by gravitational lensing. X-ray satellites have found compact binary star systems in which a neutron star or black hole is swallowing large quantities of matter shed by its companion. And gamma-ray observations have revealed mysterious bursts of energy that erupt about once a day from seemingly random directions in the sky.

The collection of observational data from astronomical systems is no longer restricted to the detection of electromagnetic radiation. Other signals reaching us from distant quarters include very energetic atomic nuclei and electrons known as cosmic rays, elusive subatomic particles called neutrinos, and gravitational waves predicted by the general theory of relativity. Although the observational techniques of cosmic-ray physics are relatively mature, the origin of cosmic rays remains obscure. Observations of individual cosmic-ray particles with energies of more than 10 J (equivalent to that acquired by a 1-kg mass dropped from a height of 0.5 m) are especially intriguing. Neutrino astrophysics has developed rapidly, starting with the Homestake Mine experiment to detect neutrinos from the Sun in the 1960s. The dearth of solar neutrinos compared to the predictions of nuclear physics, as revealed by the Homestake Mine and other experiments, has triggered an intense theoretical effort to explain the discrepancy. Neutrinos from outside the Milky Way Galaxy were first detected in 1987 from Supernova 1987A in the Large Magellanic Cloud. Although gravitational wave astronomy is still in its infancy, detectors like those now slated for construction are expected to reveal unique signatures produced when a black hole forms by stellar collapse, or when two compact objects merge.

As the range of physics required to understand astronomical systems is so vast, astrophysicists have tended to become more specialized in recent decades. Contemporary subdisciplines include high-energy astrophysics, which deals with objects that emit a large fraction of their energy at x-ray and gamma-ray wavelengths; plasma astrophysics, which deals with the peculiar properties of ionized gases in astronomical systems; and nuclear astrophysics, which explores the nuclear reaction chains that are so important in the interiors of star as well as in the early universe. Astrophysicists have also become specialized somewhat in the use of particular research techniques, most notably in the use of sophisticated computer programs to simulate complex astrophysical processes. Nevertheless, the community of astronomers and astrophysicists is still a relatively small (about 10,000 worldwide) and tightly knit one. Each of the major professional journals covers a broad range of topics, and there continues to be considerable cross-fertilization among the subdisciplines.

See also: ASTROPHYSICS, X-RAY; BLACK HOLE; COSMIC MICROWAVE BACKGROUND RADIATION, DISCOVERY OF; NEUTRINO, HISTORY OF; NEUTRON STAR; PULSAR; QUASAR; SUPERNOVA; UNIVERSE, EXPANSION OF, DISCOVERY OF

Bibliography

CHAISSON, E., and MCMILLAN, S. *Astronomy Today,* 2nd ed. (Prentice Hall, Upper Saddle River, NJ, 1996).

KAUFFMANN, W. J. *Universe,* 4th ed. (W. H. Freeman, New York, 1994).

SHU, F. H. *The Physical Universe: An Introduction to Astronomy* (University Science Books, Mill Valley, CA, 1982).

MITCHELL C. BEGELMAN

ASTROPHYSICS, X-RAY

The discovery of bright x-ray sources in the sky was a remarkable and unanticipated result from early searches for x-ray emission from the Moon. In 1962, Riccardo Giacconi and his colleagues launched a rocket from the test range in White Sands, New Mexico. The rocket, which ushered in the era of cosmic x-ray astrophysics, carried an experiment to search for x-ray emission expected from the bombardment of the Moon by the recently discovered solar wind, an outflow of high energy particles (cosmic rays) from the Sun. The experiment carried three simple x-ray detectors (i.e., Geiger counters that recorded a pulse for each x-ray detected by its ionization in a gas chamber) that together viewed a 55° cone of sky as the rocket spun about its longitudinal axis. The presence of a pronounced peak in the distribution of counts from each of the two working detectors at a spin position significantly offset from the Moon was a complete surprise since the Moon itself was not detected (and was not until viewed close-up with x-ray detectors carried on the orbiting command module of the Apollo 16 mission a decade later). The bright source was identified a few years later with a faint (12th magnitude) blue and variable star in the constellation of Scorpius and called Sco X-1. Some thirty other bright cosmic x-ray sources were discovered with a series of rocket-borne detectors in the next eight years (each detector, as with the first, viewed the sky only for the five minutes it remained above the earth's atmosphere). However, it was not until the launch of the first satellite dedicated to the new x-ray astronomy (Uhuru) that the nature of these bright x-ray sources became clear: x-ray binaries. Other intrinsically luminous cosmic x-ray sources are supernova remnants and extragalactic sources such as galaxy clusters and active galactic nuclei. Detailed studies of these objects have been carried out with the sensitive imaging x-ray telescope of the Einstein observatory, which was launched in 1978. The Einstein observatory discovered x-ray emission from virtually all classes of astronomical objects, from planets (Jupiter) to the most distant quasars. The astrophysics of these objects when studied at x-ray wavelengths is the now active and still expanding field of x-ray astrophysics.

The astrophysics of x-ray sources may be divided into two broad categories: the study of compact objects (including x-ray binaries and active galactic nuclei) and of diffuse sources (including supernova remnants and galaxy clusters). The first involve the physics of accretion as well as thermal and nonthermal processes in compact and usually high-density source regions. Accretion, or the gravitational settling of matter onto a gravitating mass, can be the most efficient source of energy in the universe. Luminosity is produced by the conversion of free fall (kinetic) energy into thermal energy and/or nonthermal particle production at the "bottom" of the potential well formed by the surface of the accreting object or its surrounding accretion disk. In contrast, the diffuse sources are predominantly low-density thermal (or nonthermal) plasmas that have acquired their energy "externally" (e.g., from the kinetic energy and shocks associated with a supernova blast wave or the motion of galaxies in a galaxy cluster).

Compact Objects

Compact objects are collapsed objects produced as the endpoints of stellar evolution—white dwarfs,

neutron stars, and black holes—or the very massive black holes thought to power the quasars and active galactic nuclei. These all are best studied in the x-ray domain since thermal x-rays are the natural radiation emitted by a neutron star or black hole (with mass perhaps 10 $M_\odot$) accreting at a rate within a factor of 10–100 of the Eddington rate. The Eddington accretion rate is that which produces an accretion luminosity ($L_{acc} = GM\dot{m}/R$, where M and R are the mass and radius of the compact object, and $\dot{m}$ is the accretion rate [e.g., in solar masses per year]) that equals the Eddington luminosity, which in turn is the maximum that can be radiated by the compact object in a steady state; at higher luminosities, radiation pressure would halt and expel the incoming accretion flow of gas. The accretion onto neutron stars and black holes in x-ray binaries is studied from the spectrum and temporal variability of the x rays produced near the compact object. The x-ray emission itself arises from the optically thick region near the neutron star surface or, in the case of black hole accretion, the inner accretion disk. A region of higher temperature x-ray emission is produced (preferentially in black hole sources) in a surrounding thin corona. Optical emission (as in the optical counterpart of Sco X-1 and the other bright x-ray binaries) is produced by the reprocessing of x rays in the outer, cooler accretion disk. This reprocessing is dramatically revealed in the optical bursts that accompany x-ray bursts, where the optical emission is delayed and smeared by amounts consistent with re-emission of the x-ray burst into lower temperature optical emission.

Active galactic nuclei, such as in Seyfert galaxies and quasars, are luminous x-ray sources. The x-ray luminosities range from approximately 10^{41} to 10^{46} erg/s (from Seyferts to the brightest quasars) and are again due to accretion onto a compact object, which must be a supermassive black hole. In order that the luminosities are not super-Eddington, and to explain the remarkable short-timescale variability seen in the x-ray emission from the active galactic nuclei, the black hole masses are constrained to be typically in the range from 10^6 to 10^9 $M_\odot$. Recent optical observations of the nuclei of several active galaxies (e.g., M87) with the Hubble Space Telescope have directly measured central masses in this range.

Diffuse Sources

Supernova remnants, such as the Crab nebula or the relatively young Casseipeia remnant Cas A, are bright diffuse sources of x rays. Here, the x-ray emission is due to synchrotron emission from high energy electrons (Crab) or thermal emission from gas shock heated to x-ray temperatures by the outward expansion of the high-velocity ejecta of the supernova (Cas A). The imaging x-ray observations begun with the Einstein observatory provide dramatic views of these endpoints of stellar evolution. Some forty-seven remnants in the Galaxy were studied with the Einstein observatory, and many more are being discovered with the follow-up Röntgen Satellite (ROSAT) x-ray observatory, which has conducted an all-sky survey. The spatial structure of these objects in the x ray versus optical (or radio) reveals the physical extent of the hot versus cold (dense) gas or high-energy electrons. The x-ray spectral lines first detected for supernova remnants with the spectrometers on the Einstein observatory provide important plasma diagnostics for the temperature and density of the gas.

Galaxy clusters, such as the Virgo cluster, and their member galaxies could be imaged in x rays for the first time with the Einstein observatory. Rather than simply crude measures of the total luminosity and integrated spectrum, such as those obtained in the initial Uhuru satellite observations of x-ray emission from galaxy clusters, the Einstein images and spectra showed that galaxy clusters contain considerable sub-structure and reveal much about the dynamical evolution of galaxy clusters. Analysis of the surface brightness profile in x rays, together with assumptions of hydrostatic equilibrium for the hot gas, allow measurements of the total mass of galaxy clusters to be made independent of the optical results that rely on Doppler velocity dispersion measures of the individual galaxies. The x-ray observations show that galaxy clusters contain more mass in their hot, diffuse intergalactic gas, bound to the cluster, than they do in the visible galaxies themselves. Spatially resolved x-ray spectra of galaxy clusters, such as those available with the Advanced Satellite for Cosmology and Astrophysics (ASCA) x-ray satellite, and expected with much higher resolution from the Advanced X-Ray Astrophysics Facility (AXAF) satellite, will allow new constraints on the distribution and mass fraction in dark matter in galaxies in clusters.

See also: ACTIVE GALACTIC NUCLEUS; ADVANCED X-RAY ASTROPHYSICS FACILITY; BLACK HOLE; COSMIC BACKGROUND EXPLORER SATELLITE; COSMIC MICROWAVE BACKGROUND RADIATION; COSMIC MICROWAVE BACK-GROUND RADIATION, DISCOVERY OF; DARK MATTER; GALAXIES and GALACTIC STRUCTURE; INTERSTELLAR and INTERGALACTIC MEDIUM; PULSAR; PULSAR, BINARY; QUASAR; RADIATION, BLACKBODY; SUPERNOVA; UHURU SATELLITE; X-RAY BINARY

Bibliography

FORMAN, W., and JONES, C. "X-Ray Imaging Observations of Clusters of Galaxies." *Annu. Rev. Astron. Astrophys.* **20,** 547–596 (1982).

GIACCONI, R.; GURSKY, H.; PAOLINI, F.; and ROSSI, B. "Evidence for X Rays From Sources Outside the Solar System." *Phys. Rev. Lett.* **9,** 439–443 (1962).

REES, M. "Black Hole Models for Active Galactic Nuclei." *Annu. Rev. Astron. Astrophys.* **22,** 471–506 (1984).

SEWARD, F. "Einstein Observations of Galactic Supernova Remnants." *Astrophys. J. Suppl. Ser.* **73,** 781–819 (1990).

JONATHAN E. GRINDLAY

ATMOSPHERIC PHYSICS

Atmospheric physics is the application of the laws of physics to processes within the atmosphere. Much of the physics used in studying the atmosphere is classical, given the spatial scales of the atmosphere (micrometers to thousands of kilometers). Atmospheric motions obey Newtonian mechanics, and cloud formation processes obey the laws of thermodynamics. Both the interaction of electromagnetic radiation with atmospheric particles and the electrical conductivity of the atmosphere follow Maxwell's equations. The only nonclassical area of atmospheric physics involves the propagation of electromagnetic radiation through the atmosphere. Study of this process depends on a knowledge of how molecules emit and absorb quanta of energy, and thus is based on quantum physics.

Meteorology is mainly concerned with understanding and predicting weather phenomena. The dynamics of atmospheric motions is the traditional purview of meteorology; historically, this arose from the early focus of meteorology on understanding and predicting middle latitude storm systems. These storms are dynamic in origin and are not strongly affected by other physical processes, such as radiative or latent heating. For this reason dynamic meteorology became associated with the motion of the atmosphere, and mainly considers adiabatic processes. All nonadiabatic processes (called diabatic), such as radiative transfer and cloud processes, constitute the field of atmospheric physics. However, since the atmosphere system involves atmospheric motions, it is appropriate to include atmospheric dynamics in the field of atmospheric physics.

Climate research considers the long-term (monthly and longer) averaged behavior of the atmosphere. On these longer time scales many of the physical processes, such as radiation and latent heating, play a major role. Climate research therefore strongly depends on the field of atmospheric physics. As with other branches of science, atmospheric physics now includes many subdisciplines, such as atmospheric radiation, cloud research, aerosol science, and atmospheric electricity.

One major area of atmospheric physics deals with the propagation of electromagnetic radiation through the atmosphere. This propagation is described by the radiative transfer equation, which is a conservation equation for the number of photons in the atmosphere. (The greenhouse effect is related to the subject of radiative transfer.) The Sun emits electromagnetic radiation in a spectral region that peaks in intensity in the visible wavelength region ($0.5~\mu$m). The total flux of radiant (shortwave) energy available to the earth (called the solar constant) is 1,370 W/m². This value is not a true constant, since fluctuations in the solar atmosphere lead to small changes in this solar flux of energy. Only one-quarter of this energy is available to the earth's climate system at any time, since the surface area of the earth is $4\pi r^2$, but only one hemisphere is exposed to this sunlight with a projected area of πr^2, where r is the radius of the earth. The fraction of the earth exposed to sunlight is the ratio of the projected area to total surface area, or $1/4$.

Upon entering the atmosphere shortwave radiation can be absorbed by molecules (e.g., ozone), or it can be scattered by molecules (Rayleigh scattering) or particulates in the atmosphere. These particulates can be either aerosol particles or cloud particles. Aerosol particles are suspended particulates composed of various chemicals (e.g., sulfur, carbon) with sizes in the submicrometer range. Cloud particles are composed of either liquid or solid water and have sizes ranging from 1 μm to hundreds of micrometers. Liquid particles are spherical and exist as either suspended cloud particles or falling rain particles. Solid particles are nonspherical; their shape is determined by a number of

factors, but temperature is perhaps the most important. Suspended ice particles constitute high altitude (altitudes of 5 km or more) cirrus clouds. Falling ice particles transform into snow. Suspended cloud particles (liquid or ice) are most efficient at scattering and absorbing incoming solar radiation. Larger particles scatter more radiation toward the earth's surface.

Satellite measurements indicate that approximately 30 percent of incoming solar radiation is reflected back to space, predominantly by clouds. Around 7 percent of this total reflected radiation is due to molecular (Rayleigh) scattering. It is this scattering, which preferentially scatters shorter wavelength solar radiation, that leads to our blue sky. Another 12 percent of the total scattered radiation comes from solar radiation reflected from the earth's surface. The total available shortwave energy to the climate system is thus $1,370 \cdot (1.0 - 0.3)/4$, or 240 W/m². This absorbed shortwave energy is used to heat the earth's surface and its atmosphere.

The earth emits thermal, or infrared, radiation (wavelengths between 4 μm and a few hundred micrometers) from the surface out to space. Infrared radiation is emitted from the earth's surface according to the Stefan–Boltzmann law. The emitted radiation is absorbed and re-emitted by various molecules in the atmosphere. The most important infrared absorbers are water vapor, carbon dioxide, and ozone; however, many other trace gases in the atmosphere also absorb infrared radiation. The strength and location of their absorption is determined by the spectroscopy of the molecule. These gases do not absorb the same amount of infrared radiation in each spectral interval. Because of the spectral variation in absorption, the altitude at which the gases absorb and emit varies with spectral location. An example of this is the water vapor pure rotation band centered at 25 μm, which absorbs and emits at 5 to 7 km in the atmosphere. One spectral region between 8 to 12 μm is fairly transparent to outgoing infrared radiation and is called the "atmospheric window." If a gas is added to the atmosphere that absorbs infrared radiation in the window region, it has a significant effect on the outgoing infrared radiation—this is why chlorofluorocarbons have a large greenhouse effect.

Clouds are also very efficient at absorbing and re-emitting infrared radiation. The amount of re-emitted radiation is determined by the local temperature of the emitting molecules. Since the temperature of the atmosphere decreases with altitude by roughly

6.5 K/km, the amount of emitted radiation decreases with altitude. High altitude clouds near 10 to 15 km absorb longwave radiation from the lower part of the atmosphere and re-emit this radiation at very low temperatures (230 to 200 K). According to the Stefan–Boltzmann law, the amount of radiation emitted at these temperatures is approximately half that emitted by a surface of 280 to 300 K. Thus, clouds act to trap outgoing longwave radiation. The trapping effect of clouds and gases is called the greenhouse effect. This term originates from the similarity to a greenhouse, which allows solar radiation to enter it, but traps the outgoing infrared radiation. The trapping of infrared radiation results from the infrared-absorbing property of glass. Similarly, the atmosphere allows a substantial amount of shortwave radiation to penetrate through it, but efficiently traps outgoing longwave radiation.

Presently, humans are increasing the amount of carbon dioxide in the atmosphere by 0.5 percent per year. This increase in carbon dioxide causes an increase in the trapping of outgoing longwave radiation. For a stable climate system, the amount of emitted infrared radiation at the top of the atmosphere must equal the amount of shortwave energy available to the climate system; that is, 240 W/m². Thus, the climate system must readjust to maintain the balance with the absorbed shortwave flux of 240 W/m². Indeed, any process causing an imbalance in either the available shortwave energy or outgoing infrared energy will cause a climate change. Examples of these processes are changes in the incoming solar radiation due to volcanic aerosols, increases in gases that efficiently absorb thermal radiation (i.e., greenhouse gases), and increases in particulates in the atmosphere that reflect shortwave radiation back to space.

Another area of research in atmospheric physics is the formation of clouds. Clouds form in the atmosphere through a conversion of water vapor to a condensed phase of water, either liquid or solid. Cloud particles initially form on small particles that are called cloud condensation nuclei. In the absence of cloud condensation nuclei, extremely high relative humidities (over 100%) are needed to form clouds. The ease with which a particle nucleates depends on the nuclei's size and chemical composition. Particles that easily take on water vapor molecules are called hygroscopic. The particles grow to a few micrometers in size through the condensation process. At this size it is much more effi-

cient for particles to continue growing by collision. Collisional growth leads to particles as large as tens of millimeters, where they are sufficiently large to overcome the upward air motions suspending them. At this point the particles fall from the cloud base in the form of rain or snow. Other types of precipitating particles, hail and grauple, can also form through more complex melting and freezing processes. Presently, the growth of liquid cloud particles is well understood compared with the processes governing ice particle growth. The growth of cloud particles is also altered by the small-scale turbulent air motions within and near clouds. Mixing of dry air into the cloud can dilute the number of cloud droplets. The air motions of clouds can transport sensible heat, moisture, and momentum, where vertical transport dominates. Clouds can also transport chemical species in the atmosphere. Many clouds have bases that are near the earth's surface. Any quantity pulled into the cloud base, called entrainment, can be quickly transported to higher altitudes and then detrained to the free atmosphere. This is how many species are quickly carried out of the boundary layer near the earth's surface.

The condensation of vapor to form cloud particles also releases latent heat. Cloud condensational heating can release hundreds of watts per square meter of energy into the atmosphere. In the tropical equatorial regions, latent heating by systems of clouds hundreds of kilometers in extent can drive vigorous upward air motions that extend for many thousands of kilometers. These circulations of air also transport heat and moisture, and form a major part of the general circulation of the atmosphere. The clouds in these systems also absorb shortwave and longwave radiation. This additional energy helps drive the large air motions in the atmosphere.

The motion of the atmosphere is governed by Newton's laws. The equations of motion are defined on a rotating sphere and thus include inertial forces like the centripetal and Coriolis forces. Based on scale analysis of the equations of motion, certain terms in this formalism are neglected. The reduced set of equations are called the primitive equations on a sphere. There are three momentum equations, one for the longitudinal direction of motion, one for the latitudinal direction, and one for the radial (vertical) direction. The solution of these equations yields all possible motions of the atmosphere. The forces acting on these motions are surface friction and mountain torques, due to flow over major mountain systems. Only simplified versions of the primitive equations yield analytic solutions. To solve the full equations on a sphere requires numerical solutions. The other equations that define the motions of the atmosphere are the mass continuity equation, the first law of thermodynamics, and the ideal gas law. These six equations form a closed set to determine the three velocity components, the temperature, pressure, and density of the earth's atmosphere. Heating by radiation and clouds are included in the first law of thermodynamics.

A new branch of atmospheric physics focuses on the dynamic behavior of the complete climate system. Thus, it views the atmosphere as a complex, nonlinear dynamic system. The first study of the chaotic behavior of a dynamic system was based on a simple reduced model of the atmosphere. It was noted that in solving this system of equations, the exact solution of this system was extremely sensitive to the initial conditions used to begin the numerical solution. Depending on the initial condition, the climate of the model atmosphere would be quite different. The concept of "strange attractor" for a dynamic system arose from these early studies in atmospheric physics. Presently, these ideas are being applied to understanding the geologic climate record, where ice ages are one climate state.

See also: CHAOS; NEWTONIAN MECHANICS; NEWTON'S LAWS; RADIATION, THERMAL; RAINBOW; STEFAN–BOLTZMANN LAW

Bibliography

HARTMANN, D. L. *Global Physical Climatology* (Academic Press, San Diego, CA, 1994).
HOUGHTON, J. T. *The Physics of the Atmosphere* (Cambridge University Press, Cambridge, Eng., 1986).
WASHINGTON, W. M., and PARKINSON, C. L. *An Introduction to Three-Dimensional Climate Modeling* (University Science Books, Mill Valley, CA, 1986).

JEFFREY T. KIEHL

ATMOSPHERIC PRESSURE

See PRESSURE, ATMOSPHERIC

ATOM

The concept that matter consists of indivisible basic quantities, the atoms, is generally attributed to the Greek philosopher Democritus. According to his ideas, the atoms are all identical and made of the same, nonspecified, basic substance. By changing the way the atoms are positioned and arranged, different kinds of materials can be formed. The concept of density appears naturally in this picture. If the atoms are more closely packed the substance is heavy (e.g., gold). If there are more empty spaces among the atoms, the substance is light (e.g., ice). The idea that the macroscopic properties are related only with the way the atoms are arranged, however, is too simple to explain the enormous variety of optical, electrical, magnetic, thermal, or mechanical properties of the materials found in nature.

It was only about 2,000 years later, in 1808, that John Dalton formulated two empirical laws setting forth the ideas of Democritus. Dalton made the important discovery that only the atoms of the same element are alike, and they are the same regardless to which other atom they may combine to form a molecule. By that time, the concept of the atom strengthened, and many of its properties and collective behavior were discovered by several investigators (Robert Boyle, Antoine Lavoisier, Daniel Bernoulli, and Dmitri Ivanovich Mendeléev, among others). However, the nature of the atoms remained hidden until the discovery of the electron in 1897 by Joseph John Thomson, establishing their electromagnetic nature.

In 1904 Thomson proposed the first model for the atom (known as the jelly model). In this model, the electrons are disposed in a ring pattern, immersed and at rest in a positive fluid of unknown nature. Because atoms do not radiate spontaneously, and, according to the Maxwell equations, an accelerated electric charge (that is, one which is not in uniform motion or at rest) radiates electromagnetic energy, the electrons were considered at rest to keep the model compatible with the laws of electromagnetism.

The way in which positive charges are distributed in an atom remained unknown until 1911, when the experiments performed by Ernest Rutherford on the scattering of alpha particles on a gold foil showed that the positive charges are concentrated in a very small volume of the atom, the nucleus. Considering that the nucleus is much heavier than the electrons (the lightest nucleus is that of the hydrogen atom, composed of a single proton with a mass 1,836 times the electron mass), Rutherford proposed the planetary model of the atom. This model consists of the nucleus at rest with the electrons moving around it, like planets moving around the Sun. Electrons, however, lose energy when accelerated and would necessarily spiral into the nucleus, leaving unsolved the question of the stability of the atom. Furthermore, the observation of the light emitted by the hydrogen atom shows a series of well-defined wavelengths (discrete lines) following a pattern systematized by Johann J. Balmer in 1884. These spectroscopic observations could not be explained with the simple planetary model of Rutherford.

Following the studies of Rutherford, Niels Bohr proposed in 1913 a radical conceptual modification of the planetary model of the atom, introducing the hypothesis of the quantization of the orbital angular momentum. Assuming that the angular momentum of the electrons are integral multiples (n) of $h/2\pi$, where h is Planck's constant, Bohr showed that the corresponding orbital radius (r_n) and energy (E_n) of the electrons are also quantized. Furthermore, to justify the atomic stability, Bohr postulated that if the electron is in one of these quantized orbits, it does not lose energy through electromagnetic radiation. If, however, the electron jumps from one orbit (a) to another (b), it emits an amount of energy (through electromagnetic radiation) equal to E_b-E_a. The concepts of quantization of angular momentum and that of nonradiative orbits (substituted by atomic states with well-defined energies, in modern quantum mechanics) introduced by Bohr, constitute a strong rupture with classical physics, indicating its inability to give a consistent model for the atom. Although the Bohr model successfully explained the frequencies of the hydrogen spectral lines, it is unable to determine their intensities. The model proves also to be inadequate to describe a two-electron system such as the He atom.

The modern theory of atomic structure begins with the advent of quantum mechanics through the Schrödinger equation established by Erwin Schrödinger in 1926 and the uncertainty principle set by Werner Heisenberg in 1927. The uncertainty principle states that the position and the momentum of the electron cannot be simultaneously measured, discarding the concept of well-defined orbits, as adopted in the Bohr model. As a result of this uncertainty, the electron can be viewed as a cloud whose density of charge distribution, obtained from the Schrödinger equation, gives the probability of finding the electron at some distance from the nucleus. Although the position of the electron is not well determined, its total energy, the angular momentum, as well as one of its components, are well defined. To each one of these three quantities is associated one quantum number (usually denoted by n, ℓ, and m_ℓ, respectively) that characterizes the state of the electron in the hydrogen atom. An electron in a definite state is stable. If the electron makes a transition into a state of lower energy, it emits electromagnetic radiation whose intensity depends on the charge distribution of the initial and final states.

The final ingredient in understanding the structure of atoms is the Pauli exclusion principle established in 1925 by Wolfgang Pauli. Studying the classification of the spectral lines of atoms in strong magnetic fields, Pauli concluded that a fourth quantum number is needed (to account for the spin of the electron) to characterize an atomic state properly. In addition, Pauli found that electrons have a strong antisocial behavior in a many-electron atom, in such a way that no two electrons can have the same set of quantum numbers. If the Pauli principle did not hold, all the ninety-two electrons of the uranium atom, for example, could be placed in the lowest energy state and the size of the uranium atom would be about $1/100$ of its actual size. Because of the Pauli principle, however, electrons must occupy, in a sequential way, the states with higher energies, which also have a larger average radius. The net electric field in the higher energy states is thus strongly reduced by the shielding of the electrons placed in the lower energy states. As a consequence, the net charge viewed by outermost electrons in any atom is close to unity, making the average size of any atom of the same order as hydrogen.

The correct understanding of the nature of atoms provided an enormous increase in our knowledge of astrophysics, atmospheric physics, chemistry, biology, genetics, and new materials. It also served as the basis for many of the technological developments of the twentieth century.

See also: ATOM, RUTHERFORD–BOHR; ATOM, RYDBERG; ATOMIC PHYSICS; ATOMIC THEORY, ORIGINS OF; BOHR, NIELS HENRIK DAVID; ELECTRON; NEUTRON; NUCLEUS; PAULI'S EXCLUSION PRINCIPLE; PROTON; QUANTUM MECHANICS, CREATION OF; RUTHERFORD, ERNEST; THOMSON, JOSEPH JOHN; UNCERTAINTY PRINCIPLE

Bibliography

BOORSE, H. A., and MOTZ, L., eds. *The World of the Atom* (Basic Books, New York, 1966).

HOLTON, G., and BRUSH, G. *Introduction to Concepts and Theories in Physical Science* (Princeton University Press, Princeton, NJ, 1985).

RELLY, J. G., and VANDER PYL, A.W. *Physical Science: An Interrelated Course* (Addison-Wesley, Reading, MA, 1970).

EDUARDO C. MONTENEGRO

ATOM, RUTHERFORD–BOHR

Toward the end of the nineteenth century, the old Democritean view that matter consists of small particles had proved its usefulness in chemistry as well as in physics to such an extent that reasonable doubts could only concern details. Therefore, opponents to atomism, such as Ernst Mach and Wilhelm Ostwald, seem to have been a minority. Especially, the theoretical investigations of Rudolf Clausius and James Clerk Maxwell into the kinetic theory of gases, and Ludwig Boltzmann's researches on the mechanical foundations of the second law of thermodynamics, had made atomism highly probable. Thus Boltzmann, in 1899, could state that the only question left open at the time regarded the way in which atoms act upon each other either by attractive or repulsive forces, or both.

The question of the inner constitution of atoms was first raised by Joseph John Thomson, Cavendish Professor of Experimental Physics in Cambridge, England, in 1897. During the period of classical atomism this question would have been meaningless, since philosophers, beginning with Leukippos (born about 470 B.C.E.), always had presupposed that atoms are solid and indivisible. Thomson, however, discovered that atoms consist of particles of negative charge, which he called electrons. Since atoms are neutral, he held the view that a second kind of matter that is electrically positive must constitute the rest of the atom, contributing nearly its complete mass. Thomson further assumed that the electrons oscillate within the atoms under the influence of positive matter, in this way offering a possibility to explain spectral lines.

In 1903 Philipp Lenard at the University of Kiel in Germany made a further important step. Lenard had realized that atomic cross sections, if calculated from the absorption coefficients of swift electrons in matter, proved to be orders of magnitude (factors of ten times) smaller than the cross sections found in the kinetic theory of gases (i.e., electrons must pass through the atoms almost freely). Hence, he concluded that the atom is nearly empty, and introduced what he called "dynamides," namely, centers of positive forces within the atoms that compensate the repulsive forces of the electrons.

Meanwhile, Ernest Rutherford at McGill University in Montreal had made experiments with a kind of ray originating from radioactive decay, which he had called α rays, and met with great difficulties in making them deflect, even by means of strong electric and magnetic fields. Yet, when he let α rays pass through thin foils of matter, he observed a small broadening of the pencil of the beam. Therefore, in 1906, he became convinced that atoms must be the seat of strong forces, either electric or magnetic. In 1908, after moving to the University of Rochester, Rutherford and Thomas Royds, proved that the α rays consist of doubly ionized helium atoms. In the spring of 1909 Rutherford asked Ernest Marsden and Hans Geiger to examine the question of whether it was possible to reflect α particles from the surface of metals. Geiger and Marsden prepared a strong source of radium, and were able to observe the effect of back-scattering of α particles from thin foils of, for example, gold. In his last public lecture, delivered in 1937, Rutherford had pronounced that the observation of reflected α particles had been the most incredible event in his life. It was as surprising as if someone had fired a 15-in. shell at a piece of tissue paper and the shell had returned to hit the gunner.

These observations were strong evidence of the fact that atoms must exert enormous forces on the α particles. However, the theoretical explanation was in suspense for about two years. Finally, at the end of 1910, Rutherford could demonstrate that the angular distribution of scattered α particles follows from mechanical (and statistical) considerations, under the assumption that the mass and the charge of the atom are concentrated in a point that interacts with the α particle by forces of the Coulomb type. The discovery was published in April 1911. The term "nucleus" did not appear yet and was first introduced by Rutherford in 1912. General interest in the nuclear atom did not arise earlier than the summer of 1913.

Originally, Rutherford had left open the question as to whether the nucleus carries a negative or a positive charge, but further consideration showed him that the nucleus must be positive and surrounded by a number of electrons, compensating the nuclear charge. This was what was called Rutherford's atom model. It was similar to the solar system, except that the forces among the electrons were assumed to be repulsive, whereas in Newton's theory of gravitation the forces are attractive.

Rutherford's experiments not only shed light on the scattering phenomena, but also yielded information about the velocity losses that α particles suffer when passing through matter. In 1912, Charles Galton Darwin had propounded a theory of velocity losses under the assumption that α particles interact with electrons when passing the interior of the atom, and remain uninfluenced outside. Atomic cross sections that he obtained in this way proved to be one order of magnitude (ten times) too high in elements of small nuclear charge and one order of magnitude smaller than kinetic cross sections in heavy elements.

Darwin's partial failure immediately aroused the interest of Niels Bohr who, in the spring of 1912, had joined Rutherford's laboratory at Manchester. Bohr took into account the forces that atomic electrons exert on α particles passing ouside the scattering atoms. Presupposing that the disturbing forces are simple harmonic, he succeeded in making theory agree with observations, if the number of electrons in hydrogen is 1, and 2 in helium, referring to experiments on the dispersion of light.

Probably in pursuit of these investigations, Bohr made two important observations:

1. An atom of the Rutherford type is mechanically unstable in general (the Japanese physicist Hantaro Nagaoka had obtained the same result in 1904).
2. Classical mechanics does not offer any means to determine the radius of the hydrogen or any other atom; for, in the case of the one-electron system, classical mechanics only yields the equation relating centrifugal to Coulomb force

$$ma\omega^2 = \frac{e^2}{a^2} \qquad (1)$$

if the orbit of the electron is assumed to be circular, a being the radius, ω the frequency, m the mass of the electron, and e the charge of nucleus and electron.

Obviously, there are two unknown quantities, a and ω, connected by only one equation, Eq. (1). By contrast, experience seems to show that chemical elements are stable in general and that atomic cross sections and spectral lines have definite values. The central problem of Bohr's theory of the Rutherford atom, therefore, was to solve the two observed dilemmas by appropriate assumptions.

Bohr settled the problem of instability by introduction of the hypothesis that the angular momentum of the electron remains constant, whatever force may disturb the electron.

Bohr attacked the problem of indefiniteness by taking up Max Planck's theory of radiation, which he had been acquainted with at least since 1910, when he was writing his doctoral dissertation on electron theory. At that time already, Bohr was convinced that it was impossible to explain the phenomena of radiation and of diamagnetism and paramagnetism without referring to Planck's theory of quantization of energy. Now, the theory of the Rutherford atom again confronted him with a similar problem that seemed to be solvable by Planck's hypothesis, developed in his theory of radiation, that the energy E of a harmonic oscillator of frequency ν is given by

$$E = h\nu, \qquad (2)$$

where h is Planck's constant.

Bohr therefore introduced an analogous equation, namely

$$E = K\nu, \qquad (3)$$

where E meant the kinetic energy of the electron and K a certain constant (nearly equal to $\frac{1}{2}h$), and he was able to calculate the diameter of the hydrogen atom and its dispersion frequency in satisfying agreement with observations. Moreover, he could explain why molecular hydrogen is more stable than atomic hydrogen, whereas in helium the situation is the reverse. Finally, he endeavored to give a rudimentary theory of the system of chemical elements.

At the same time (1912), John William Nicholson had also taken up Planck's quantum theory and had applied it to spectral lines. The most striking feature of Nicholson's theory was the idea that spectral lines originate from different states of the atoms that can be characterized by whole numbers τ. Thus, Bohr became interested in the laws of spectral lines and learned the Balmer formula of the hydrogen spectrum,

$$\lambda = H\frac{\tau^2}{\tau^2 - 4},\qquad(4)$$

where H is a certain constant and λ the wavelength. In the spring of 1913 he replaced Eq. (3) by the "quantum condition"

$$E = \tau\left(\tfrac{1}{2}\,h\,\nu\right).\qquad(5)$$

From Eqs. (1) and (5) Bohr obtained the following expression for the total energy W of the electron:

$$W = -2\pi^2\,\frac{m\,e^4}{h^2}\,\frac{1}{\tau^2}.\qquad(6)$$

A comparison with the Balmer formula led him to the equation

$$h\nu_{nm} = W_n - W_m,\qquad(7)$$

which later became known as Bohr's frequency condition. Equation (7) is the fundamental law governing spectra, and Bohr gave the following interpretation of the formula: When the electron passes from a state of total energy W_n to another state of total energy W_m, the energy difference $W_n - W_m$ is given off as electromagnetic radiation.

An immediate consequence of Bohr's theory of radiation—and perhaps its most convincing result—was the discovery that Rydberg's constant R, determining the spectra of hydrogen, is related to Planck's constant h by the equation

$$R = 2\pi^2\,\frac{me^4}{h^3 c},\qquad(8)$$

where $R = 4/H$, H being Balmer's constant, and c is the velocity of light. The experimental value of R and the theoretical result agreed within an error of about 5 percent.

Bohr published his theory in 1913 in a three-part paper entitled "On the Constitution of Atoms and Molecules" in the *Philosophical Magazine.* In his first paper of July 1913 he based his considerations on two principles:

1. That the dynamical equilibrium of the systems in the stationary states can be discussed by help of the ordinary mechanics, while the passing of the systems between different stationary states cannot be treated on that basis.
2. That the latter process is followed by the emission of a homogeneous radiation, for which the relation between the frequency and the amount of energy emitted is the one given by Planck's theory.

Another important implication of Bohr's theory of the Rutherford atom was that certain lines with half quantum numbers, hitherto attributed to hydrogen, are due to ionized helium. When the measurements of Evan Jenkin Evans in September 1913 proved this result, Bohr's theory won considerable support.

However, there were many physicists who remained reluctant and hesitated to accept Bohr's theory of the Rutherford atom. The most prominent difficulties were Bohr's assumptions that the electrons do not radiate when moving in the stationary states (although being accelerated in their orbits), and that the radiation frequencies are different from the frequencies of revolution of the electrons. Both assumptions were in striking conflict with classical electromagnetic theory.

Another difficulty was that Bohr failed to deduce the characteristic features of the next simple element, the two-electron atom helium, from the premises of his theory, let alone of more complex atoms.

Moreover, in his theory of the hydrogen atom he had introduced one single quantum condition and only one quantum number. In 1915 Arnold Sommerfeld published a more general theory employing several quantum numbers. The energy states turned out to be the same as in Bohr's theory, but Sommerfeld made considerable progress in the relativistic theory of the hydrogen atom. Nevertheless, even his theory of phase integrals could not solve the helium problem.

Further development proceeded from the correspondence principle, first introduced by Bohr in 1913, who had realized that, in the case of large quantum numbers, the radiation frequency, as calculated from Bohr's frequency condition [Eq. (7)], and the frequency of revolution of the electron, co-incide. The correspondence principle finally led to Werner Heisenberg's matrix mechanics in 1925.

Erwin Schrödinger, by contrast, never accepted Bohr's quantum jumps and took a different approach to the theory of atomic constitution, starting from Louis de Broglie's dualistic views concerning the properties of light and matter, and combining corpuscular and wave aspects in his theory of matter. In 1926 he conceived partial differential equations that allowed the interpretation of Bohr's discrete quantum states as mathematical eigenvalues. Schrödinger believed he had replaced classical mechanics by what was called wave mechanics and that he therefore, to some extent, had reached the aim that Bohr had in mind when, in 1913, he applied Planck's quantum theory to the theory of the Rutherford atom; for Bohr, from the beginning of his career, was convinced that classical mechanics as well as classical electrodynamics do not suffice to account for atomic phenomena.

See also: ATOM; ATOM, RYDBERG; BOHR, NIELS HENRIK DAVID; CORRESPONDENCE PRINCIPLE; HEISENBERG, WERNER KARL; PLANCK, MAX KARL ERNST LUDWIG; RUTHERFORD, ERNEST; SCHRÖDINGER, ERWIN; THOMSON, JOSEPH JOHN

Bibliography

BOHR, N. *Collected Works,* Vol. 2: *Work on Atomic Physics (1912–1917)* (North-Holland, Amsterdam, The Netherlands, 1981).

RUTHERFORD, E. *The Collected Papers of Lord Rutherford of Nelson,* 3 vols. (Allen and Unwin, London, Eng., 1962–1965).

ULRICH HOYER

ATOM, RYDBERG

An atom in a highly excited state of large principal quantum number n is known as the Rydberg atom. After studying the long wavelength line-spectra of various elements, Johannes Robert Rydberg discovered that the spectra could be classified into different series. The observed lines in each series were expressed as

$$\nu_{n'n} = R_\infty \left(\frac{1}{n'^2} - \frac{1}{n^2} \right),$$

$$(n' = 1,2,\dots, n = n' + 1, n' + 2,\dots),$$

where $\nu_{n'n}$ is the wave number (inverse of the wavelength), R_∞ is a universal constant known as the Rydberg constant.

The discovery of the Rydberg series played a pivotal role leading to the establishment of the quantized model of the atom, such as the semiclassical Bohr model of the H atom. The Bohr model predicted discrete energy levels proportional to $-n^{-2}$. The Rydberg series were explained as transitions between the states n' and n.

The properties of Rydberg atoms are made more explicit with the advent of quantum mechanics. Several properties for the H atom are listed in Table 1, where ℓ refers to the angular momentum quantum number and a_0 is the Bohr radius.

Rydberg atoms show extraordinary properties. They are weakly bound and fragile. They can be very large on the atomic scale for even moderately high n. It is therefore not surprising that Rydberg states of nonhydrogenic atoms (such as Na) are similar to those of H atoms with one exeption: Electrons in Rydberg states of low angular momentum, $\ell \leq 2$, can penetrate into the noncoulombic core of nonhydrogenic atoms. This causes a slight modification in the binding energy as $-(n - \delta_\ell)^{-2}$, where δ_ℓ is known as the quantum defect. It depends strongly on ℓ and is usually negligible for $\ell > 2$.

Rydberg atoms may be produced by collisional or optical excitation and by electron capture. In collisional excitation, an atom is excited by collisions with electrons, ions, or atoms. Optical excitation occurs through the absorption of a photon. In electron capture, a positively charged ion picks up the electron in the collision. Only the optical excitation is state-selective as determined by the photon energy. The other two methods result in a distribution of Rydberg states and are not state selective. It is also possible to produce Rydberg atoms by optical excitation in the presence of electric and/or magnetic fields to take advantages of Stark and Zeeman effects. Rydberg atoms can be detected by fluorescence emission, collisional ionization, or field ionization.

Besides spectral analysis, Rydberg atoms also have applications in astrophysics (such as radio emission lines) as well as in a hot plasma, where Rydberg states are formed. Due to the weak binding of the

Table 1 Properties of Rydberg States of the H Atom

Quantity	Expression	Value at $n, \ell = 30, 29$
Binding energy (eV)	$-13.6n^{-2}$	-0.0151
Average potential energy (eV)	$-27.2n^{-2}$	-0.0302
Average kinetic energy (eV)	$13.6n^{-2}$	0.0151
Average orbital radius (a_0)	$\frac{1}{2}[3n^2 - \ell(\ell + 1)]$	915

Rydberg atom, laboratory fields of strength comparable to or greater than the atomic field may be easily achieved, which allows the study of nonlinear effects in the strong perturbation regime. An interesting property of the Rydberg atom is its semiclassical character owing to the correspondence principle as $n \rightarrow \infty$. This facilitates fundamental studies such as quantum behavior of classically chaotic systems, as well as semiclassical aspects of double Rydberg atoms where two electrons are in high Rydberg states.

See also: ATOM; BOHR'S ATOMIC THEORY; ELECTRON; EXCITED STATE; IONIZATION; QUANTUM NUMBER; RYDBERG CONSTANT; SPECTRAL SERIES; STARK EFFECT; ZEEMAN EFFECT

Bibliography

BETHE, H. A., and SALPETER, E. A. *Quantum Mechanics of One and Two Electron Atoms* (Academic Press, New York, 1957).

GALLAGHER, T. F. *Rydberg Atoms* (Cambridge University Press, Cambridge, Eng., 1994).

RYDBERG, J. R. "On the Structure of the Line-Spectra of the Chemical Elements." *Phil. Mag. and J. Sci. 5th Ser.,* **29,** 331 (1890).

STEBBINGS, R. F., and DUNNING, F. B., eds. *Rydberg States of Atoms and Molecules* (Cambridge University Press, Cambridge, Eng., 1983).

JIANYI WANG

ATOMIC BOMB

See FISSION BOMB

ATOMIC CLOCK

Atomic clocks use natural resonances within atoms (or molecules) to keep time. These "quantum mechanical oscillators" are vastly less sensitive to environmental effects, such as temperature, pressure, humidity, vibration, acceleration, and so on, than are macroscopic oscillators such as pendulums and quartz crystals. The result is that atomic clocks are the most stable and accurate clocks known.

History

Clocks are instruments to measure the passage of time. In the period from the earliest human awareness of time until 1955, the measure of time and, subsequently, even its legal definition were based on the rotation of the earth. Clocks were first developed to help interpolate this motion and to maintain a record during intervals of darkness. During the "age of exploration" the problem of navigating during long sea voyages led to a great need for better clocks (Fig. 1). In 1759, after 35 years of work, Richard Harrison developed a "chronometer," which could keep time to ± 0.4 s per day even in the harsh environment of a ship at sea. Subsequent developments to laboratory pendulum devices culminated in 1921 in the Shortt "free" pendulum in which the master pendulum was separated from the escapement mechanism. It was able to keep time to within about one millisecond per day.

In the early 1920s the electronic quartz crystal oscillator was developed. It rapidly improved to the point where it was more stable than the rotating earth. This necessitated a redefinition of the second. In 1956, the definition of the second was changed to 1/31,556,925.9747 of an orbital period of Earth around the Sun. However, advances in atomic time-keeping quickly made that definition obsolete and, in 1967, the SI (*Le Système International d'Unités;* the metric system) second was defined as ". . . the duration of 9,192,631,770 periods of the radiation corresponding to the transition between the two hyperfine levels of the ground state of the cesium-133 atom."

Basic Operation

Atomic clocks and atomic frequency standards (a clock is just a constant frequency generator with a counter added) are based on transitions between two energy levels in an atom or molecule. The frequency ν corresponding to a transition is given by

$$h\nu = |E_1 - E_2|,$$

where E_1 and E_2 are the energies of the two levels and h is Planck's constant. The "hydrogenic" atoms that form the basis of our atomic clocks can be viewed as an electron in orbit about a nucleus. Both particles have spin angular momentum and, hence, a magnetic moment. Just as two magnets attract each other when oriented one way and repel when oriented differently (an energy difference), so too do the magnetic moments of the nucleus and the electron interact. However, according to the rules of quantum mechanics, only two precisely defined orientations are allowed: spin parallel and spin antiparallel. This leads to two energy states whose energy difference is established by the magnetic coupling between the electron and the nucleus. These states are called "hyperfine" states from their first observation as very small splittings in optical spectra. Their energy difference falls in the microwave region of the electromagnetic spectrum.

The energy difference between hyperfine states is much smaller than thermal energy at room temperature. According to the rules of a Boltzmann energy distribution, atoms are nearly equally distributed in the two energy states at room temperature. If we irradiate such an ensemble of atoms with resonant radiation, some atoms in the lower energy state will absorb a photon (the basic quantum of radiant energy) and make the transition to the upper energy state. However, an equal number of atoms in the upper state will be stimulated to emit a photon and make the transition to the lower state. There is no

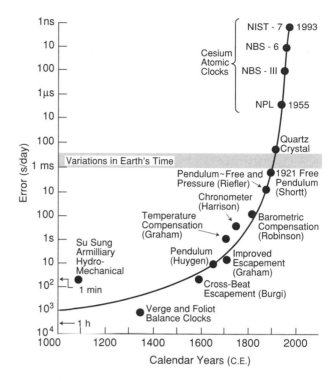

Figure 1   Historical growth in timekeeping accuracy.

net change in either the atomic state populations or the number of photons; we have no way of knowing if the radiation was the right frequency to excite the resonance.

This, then, sets the stage for describing the operation of an atomic clock; three basic operations must happen. First, the initial population distribution must be modified. Second, the atoms must be exposed to resonant radiation to cause the transition to happen. Finally, the degree to which the atoms made the transition must be determined. Any given atom must either make the transition or not; things are quantized. However, the number of atoms that make the transition is a function of how nearly the applied radiation matches the resonant frequency of the transition. Hence the strength of the signal at the detector is a measure of the correctness of the applied frequency. In this way a "control signal" is generated and used to lock the applied frequency to the resonant frequency of the atomic transition.

Three types of atomic clocks are available commercially. We will discuss briefly the operation of each in light of the foregoing model to explain

67

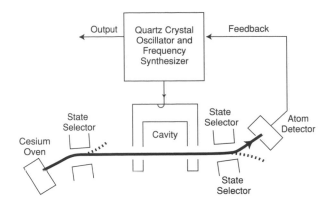

Figure 2 Schematic of a cesium atomic beam clock.

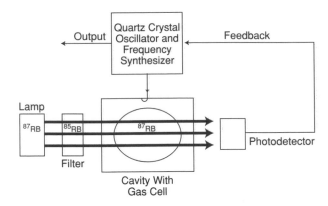

Figure 3 Schematic of a rubidium gas cell clock.

their stability and accuracy. In this field, we call "accuracy" the degree to which a device can realize the frequency of the free and unperturbed atomic transition. "Stability" has two types: Short-term stability is dominated by random noise processes and improves with increased measurement time; long-term stability, including drift, results from changes in the perturbations to the natural atomic resonance frequency and becomes worse at longer times. Long-term stability, by this definition, must be better than accuracy.

Examples

Cesium beam. The cesium clock is the first and still most accurate type of atomic clock (Fig. 2). Cesium is a gold-colored metal that melts at body temperature and has a hyperfine frequency that is set by the international definition of the second to be 9,192,631,770 Hz. The metal is contained in a small oven, where it is heated to about 100°C. While still well below its boiling point, a vapor of the atoms exists in equilibrium with the liquid. A tube leading from the oven allows some of this vapor to escape into a vacuum chamber, forming a beam of atoms traveling with thermal velocity ($\approx$ 200 m/s for cesium). The beam of atoms first passes through a strongly inhomogeneous magnetic field (a Stern–Gerlach magnet) where the trajectories are altered according to the internal magnetic state of the atom. The state-selected atomic beam then passes through a microwave cavity where it is exposed to the microwave radiation to excite the atomic transition. Finally, the atomic beam passes through a second Stern–Gerlach magnet where atoms that made the transition are directed toward a hot wire similar to the filament in a lightbulb.

Cesium atoms that reach the hot filament are ionized, causing an electric current proportional to their number. The microwave frequency is controlled by this signal and is divided down to a lower frequency by electronic circuits to run a clock.

Magnetic and electric fields in cesium clocks are easily controlled, and the atoms do not touch anything during their flight. Perturbations to the "clock" transition are small and well-understood. The result is both high accuracy and long-term stability. Commercial cesium clocks are the size of a personal computer, weigh 10 to 20 kg, consume several tens of watts, and cost the equivalent of a luxury car. The annual production is several hundred units, and the lifetime of the cesium tube is 3 to 10 years. The long-term stability of the best commercial cesium clocks is of the order of 1 part in 10^{14}, or 1 s in 3 million years. Laboratory standards have accuracy of this order.

Rubidium cell. Rubidium is a silvery metal that has two naturally occurring isotopes: ^{85}Rb and ^{87}Rb. The process of state selection and detection in this type of standard uses a technique called "optical pumping" and takes advantage of a natural coincidence in the spectroscopy of the two isotopes. The two hyperfine levels that form the clock transition exist within the lowest energy orbit of the electron, the "ground state." In the optical pumping process, atoms in one of the hyperfine states preferentially absorb light, which causes the outer electron to be excited to a higher orbit. When an electron returns to the ground state nanoseconds later, it may enter either hyperfine state. In this way, after repeated interactions with the light, all the population can be "pumped" from the absorbing hyperfine level to the other. In the rubidium cell–type standard (Fig. 3),

^{87}Rb atoms are held in a glass cell and the resonant, optical-pumping light is generated by a lamp that passes an electric discharge through a vapor of ^{87}Rb atoms. Such a simple set-up, however, would not produce optical pumping because the lamp puts out light resonant with atoms in both hyperfine levels. The magic in the case of rubidium is that an additional cell filled with ^{85}Rb can be inserted between the lamp and the clock cell to filter out the light that is resonant with one hyperfine level. As the clock cell is optically pumped, the number of atoms absorbing light decreases and the amount of light transmitted through the cell increases. However, if the clock transition is excited by resonant microwaves, atoms are returned to the light-absorbing state, and the transmitted light decreases proportionally. This process is called microwave–optical double-resonance and it allows the absorption of microwave photons to be detected by their action on much more easily detected visible photons. As before, the signal from the detector is used to lock the frequency of the microwave generator to that of the atomic resonance.

This type of atomic clock can be quite small; typical units are the size of a tea cup, with advanced prototype units only 25 cm^3. They are relatively inexpensive and are widely used in telecommunications. The annual production is thousands of units that typically have a 5- to 10-year lifetime. Because the atoms are perturbed by both the optical pumping light and collisions with the buffer gas in the clock cell, these clocks are not as accurate or long-term stable as cesium beam devices; typical stability at a year is parts in 10^{11}. Their short-term stability, however, is as good as the best cesium beam devices because the number of atoms in even a small cell is larger than the number of atoms passing down an atomic beam. The short-term stability is on the order of 1 part in 10^{11} at 1 s and improves linearly with the square root of the measurement time.

Hydrogen masers. This type of atomic clock (Fig. 4) differs from those just discussed in that it usually operates as an "active maser." That is, the clock transition is not probed by externally generated radiation but rather by radiation generated by the atoms themselves. It starts much like the cesium beam clock. Molecular hydrogen is contained in a bulb where an electric discharge breaks the molecular bond and creates atomic hydrogen. This is allowed to escape through a tube to make an atomic beam, which is directed through a state-selecting magnet. The magnetic state-selector is designed to focus

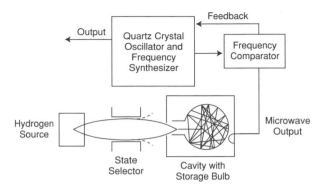

Figure 4   Schematic of a hydrogen maser clock.

atoms in the higher energy state into a storage bulb contained within a microwave cavity. The bulb has a specially coated wall that allows the atoms to bounce around for up to one second without destroying the carefully prepared initial atomic state. In operation, the microwave cavity is filled with photons resonant with the hyperfine transition. A state-selected atom interacting with these photons can be "stimulated" to emit a similar photon and make the transition to the lower hyperfine level. In this way, a kind of chain reaction is established where one photon stimulates the creation of another, and they in turn stimulate more. The rate at which new atoms are supplied to the cavity is adjusted to balance the loss of photons in the cavity walls. A very small amount of the microwave power (picowatts) is coupled out of the cavity and used to steer the frequency of a more powerful oscillator that runs the clock output.

The advantage of this type of clock is its exceptional short-term stability; parts in 10^{15} for hours to days. Hydrogen masers are extensively used in a type of astronomy called "very-long-baseline interferometry," where the exact time is less important than very high stability over hours and days. However, the hyperfine transition frequency at 1.4 GHz is 43 times more sensitive to magnetic fields than is that of cesium. Furthermore, an aspect of the active maser action is that the atomic resonance is much more strongly coupled to the microwave cavity resonance (coupled harmonic oscillators) and is more seriously perturbed by it. We call this effect cavity pulling. These effects make the technology very difficult, requiring state-of-the-art temperature and magnetic field control. As a result, hydrogen masers are the size of a small desk and cost more than a typical house. Only a few are made each year. The cavity pulling and collisions of the atoms with the walls cause frequency biases that degrade the accuracy and lead to long-term frequency drift.

Future Devices

Research into laser-cooled atoms and ions is pointing the way to future atomic clocks in which the perturbations to the atomic transition can be controlled to parts in 10^{18}. It follows that the stability will be correspondingly better than the standards of today.

See also: ATOMIC PHYSICS; BOLTZMANN DISTRIBUTION; ENERGY LEVELS; MASER; OSCILLATION; OSCILLATOR; PENDULUM; PHOTON; QUANTUM MECHANICS; RESONANCE; STERN–GERLACH EXPERIMENT

Bibliography

HELLWIG, H. "Time and Frequency Applications." *IEEE Transactions on Ultrasonics, Ferroelectronics, and Frequency Control* **40,** 538–542 (1993).

HELLWIG, H.; EVENSON, K. M.; and WINELAND, D. J. "Time, Frequency and Physical Measurement." *Phys. Today* **31**(12), 23–30 (1978).

ITANO, W. M., and RAMSEY, N. F. "Accurate Measurement of Time." *Sci. Am.* **269,** 56–65 (1993).

McCOUBREY, A. O. "A Survey of Atomic Frequency Standards." *Proceedings of the IEEE* **54,** 116–135 (1966).

RAMSEY, N. F. "Precise Measurement of Time." *Am. Sci.* **76,** 42–49 (1988).

ROBERT DRULLINGER

ATOMIC MASS UNIT

The unified atomic mass unit (u) is defined as one-twelfth the mass of an atom of the carbon isotope $^{12}_{6}C$, that is, an atom comprising six electrons and protons and six neutrons. Measurements give its value as 1.66054×10^{-27} kg. An Avogadro number (1 mole, about 6.022137×10^{23}) of $^{12}_{6}C$ atoms has a mass of exactly 12 g. In energy units, 1 u = 931.494 MeV/c^2, where $c = 299\ 792\ 458$ m/s is the speed of light.

The International Union of Pure and Applied Chemistry adopted the current carbon standard for the atomic mass unit in 1961. This action resolved a conflict between two slightly different definitions of the atomic mass unit, both based on oxygen. Physicists employed the oxygen isotope $^{16}_{8}O$ as the standard and defined each atom of this isotope to have a mass of 16 u. Chemists, on the other hand, defined the natural isotopic mixture of oxygen to have an average mass of exactly 16 u, which some called 16 avograms. Because about 99.76 percent of oxygen on Earth are atoms of the isotope $^{16}_{8}O$, the chemical and physical scales were for most purposes essentially equal: The average isotopic mixture of oxygen has a mass of 16.00445 u on the $^{16}_{8}O$ physical scale, and the ratio of the atomic masses of the isotopic mixture of oxygen to that of the pure isotope $^{16}_{8}O$ is the Mecke–Childs factor, 1.000275. On the current $^{12}_{6}C$-based scale, the mass of an atom of $^{16}_{8}O$ is 15.99491 u and the average mass of a natural isotopic mixture of O is 15.9994 u.

The history of the atomic mass unit goes back to the beginning of the nineteenth century, when chemists were making tables of definite proportions for the formation of compounds from the elements. The English physician William Prout noted in 1815 that the weights of most atoms could be taken as multiples of the weight of the hydrogen atom. He went so far as to suggest (erroneously) that all matter consisted of hydrogen in varying amounts. A full understanding of the atomic mass unit and its application had to wait almost a century for J. J. Thomson's demonstration by means of ion-beam analysis in 1913 that some elements exist in different isotopes.

Chemists can determine average relative atomic masses by weighing the relative amounts of chemical elements needed to form compounds of known compositions, and where there are ambiguities in chemical compositions, these can often be resolved by comparing volumes of vaporized constituents or by measuring the heat capacities of gases and solids. However, since the early work of Thomson and the subsequent measurements of A. J. Dempster and F. W. Aston with mass spectrometers in 1918 and 1919, scientists have been able to determine absolute atomic masses of individual isotopes from the deflection of beams of charged ions of these isotopes in electromagnetic fields. The deflection depends both on the initial velocity and on the ratio of the charge to the mass. The velocity dependence can be eliminated since it is different for electric and magnetic fields, and the charge is found independently, for example, by balancing the force in an electrostatic field against an oppositely directed gravitational force as in the Millikan oil-drop experiment. The charge neutrality of atoms has established that the charge on a proton is precisely opposite to that on an electron, and the neutron is neutral, as its name suggests. Taken together, the charge divided by the charge-to-mass ratio gives the atomic mass.

The atomic mass unit should not be confused with the unit mass in atomic units. Atomic units (a.u.) constitute a unit system in common use in atomic physics in which the electron mass $m_e = 9.10939 \times 10^{-31}$ kg, is taken as the unit of mass, the Bohr radius $a_0 = 5.2917725 \times 10^{-11}$ m is the unit of length, and $\hbar$ ($= 1.054573 \times 10^{-34}$ J·s, the Planck constant h divided by 2π) is the unit of angular momentum. Other atomic units can be derived in terms of m_e, a_0, $\hbar$, and the dimensionless fine-structure constant $\alpha = e^2 (4\pi\varepsilon_0\hbar c)^{-1} = 1/137.036$, where $e = 1.602177 \times 10^{-19}$ C is the magnitude of the electron charge and $\varepsilon_0 = 10^7 (4\pi c^2)^{-1}$ A^2/N is the permittivity of the vacuum. Thus the unit of energy in a.u. is the hartree: $e^2/(4\pi\varepsilon_0 a_0) = \alpha^2 m_e c^2 = 27.211396$ eV $= 4.359748 \times 10^{-18}$ J, and the unit of time is $\hbar$ divided by the unit of energy, which is equivalent to the unit of distance a_0 divided by the unit of velocity αc: $a_0/(\alpha c) = 2.418884 \times 10^{-17}$ s. In a.u., the atomic mass unit 1 u is about 1822.8885 m_e.

In simple chemical calculations, one often uses atomic mass numbers in place of the more accurate atomic masses (usually called atomic weights) because the mass numbers are all integers, representing the number of nucleons in a given nucleus. However, there are several reasons that the mass numbers only approximate the true atomic masses. First, there is the difference between the masses of the neutron and the proton, as well as the presence of the electron mass. However, at least as important is the fact that the mass of an atom is slightly less than the sum of the masses of its isolated constituents. The difference, as predicted by the theory of relativity, is equal to the total binding energy of the atom divided by c^2. Thus, free protons and neutrons have masses $m_p = 1836.1527 \, m_e$ and $m_n = 1838.6837 \, m_e$, respectively, so that the total mass of the particles in a $^{12}_{6}$C atom is $6 \cdot (1 + 1836.1527 + 1838.6837) \, m_e = 2.009077 \times 10^{-26}$ kg, whereas the mass of the atom is $12/(6.022137 \times 10^{23})$ g $= 1.992648 \times 10^{-26}$ kg. The difference times c^2 is equal to the binding energy per atom: $(2.009077 - 1.992648) \times 10^{-26}$ kg $c^2 = 1.4766 \times 10^{-11}$ J $= 92.16$ MeV.

Practically all of the binding energy is attributable to the nucleons, and in particular, to the confinement of positively charged protons into the tiny nu-

cleus of about 1 fm (10^{-15} m) radius by the strong force; the total binding energy of the electrons by electromagnetic forces in a carbon atom is less than 0.5 keV. Differences among the various isotopes in the binding energy per nucleon make nuclear energy possible and ensure that there is no linear mass scale in which the atomic masses of isotopes are precisely equal to the number of nucleons. The largest relative deviation of the atomic mass from the atomic mass number is nevertheless small: even at its largest, for $^1_1$H, it is less than 0.8 percent. Consequently, errors made by approximating atomic masses by atomic mass numbers are also usually small.

See also: ATOMIC NUMBER; ATOMIC WEIGHT; ENERGY, NUCLEAR; FINE-STRUCTURE CONSTANT; ISOTOPES; NUCLEAR BINDING ENERGY

Bibliography

COHEN, E. R., and TAYLOR, B. N. "The 1986 Adjustment of the Fundamental Physical Constants." *Rev. Mod. Phys.* **59**, 1121 (1987).

JEANS, J. *The Growth of Physical Science* (Macmillan, New York, 1948).

MOELLER, T. *Inorganic Chemistry* (Wiley, New York, 1952).

WILLIAM E. BAYLIS

ATOMIC NUMBER

The atomic number Z of an element specifies the number of protons in the nucleus of each of its atoms. Since each proton carries a positive atomic unit of charge $e = 1.6022 \times 10^{-19}$ C, the total charge on the nucleus is Ze. If the atom is neutral, the positive charge on the nucleus is balanced by Z negatively charged electrons, which determine the chemical properties of the element and its place on the periodic table.

Isotopes

All elements exist in different isotopes, distinguished by their atomic mass numbers A. The difference $A - Z$ gives the number of neutrons in each nucleus of the isotope. The majority of nearly 2,000 known isotopes are radioactive and decay with a definite mean lifetime to isotopes of other elements. There are ninety-one elements that have been found naturally on Earth. These range from hydrogen ($Z = 1$) to uranium ($Z = 92$). In addition, an isotope of plutonium ($Z = 94$) with $A = 244$ was discovered in minute amounts in rocks in California. However, the elements technetium ($Z = 43$) and promethium ($Z = 61$) have not been seen naturally on earth, although they have been created artificially and observed on stars. The naturally occurring elements on Earth include 347 isotopes, of which 67 are radioactive.

The values of Z and A for an isotope are often indicated by lower and upper prefixes to the chemical symbol. For example, the radioactive isotope carbon-14, with $Z = 6$ and $A = 14$, has the symbol $^{14}_{6}C$. An older notation, with the mass number as a superscript after the symbol, as in $_{6}C^{14}$, is also common. Of course, the notation is somewhat redundant, since all isotopes of carbon (C) have the atomic number $Z = 6$.

Transuranium Elements

All isotopes of transuranium elements ($Z > 92$) are radioactive, and as a general rule, those transuranium elements with higher Z tend to have shorter lifetimes. However, the lifetime of an isotope depends on both Z and A, and there are islands of stability, that is, small regions of A and Z values in which isotopes have longer mean lifetimes than in neighboring regions. There has been speculation that a sufficiently strong island of stability may exist around $Z = 114$ and $A = 298$ that, if such an isotope can be produced, it may last for billions of years. There are also unstable regions. For example, there is no known isotope of promethium with a mean lifetime greater than 25.5 years (corresponding to a half-life of 17.7 years).

Transuranium elements can be produced in nuclear reactors by neutron absorption and conversion and in accelerator laboratories by colliding ions into heavy isotopes together with sufficient energy to overcome the coulomb repulsion and fuse their positively charged nuclei. Between 1940 and 1955, Glenn T. Seaborg, Albert Ghiorso, and their colleagues, mainly at the radiation laboratory of the University of California at Berkeley, discovered elements from plutonium ($Z = 94$) to Mendelevium ($Z = 101$) in nuclear reactors, accelerator targets, and in the debris of nuclear explosions. Ghiorso and his group have continued the search to higher Z. Elements of atomic number less than 100 can be produced in macroscopic quantities to allow chemical separation and characterization. In most cases of larger Z, only a few nuclei are produced and then detected in sensitive mass spectrographs. Elements with Z up to 107 have been made and verified by at least one independent group. The Gesellschaft für Schwerionenforschung (Society for Heavy Ion Research) in Darmstadt, Germany, has reported the production of elements up to $Z = 111$.

Historical Origins

John Dalton, the son of a Quaker weaver in Cumberland County in the northwest of England, was a largely self-taught professor of mathematics and physics at New College in Manchester during the late eighteenth and early nineteenth centuries. His studies of atmospheric gases and how they combine laid the foundation for modern atomic theory and for the significance of atomic numbers. The concept of atoms as the smallest, indivisible units of matter goes back at least to the Greek philosophers Democritus and Leucippus in the fifth century B.C.E. However, it was Dalton who recognized by 1804 that each element is composed of identical atoms of a characteristic mass, that atoms of different elements combine in specific ratios to form chemical compounds, and that chemical reactions changed the combinations of atoms while leaving the atoms themselves intact. He published his findings in *A New System of Chemical Philosophy* (1808). Dalton's atomic model was strongly influenced by Isaac Newton's philosophy of atomism. It extended earlier work by such scientists as Robert Boyle, who introduced the concept of an element; Antoine L. Lavoisier, who showed that mass is conserved in chemical reactions; and Joseph L. Proust, who showed that chemical compounds contain elements in fixed ratios by weight. Dalton, by assuming that atoms combined in the simplest small ratios compatible with observations, was able to establish ratios of atoms in many compounds. However, he also ran into inconsistencies because he failed to recognize, even after the suggestion made by Amadeo Avogadro in 1811, that many gaseous elements exist primarily as diatomic molecules.

Dalton is also well known for his law of partial pressures, according to which the total pressure exerted by a gaseous mixture is the sum of the partial pressures of all its components. The law is easily understood in the context of the atomic theory when the molecules, whose collisions are responsible for the pressure, are sufficiently well separated that their mutual interactions can be ignored. Dalton also proposed and used a set of chemical symbols in which each atom was represented by a decorated circle. He continued to use his graphical symbols long after most of the scientific world had adopted the more modern chemical symbols along the lines proposed by J. J. Berzelius.

The law of atomic numbers was discovered by the British physicist Henry Gwyn-Jeffreys Moseley in his experimental studies of about fifty elements and of the x rays emitted when the elements are bombarded with electrons ("cathode rays"). The law states that the frequency of the x rays is proportional to the square root of the atomic number of the element. After its discovery, Moseley used the law to revise the periodic table. His new table accurately placed all elements with $Z \leq 92$.

Modern Importance

The atomic number Z is an important parameter in modern atomic physics and quantum electrodynamics. The properties of atomic states and of the radiative transitions between them are closely related for atoms and ions with the same number of electrons (isoelectronic) but with different atomic number. A series of isoelectronic ions with different values of Z is called an isoelectronic sequence. The gradual change of properties along such a sequence are often expressed as an expansion in inverse powers of Z or in powers and logarithms of $Z\alpha$, where $\alpha = 1/137.036$ in the fine-structure constant. For example, hydrogenic ions form an isoelectronic sequence with a single electron bound to a positive nucleus of charge Ze. The binding energy of the electron in an ion of this sequence is $\frac{1}{2}(Z\alpha)^2 mc^2$, where m is the (reduced) mass of the electron.

See also: ATOMIC MASS UNIT; ATOMIC PHYSICS; ATOMIC THEORY, ORIGINS OF; DECAY, NUCLEAR; ELEMENTS, TRANSURANIUM; NEUTRON; PROTON; RADIOACTIVITY

Bibliography

EMSLEY, J. *The Elements*, 3rd. ed. (Oxford University Press, Oxford, Eng., 1996).

HAMMOND, C. R. "The Elements" in *CRC Handbook of Chemistry and Physics,* 76th ed., edited by D. R. Lide and H. P. R. Frederikse (CRC, Boca Raton, FL, 1995).

NASH, L. K. "The Atomic–Molecular Theory" in *Harvard Case Histories in Experimental Science,* edited by J. B. Conant (Harvard University Press, Cambridge, MA, 1957).

WILLIAM E. BAYLIS

ATOMIC PHYSICS

Atomic physics is the knowledge of the structures of atoms, of their interactions with electromagnetic fields, and with other atoms, molecules, and matter in general. Also it concerns other particles that may be found outside the nucleus, for example, electrons, positrons, photons, muons, and combinations of those particles, such as positronium and other exotic atoms. It helps us to understand the fundamental laws of nature and how it forms the basis for other areas of physics such as condensed matter and other fields of science, engineering, chemistry, biology, and medicine.

Many of the advances of modern technology and of new instruments, for example, lasers, magnetic resonance imaging, fusion plasmas, high precision clock, and even computers, have part of their origin in atomic physics. Even one of the first discoveries of atomic physics, x rays, are still used as one of the key diagnostic tools of modern medicine.

Basic Model of the Atom

Knowledge of the structure of an atom has advanced significantly from the 1913 Bohr model of planetary-like atoms with negatively charged electrons revolving around a positively charged point nucleus. These particles interact through the electromagnetic Coulomb force and their motion is described by the quantum mechanical Schrödinger equation in terms of their kinetic and potential energies. However, the magnetic interactions have far-reaching consequences. In classical physics a current circulating in a loop creates a magnetic field; so too in quantum physics, an electron circulating with a quantized motion around a nucleus generates a magnetic field that is represented by a quantized orbital magnetic moment. The interaction of this magnetic moment with external fields and other magnetic moments, from the electron intrinsic spin and the nucleus, is important in determining atomic interactions and structure.

In the simplest model, the energies in the Schrödinger equation of motion take account of the electrostatic interaction of the electrons with a heavy point nucleus. The solutions to the equation of motion show that the energy levels are quantized and are described by a series of quantum numbers: n, the principle quantum number; l (equal to $n - 1$), the orbital angular momentum and its component m_l along a quantization axis; the electron spin s and m_s, the component of electron spin; and j, the total orbital angular momentum. Every electron moves in an orbit, with a given quantum number determining its energy, shape, size, angular momentum, and spin.

The electromagnetic interaction between the electrons in heavy atoms is more difficult to treat. A central field approximation, in which the innermost electrons move in the average field of all the other electrons, explains simply that electrons with the same n are situated at similar average distances from the nucleus and form an electronic shell. Electrons with the same n and l form a subshell. The Pauli Principle states that the wave functions of the individual electrons must be different from one another. This principle has fundamental consequences for all atomic physics. It places a limit on the number of electrons that can have the same energy and specifies the quantum numbers of an electron. Consequently there are $2(2l + 1)$ electrons in a given subshell of quantum number l where $l = 2$, 6, 10, 14, and so on, with $2n^2$ (i.e., 2, 8, 18, 32, . . .) electrons in a given shell from which the periodic classification of the elements and their general physical and chemical properties can be explained. The valence electrons and excited states have binding energies corresponding to transitions with visible photons of several eV energy. The inner shells have binding energies that increase as the nuclear charge and hence the number of electrons increase and are attracted deeper into the atom; then the transitional energies (keV) correspond to wavelengths in the x-ray region. Note that 1 eV corresponds to light of frequency of 2.42×10^{14} Hz and a wavelength of 1.24×10^4 Å.

Finer details are included by considering the magnetic interaction of the electron intrinsic spin with the orbital magnetic moment and the interaction with other spin magnetic moments. The description can be extended to include the even smaller nuclear effects, that is, the interaction of the orbital and spin magnetic moments of the electrons with the magnetic moments of the nucleus and additionally the motion of the nucleus, its finite size, and the distribution of the nuclear charge when it is not spherically symmetric as well as isotope effects.

Magnetic properties of atoms yield a deeper insight into the shell structure. Using the simplest spectroscope, the observation of photons from salt sprinkled into a gas flame shows strong double (or split) yellow lines. The splitting is caused by the magnetic interaction of the orbital magnetic moment with the intrinsic spin magnetic moment of the electron. It is called fine structure splitting, typically about 10^{-3} eV. Similarly, the nuclear magnetic moment causes a hyperfine structure of about 10^{-6} eV.

As the orbital angular momentum l and spin s are quantized, so too are the orbital and spin magnetic moments. An external magnetic field interacts with the magnetic moment of the atom such that the average moment points either parallel or antiparallel to the field and each direction corresponds to a different energy of the atom. Then external energy can perturb the atom and cause transitions between states, which enables studies of their structure. Also an external electric field displaces the electrons according to their electronic configuration and causes a change in energy levels.

Another very small energy shift (the Lamb shift) of about 10^{-6} eV of the $2S_{1/2}$ from the $2P_{1/2}$ (i.e., same j, total angular momentum) occurs. To explain this shift, it is necessary to assume that electrons are continually emitting and absorbing light quanta. This fundamental concept was a major factor in the development of quantum electrodynamics from which the explanation arose that the quantum mechanical "zero point fluctuations" of the electromagnetic field act statistically on the electrons and so change their potential energy. Many ingenious experiments have been devised to test such ideas and, similarly, theories to model the observations.

An exciting and wonderful consequence of the above structures is that transitions may occur between these energy levels such that the energy difference between those levels is observed as photons. Their wavelengths correspond to x-ray, visible, microwave, and radio-frequency spectra. All of the features of the quantum description can be tested by observations.

Contemporary Interests

The apparent simplicity of atoms invites a detailed study of their characteristics. New challenges are made possible by advancing technology and new instruments as well as by the enormous computing power available for theoretical modeling. Modern studies include, for example, how the properties of space and time can test the elementary interactions and symmetries of nature; how the electronic motion is correlated in both few- and many-electron systems; and how the exchange of energy, angular momentum, and particles between atoms occurs in collisions.

Atomic structure is one of the best examples of the success of quantum mechanics. Different structures can be made by exciting one or more electrons to larger orbits, by adding or removing electrons, or by joining large numbers of atoms into clusters. Now studies are made, for example, with a uranium atom, stripped of all but one of its ninety-two electrons, to explore nuclear and otherwise small effects in a simple hydrogen-like atom. Also the states of free atoms are well described but when the atom moves in external electric and magnetic forces, which are comparable with atomic forces, only approximate theoretical descriptions of electronic motion can be made.

Great interest has been shown in Rydberg atoms, which have one electron very far from the ionic core in a highly excited state with the electron effectively moving in a Coulomb field of unit charge. These states have the near-zero bonding energies, proportional to R/n^2 where R is the Rydberg constant, huge radii (na_0^2) up to 10^{-3} mm and extremely long lifetimes up to 1 s, compared with lower excited state lifetimes of 10^{-8} s. Such atoms have been observed, for example, in interstellar space, by detecting radio frequency photons with energies that correspond to transitions between states with n up to 350. The outer electron is essentially free yet has a momentum distribution characterized by its quantum state so that it offers new opportunities for experiments with near-zero energies.

In contrast to Rydberg atoms, muonic atoms, first prepared in 1952 and containing the 207-times-heavier muon instead of an electron, are extremely small and can even be of nuclear size but with a structure similar to atoms with electrons.

Positron (i.e., antielectron) studies are opening new applications because of their opposite charge to electrons. Exotic atoms, containing positrons or muons or other elementary particles, are important since they allow better tests of certain aspects of theory and their application, for example, in muon catalyzed fusion.

Modern light sources, such as tunable dye lasers and synchrotron radiation, have been used for detailed and precise studies of atomic structure. Their properties of high directionality, brightness, and monochromaticity are indicated by several examples. The smearing of spectral lines arising from random atomic velocities in a gas at room temperature, that is, Doppler broadening, can be avoided by using two counterpropagating lasers beams interacting successively with the same atom so that the atom has effectively zero axial velocity. Individual atoms can be counted by laser-induced fluorescence or ionization. Consequently, complex spectra can be studied by labeling the absorption or emission lines from a chosen energy level.

Many-body interactions are attracting considerable interest because of the prospect of explaining the transitions from single particle interactions to the solid state. Studies of clusters of up to 4,600 argon atoms have observed diffraction oscillations and vibrational excitations of meV energies. A most exciting discovery of a new form of carbon, C_{60} (Buckminsterfullerene), is exceptionally stable and has the structure of a soccer ball. The carbon atoms occupy the vertices of twenty regular hexagons and twelve pentagons that cover the surface of a sphere in such a way that all pentagons do not have a common boundary. The cage diameter of the molecule is 7.1 Å, which is large enough that its hollow center can encapsulate other atoms or molecules useful, for example, for medical diagnostic studies and superconductors.

Techniques, Instruments, and Applications

Many experiments are designed to give physicists information from which they can construct models either to describe a phenomenon or from which theory can be developed. Atomic physics continues to provide many new challenges.

The techniques of optical, radio frequency, microwave, and mass spectroscopies are well developed but new prospects continually emerge. Increasing manipulation of atomic states using lasers, and more precise detection of them using position-sensitive detectors, are making experiments more interesting and challenging. The ability to use external electric or magnetic fields to perturb atomic energy levels, such that the electronic motion moves coherently in phase, has given rise to interfering waves and interference spectroscopy. The ability to produce and detect hydrogen atoms, since they exist normally only bound together as hydrogen molecules, as well as their transitions between the hyperfine states separated in a magnetic field, led to the development of the hydrogen maser and subsequently its use as a frequency and time standard of $1,420,405,751.786 \pm 0.028$ Hz.

The deflection of an uncharged atom, through the interaction of its magnetic moments with an external magnetic field gradient, led to the separation of fine and hyperfine states of atoms and hence study of their properties. Transitions between hyperfine levels in which the orientation of the nuclear spin changes led to the development of nuclear magnetic resonance (NMR) and subsequently imaging (magnetic resonance tomography) of bodies in diagnostic medicine to complement x-ray imaging. When an atom is chemically bonded, an external magnetic field perturbs the electronic shells and causes a characteristic shift of the NMR frequency. Using magnetic field gradients and pulsed signals, the characteristic time decay and NMR images of body parts can be obtained.

Conclusion

The knowledge gained from atomic physics is used in many ways and in many places. For example, it provides the basis for the structures of solid state physics; the formation of molecules and chemical reactions in chemical physics; atomic spectroscopy for astronomy and astrophysics; excitation mechanisms for plasma physics; and complex molecular structures for biophysics. In technology, atomic physics provides knowledge such as radiation effects for medical technology and laser techniques for ionospheric studies. In interdisciplinary studies, its applications are widespread and include space research, meteorology, and new methods of energy production.

See also: ATOM; ATOM, RYDBERG; BOHR'S ATOMIC THEORY; HYPERFINE STRUCTURE; LAMB SHIFT; NUCLEAR MAGNETIC RESONANCE; QUANTUM; QUANTUM NUMBER; SCHRÖDINGER EQUATION

Bibliography

CAGNAC, B., and PEBAY-PEYROULA, J. C. *Modern Atomic Physics: Quantum Theory and Its Applications* (Macmillan, London, 1975).

HAKEN, H., and WOLF, H. C. *Atomic and Quantum Physics*, 3rd ed. (Springer-Verlag, New York, 1994).

JAMES F. WILLIAMS

ATOMIC SPECTROSCOPY

See SPECTROSCOPY, ATOMIC

ATOMIC THEORY, ORIGINS OF

The idea that matter may not be infinitely subdivided, that every homogeneous substance may consist of aggregations of unimaginably tiny, smallest parts ("atoms"), originated with the pre-Socratic Greek philosophers. For more than 2,000 years this belief was more metaphysical than scientific, and many contested it. Atomistic ideas gained popularity in the seventeenth century, especially among physical scientists, but little supporting evidence could be adduced; the supposed atoms were well beyond the reach of the senses, even when aided by such new instruments as the microscope.

Theories about atoms finally entered the world of empirical research through the efforts of chemists. Once Antoine Lavoisier had convinced his eighteenth-century contemporaries to agree on a list of chemically simplest substances called "elements" (oxygen, nitrogen, lead, tin, sulfur, and so on), and had emphasized that the way to follow chemical transformations was to study the weights of the substances that take part in reactions, the conceptual route to a chemical atomic theory was cleared. The creation of the theory, however, was due to the unlikely figure of the English Quaker schoolmaster John Dalton.

In a paper read in 1803, Dalton referred obliquely to an "enquiry into the relative weights of the ultimate particles of bodies" that he had "lately been prosecuting," and then without further explanation presented the first table of provisional relative atomic weights for many of the elements known in his day. Subsequent historical research has revealed how Dalton actually computed these numbers; some examples will illustrate the principle. Lavoisier had determined that water consists of 85 percent oxygen and 15 percent hydrogen by weight. Dalton made the simple and reasonable assumption that every water molecule consists of one atom of oxygen and one atom of hydrogen; it follows from this that oxygen atoms must be 5.66 times as heavy as hydrogen atoms, for $85/15 = 5.66$ (using modern analytical results, the ratio would be more like $89/11 = 8$). Therefore, conventionally setting the weight for a hydrogen atom as 1 exactly (the unit of weight deliberately left unspecified, since there was no way to measure the real sizes or weights of the atoms), the relative atomic weight for oxygen must be 5.66.

Similarly, William Austin had found that ammonia consists of 80 percent nitrogen and 20 percent hydrogen. If every ammonia molecule is "binary" (i.e., if one atom of each element combines to form the ammonia molecule NH), then the atoms of nitrogen must weigh 4 relative to hydrogen = 1. The simplest compound of nitrogen and oxygen, whose formula was also assumed to be binary (NO), must therefore consist of molecules that weight 9.66 on the scale of hydrogen = 1, for that is the sum of 5.66 and 4. This last conclusion could be tested empirically, by examining analytical results for this compound (Dalton's "nitrous gas," today known as nitric oxide), and asking whether those results are compatible with the atomistic prediction. Another sort of test would be to examine a second oxide of nitrogen such as nitrous oxide (assumed by Dalton, as it is today as well, to be N_2O): Is the analysis of this substance consistent with the atomistic prediction that the relative weight of nitrogen to oxygen is $(4 \times 2) = 8$ to 5.66? An affirmative answer increases confidence that the weights are correct. Once relative atomic weights for a few elemental substances are provisionally fixed in this way, these elements can be used in a "bootstrap" fashion to determine the atomic weights for other elements in a similar way, from chemical analysis of the compounds in which they occur (always assuming appropriate molecular formulas for such compounds).

Dalton unveiled the details of his theory in his *New System of Chemical Philosophy* (Vol. 1, 1808; Vol. 2, 1810). In the years since his first sketch of the theory, the laws of stoichiometry had become well-established, increasing Dalton's confidence that the theory was valid. (The laws of definite proportions, equivalent proportions, and multiple proportions represent empirical regularities regarding the weights of elemental substances that combine to form compounds; these stoichiometric laws become self-evident once one accepts the principles of the chemical atomic theory, or, as it happened historically, their prior confirmation by experiment justified the subsequent development of chemical atomism.) However, the atomic theory was by no

means yet mature. One issue was the uncertainties of early analyses; by 1810 Dalton had made considerable adjustments in all of his numbers (e.g., increasing the estimated weight of his oxygen atoms to 7 and of nitrogen to 5), and this was only the beginning of a process that has continued to the present.

A much more fundamental problem was raised by the necessity always to assume a definite molecular formula before one could calculate any proposed atomic weight. Again, let us look at a case in point. Dalton had assumed, in the absence of any empirical evidence on the question, that water molecules are binary (water = HO). However, in the formation of water, hydrogen combines with oxygen in a two-to-one proportion by volume, and one might reasonably infer that two hydrogen atoms, rather than one, always combine with each oxygen atom (water = H_2O). But if that is the case, then each oxygen atom must weigh twice as much relative to hydrogen as under the first scenario. This was just how the Swedish chemist Jacob Berzelius reasoned, soon after reading Dalton's book. Berzelius was a master theoretician, and his patience, ingenuity, and skill at chemical analysis have perhaps never been equaled. He adjusted Dalton's analytical data for the atomic weight of oxygen, and then doubled the number to accord with his new presumed formula for water, H_2O. The weight he calculated is very close to the currently accepted value for oxygen of about 16.

Berzelius's inference that two volumes of hydrogen gas contain twice as many ultimate particles as one volume of oxygen gas, hence that water is H_2O, follows from a more general assumption that the gaseous volume of every atom or molecule is the same, whatever the substance. This general assumption that equal volumes of gases under identical physical conditions contain equal numbers of ultimate particles (equal volumes, equal numbers, or EVEN) appeared to some scientists to be justified from Joseph L. Gay-Lussac's law of combining volumes of gases (1809), which established that elements and compounds that undergo reactions in the gaseous state always do so in small integral proportions by volume. But there were still problems with this viewpoint. Take again the case of water. Experiment shows that two volumes of hydrogen combine proportionally with one volume of oxygen to form two volumes of gaseous water (water vapor or steam). But if one volume of hydrogen be symbolized by H, one of oxygen by O, and one of water by H_2O, and if EVEN holds, then the product of the reaction should occupy only one proportional volume, not two: $2H + O = H_2O$. A way out of this impasse was offered as early as 1811 by the Italian physicist Amedeo Avogadro. Avogadro adopted the EVEN assumption (now known as Avogadro's hypothesis), and further suggested that the elementary "atoms" of gaseous hydrogen and oxygen may actually be molecules that can split in two when they react to form compounds; in combining, hydrogen and oxygen molecules could thereby form two molecules of water in every simple reaction: $2H_2 + O_2 = 2H_2O$.

Much as this reasoning may immediately appear to provide the basis for modern chemical atomism, Avogadro's contemporaries found it very difficult to accept—and for good reason. Aside from the apparent internal contradiction in positing "compound atoms" of elements, there was no real evidence for either of Avogadro's hypotheses—EVEN and compound "atoms" of elements —other than the apparent regularity revealed in Gay-Lussac's law. The entire scheme appeared to be a castle of cards built upon unsubstantiated hypotheses. Even worse, the suggestion of diatomic elementary gases such as H_2, O_2, N_2, or Cl_2 violated then well-accepted physical and chemical principles. It became ever more firmly accepted that the basis of chemical combination must be electrical in character, specifically, Coulombic attraction. From this it follows that any two identical atoms must bear identical charges, hence must repel each other, hence can never combine to form compound elementary molecules. Also, for complicated physical reasons Dalton deduced that the gaseous volumes of all species of atoms and molecules must be as distinct as their weights.

For these and other reasons, Avogadro's hypotheses were not soon accepted, though some people adopted some parts of the scheme—Berzelius, for example, accepted EVEN for elements but not for compounds. Between 1813 and 1826 Berzelius provided the fullest development of a system of chemical atomism yet seen. In addition to his partial adoption of Avogadro's hypothesis on the basis of Gay-Lussac's law, he also used Eilhard Mitscherlich's studies of crystalline salts, Alexis Petit and Pierre Dulong's law of specific atomic heats, and sensitive application of analogies across the entire spectrum of known chemical reactions, all combined with his own extraordinarily precise analyses, in order to deduce the most accurate relative atomic weights from the most probable molecular formulas for all known elements. Berzelius's set of relative atomic weights was widely adopted, especially in the rising school of German chemistry.

Nonetheless, there was also a current of skepticism regarding atomic theory, for it was apparent to everyone that the system relied on certain assumptions, the empirical warrants of which were distant, unclear, or even nonexistent. Above all, the assignment of atomic weight multiples was, as we have seen, directly dependent on assumed molecular formulas, and there were as yet no unambiguous or direct methods to determine those formulas. Consequently, most scientists could be relatively confident that such things as chemical atoms existed, but disagreements over the particular set of relative atomic weights could not be decided conclusively by experiment. There was also the knotty problem of the precise physical nature of the presumed atom. Was it a dense, solid sphere, an area of immaterial force, a complex of smaller particles, or something entirely different from any of these? Neither chemists nor physicists had any empirical purchase on such questions of atomic detail, and some even questioned the very existence of atoms (though chemists continued to make fruitful use of them in their science).

This unsatisfactory situation began to change soon after the middle of the nineteenth century. For one thing, chemists began to develop direct evidence regarding the formulas of some important compounds (water as H_2O, ammonia as NH_3, sulfuric acid as H_2SO_4, and so on) from astute interpretations of novel reactions designed for this purpose. Once the formulas were thus established from such chemical evidence, the ambiguity of atomic weights could be resolved. At about the same time, innovations pertaining to these problems also began to appear from the work of physicists. In particular, the early development of kinetic gas theory by Rudolf Clausius and James Clerk Maxwell between 1857 and 1860 led to a reassertion of Avogadro's hypothesis (or the Avogadro number) in the most general case, as a direct consequence of kinetic assumptions. In turn, the establishment of Avogadro's hypothesis for compounds as well as elements led almost inexorably to Avogadro's second hypothesis, that elements consist of "compound atoms." Finally, the older Coulombic interpretation of chemical affinity had considerably weakened over the years, and therefore could provide less of a psychological barrier to acceptance of polyatomic elementary molecules. By around 1860 a new set of relative atomic weights, similar to those of Berzelius but with some appropriate adjustments—essentially the system we use today—was generally adopted throughout the world.

The capstone of the chemical atomic theory was provided by the development of the periodic system of the elements. Several early nineteenth-century chemists had noted examples of periodicities in the chemical properties of the elements when they are schematically arrayed according to increasing atomic weight. Simultaneously and independently during the late 1860s, the German physical chemist Lothar Meyer and the Russian chemist Dmitrii Mendeleev developed a satisfactory understanding of chemical and physical periodicity. Arraying the elements in a systematic fashion (today usually left to right in horizontal "periods" of increasing atomic weight) can reveal "groups" of elements (vertical columns) that show family resemblances of physical and chemical properties. Such a "periodic table of the elements" is not only pedagogically useful, but has also spurred further development of the understanding of atomic chemistry and physics well into the twentieth century.

Despite this highly satisfactory development of atomic theory within the science of chemistry, there continued to be a current of skepticism among some late nineteenth-century physicists. These doubts were gradually dispersed by a variety of discoveries toward the end of the nineteenth and the beginning of the twentieth centuries. A number of estimates of the sizes of molecules inferred from experimental data emerged in the late 1860s and early 1870s, all reasonably consistent among themselves. The kinetic theory of gases and of heat became ever more fully developed, and connection was made to the science of thermodynamics with the creation of statistical mechanics. Around the turn of the twentieth century a number of additional developments, including the discoveries of x rays, radioactivity, the electron, light quanta, and a kinetic-molecular interpretation of Brownian motion, gave further support to an atomistic worldview. Finally, the discovery of the rare gases (argon, helium, neon, krypton, and xenon) by William Ramsay and Lord Rayleigh not only provided a new group in the nearly complete periodic chart of the elements, but also was itself motivated on the basis of atomistic reasoning. By about 1910 physicists and chemists had essentially closed the circle of experimental and theoretical reasoning; the atomic theory had entered the very lifeblood of physical science.

See also: ATOM; ATOMIC WEIGHT; AVOGADRO NUMBER; ELEMENTS; MOLECULE

Bibliography

BROCK, W. H., ed. *The Atomic Debates* (Leicester University Press, Leicester, Eng., 1967).

BRUSH, S. G. *The Kind of Motion We Call Heat: A History of the Kinetic Theory of Gases in the Nineteenth Century* (North-Holland, Amsterdam, The Netherlands, 1976).

NYE, M. J. *The Question of the Atom: From the Karlsruhe Congress to the Solvay Conference, 1860–1911* (Tomash, Los Angeles, 1983).

ROCKE, A. J. *Chemical Atomism in the Nineteenth Century: From Dalton to Cannizzaro* (Ohio State University Press, Columbus, 1984).

ALAN J. ROCKE

ATOMIC WEIGHT

In the past, chemists used a chemical atomic weight scale in which naturally occurring oxygen was assigned by definition the atomic weight 16 exactly. The atomic weight of hydrogen was thus defined as

$$\text{atomic weight} \atop \text{of hydrogen} = 16 \times \frac{\text{mass of hydrogen atom}}{\text{mass of oxygen atom}},$$

where the word "atom" refers to the element as it occurs in nature. The atomic weights were determined by chemists through careful weighing operations. For example, the atomic weight of hydrogen may be obtained by determining the amount, in grams, of naturally occurring hydrogen that will combine with 16 g of naturally occurring oxygen to form water with nothing left over. The resulting number divided by two is the atomic weight of hydrogen on the chemical scale.

Many elements have atomic weights that are close to integers, but there are notable exceptions; the atomic weight of magnesium is 24.3, and that of chlorine is 35.5. The mass of an atom is mainly concentrated in the nucleus. Nuclei are built of protons and neutrons, which have approximately the same mass. The number of protons plus the number of neutrons is known as the mass number of the nucleus. This integer is commonly denoted by A. The number of protons is called the atomic number of the nucleus and is denoted by Z. There are many instances of families of nuclei with the same charge but different mass numbers. These different nuclei are referred to as different isotopes of the element. The explanation for the occurrence of markedly nonintegral atomic weights is that many naturally occurring chemical elements are mixtures of two or more isotopes. Furthermore, different isotopes have, for all practical purposes, identical chemical properties; that is, the chemical properties of an atom are determined almost exclusively by the nuclear charge. The chemical atomic weight scale does not agree exactly with the Aston mass-spectroscopic scale, used primarily by physicists, and which is based on the mass of the oxygen isotope ^{16}O.

At an international convention in 1960, scientists agreed upon a new standard for atomic weights based on the mass of the carbon isotope ^{12}C, which is assigned a mass of precisely 12 u (or 12 amu). This convention gives rise to the unified new scale in which the mass unit m_0 is defined in terms of the mass m_C of an atom of the particular carbon isotope ^{12}C by

$$m_0 = \frac{m_C}{12}.$$

On this scale, the atomic weight of naturally occurring oxygen is 15.9994 u. This is calculated from the atomic masses and natural abundance percents of three isotopes of oxygen: ^{16}O, 15.99491 u, 99.759 percent; ^{17}O, 16.99914 u, 0.037 percent; ^{18}O, 17.99916 u, 0.204 percent. Thus the atomic weight of naturally occurring oxygen is (15.99491 $\times$ 0.99759 + 16.99914 $\times$ 0.00037 + 17.99916 $\times$ 0.00204) u = 15.99937 u.

The relative numbers of atoms or molecules of each species taking part in a chemical transformation can be stated by expressing the weighed amount of each participating substance in terms of a standard quantity that contains a fixed number of atoms or molecules. This standard, called a mole, weighs a number of grams equal to the atomic or molecular weight of interest. The number of atoms or molecules in a mole is called Avogadro's number. A convenient macroscopic number of atoms is the number N of atoms of mass m_0, which would have a total mass of 1 g (i.e., the number N is defined by $N = 1/m_0$).

Another method of determining atomic weights is based on Avogadro's law (1811) that equal volumes of ideal gases at the same pressure and temperature contain equal numbers of molecules.

Ratios of molecular weights may be determined from extrapolating to zero density the ratio of the densities of two real gases at a series of decreasing pressures.

Atomic masses determined from chemical data are generally subject to considerable uncertainty because of the fact that ordinary chemical elements are usually mixtures of several isotopes. A mass spectroscope can be used to determine the abundances of different isotopes in a naturally occurring chemical element. The mass spectroscope is an instrument developed initially by J. J. Thompson and F. W. Aston to determine the ratios of charge to mass for ions through deflection experiments in combined electric and magnetic fields. A gas sample to be analyzed is ionized by electron bombardment in an ion source. The ions are accelerated and deflected by a magnet. Different isotopes are deflected by different amounts and, by varying the magnetic field strength, the current passing through a collector slit can be measured for each isotope in turn. The abundance of the isotope is proportional to the current. In this case of a spectrum swept across a slit in front of an electrical detector, the instrument is called a mass spectrometer. If the deflected ions fall on a photographic plate, which after development shows a mass spectrum, the instrument is called a mass spectrograph.

In cases in which a mass spectrometer cannot be applied readily, measurement of the relative intensities of isotopic molecular bands may serve as a convenient method of determining the abundance ratios of isotopes. The first rare isotope discovered by the use of band spectra was ^{18}O in 1929 by W. F. Giauque and H. L. Johnson who explained some anomalies in observations of the solar spectrum by G. H. Dieke and H. D. Babcock in 1927 as being due to the presence of ^{18}O.

Mass spectroscopy has developed in many sophisticated ways in recent years with the coupling of an accelerator to a mass spectrometer to form the tandem accelerator mass spectrometer, the coupling of mass spectrometers in series, or the coupling of a laser to a mass spectrometer. This has extended the range and sensitivity of mass spectrometry as an analytical tool.

Mass spectrometers are widely used in industries such as oil and petroleum where analysis of complex hydrocarbon mixtures are required, or in environmental science and pollution control where a trace analysis of organic compounds is needed. Agricultural and pesticide research, biological and drug design laboratories, and companies involved in new materials and surface science phenomena rely heavily on mass spectrometers as do other areas such as archeology, anthropology, and geology. Radioactive dating can be achieved with relatively small samples of material. For example, a tandem mass spectrometer can measure ^{14}C using only 10 μg to a few milligrams of carbonaceous material, whereas it takes 1 to 10 g of the material to make such a measurement by radioactive decay counting techniques. This is because of the long half-life of 5,730 years for ^{14}C. Isotopic analysis of elements such as lead that may result from radioactive decay of other elements are of particular interest because they make possible the determination of the geologic age of minerals from which the elements are extracted.

See also: ATOMIC MASS UNIT; ATOMIC NUMBER; AVOGADRO NUMBER; ISOTOPES; MASS SPECTROMETER; NUCLEAR STRUCTURE; RADIOACTIVE DATING; SPECTROSCOPY, MASS

Bibliography

DELGASS, W. N., and COOKS, R. G. "Focal Points in Mass Spectrometry." *Science* **235**, 545–552 (1987).

EISBERG, R., and RESNICK, R. *Quantum Physics: Of Atoms, Molecules, Solids, Nuclei, and Particles*, 2nd ed. (Wiley, New York, 1985).

WATSON, J. T. *Introduction to Mass Spectrometry* (Raven Press, New York, 1985).

RONALD J. W. HENRY

ATOM TRAP

An atom trap is a device that localizes an atom at or near a specific point in space. The trapping action is accomplished through an electric and/or magnetic interaction between the atom and an applied field. The trap is designed so that atom-field interaction gives rise to a spatially dependent net average force $F_i(\mathbf{r}) \sim -k_i(r_i)$—a restoring force—or equivalently to a quadratic trapping potential $U = -\frac{1}{2}k_i x_i^2$ ($i = x,y,z$). The atom traps that have been experimentally realized can be classified into two categories: traps for neutral atoms and traps for ions.

Neutral Atom Traps

The neutral atom traps that have been successfully demonstrated are of two distinct types: those using time-dependent electromagnetic fields (fields that are best described as waves or photons) and those using static or quasi-static magnetic fields. The former are radiation traps and the latter are magnetic traps.

Magnetic traps. Magnetic traps were considered as early as 1963, but it was not until 1985 that magnetic trapping was first experimentally demonstrated in the laboratory of William D. Phillips at the National Institute of Standards and Technology. In these experiments, a pair of circular, current-carrying coils were used to create a static, spheroidal-quadrupolar magnetic field. Due to the Zeeman effect, the energy of an atom in an external magnetic field can either increase or decrease, depending on the internal state of the atom and the orientation of the magnetic moment of the atom with respect to the field. In a spatially varying magnetic field there is a gradient in this energy, and hence there is a force on the atom. When the magnetic moment has a component that is aligned along the applied field, the force is toward the direction of increasing field, whereas when the moment is aligned against the field, the force is directed to expel the atom from the field. To form a stable three-dimensional magnetic trap, there must therefore be either a maximum or a minimum in the trapping field. This position will correspond to the minimum of the potential well that defines the trap. Starting from the Maxwell equation $\nabla \cdot \mathbf{B} = 0$, one can show that a static magnetic field cannot have a three-dimensional maximum but that it can exhibit a three-dimensional minimum. As a result, magnetostatic traps can only work for atoms prepared in specific internal states, states for which the magnetic moment is aligned against the local trapping field. These states are referred to collectively as low-field-seeking states. The depth of the trapping well depends both on the strength and profile of the applied magnetic field and on the atomic species being trapped. In Phillips' experiment a 0.025 T field was used to trap atomic sodium (magnetic moment approximately one Bohr magneton) yielding a trap depth of 17 mK, a trap volume of ~20 cm^3 and a trapped atom density of ~10^3 cm^{-3}. More recent magnetostatic traps have captured more than 10^{12} atoms and have achieved densities in excess of 10^{13} cm^{-3}.

One attractive feature of a magnetic trap is that the trapped atoms can be evaporatively cooled. In this technique, the temperature of the trapped atoms is decreased by decreasing the height of the trap potential, thus allowing the most energetic trapped atoms to escape or "evaporate." A variation on this cooling technique was recently used by Carl E. Weiman, Eric A. Cornell, and coworkers to achieve Bose–Einstein condensation in a dynamic version of the magnetic trap.

To trap the high-field-seeking states, time varying magnetic fields can be used. Magnetic traps with time varying fields are referred to as ac or dynamic traps, although they are quasi-static by comparison with radiation traps. In the dynamic trap scheme an unstable trapping field is rotated or modulated so as to generate a stable, time-averaged, three-dimensional trapping well. A simple mechanical model can be used to understand this trap: Consider a mass placed at a saddle point of a two-dimensional, gravitational potential surface. This point represents an unstable equilibrium; any weak perturbation can destabilize the mass. If the potential is rotated about an axis that is perpendicular to the surface and passes through the saddle point, the mass can be dynamically stabilized. It is important to recognize that the time dependence of the field in a dynamic trap is sufficiently slow that the electric field associated with the time variation of the magnetic field is unimportant.

Radiation traps. The second class of neutral atom traps derives the trapping force from the interaction of the atom with an electromagnetic wave. Here, the trapping force is known as the radiation or light pressure force. It is convenient to divide this force into two contributions: the scattering or spontaneous force and the reactive or dipole force. The force derives from the fact that photons carry linear momentum and that that momentum is transferred, through recoil, to the atom on absorption and emission; this is the time-dependent transfer of momentum that defines the radiation pressure force. The spontaneous force is the working force of the magneto-optical trap. The essentials of this trap were suggested independently by David E. Pritchard and by Jean Dalibard and it was first demonstrated in 1986 in a collaboration between the groups of Pritchard and Steven Chu. In this trap, three pairs of mutually-orthogonal, counterpropagating laser beams are coincident on the center zero of a spheroidal magnetic quadrupole field. By comparison with the magnetic trap, the

magnetic field strength in the magneto-optical trap is too weak to provide any significant trapping force. Instead, the field is used to provide a spatially dependent frequency shift of the atomic energy levels with respect to the laser frequency. The static field is arranged so that as an atom moves away from the trap center, a transition between two of its levels is Zeeman shifted closer to resonance with respect to a counterpropagating laser beam, a beam that is necessarily directed back toward the trap center. As a result, the atom preferentially absorbs photons from this beam and experiences a restoring force directing it back toward the trap center. The depth of the trapping well depends on the experimental parameters, but is typically between 0.1 and 1 K. Magneto-optical traps have been demonstrated that hold more than 10^{10} atoms at densities in excess of 10^{13} cm^{-3}. Atomic species that have been successfully trapped include many of the alkalis (Li, Na, Rb, Cs) as well as some transition species such as Ca. Some noble gases prepared in metastable states have also been trapped, such as He, Ne, and Xe. Several efforts also exist to trap radioactive species for experiments in nuclear physics. Other laser beam and magnetic field configurations can be used to create the magneto-optical trap; for example, a tetragonal laser beam geometry is known to work.

Another important optical trap is based on the dipole force. The dipole force is associated with photon absorption and emission processes in which stimulated emission plays an important role. The dipole force can be understood using a Lorentz model of the atom in a description analogous to that used to describe the magnetostatic trap: When the laser beam interacts with the atom, it induces an electric dipole moment in the atom, which in turn interacts with the applied electric field through the Stark effect. The interaction causes the energy of the atom+laser−field system to be either increased or decreased depending on the phase, or instantaneous orientation, of the moment with respect to the driving field. When the field frequency is less than a nearby transition in the atom, the phase of the induced dipole moment follows the driving field, and the net energy of the atom+field system is reduced by the interaction. The atom is described as a strong-field seeker because if the intensity of the field varies in space, then the atom experiences a force directed toward the intensity maximum of the field. Similarly, if the field frequency is larger than the atom transition frequency (above resonance), there is a π phase shift between the induced dipole and the driving field and the force is directed to expel the atom from the field. When the field is on resonance, the time averaged dipole force is zero.

The simplest version of the dipole trap is formed at the focus of a laser beam, tuned below the resonance frequency of the atomic transition. The stable point of the trap well is located at the beam focus. This type of trap was treated theoretically in some detail by Arthur Ashkin and demonstrated experimentally by Chu in 1986. The depth of the confining potential in the focused-beam dipole trap depends on the atomic parameters, the laser frequency, and the laser intensity and, in principle, can be arbitrarily deep. In Chu's experiment, a depth of ~5 mK was achieved and about 500 atoms were trapped in a volume of 10^{-7} cm^{-3}. The holding time of this trap was quite limited because quantum fluctuations associated with the atom-field coupling caused the kinetic energy of the trapped atoms to grow in time until finally the atoms escaped from the dipole potential well. More recent dipole traps have been built that are capable of holding larger numbers of atoms (more than 10,000) for much longer times. The longest holding times have been achieved by using a trap beam that is detuned far from any atomic resonance, thus minimizing the effect of quantum fluctuations. As detuning is increased, however, laser power must be increased if the trap depth is to remain constant. Dipole traps have also been operated in other geometries and in combination with other trapping and cooling fields.

Neutral atom traps have been used extensively in combination with various forms of atom cooling to construct sources of ultracold atoms. The interest of these cooled and trapped atoms is largely centered around two of their properties: (1) they have exceptionally low temperatures and therefore move very slowly (from less than 0.01 m/s to 1 m/s) and (2) their de Broglie or matter waves are characterized by a very long wavelength (0.01 μm to over 1 μm). Neutral atom traps have been used to carry out research on atomic collisions, to construct high-precision clocks and gravitometers as well as to investigate atom-optical devices—devices where the wave-like properties of the atomic center-of-mass motion allow manipulation of the atomic trajectories in direct analogy with the manipulation of photons in traditional optics.

Ion Traps

A trap for ions differs from a neutral atom trap in that trapping forces can be exerted on the atom through the interaction of an applied field with the static charge of the ion (the monopole term of the field interaction). The situation is also different because the individual ions interact strongly through the long-range ion-ion Coulomb potential. Two very well-known ion trap designs are the Paul trap and the Penning trap. Experimental realization of ion traps preceded the neutral atom traps by several decades, having been first demonstrated in 1955. Indeed, experiments with these traps played a central role in the work of Nobel Laureates Wolfgang Paul and Hans Dehmelt.

The Paul trap consists of one hyperbolically shaped ring electrode and two hyperbolically shaped cap electrodes. The three components are arranged coaxially about their common axis of rotational symmetry. A dc voltage and an ac voltage are applied between the ring and the two caps (the caps are held at identical potentials with respect to the ring) and an ion is trapped in the region at or near the center of the ring. The trap is similar in principle to the ac magnetic trap in that the time-dependent potential is introduced to stabilize in three dimensions, an otherwise unstable situation. If only a dc field is applied the trap can be made stable along the z direction, but it is not stable in the transverse direction.

The Penning trap uses only static fields, yet it can still be viewed as a dynamic trap because stabilization is achieved through atomic motion. Consider first an electric quadrupolar field created using four symmetrically spaced electrodes, each held at an identical potential, selected to attract the ion to the electrode. In the absence of other applied fields, an ion placed at a point equidistant from the four electrodes will be in equilibrium but will be unstable against (transverse) displacement toward any of the four electrodes. The ion will, however, be stable for motion along the (longitudinal) axis of symmetry of the quadrupole, being confined along this axis by a harmonic potential. To stabilize the transverse motion, a magnetic field is applied along the longitudinal axis such that the ion undergoes cyclotron motion in the transverse plane. The magnetic component of the Lorentz force that generates the cyclotron motion acts to compensate the radial electric force due to the electrodes, and stability is thereby achieved. A similar trap design can be used to trap other charged particles such as electrons. In addition, other variations on these geometries can be used; for example, rows of ions can be trapped in a linear variation of the Penning trap.

The strong Coulomb interaction between the ions plays an important role in limiting the trapped atom number and density. Hence ion traps cannot compete with neutral atom traps as high-brightness (high-phase-space density) sources. However, the ion-ion interaction in the ion trap gives rise to a rich spectrum of ion dynamics. For example, the ions can order to form novel crystalline states or can be made to follow complex chaotic trajectories. The small trapped particle number is also ideal for experiments in fundamental quantum physics, for example in the realization of Young's double-slit experiment using two trapped ions.

Two important applications of ion traps are (1) the construction of time standards (clocks) based on single trapped ions and (2) ultrahigh-precision mass spectroscopy. The trapped ion has many advantages for high-resolution spectroscopy. In the radiation trap and in the magnetostatic trap, the trapping forces cause very strong perturbations of the optical transitions of the trapped atom. Ion traps by contrast cause little perturbation of the optical transitions of the atoms because they trap the ion by interacting with the total ion charge (the monopole moment), not by coupling to the internal states (through the dipole or other higher order moment). Moreover, ion traps also offer very long single-ion storage times and hence very long periods of interrogation are possible. The real disadvantage to ion traps for spectroscopy is that they hold only a small number of atoms and hence have reduced signal-to-noise.

See also: Bohr Magneton; Broglie Wavelength, de; Condensation, Bose–Einstein; Dipole Moment; Double-Slit Experiment; Electromagnetic Wave; Energy Levels; Ion; Laser; Laser Cooling; Lorentz Force; Magnetic Moment; Maxwell's Equations; Resonance; Stark Effect; Zeeman Effect

Bibliography

Anderson, M. H.; Ensher, J. R.; Matthews, M. R.; Wieman, C. E.; and Cornell, E. A. "Observation of Bose–Einstein Condensation in a Dilute Atomic Vapor." *Science* **269**, 198–201 (1995).

Arimondo, E.; Phillips, W. D.; and Sturmia, F., eds. *Laser Manipulation of Atoms and Ions* (North-Holland, Amsterdam, 1992).

Chu, S.; Bjorkholm, J. E.; Ashkin, A.; and Cable, A. "Experimental Observations of Optically Trapped Atoms." *Phys. Rev. Lett.* **57**, 314–317 (1986).

Dehmelt, H. "Experiments with an Isolated Subatomic Particle at Rest." *Rev. Mod. Phys.* **62**, 525–530 (1990).

Migdall, A. L.; Prodan, J. V.; Phillips, W. D.; Bergeman, T.; and Metcalf, H. J. "First Observation of Magnetically Trapped Neutral Atoms." *Phys. Rev. Lett.* **54**, 2596–2599 (1985).

Paul, W. "Electromagnetic Traps for Charged and Neutral Particles." *Rev. Mod. Phys.* **62**, 531–540 (1990).

Raab, E.; Prentiss, M G.; Bjorkholm, J. E.; Cable, A.; Chu, S.; and Pritchard, D. E. "Trapping of Neutral Sodium Atoms with Radiation Pressure." *Phys. Rev. Lett.* **59**, 2631–2634 (1987).

Raizen, M. G.; Gilligan, J. M.; Berquist, J. C.; Itano, W. M.; and Wineland, D. J. "Ionic Crystals in a Linear Paul Trap." *Phys. Rev. A* **45**, 6493–6501 (1992).

Nicholas P. Bigelow

ATTRACTION

See Electrostatic Attraction and Repulsion; Gravitational Attraction

AUGER EFFECT

The process of radiationless rearrangement of an atom or an ion that has been ionized in an inner shell by the emission of a free electron is usually known as the Auger effect. The free electron is called the Auger electron, in recognition of Pierre Auger who, in 1925, interpreted his experiments on the ionization of neon, argon, krypton, and xenon by incident x rays. The experiments consisted of observing the tracks of two electrons in a Wilson expansion cloud chamber; the length of the track is directly proportional to the energy of the free electron. The energy of one electron (the photoelectron) increased with increasing x-ray energy, whereas the energy of the second electron remained constant. Auger proposed that the photoelectron is produced by the ionization of the inner shell. The other electron with a constant energy results from the rearrangement of the ionized atom and, therefore, its energy is a characteristic of the ionized atom. It should be noted that the ionized atom with an inner-shell vacancy also produces characteristic x rays. The Auger process is dominant as compared to x-ray emission for low atomic numbers.

The general notation used in the classification of the Auger electron is *V-AB,* where *V* specifies the inner shell with a vacancy. *A* and *B* are the shells of the two bound electrons that participate in the radiationless rearrangement, with one filling the vacancy and the other ejected from the ion.

The energy of the Auger electron is the difference between the total energies of the initial atom/ion/molecule with an inner-shell vacancy and of the final state after Auger emission. These energies can be measured very precisely using electron spectrometers. This has allowed the use of Auger spectroscopy not only in atomic and molecular studies but also in the investigation of the surfaces of solids.

See also: ATOM; ELECTRON, AUGER; EMISSION; ION; IONIZATION; SPECTROSCOPY

Bibliography

AUGER, P. *J. Phys. Radium* **6,** 205 (1925).
AUGER, P. *J. Ann. Phys. Paris* **6,** 183 (1926).
CHATTARJI, D. *The Theory of Auger Transitions* (Academic Press, San Diego, CA, 1976).

CHANDER P. BHALLA

AUGER ELECTRON

See ELECTRON, AUGER

AURORA

The aurora, sometimes called the Northern Lights or Southern Lights, has fascinated people throughout the centuries. The native peoples of the polar regions have a great oral tradition associated with the aurora. The Eskimos tell stories of the aurora being the spirits of dead relatives who have gone to a wonderful afterlife. In Scandinavia the aurora was believed to be a portent of great events. In Europe and North America the appearance of the aurora at lower latitudes, where it does not frequently occur, was sometimes interpreted as the end of the world. All the stories and folklore associated with the aurora have in common their attempt to explain the beautiful lights that lie like a crown on the top and bottom of our globe.

The aurora can have many different appearances. The aurora that is visible to the human eye usually is either diffuse or discrete. Both types of aurora can be visible together at the same time, and distinguishing between the diffuse and discrete auroras is sometimes a matter of definitions. The diffuse auroras appear as large, luminous areas in the sky with no, or little, observable structure. The diffuse auroras normally are greenish white, green, or, occasionally, red in color. The discrete auroras appear as great arcs or curtains when viewed from a distance. When directly overhead, the lower edges of these arcs or bands are visible, and the arcs extend up into the zenith. The discrete arcs can fold back over on themselves, in the way that a piece of paper can be folded, and can even form spirals or swirls. The colors of the discrete auroras are generally more varied and can include blue and violet as well as green, red, and orange.

The different colors of the aurora are linked to the process that generates the light, and to understand this process requires some knowledge of the atmosphere of the earth. The aurora occurs at very high altitudes in the upper atmosphere of the earth. Typically, the lower height of the aurora is 90 to 100 km (approximately 55 to 60 miles) above the surface of the earth. Solar-wind particles from the Sun, mainly composed of electrons and protons, can penetrate to this altitude into the atmosphere. As these particles penetrate into the atmosphere they lose energy due to collisions with the atoms and molecules that compose the upper atmosphere. Sometimes the energy transferred in these collisions is large enough to tear apart the atoms and molecules. This is called ionization, and results in the presence of free electrons and nuclei of the atoms and molecules. This is why the upper atmosphere at these altitudes is sometimes called the ionosphere. When the energy of the collisions is not large enough to ionize, the energy is instead absorbed by the atoms and molecules in the ionosphere. This increases the energy of the bound electrons in the atoms or molecules, a process known as exciting the energy states of the bound electrons. These excited states in the atoms or molecules can return to their lower energy states by giving off (emitting) a photon, and these emitted photons are the light seen in the aurora.

One of the major achievements of modern physics in the twentieth century is the understanding that the amount of energy that can be gained or lost by the bound electrons in atoms and molecules is quantized. This means that the amount of energy that can be exchanged comes in integer units of a basic quanta of energy. More importantly, for each individual atom or molecule only particular transitions of their bound electrons are allowed. A transition means the change of the bound electron from one energy state to another. The energy of the emitted photon (light) is inversely related to the wavelength of the light. Different wavelengths of light are seen as different colors by the eye, and so each different transition will give off light of a different color. This explains the diverse colors seen in the aurora. The different colors come from transitions in the many different atoms and molecules that make up the ionosphere. Since not all the energy of the incident electrons and protons can be lost in one collision, many collisions are required before all the energy from the incident electrons and protons is lost to the ionosphere. The large numbers of required collisions explains why the aurora can be very tall. The height is simply a record of where the collisions between the incident particles and the constituents of the ionosphere occurred. Which colors are seen where, and in which type of aurora, require a detailed understanding of how the energy of the incident particles is partitioned in the ionosphere. However, the important idea to understanding these details is that keeping track of where the energy goes, and how it moves around, explains what is seen in the auroral light.

With this explanation in mind, we turn to a description of where the auroras are seen, and some ideas of where the incident particles described above come from. We currently know that the auroras occur around both magnetic poles of the earth. When seen in the Northern Hemisphere the aurora is called the aurora borealis, when viewed in the Southern Hemisphere it is called the aurora australis. The auroras are aligned with the magnetic

poles because the charged electrons and protons that cause the aurora are guided along the magnetic field lines of the earth. This guiding is caused by the force of the magnetic field on the particles, which causes the particles to move in helices with the center of the helix being the magnetic field lines of the earth. Since the magnetic field of the earth resembles that of a bar magnet, called a dipole field, the field lines, and hence the particles, enter the earth around its magnetic poles.

The aurora occurs in a fairly narrow band, shaped like an oval, that is aligned with the magnetic poles of the earth. The oval is narrower on the side close to the daylit portion of the earth and is correspondingly wider on the side away from the Sun. In addition, the intensity of the aurora is not uniform around the oval. The side of the auroral oval away from the Sun is called the nightside auroral oval, and the side closest to the Sun is called the dayside auroral oval. Very intense auroral emissions seen on the nightside of the oval are common, and they are key to understanding the temporal morphology of the auroral oval.

Although it is known that energetic electrons and protons are responsible for the light from the aurora, exactly how these particles gain their energy is one of the major mysteries in auroral physics. While in the solar wind, the particles have a speed of 300 to 600 km/s at the earth; this is approximately 1,000 times the speed of sound at sea level. This gives an electron in the solar wind an energy of about 0.25 to 0.5 eV, where $1 \text{ eV} = 1.6 \times 10^{-19}$ J. This is not much energy! By the time this same electron comes into the ionosphere, and creates the aurora, it has approximately 10,000 eV of energy. The current explanation of this increase in energy is that some amount of the kinetic energy (energy of motion) of the solar wind is transferred into electric and magnetic field energy in the magnetosphere of the earth. The magnetosphere is simply defined as the locations where the magnetic field strength of the earth is large compared with the magnetic field of the sun. An analogy for this transfer is a generator. A generator takes kinetic energy and transforms it into electromagnetic field energy. This generator is located on the dayside of the magnetosphere where the solar wind interacts with the magnetosphere of the earth. The electromagnetic field energy is then transferred back into kinetic energy of the electrons and ions on the nightside of the magnetosphere. The analogy for this region is an electric motor, which transfers electromagnetic field energy into kinetic energy. Since this motor is on the nightside this explains why the auroral oval is much larger on the nightside when compared with the dayside. This process basically takes energy from many solar-wind particles and uses it to accelerate, or speed up, a fewer number of electrons and ions that cause the beautiful aurora.

See also: IONOSPHERE; MAGNETOSPHERE

Bibliography

DAVIS, N. *The Auroral Watcher's Handbook* (University of Alaska Press, Fairbanks, 1992).

KIVELSON, M. G., and RUSSELL, C. T., eds. *Introduction to Space Physics* (Cambridge University Press, Los Angeles, 1995).

PARKS, G. K. *Physics of Space Plasmas* (Addison-Wesley, Reading, MA, 1991).

MATTHEW G. MCHARG

AVALANCHE

Loose materials on steep slopes are inherently unstable. The materials may include ice, snow, water, soil, and rock. They may move extremely slowly or suddenly break loose and rapidly cascade down slopes, creating an avalanche. More generally, an avalanche is a large mass flow, at high velocity, of either rocks, soil, ice, snow, or a mixture of these, under the influence of gravity and usually beginning on a steep slope. The terms "rockslide" and "avalanche" are often used interchangeably. An avalanche may be initiated by a rockfall from collapse of a cliff or other nearly vertical rock wall, such that the rocks cascade at nearly free-fall velocities. Many large avalanches are facilitated by unusually high rainfall, decreasing the coefficient of friction along the basal plane of the loose materials. Earthquakes also can initiate avalanches. Material accumulates on steep slopes by various processes of weathering/rock disintegration. Of these, frost wedging (the freezing and expansion of water in surficial fractures) is important in cold climates or on high mountains.

Rock avalanches can be devastating. One of the most destructive was triggered by a 1970 earthquake in Peru. A gigantic block of summit rock, ice, and snow broke loose on the 6,663-m Nevado Huascaran, about 15 km east of the town of Yungay. It fell in nearly free-fall for 1,000 m and on hitting the slopes mobilized even more material and reached velocities as high as 210 to 280 km/h, destroying much of Yungay and then the smaller town of Ranrahirca, and entombing about 20,000 inhabitants. Avalanches such as these eventually entrain water, boulders, and a tremendous amount of mud and are termed "debris flows." Debris flows can be triggered by volcanic eruptions and associated earthquakes. During March, April, and early May of 1980, a large bulge developed on the north slope of Mount St. Helens volcano in Washington as molten rock slowly intruded the cone. At 8:32 A.M. on May 18, a sharp earthquake triggered the collapse of the over-steepened bulge. About 1,313 ft (400 m) of the formerly 9,677-ft (2,950-m) volcano collapsed, forming an avalanche of rock, snow, and ice from the glaciated summit that moved rapidly down the north slope. It then turned eastward along the valley of the Toutle River, destroying or burying all in its path.

It is not unusual for an avalanche to pick up diverse materials as it descends, and, thus, to change its composition along its course. For example, a rock avalanche may incorporate so much water along with large rock fragments as it descends that it may grade into a debris flow. As the amount of water increases and grain size decreases, a debris flow can grade into a mudflow, in which the water content may reach about 60 weight percent. The high density and low viscosity of many mudflows make them powerful and destructive erosive agents.

Snow avalanches are common and sometimes destructive phenomena on steep snow-covered mountain slopes. These are particularly frequent after unusually thick accumulations of winter snows. One of the most disastrous snow avalanches occurred in December 1916, killing about 6,000 Austrian soldiers in the Tyrolean Alps. Many snow avalanches consist of dry, pulverized snow. Avalanches that occur during times of melting may be mixtures of snow, ice, and water; these are far more dense and more erosive than dry snow avalanches.

"Hot avalanches" refers to the fall-back and rapid flow of hot volcanic blocks and bombs from a volcanic eruption, or from the collapse of a growing, extremely viscous mass of lava, either a volcanic spine or dome. The resulting hot avalanche is also one type of a pyroclastic flow (a flow of hot volcanic materials). Hot avalanches, like all avalanches, travel extremely rapidly, reaching velocities in excess of

100 km/h, and are directly responsible for some of the worst volcanic catastrophes that have occurred, such as the 1902 destruction of the city of St. Pierre on the island of Martinique. Cool air entrained in a hot avalanche rapidly expands, lowering the coefficient of friction within the flowing mass, leading to higher velocities than observed in ordinary, ambient temperature avalanches. Typically, hot avalanches emit billowing clouds of volcanic ash lofted upward by the escaping, hot, expanding gases (some from entrained air and others emitted from the impacting lava within the avalanche).

See also: EARTHQUAKE; VOLCANO

Bibliography

SANDERS, J. E. *Principles of Physical Geology* (Wiley, New York, 1981).

WILLIAM G. MELSON

AVOGADRO NUMBER

The number of atoms in exactly 12 g of carbon 12 (^{12}C; the most common isotopic form of elemental carbon) is the definition of the Avogadro number. Its value, which was not determined by Avogadro, has been measured experimentally with varying degrees of precision since 1865. Its most recent value is 6.02213×10^{23}. In chemistry and physics, this number is also identified as the definition of the "mole," so that 6.02×10^{23} "things" constitute one mole of things (sort of a "chemist's dozen"). Normally defined in this way, the number becomes a constant with dimensions 6.02×10^{23} mol^{-1} (things/mol), and is symbolized by N_A, N_0, or L.

The Avogadro number was named after a scholarly and humble physics professor working in Turin, Italy, named Lorenzo Romano Amedeo Carlo Avogadro di Queregna e di Cerreto. Avogadro is known to the scientific world mainly for his paper of 1811, today called Avogadro's hypothesis. Originally, Avogadro wrote his paper to resolve a conflict between the atomic theory of John Dalton and the experimental results of Joseph Louis Gay-Lussac. For example, when Gay-Lussac combined one volume, each, of nitrogen and oxygen, he produced two vol-

umes of nitric oxide. Dalton's theory predicted that only one volume should be produced, and, hence, Dalton rejected the measurements of Gay-Lussac (who, in turn, rejected Dalton's atomic theory). On the basis of Gay-Lussac's law of combining volumes, Avogadro suggested that some gases, such as nitrogen and oxygen, exist as diatomic molecules, and maintained that equal volumes of all gases under the same conditions of temperature and pressure contain the same number of molecules. Unfortunately, this brilliant hypothesis was rejected by both Gay-Lussac and Dalton and was ignored by the scientific community for fifty years (the rest of Avogadro's life). Avogadro's paper was read in French, German, and English, but it lacked evidence, tests, and verification. In addition, the scientific community did not accept the ad hoc nature of the proposition that some gases existed as diatomic molecules. At a meeting in 1860, however, Avogadro's hypothesis was reintroduced by Stanislo Cannizzarro as the basis of a coherent scientific theory, and it was accepted by his peers. Later, the hypothesis was derived from the kinetic theory of gases.

During this time, it was also shown that gram molecular weights of different gases under the same conditions of temperature and pressure occupy equal volumes. (A gram molecular weight is the weight in grams of a substance equal to its molecular weight. A similar definition holds for gram atomic weights.) For example, 2 g of hydrogen and 32 g of oxygen occupy equal volumes under the same conditions of temperature and pressure, and, thus, they contain the same number of molecules. That number, determined by experiment, was named in honor of Avogadro (and is called the mole). Thus, 18 g of water (H_2O), which is 1 mole of water, contains 6.0×10^{23} molecules, 6.0×10^{23} atoms of oxygen, and 1.2×10^{24} atoms of hydrogen.

One can also use the Avogadro number (in the form of a constant) to calculate the mass in grams of a single atom or molecule. For example, the molecular weight of water is 18 g/mol, that is, 18 g contains one mole of molecules. Therefore, the mass of one water molecule is 18 g/mol divided by 6×10^{23} molecule/mol, or approximately 3×10^{-23} g/molecule. [Note: Dividing one mole of atoms or molecules into two equal parts seventy-nine times results in a single molecule (since $2^{79} = 6 \times 10^{23}$).]

In 1865 Josef Loschmidt used the kinetic theory equation for gas viscosity and estimated that the number of gas molecules in a cubic centimeter (cc) at 0°C and 1 atm is 2.7×10^{19} (approximately $6.0 \times$

10^{23} mol^{-1}). Jean Baptiste Perrin in 1909 used his observation of Brownian motion and the Boltzmann distribution law to determine Boltzmann's constant k and, thereby, the Avogadro constant $N_A = R/k$ (where R is the ideal gas constant). Perrin's value was 7×10^{23} mol^{-1}. In 1917 Robert Andrew Millikan found that the changes in electric charge of his oil drops were always multiples of 1.60×10^{-19} C. Using $N_A = F/e$ (where F is the Faraday constant equal to 96,500 C/mol and $e = 1.6 \times 10^{-19}$ C), Millikan obtained 6.06×10^{23} mol^{-1}. The most precise method to measure the Avogadro constant used x-ray diffraction coupled with optical interferometry studies on a crystal of pure silicon. By measuring the lattice spacing a of the crystal, its density d, and atomic weight M, Richard Deslattes in 1980 used the equation $N_A = MZ/da^3$ (where Z is the number of atoms per unit cell) to determine $N_A = 6.02213 \times 10^{23}$ mol^{-1}.

Today, a popular classroom technique for estimating the Avogadro constant uses an idea described by Benjamin Franklin in 1773: measuring the size of an oil film on the surface of a lake. The experimental method was refined by Lord Rayleigh, Agnes Pockels, and Irving Langmuir. One can perform this experiment on a small scale by measuring the diameter of the film formed from one drop of a dilute solution of oleic acid and by counting the number of drops of the solution in 1 cc. Assuming the film is one molecule thick, and using the density and molecular weight of oleic acid, its molecular mass (assuming a cubic shape) and the Avogadro constant can be estimated. (One can make a more accurate determination by assuming the molecules are rods, with each rod having a length ten times its width and standing on end in the film.) The prob-

lems in measuring the Avogadro constant, however, are never simple. With current technologies such as atomic field microscopy, one can actually "see" that the oleic acid molecules lie at an angle to the surface of the film.

The Avogadro constant is accepted today as a "fundamental" relationship between atomic mass units and the arbitrarily defined mass unit of 1 g. It is an indispensable quantitative tool for relating macroscopic measurements of all matter (gas, liquid, and solid) to its microscopic constituents.

See also: ATOMIC THEORY, ORIGINS OF

Bibliography

DESLATTES, R. "The Avogadro Constant." *Annu. Rev. Phys. Chem.* **31**, 435–461 (1980).
JAFFE, B. *Crucibles: The Story of Chemistry* (Simon & Schuster, New York, 1951).
KOLB, D. "The Mole." *J. Chem. Educ.* **55**, 728–732 (1978).
STAUFER, F. "An Estimate of Avogadro's Number." *Physics Teacher* **29**, 252–254 (1991).
TANFORD, C. *Ben Franklin Stilled The Waves* (Duke University Press, Durham, NC 1989).

FREDRICK M. STEIN

AXAF

See ADVANCED X-RAY ASTROPHYSICS FACILITY

B

BACKGROUND RADIATION

See COSMIC MICROWAVE BACKGROUND RADIATION

BARDEEN, JOHN

b. Madison, Wisconsin, May 23, 1908; *d.* Boston, Massachusetts, January 30, 1991; *solid-state physics, condensed matter physics.*

Bardeen was a uniquely gifted scientist who combined superb physical intuition with analytic abilities of the highest order and a remarkable instinct for invention. He was the first scientist to receive two Nobel Prizes in the same field, physics: in 1956 (with Walter Brattain and William Shockley) for the invention of the transistor, and in 1972 (with Leon Cooper and Robert Schrieffer) for the development of the microscopic theory of superconductivity, while, as a consultant to the Xerox Corporation, he played a major role in the development of the xerox process. Since the information and communication revolution is built on the transistor and xerography, Bardeen arguably did more to change the daily lives of Americans than any other scientist of the twentieth century.

One of five children, Bardeen grew up in Madison, Wisconsin, where his father was Dean of the Medical School of the University of Wisconsin. He was a child prodigy whose early talents for mathematics and learning led to his being promoted directly from the third grade of his elementary school to the seventh grade of the University High School in Madison. Because the school lacked laboratory facilities, Bardeen switched to the public high school in his junior year, graduating at the age of fifteen. At the University of Wisconsin, he majored in electrical engineering, and despite spending a year away from campus working for a company in Chicago, obtained his master's degree there in 1930 at the age of twenty-two.

After graduation, Bardeen followed his engineering professor, Leo Peters, to Gulf Research Laboratories in Pittsburgh, where he invented a novel electromagnetic method for oil prospecting which the company found so important that it waited thirty years to file a patent application. After three years as an oil-exploration engineer, Bardeen decided to become a theoretical physicist, leaving his highly paid position at Gulf Research Laboratories to begin graduate work in physics at Princeton University in 1933. There, he carried out research under the supervision of Eugene Wigner, receiving his Ph.D. in 1936 for his theoretical work on the physics of a metal surface in which he showed that exchange and correlation effects play a significant role in determining the charge distribution near the surface. At Harvard, as a Junior Fellow of the Society of Fellows from 1936 to

1938, he carried out the first self-consistent account of the influence of electron-electron interactions in metals on phenon energies and the coupling of electrons and phonons.

Bardeen joined the faculty of the University of Minnesota in 1938; it was there that he became interested in superconductivity and conceived the idea that a sufficently strong electron–phonon interaction could bring about superconductivity. During World War II he worked at the Naval Ordinance Laboratory in Washington, D.C., on magnetic mines, torpedoes, and counter measures against them. In 1945 he joined the technical staff of Bell Laboratories, where, in part because he shared an office with Walter Brattain, he began theoretical work on the transport of current through semiconductor junctions. In the course of explaining the results obtained in Brattain's experiments, Bardeen concluded that a new phenomenon, hole injection, must be occurring at metal-semiconductor junctions, and that holes injected by a forward current through a metal point contact could lower the resistance of a nearby point contact. This led within a week to the demonstration by Bardeen and Brattain of the point-contact transistor.

Bardeen continued his research on the physics of semiconductors for the next few years until word reached him, in early 1950, of the discovery of the isotope effect, the influence of isotopic mass on the superconducting transition temperature, which demonstrated that electron-phonon interactions must play a significant role in bringing about superconductivity. He quickly returned to his earlier work on electron-phonon interactions, and soon developed a theory based on the changes this brings about in single-electron states at low temperatures. Neither Bardeen's theory, nor a similar one developed independently by Herbert Fröhlich, turned out to explain the perfect diamagnetism characteristic of superconductors, but these developments led Bardeen to focus his major scientific efforts on superconductivity. However, his supervisor at Bell Laboratories, William Shockley, actively attempted to persuade Bardeen to abandon his work on superconductivity in favor of continuing theoretical work relating to transistors. As a result, Bardeen, who was passionately interested in finding an explanation for superconductivity, decided to leave Bell in 1951 for a joint professorship in physics and electrical engineering at the University of Illinois at Urbana-Champaign.

At Illinois, where he remained a faculty member for the next forty years, he established two major research groups: one in physics on theoretical work on superconductivity and, more generally, the properties of quantum liquids; the other in electrical engineering for experimental work on semiconductors. During the period from 1952 to 1955, he devoted himself to developing a phenomenological description of superconductivity that could explain the emerging experimental facts, and his review article for the *Handbook on Physics* represents the distillation of this effort. With his first postdoctoral research associate, David Pines, he developed a self-consistent description of the combined consequence of electron-phonon and electron-electron interaction in metals, which showed that for pairs of electrons near the Fermi surface whose energies differed from each other by a typical phonon energy, the effective screened electron interaction would be attractive, while for larger energy differences the interaction would be repulsive. This interaction was used by his second postdoctoral research associate, Leon Cooper, to demonstrate in the presence of an attractive interaction the instability of the Fermi surface for the formation of pairs, and subsequently by Bardeen, Cooper, and Bardeen's first physics graduate student, Robert Schrieffer, in their development of the microscopic theory of superconductivity, the Bardeen–Cooper–Schrieffer (BCS) theory.

The microscopic theory of superconductivity not only explained all existing experiments but also made a number of predictions that were soon verified. It had a significant impact on nuclear physics (pairing phenomena in nuclei), astrophysics (the superfluidity of neutron stars), and low-temperature physics (superfluid ^{3}He). From 1959 until his retirement from the University of Illinois faculty in 1975, Bardeen's physics research interests were primarily directed toward understanding superconductivity and, more generally, macroscopic quantum phenomena in condensed matter. In addition to carrying out seminal work on the motion of flux lines in type-II superconductors, he made significant contributions to our understanding of charge-density waves in quasi-one-dimensional metals and to our understanding of the quantum-liquid behavior of dilute mixtures of ^{3}He in liquid ^{4}He. He also investigated the possibility that an excitonic mechanism might give us higher temperature superconductors.

Bardeen had a keen interest in issues linking science and society; he was a member of President Kennedy's Presidential Science Advisory Council, and was elected president of the American Physical

Society. He loved to travel, and his passion for science was almost matched by his passion for golf. A devoted father and grandfather, he took great pleasure in the fact that his three children were all highly successful in their pursuit of careers in physics and in science-based management. His remarkable accomplishments were recognized widely in the United States, Europe, and the former Soviet Union, but nowhere more than in Japan. On his eightieth birthday, Sony established the Bardeen Chair at the University of Illinois at Urbana-Champaign.

Throughout his career, Bardeen had a rare ability to see into the heart of the problem, to pluck out the significant phenomena from a myriad of (often conflicting) experimental results, and to isolate the elements required to develop a coherent description. His approach to scientific problems went something like this:

- Focus first on the experimental results, by careful reading of the literature and personal contact with members of leading experimental groups.
- Develop a phenomenological description that ties the key experimental facts together.
- Avoid bringing along prior theoretical baggage and do not insist that phenomenological description map onto a particular theoretical model; explore alternative physical pictures and mathematical descriptions without becoming wedded to a specific theoretical approach.
- Use thermodynamic and macroscopic arguments before proceeding to microscopic calculation.
- Focus on physical understanding, not mathematical elegance; use the simplest possible mathematical descriptions.
- Keep up with new developments and techniques in theory, for one of these could prove useful for the problem at hand.
- Do not give up; stay with the problem until it is solved.

See also: CONDENSED MATTER PHYSICS; FERMI SURFACE; PHONON; QUANTUM; SUPERCONDUCTIVITY; SUPERCONDUCTIVITY, HIGH-TEMPERATURE; THEORETICAL PHYSICS; TRANSISTOR

Bibliography

HERRING, C. "Recollections from the Early Years of Solid-State Physics." *Phys. Today* **45** (4), 26 (1992).

HOLONYAK, N., Jr. "John Bardeen and the Point-Contact Transistor." *Phys. Today* **45** (4), 36 (1992).

LUBKIN, G. B. "Special Issue: John Bardeen." *Phys. Today* **45** (4), 23 (1992).

PAKE, G. "Consultant to Industry, Adviser to Government." *Phys. Today* **45** (4), 56 (1992).

PINES, D. "An Extraordinary Man: Reflections on John Bardeen." *Phys. Today* **45** (4), 64 (1992).

SCHRIEFFER, J. R. "John Bardeen and the Theory of Superconductivity." *Phys. Today* **45** (4), 46 (1992).

DAVID PINES

BARRIER PENETRATION

It was known as early as 1910 that the lifetime of heavy nuclei that decayed by the emission of alpha particles (helium nuclei) was strongly dependent on the energy of the alpha; the half-life was much greater if the energy was small. Hence the most common isotope of uranium, ^{238}U, which decays through the emission of a 4.25 MeV alpha particle, has a half-life of about 4.5 billion years, whereas radon, ^{222}Rn, which decays through the emission of a 5.59 MeV alpha, has a half-life of only 3.8 days. Overall, the half-life was found to vary approximately as a Geiger–Nuttall relation,

$$\log_{10} \tau \approx 39 - 7 \cdot T$$

where τ is the half-life in years and T the kinetic energy of the alpha in MeV.

Moreover, the alpha energies seemed anomalously small. The alpha carries a charge of $+2e$, where e is the electronic charge, while the residual nucleus left from uranium decay held a charge of $+90e$ and, hence, the electrostatic repulsive forces are strong. Indeed, the forces are so strong that an alpha particle placed at the edge of the nucleus would gain an energy of about 25 MeV as it was pushed away. Then, how could an alpha particle ejected from the nucleus have as little energy as 5 MeV? With so little energy the particle should never have been able to get out.

We illustrate these considerations with the help of Fig. 1, which shows a puck sliding back and forth under gravity on the frictionless surface of a hole surrounded by a raised barrier that acts to repel pucks near to, but not in, the hole. At the high-point

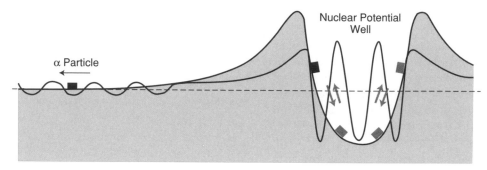

Figure 1   The diagram shows a puck sliding on the surface of a hole surrounded by a barrier or, equally, an alpha particle held in a nucleus where the barrier consists of an electrostatic potential. The wave forms suggest symbolically the wave that describes the particle.

of its oscillation, the puck is higher than the puck on the flat surface to the right of the hole. Hence, energy will be conserved if the puck were to escape the hole and move away. But the hill around the hole will contain the puck; in this classical picture, the puck will never escape. Any puck that does escape must have enough energy to reach the top of the barrier and will then accumulate a large kinetic energy as it slides down the hill.

The classical picture of the alpha particle is similar on a much smaller scale if we consider just one direction and one dimension of the three-dimensional system of the alpha particle and the residual nucleus. Now the hole represents the nucleus where nuclear forces hold the alpha tightly, and the barrier represents the effect of the electrostatic forces that repel the alpha outside of the nucleus. Like the puck, in a classical world the alpha particle will never escape the nucleus, though such an escape is allowed energetically.

In the macroscopic world in which we live, nature differentiates sharply between the properties of particles and of waves, between the characteristics of a stone we might toss into a pond and the ripples on the water initiated by the falling stone. However, when the scale is reduced enormously, such that the momentum of the particle (e.g., the alpha particle) times the length of the region over which that momentum is relevant is not much larger than Planck's constant ($h = 6.626 \times 10^{-34}$ J·s), the particle will exhibit wave-like properties. In 1928 George Gamow, and Ronald W. Gurney and Edward U. Condon, showed that the wave-like properties of alpha particles lead to the leakage of the particle through the barrier of electrostatic potential about heavy nuclei that we call barrier penetration and, hence, to alpha-decay.

According to the wave form, or Schrödinger form, of quantum mechanics, the wave function ψ describing a particle of mass m moving in an x direction with a velocity v and momentum $p = mv$, accelerates toward zero, or curves toward zero, at a rate described by the differential equation

$$\frac{d^2\psi}{dx^2} = -k^2\psi,$$

where $k = p/2\pi h$. The solutions to the wave equation take the trigonometric wave forms $\psi(x) = A \sin kx$ and $\psi = A \cos kx$, where the square of the amplitude $\psi^2(x) \propto A^2$ is proportional to the probability that the particle is near the position x. Here the wavelength is $\lambda = p/h$.

The kinetic energy of the particle, $T = p^2/2m$, is the difference between the total energy W and the potential energy V of the particle; $T = W - V$. Hence, $k^2 = 2m(W - V)/(2\pi h)$. In classical physics, the kinetic energy can never be negative; a negative kinetic energy has no meaning. But in the wave form of quantum mechanics, a negative kinetic energy simply means a change of sign of k^2 in the wave equation, which then has solutions that curve, or accelerate, away from the origin, taking the form of $\psi(x) = B \exp(-\kappa x)$ and $\psi(x) = B \exp(\kappa x)$, where $\kappa^2 = -k^2 = 2m(V - W)/(2\pi h)$. Again, $\psi^2(x) \propto B^2$ is the probability of the particle being near x.

Therefore, in the nucleus, where the potential energy of the alpha particle is negative, the kinetic energy is large and positive and the wave function

94

describing the alpha varies as a sine or cosine function. But just outside of the edge of the nucleus and the effects of the short-range nuclear forces, the alpha will have a negative kinetic energy even as the potential energy is greater than the total energy and the wave function decays exponentially, as suggested by the symbolic drawing of the wave function in Fig. 1. The smaller the energy of the particle W—which is equal to the final kinetic energy of the escaping alpha particle—the larger the factor κ and the stronger the attenuation. Then, when the alpha particle is far from the nucleus, and the electrostatic potential energy is negligible, the kinetic energy of the alpha is positive and equal to the total energy, and the wave function varies as a sine or cosine function as shown.

The amplitude of the wave is very much reduced by the exponential reduction in the barrier, hence the probability of finding the alpha outside of the nucleus is very small. This means that the decay probability is small and the lifetime for decay is long. For radium, with a half-life of about 1,600 years, the amplitude of the wave function at a distance of 1 mm from the nucleus is reduced by a factor of about 10^{-26} from that inside of the nucleus. For uranium, which emits a lower-energy alpha particle, the amplitude is smaller by another factor of about 1,500.

Barrier penetration of other particles, especially electrons on an atomic scale, is also important in other areas of physics, and in chemistry.

See also: DECAY, ALPHA; EMISSION; WAVE FUNCTION

Bibliography

BORN, M. *The Restless Universe* (Dover, New York, 1951).
DAS, A., and FERBEL, T. *Introduction to Nuclear and Particle Physics* (Wiley, New York, 1993).
RICE, F. O., and TELLER, E. *The Structure of Matter* (Wiley, New York, 1949).

ROBERT K. ADAIR

BARYON NUMBER

Each fundamental particle carries an attribute known as its baryon number, *B,* which, like electric charge, takes only discrete values. Protons and neutrons and their excitations have $B = 1$, their antiparticles have $B = -1$, while mesons (e.g., the pions, which are partially responsible for mediating the nuclear force) and the leptons (i.e., particles with no strong interactions, such as the electron and neutrino) are neutral ($B = 0$). The baryon number of an atom is the number of protons and neutrons in its nucleus. To a good first approximation it coincides with its atomic weight. *B* is a consequence of the number of quarks that make up each type of particle. Baryons or antibaryons are bound states of three quarks or three antiquarks, respectively, and mesons consist of a quark-antiquark pair. Thus baryon number is the number of quarks (n_q) minus the number of antiquarks ($n_{\bar{q}}$) times a conventional factor of $1/3$, that is, $B = (n_q - n_{\bar{q}})/3$. Equivalently, quarks and antiquarks carry $B = \pm 1/3$, respectively.

The total baryon number is conserved in all known reactions. For example, the process $p + \pi^- \rightarrow n + \pi^0$ is allowed, because the net baryon number of $+1$ is unchanged. Similarly, a proton and antiproton can annihilate into pions, because the total initial and final baryon number is zero. However, the decay $p \rightarrow e^+ \pi^0$, which would involve a change of baryon number from $+1$ to 0, has not been observed. Thus *B* conservation is responsible for the stability of protons and ordinary matter. At the quark level, *B* conservation holds because the known interactions can transform one type of quark into another, or can allow the annihilation of a quark and antiquark into radiation (or the inverse creation of a $q\bar{q}$ pair), but do not allow the creation or disappearance of a single quark or a quark pair.

Nevertheless, there are several reasons to believe that there are very weak *B*-violating processes in nature. The only strong theoretical motivation for an exactly conserved quantum number is the existence of an associated long-range force, such as the electromagnetic force associated with electric charge. However, there appears to be no analogous force coupling to baryon number. Experimentally, any such force would have to have a strength less than 10^{-45} that of electromagnetism.

Similarly, black holes retain information only for quantities associated with long-range forces. If a neutron were to drop into a black hole its quantum numbers would disappear from the universe, violating baryon number.

The weak interactions are predicted to involve a small *B* nonconservation, because there are many possible lowest energy (ground) states of nature differing in their baryon number. Quantum mechanical tunneling from one to another can lead to the creation of baryon number. However, the rate is suppressed by an incredibly tiny factor, 10^{-172}, and thus this process is irrelevant for ordinary purposes.

Grand unified theories, which attempt to unify the strong, weak, and electromagnetic interactions as aspects of an underlying unified theory, generally involve new interactions that can mediate proton decay and thus violate baryon number conservation. The simplest such theories predict $p \rightarrow e^+ \pi^0$ with a lifetime around 10^{30} years. Although very rare, such decays could be detected by dedicated experiments observing around a kiloton of matter. Existing experiments have not observed proton decay at the predicted level, thus ruling out this simplest grand unified theory. However, theories involving additional particles, known as supersymmetric grand unified theories, are motivated by the inclusion of gravity and by technical problems involving the energy scale of the weak interactions. These predict a slower proton decay rate, typically into different final particles. The decays predicted are consistent with existing limits, but possibly observable in the future. The even more ambitious superstring theories generally predict proton decay at a very slow rate or not at all.

The universe, or at least our region of it, consists of matter (baryons) and not antimatter. This is necessary for the existence of stars, planets, and life; otherwise, the matter and antimatter would rapidly annihilate into radiation. This suggests B violation played an important role at some time in the history of the universe, in which a small excess (by one part in 10^{10}) of quarks was created compared to antiquarks. Later, the antiquarks and most of the quarks annihilated, leaving the relatively small quark excess to force the matter of all known objects. The period at which this excess developed may have been associated with the grand unification scale, when the universe had a temperature around 10^{29} K. Alternately it may have developed later by thermal processes related to the weak ground state tunneling, at a temperature around 10^{16} K. It is also possible that there was a small initial asymmetry at the start of the big bang, or that there is a large-scale separation between baryons and antibaryons, but these possibilities meet with other difficulties. Thus the dominance of matter over antimatter in the universe is one of the outstanding puzzles of cosmology and particle physics.

See also: ANTIMATTER; ELEMENTARY PARTICLES; GRAND UNIFIED THEORY; INTERACTION, ELECTROMAGNETIC; INTERACTION, STRONG; INTERACTION, WEAK; LEPTON; PARTICLE PHYSICS; QUARK

Bibliography

GOLDHABER, M.; LANGACKER, P.; and SLANSKY, R. "Is The Proton Stable?" *Science* **210,** 851 (1980).

KOLB, E. W., and TURNER, M. S. *The Early Universe* (Addison-Wesley, Reading, MA, 1990).

LANGACKER, P. "Grand Unified Theories and Proton Decay." *Physics Reports* **72,** 185–385 (1981).

PAUL LANGACKER

BASIC, APPLIED, AND INDUSTRIAL PHYSICS

Physics—like biology, chemistry, and other disciplines of science—is readily described in terms of its research topics, its research activities, and its research findings. Physics topics are centered around studies of the fundamental principles and laws that govern the behavior of energy and matter in nature. The activities of physicists are often organized within the three branches of physics research: basic, applied, and industrial. Although each physics branch has its own set of topics, activities, and findings, all three are connected through fundamental physics findings; such as the behavior of energy and matter under similar physical conditions is the same in laboratories on earth as it is in the solar system or anywhere else in the universe.

Historically, physics developed from empirical observations of nature that were generalized with the aid of experimental verification under carefully controlled conditions. In time, physicists were aided in their pursuit of knowledge about natural phenomena by theoretical models that were developed from mathematical descriptions of the observational data. Through the use of mathematics as the language for communicating between physicists, analytical theories were able to predict observations from controlled laboratory experiments.

More recently with the aid of computers, physics research is being explored simultaneously with the use of theory, experiments, and computations.

Physicists are often categorized as either a theoretical, experimental, or computational physicist depending on how or where they spend the majority of their research time—thinking about mathematical models, collecting data in the laboratory, or computing in front of a monitor, respectively. Nevertheless, most modern-day physicists cross over these descriptive boundaries during the course of their research careers.

The acquisition of physics knowledge is generally referred to as "basic physics." The application of that knowledge is commonly referred to as "applied physics." The commercialization of physics applications is ordinarily referred to as "industrial physics." These three branches collectively span the equivalent activities that fall under the broad descriptive phrase of science and technology. Even though the separation between basic, applied, and industrial physics becomes less and less every day as physicists are motivated to share their collective knowledge in shorter and shorter time frames, these three branches lead to a natural organizational breakout for further discussion of the topics, activities, and findings of physics and physicists.

Basic Physics

Basic physics topics can be summarized by looking at the table of contents of an introductory physics book for a high school or lower-division college class. The textbook would cover classical physics topics such as energy, matter, motion, forces, gravity, electricity, magnetism, and light; and modern physics topics such as relativity, atomic physics, and quantum mechanics. Physicists whose activities try to further the understanding of these examples of energy and matter topics are doing basic research. Physicists who concentrate their research in basic physics tend to do their work within individual fields of physics that are defined as physics subdisciplines.

Examples of physics subdisciplines with ongoing basic physics research activities include astrophysics, atomic physics, computational physics, condensed matter physics, fluid dynamics, high-polymer physics, materials physics, molecular physics, optical physics, nuclear physics, particle physics, plasma physics, and many other headings scattered throughout the letters of the alphabet. Numerous upper-division college textbooks exist on every one of the subdisciplines of basic physics, as well as refereed scientific journals.

Plasma physics is a physics subdiscipline that requires the application of findings in several basic physics topics at the same time (motion mechanics, electromagnetism, and atomic physics). Plasma physics grew from turn-of-the-century basic research and investigations of such natural plasma phenomena as lightning, the Northern Lights, the magnetic field of the earth, and sunspots. Plasmas are commonly referred to as the "fourth state" of matter (solid, liquid, and gas being the first three states) because they occur at temperatures greater than a few thousand degrees kelvin, where matter ionizes from a gaseous state of atoms and molecules into a plasma state of negatively charged electrons and positively charged ions. The motion of the ions and electrons within the plasma is influenced by collective behavior arising from the electric and magnetic forces that are produced from charges and currents generated throughout the plasma.

Modern-day plasma physicists find themselves making contributions in basic physics research dealing with space plasmas and astrophysical plasmas. Space plasma research includes studies of phenomena occurring at the surface of the sun, in the solar wind, and in the magnetospheres and upper atmospheres of the planets. Astronomical plasma physics research has revealed that many of the unsolved astrophysical problems involve plasma physics topics that have not yet been fully understood. Approximately 99 percent of the matter in the universe is in the plasma state.

Applied Physics

Applied physics topics differ from basic physics topics due to the research emphasis on the development of applications that could lead to solutions to problems that have a measurable effect on the daily lives of people. In many instances, applied physics research shares common topics of interest with other disciplines of the sciences, such as biological physics and chemical physics. Applied physics tends to be more cross-disciplinary in its research approach than the single subdisciplinary approach to basic physics research. For example, biophysicists use the physical principles of physics to explain the mechanics of biological processes.

Continuing with the plasma physics example from the basic physics discussion will lead naturally into an example of an applied physics topic for further description: plasma fusion energy research. Fusion energy research began in the late 1950s with the attempt to use controlled plasma confinement at high enough temperatures and densities in order to harness the energy from the nuclear fusion process. Nuclear fusion is the process by which two light nuclei combine to form a heavier nucleus with the simultaneous release of nuclear binding energy. The fusion process is the source of energy for the sun, the stars, and present-day thermonuclear weapons. Efforts to achieve controlled thermonuclear fusion in the laboratory have been the primary stimulus for plasma experiments for the last forty years.

Modern-day plasma physicists find themselves making contributions in applied physics topics such as nuclear science, materials science, and laser science. Studies indicate that a fusion system may be useful for eliminating, via nuclear transmutation, the radioactive wastes from fission reactors. Advanced materials studies will be crucial to the success of fusion energy power plants because the first fusion reactor is likely to be based on the deuterium–tritium fusion reaction, which produces helium nuclei as well as 14-MeV neutrons. Laser science research led to an advanced nonlinear optical material being combined with a plasma electrode in order to develop an optical switch. This advanced optical switch is used in high-power glass laser fusion systems to redirect a laser beam after multiple passes through a laser amplifier.

Industrial Physics

Industrial physics topics are defined by the concern to produce a product from a new technology or an improvement in an existing technology, rather than understanding new physical phenomena. Because of the customer-driven forces defined by the marketplace, industrial physics activities tend to be more multidisciplinary in nature than applied physics activities (cross-disciplinary emphasis) or basic physics activities (single-subdisciplinary emphasis). Industrial physicists can be found in all the major research and development laboratories of industry from all sectors of the marketplace, such as transportation, communication, electronics, computers, and so on.

The plasma physics example from the basic physics discussion also leads naturally into an example of an industrial physics topic for further description: industrial plasmas. Industrial plasmas owe their existence in the last several decades to the theoretical, experimental, and computational understanding of plasma physics from plasma fusion research that has been applied to plasma processes in industrial applications.

Modern-day plasma physicists find themselves making contributions in industrial physics topics such as those of interest in the distribution of electric power, electronics manufacturing, and materials-processing industry. Because plasmas exhibit characteristics of a gas, fluid, and electrical conducting medium, the electric power industry continues to use plasmas for high-power switching equipment. Plasma processes currently constitute about one-third of the procedures needed to manufacture microelectronics elements. Industrial plasmas are also commercially available for tool hardening, anticorrosion coatings, and the vapor deposition of superconducting films and diamonds.

The Physicists of Today

The traditional roles of the physicists working in basic research, applied development, and industrial commercialization are expanding. A general one-to-one mapping exists between the activities within these three branches of physics and the primary function of physicists within universities, government laboratories, and industry. Physicists today tend to cross over more and more the boundaries once defined as theoretical, experimental, and computational, as the required skills and abilities expand to meet the increased overlap in research activities in the universities, national laboratories, and industry.

Physicists are best known by the research they do. If teaching is combined with basic research, then the latest tally of the 44,000 members of The American Physical Society gives equal numbers of physicists employed in research/teaching and in applied development/industrial commercialization. The American Physical Society and the American Association of Physics Teachers publish monthly scientific journals on the latest research findings and teaching methods within all physics subdisciplines. Trade journals are available from other physics or science organizations for keeping track of recent developments within the research and development (R&D) activities of industry.

The American Institute of Physics publishes a monthly magazine, *Physics Today*, which contains physics updates, articles, book reviews, and so on, that can be understood by a college undergraduate in any field of science. Undergraduates outside of the sciences can still get an appreciation of physicists and their work by reading the letters, reports from Washington, and opinion pieces that are also contained in *Physics Today*. In recognition of the growth in the number of physicists working in modern industry, *Physics Today* has a semiannual supplement titled *The Industrial Physicist.*

Conclusion

Physics can be described as a collection of research topics with activities focused on the acquisition, application, and commercialization of scientific findings. For example, there exists a basics physics theory acquired more than 100 years ago that describes how electrical forces exist and behave between charged particles. When this electrical force theory was combined with chemistry and materials science for an applied physics application, battery technology was developed for converting chemical energy to electrical energy. Industrial physics research currently continues to explore technological improvements on novel battery concepts for commercial applications ranging from electric vehicles to artificial organs.

A dimensional analysis that uses time, distance, and speed can help explain the importance of understanding the topics that are of interest to physicists, the activities of physicists, and the findings of physics research. The time necessary to cover a fixed distance is given by the distance to be traveled divided by the speed at which one is traveling (time = distance/speed). The distance between basic physics research, applied physics development, and industrial physics commercialization is decreasing. The speed at which information is shared between theoretical, experimental, and computational physicists is increasing. Thus with a simultaneous decrease in distance and increase in speed, the time it takes for physics research to be acquired, developed into technological applications, and commercialized into the marketplace is getting shorter and shorter. As exemplified by the extremely short time it took between the first experimental demonstration of laser technology in 1960 and the commercial use of lasers in medicine, communication, entertainment, and various other industries, the future impact of physics will be defined by even shorter time scales between basic research, applied development, and industrial commercialization.

In life as in physics, time is a parameter not to be ignored. Today is the time to broadly increase the knowledge of physics topics, activities, and findings. Physics is what physicists do; and what follows will lead to profound changes in the lives of people.

See also: COMPUTATIONAL PHYSICS; EXPERIMENTAL PHYSICS; INDUSTRY, PHYSICISTS IN; PHYSICS, HISTORY OF; PHYSICS, PHILOSOPHY OF; THEORETICAL PHYSICS

Bibliography

RUTHERFORD, F. J., and AHLGREN, A. *Science for All Americans* (Oxford University Press, New York, 1990).

DONALD CORRELL

BATTERY

A battery is a source of electrical potential energy that results from a transformation of another form of energy to the electrical energy available at the terminals of the battery. The energy per unit charge available, in units of volts (1 V = 1 J/C), at the terminals is called the electromotive force (emf), E, of the cell. The positive terminal of the battery is called its anode and the negative terminal is called its cathode.

All batteries have an internal resistance r. When a current I is drawn from a battery, the terminal voltage V at the terminals of the battery becomes $V = E - Ir$. The normal energy transformation process inside of a battery results in an increase of the internal resistance over a period of time. This buildup of internal resistance depends on the nature of the particular battery. A typical "dead" battery is one that has an internal resistance so large that its terminal voltage drops rapidly once a current is drawn from the battery, making it no longer capable of providing useful electrical energy. In some batteries it is possible to "recharge" the battery by passing current through the battery in the direction opposite from current flow during normal battery use. In the recharging process, the internal resistance is reduced and the terminal potential energy of the battery is restored. Batteries have a limited number of recharging cycles due to irreversible processes in the transformation processes.

Two examples of the energy transformation processes that are used for batteries are as follows: For the standard dry cell battery, chemical energy is converted to electrical energy in the chemical reaction between the anode material (+ terminal), the cathode material (- terminal), and the electrolyte forming the medium between the anode and cathode; for a solar battery, electromagnetic energy of the light striking the solar cell is converted to electrical energy in the process that results in separating the positive and negative charges inside the solar battery.

The power (watts) supplied by a battery is equal to the product of its emf (E in volts) and the current (I in amps) supplied. High-power applications, such as powering starter motors in automobiles, provide power levels of over 1,000 W for brief intervals of time. Low-power applications in small electronic applications, such as hearing aids and digital watches, involve power levels in the milliwatt (1/1,000 W) range. It is interesting to note that we can now find batteries that have a 5-year lifetime for almost every use. The battery has played a very important part in our expanding use of electronics. Current basic research in materials science suggests that we can expect continuing improvements in batteries, our portable power packs for everything from portable stereo systems to electric cars.

See also: ELECTRIC GENERATOR; ELECTROLYTIC CELL; ELECTROMOTIVE FORCE; SOLAR CELL

Bibliography

HINRICHS, R. A. *Energy* (Saunders, Philadelphia, 1991).

RICHARD M. FULLER

BEATS

Two mechanical oscillators even weakly coupled will exchange energy back and forth. An example is two pendulums suspended from a common taut cord. If the difference in their natural frequencies $\Delta\omega = \omega_1 - \omega_2$ (radians per second) is small and if damping is slight so that energy is nearly conserved, the motion of each oscillator will be essentially sinusoidal with frequency $\omega_0 = \frac{1}{2}(\omega_1 + \omega_2)$ and have a slowly varying sinusoidal amplitude with frequency $\Delta\omega$. As time goes on each oscillator in turn will periodically slow down at the rate $\Delta\omega$, and could even

come to rest with zero energy given certain initial conditions, for example, if one of the oscillators was initially at rest. When one oscillator reaches its minimum amplitude and therefore minimum energy, the other oscillator will have achieved its maximum amplitude and energy. Thus, after passing through a minimum, the amplitude of an oscillator will slowly increase with time, while the amplitude of the other oscillator slowly decreases. This oscillatory phenomenon is known as beats and $\Delta\omega$ the beat frequency.

The notion of beats extends to the superposition of any two sinusoidal signals—electrical, electromagnetic, sound, and so on—with slightly different frequencies ω_1 and ω_2. The oscillations in the resulting interference are described by a sinusoidal wave with frequency ω_0, called the carrier signal, which is said to be modulated by a time-dependent amplitude

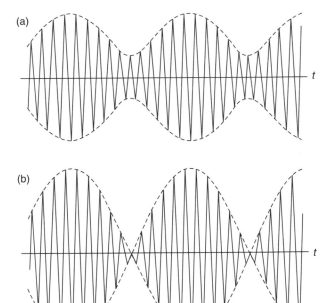

Figure 1 Part (a) shows a plot of the two-component sinusoidal function $\sin 1.1t + \frac{1}{2}\cos t$. The relatively slow modulation (dashed curves) of the rapidly varying carrier signal (solid curve) is evident. Since the amplitudes of the component functions are not equal, the destructive interference is less than 100 percent. Part (b) shows a plot of $\sin 1.1t + \cos t$ with equal-amplitude component functions and therefore 100 percent destructive interference at regular intervals.

that minimizes with the beat frequency $\Delta\omega$. If the original signals have equal amplitude, their superposition will exhibit complete destructive interference in time at the beat frequency (see Fig. 1). This behavior is fully analogous to the motion of two coupled oscillators.

The effect can be used in tuning musical instruments. For example, if two nearly identical tuning forks have natural frequencies of $\omega_1/2\pi = 440$ Hz and $\omega_2/2\pi = 441.2$ Hz, the combination of these tones will produce a beat frequency of 1.2 Hz, or one beat every 0.83 s. Beat frequencies up to about 7 Hz are discernible by the human ear, and successful tuning is achieved by adjusting the frequency of a note to that of say a standard tuning fork until the beats are eliminated.

If the interfering signals involve excitation and response of a quantum mechanical system, the resulting oscillations in their superposition are generally referred to as quantum beats. Examples occur in the observation of radiation from coherently excited states of atoms and molecules. The effect is fundamental because any attempt to determine from which state the photon is emitted suppresses the interference and therefore the origin of the beats.

See also: INTERFERENCE; OSCILLATION; OSCILLATOR; RADIO WAVE; RESONANCE; SOUND

Bibliography

HALLIDAY, D.; RESNICK, R.; and WALKER, J. *Fundamentals of Physics*, 4th ed. (Wiley, New York, 1993).

JAMES M. FEAGIN

BELL'S THEOREM

Bell's theorem was derived in 1964 by John Stewart Bell. It is a major milestone in studies of the conceptual foundations of quantum mechanics (QM).

In the early years of the development of QM a great deal of excitement was generated due to the capabilities it provides for making new predictions. However, these predictions applied to the statistical results of a large number of measurements; for a single measurement, QM could only give a probability for the outcome. For example, consider Young's double-slit experiment. A beam of photons is observed after passing through a pair of slits; if one detects the arrival positions of a large number of the photons on a distant screen, the statistical average produces an interference pattern with light and dark bands. However, the exact position at which any single photon will be detected cannot be predicted; one can, at best, specify the probabilities for detection at various positions.

Some physicists were uncomfortable with this limitation, believing there must be a far more encompassing theory in which every parameter has a precise value. An analogy is found in the relation of thermodynamics to statistical mechanics: the macroscopic laws of thermodynamics are a result of the statistical average of the precise values of microscopic molecular parameters that are the domain of statistical mechanics. In a classic 1935 paper, Albert Einstein, Boris Podolsky, and Nathan Rosen expressed their discomfort with QM by arguing that QM was an incomplete theory and introducing what is called the Einstein–Podolsky–Rosen *gedanken experiment* (EPR).

EPR considered a single quantum mechanical system consisting of two parts spatially separated by a large distance. They showed that by making a measurement on only the first part, they could predict with certainty the result of a measurement on the distant second part. Clearly, if that result can be known with certainty, then it must be a "real" property of the second part, that is, it must be a preexisting property that was revealed, but not created, by the measurement on the first part. (In quantum mechanics, the measurement on the first part actually changes the wave function for the distant second part.) Presumably, a complete theory would include additional parameters that would enable one to precisely predict all such preexisting properties; these additional parameters were called "hidden variables." In his 1951 book, *Quantum Theory*, David Bohm introduced a variant of EPR involving correlations between two spin one-half particles; it still provides a focus for much of the discussion.

An argument carrying considerable weight was a 1932 proof by John von Neumann that hidden variables were impossible in the context of QM. But, in 1952, Bohm did the "impossible" by developing a hidden variable model that reproduced the predictions of QM, a direct contradiction of von Neumann's "proof." In 1966 Bell resolved the contradiction by showing that one of von Neumann's axioms for QM was unreasonable. Bell also observed that Bohm's model did not satisfy locality. (Locality means that the experimental result produced by an apparatus depends on the local settings of that apparatus but does not depend on the settings of any other distant apparatus. In practice, this would be ensured by changing apparatus settings in a time that was short compared to the time it takes a light signal to travel from one apparatus to the other.) Bell asked if *any* theory based on hidden variables and satisfying locality could reproduce all the statistical predictions of QM. The answer, in the form of Bell's theorem, was that it could not.

Bell's theorem takes the form of an inequality. If any theory satisfying locality can assign precise values to every quantity (a "complete" theory), then its statistical predictions for correlated phenomena are limited by an inequality; QM predicts strong correlations that can violate that limit. The surprise is that knowing precise values for each single event weakens the statistical correlation. The great significance is that after three decades of philosophical arguments, Bell's theorem led to the first experimental tests. Specifically, if appropriate experimental data violate the inequality, a completely classical description of nature (hidden variables) is ruled out (i.e., a classical description would require instantaneous transfer of information between distant apparatuses). It is ironic that locality appears as a major factor in Einstein's argument that QM was incomplete; yet, it is locality that enabled Bell to prove it is impossible to "complete" QM and still retain all its statistical predictions.

Since Bell's original argument was idealized, additional assumptions were made to derive a weaker inequality that could be experimentally tested. Early experiments studied polarization correlations between two spatially separated photons that were emitted successively in an atomic cascade. In 1972 Stuart J. Freedman and John F. Clauser used a $J = 0 \rightarrow 1 \rightarrow 0$ transition in Ca (J is the angular momentum of the atomic level). In 1976 Edward S. Fry and Randall C. Thompson used a $J = 1 \rightarrow 1 \rightarrow 0$ transition in ^{200}Hg and employed lasers to improve data acquisition rates. In 1981–1982 Alain Aspect and coworkers used significantly improved technology to repeat the original Ca experiment as well as several important variants. All these experiments found agreement with QM and violated the Bell inequality. The correlated pair of photons produced by nonlinear down conversion in a crystal is the source of choice for recent experiments; these include polarization correlations, position–time correlations, and phase–momentum correlations.

Two loopholes exist in all the experiments: detector efficiency and locality. Strong Bell inequalities involve coincidence rates (rates for detection of both photons of a pair) that are normalized to single photon counting rates at each detector. For low detector efficiency, singles rates are so large that the normalized rates cannot violate a Bell inequality, and all existing experiments have required an additional assumption in order to obtain a weaker, but testable inequality (normalization to coincidence rather than singles rates). The only experiment to address locality is one by Aspect and coworkers. They have made the only space-like separated measurements and were able to rule out any ordinary transfer of information, but locality was still not rigorously enforced.

Two experiments that would close both loopholes have been proposed. Paul G. Kwiat and coworkers solve the efficiency problem by using a new two crystal down conversion interferometer together with new solid state photon counters (efficiency greater than 90 percent). Fry, Thomas Walther and Shifang Li described an exact experimental realization of Bohm's version of EPR; correlations are observed between the spin one-half nuclei of two ^{199}Hg atoms produced by dissociation of a two-atom molecule in a nuclear spin singlet state. Detection efficiencies are greater than 95 percent.

Bell's theorem is based on two particle quantum systems. Recent extensions by Daniel M. Greenberger, Michael A. Horne, and Anton Zeilinger to multiparticle systems provide direct tests without inequalities; Lucien Hardy has now shown this can also be done for two particles.

Although Bell's theorem is rather esoteric, its study has led to an important application, quantum cryptography, in which EPR correlations lead to a shared, unbreakable code that can be used to encrypt and decrypt messages. In another application, the two photon sources for Bell inequality experiments are used to measure the absolute efficiency of photon detectors.

See also: DOUBLE-SLIT EXPERIMENT; EINSTEIN, ALBERT; EINSTEIN–PODOLSKY–ROSEN EXPERIMENT; LOCALITY; QUANTUM MECHANICS; QUANTUM THEORY, ORIGINS OF; QUANTUM THEORY OF MEASUREMENT

Bibliography

BELL, J. S. *Speakable and Unspeakable in Quantum Mechanics* (Cambridge University Press, Cambridge, Eng., 1987).

Bohm, D. *Quantum Theory* (Prentice Hall, Englewood Cliffs, NJ, 1951).

Clauser, J. F., and Shimony, A. "Bell's Theorem: Experimental Tests and Implications." *Rep. Prog. Phys.* **41,** 1881 (1978).

d'Espagnat, B. "The Quantum Theory of Reality." *Sci. Am.* **241** (Nov.), 158 (1979).

Fry, E. S.; Walther, T.; and Li, S. "Proposal for a Loophole Free Test of the Bell Inequalities." *Phys. Rev. A* **52,** 4381–4395 (1995).

Kwiat, P. G.; Eberhard, P. H.; Steinberg, A. M.; and Chiao, R. Y. "Proposal for a Loophole-Free Bell Inequality Experiment." *Phys. Rev.* **A49,** 3209 (1994).

Mermin, N. D. "What's Wrong With This Temptation?" *Phys. Today* **47** (June), 9 (1994).

Wheeler, J. A., and Zurek, W. H., eds. *Quantum Theory and Measurement* (Princeton University Press, Princeton, NJ, 1983).

Edward S. Fry

BERNOULLI'S PRINCIPLE

Bernoulli's principle states that for non-turbulent, laminar flow, the fluid's pressure increases with decreasing velocity. Conversely, regions of higher velocity have the lower pressure.

No such generalization can be made for turbulent flow, when the fluid motion is irregular, churning, and disorderly. Bernoulli's principle applies to laminar flow, when each layer of fluid flows smoothly alongside its neighboring layers.

Bernoulli's principle is illustrated in Fig. 1. A laminar flow of air enters the pipe at location 1, where the pipe's cross-sectional area is A_1. The pipe narrows at location 2 to cross-sectional area $A_2 < A_1$. Since the velocity v_2 is greater through region 2 than the velocity v_1 through region 1 (for reasons that will be explained below), the pressure P_2 at location 2 is less than the pressure P_1 at location 1. That is, since $v_2 > v_1$, $P_2 < P_1$. This relationship between P_2 and P_1 can be verified with a manometer whose ends connect locations 1 and 2.

The pressure differences accounted for in Bernoulli's principle are due to velocity changes only. This means the elevations of the two points in question are close enough together that their differences in static pressure due to gravity is negligible.

The physics underlying Bernoulli's principle, and the conditions under which it holds, are most easily approached through the work-energy theorem applied to a laminar, nonviscous, steady, irrotational, incompressible fluid flow. These restrictions mean that no energy is dissipated due to various kinds of friction. In Fig. 2, the fluid of density ρ initially fills the pipe between surfaces A and C. At location 1 at surface A, the pressure is P_1 and the pipe's cross-sectional area is A_1, giving a force P_1A_1 at A on the fluid, exerted to the right. Similarly, at location 2, there is a force at surface C acting to the left with magnitude P_2A_2. Suppose the force P_1A_1 is greater than the force P_2A_2, so that the fluid at surface A moves to B, and the surface C moves to D. The net effect is the transfer of mass $m = \rho A_1 x_1 = \rho A_2 x_2 \equiv \rho V$, where V is the volume of the fluid between AB and CD. Applying the work-energy theorem to m,

$$W_{nc} = \Delta(K + U), \qquad (1)$$

where W_{nc} is the work done on the system by non-conservative forces, and K and U are the system's kinetic and potential energies, respectively, one obtains

$$P_1A_1x_1 - P_2A_2x_2 = \left[\tfrac{1}{2}(\rho V)v_2^2 + (\rho V)gy_2\right]$$
$$- \left[\tfrac{1}{2}(\rho V)v_1^2 + (\rho V)gy_1\right], \quad (2)$$

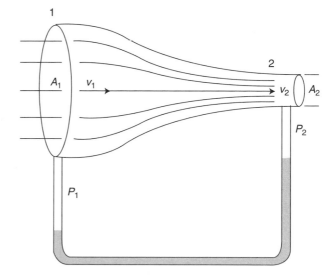

Figure 1 Illustrating Bernoulli's principle. Since the velocity at location 2 is greater than the velocity at location 1, the pressure at location 2 is less than the pressure at location 1.

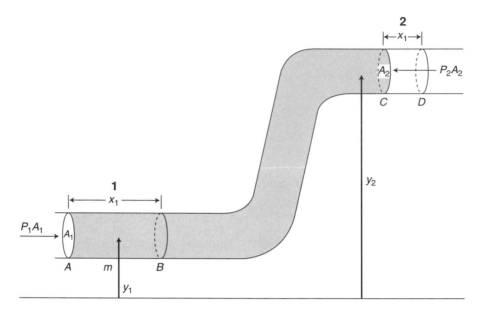

Figure 2 The system used to derive Bernoulli's equation, and from that Bernoulli's principle, via the work-energy theorem. The effect of the work $P_1A_1x_1 - P_2A_1x_2$ is in effect to move fluid AB (near location 1) to CD (near location 2), with the resulting change in kinetic plus potential energy, $\Delta[\frac{1}{2}(\rho V)v^2 + (\rho V)gy]$.

where v_2 and v_1 are the fluid's velocity at locations 2 and 1, respectively. Dividing out the volume V of m, we have Bernoulli's equation:

$$P_1 + \rho gy_1 + \tfrac{1}{2}\rho v_1^2 = P_2 + \rho gy_2 + \tfrac{1}{2}\rho v_2^2, \qquad (3)$$

which is conservation of mechanical energy (per unit volume),

$$P + \rho gy + \tfrac{1}{2}\rho v^2 = \text{const.} \qquad (4)$$

Bernoulli's principle follows at once from Bernoulli's equation. If the fluid density ρ and/or the difference in elevation of the two points in question are sufficiently small that $\rho g(y_2 - y_1)$ is negligible, then from Eq. (4),

$$P + \tfrac{1}{2}\rho v^2 = \text{const.}, \qquad (5)$$

is a mathematical expression of Bernoulli's principle: As the velocity increases, the pressure decreases; as the velocity decreases, the pressure increases. The slowest fluid has the greatest pressure. We now see that Bernoulli's principle is a consequence of the conservation of mechanical energy applied to an idealized fluid. Thus Bernoulli's principle holds, as an approximation, for real fluids that have viscosity and are compressible.

In deriving Eqs. (2), (3), and (4), it was assumed that there were no dissipative losses of energy due to viscosity and turbulence. It was also assumed that there were no sources or sinks of fluid between locations 1 and 2 of Fig. 2. This means that the amount of mass flowing between these locations is conserved. The amount of mass Δm moving with velocity v that flows through a pipe of cross-sectional area A in time Δt is $\Delta m = \rho Av\Delta t$. Thus when the fluid's density is constant, as is the case for an incompressible fluid,

$$A_1v_1 = A_2v_2. \qquad (6)$$

Equation (6) is the simplest version of the equation of continuity for fluid flow. It predicts that as the cross-sectional area decreases, the velocity increases. We can see why this is so: As the passage becomes narrower, the fluid must move faster to get the same number of kilograms per second through the pipe at the constriction as elsewhere. From Eq. (6) we see why the fluid's speed increases at a constriction, and from Bernoulli's principle, Eq. (5), we see why the pressure decreases there as a result. The increased wind speeds in hallways, doorways, and mountain passes are examples of applications of Eq. (6).

An important example of the application of Bernoulli's principle is the carburetor, which is used on many gasoline engines. The air flows into the engine through the large throat of the carburetor (location 1 of Fig. 1), but the throat narrows (location 2 of Fig. 1), so the air speeds up and the pressure drops. At the constriction, gasoline is allowed to enter the carburetor. Since the gasoline reservoir is at atmospheric pressure but the carburetor's constriction has a lower pressure, gasoline is pushed into the carburetor at the constriction, where it then mixes with air to form the combustible gasoline-air mixture.

Another familiar example of Bernoulli's principle in action is the "ballooning" of the fabric convertible top on an automobile traveling at highway speeds with the top up and the windows closed. In the reference frame of the car, the air outside flows over the top at a high speed, while the air inside the car is at rest. Thus the air outside has a much lower pressure than the air inside, so the fabric convertible top is ballooned out.

Bernoulli's principle also provides a simple explanation of aircraft lift. The wing shape and its "angle of attack" (i.e., how the wing is tipped relative to the flow of air) are designed such that the air flowing *over* the wing moves faster than the air *under* the wing. The air pressure beneath the wing is therefore greater than the air pressure above it. This pressure difference is responsible for the lift.

See also: AERODYNAMIC LIFT; FLUID DYNAMICS; FLUID STATICS; PRESSURE; PRESSURE, ATMOSPHERIC; TURBULENCE; TURBULENT FLOW; VISCOSITY

Bibliography

ACHESON, D. J. *Elementary Fluid Dynamics* (Clarendon Press, Oxford, Eng., 1990).

FEYNMAN, R. P.; LEIGHTON, R. B.; and SANDS, M. *The Feynman Lectures on Physics,* Vol. 2 (Addison-Wesley, Reading, MA, 1963).

HALLIDAY, D.; RESNICK, R.; and KRANE, K. S. *Physics,* 4th ed. (Wiley, New York, 1992).

SERWAY, R. A., *Physics for Scientists and Engineers with Modern Physics,* 3rd ed. (Saunders, Fort Worth, TX, 1990).

SHEPHERD, D. G. *Elements of Fluid Dynamics* (Harcourt Brace & World, New York, 1965).

TIPLER, P. A. *Physics,* 2nd ed. (Worth, New York, 1982).

DWIGHT E. NEUENSCHWANDER

BETA DECAY

See DECAY, BETA

BIG BANG THEORY

The term "big bang" is used to describe cosmological models for the origin and evolution of the universe in which it emerged from a state of high density a finite time in the past and evolved to the presently observed universe as the temperature and mass-energy density decreased in a universal expansion.

The modern idea of the big bang cosmology was pioneered in the 1920s by the Russian physicist Aleksander Friedmann and the Belgian physicist Georges Lemaître. Earlier investigations by Albert Einstein, Wilhelm de Sitter, and others into the application of general relativity to cosmology had demonstrated that a homogeneous (the same at every point), isotropic (the same in every direction) universe filled with matter could not be static; it must either expand or contract. While others tried to obtain static solutions by introducing other forms of mass-energy density (e.g., a cosmological constant), Friedmann and Lemaître worked out the dynamics of the expanding universe. After Edwin Hubble's 1929 discovery of the expansion of the universe, the cosmological constant was dropped in the standard big bang cosmology.

The Cosmological Principle and the Evolution of the Universe

The standard big bang theory is based upon the cosmological principle, which states that on a sufficiently large scale the universe is homogeneous and isotropic. Although at any instant of cosmic time the universe appears the same for any observer in the universe, the universe may evolve in time and still be consistent with the cosmological principle.

The cosmological principle restricts the possible geometry of the universe. Spaces consistent with the cosmological principle are unbounded (have no edge) and have no center. There are three possibilities for the local geometry of such spaces. The space may be flat, with no spatial curvature, positively curved, with spatial curvature like that of the surface of a sphere, or negatively curved, like a hyperbola.

The best way to visualize the expansion of the universe is to imagine a coordinate grid with fixed relative spacing, but with the overall scale increasing in time. For example, imagine two galaxies in flat space fixed at coordinates (x_1, y_1, z_1) and (x_2, y_2, z_2). In the expansion of the universe the galaxies would remain at fixed coordinates, and the distance between them would change according to a scale factor, $a(t)$, as

$$D^2 = a^2(t)\left[(x_1 - x_2)^2 + (y_1 - y_2)^2 + (z_1 - z_2)^2\right]. \quad (1)$$

The dynamics of $a(t)$ are found from the Einstein field equations once the form of the mass-energy density of the universe is specified. The simplest possibility consistent with the cosmological principle is that of a fluid with a time-dependent mass-energy density ρ and a time-dependent pressure p. One of the Einstein field equations results in the Friedmann equation

$$\frac{1}{a^2}\left(\frac{da}{dt}\right)^2 + \frac{k}{a^2} = \frac{8\pi G}{3}\,\rho, \quad (2)$$

where G is the gravitational constant, and $k = +1$ for the spherical space, -1 for the hyperbolic space, and 0 for flat space.

With the assumption of a relationship between the energy density and pressure, the rate of decrease of the energy density during expansion can be found. For instance, for a "matter-dominated" universe ($p \ll \rho$), $\rho(a)$ scales as a^{-3}, while for a "radiation-dominated" universe ($p = \rho/3$), $\rho(a)$ scales as a^{-4}. The difference in scaling guarantees that, in an expanding universe containing both matter and radiation, the early universe will be radiation dominated and the late universe will be matter dominated. Once $\rho(a)$ is known, the solution for $a(t)$ follows from the Friedmann equation.

The evolution of the scale factor as a function of time is shown in Fig. 1. All solutions start from $a = 0$ at $t = 0$ in a state of infinite density, and increase with time. If $k = +1$ (a closed universe), the scale factor reaches a maximum and then re-collapses. If $k = -1$ (an open universe) or $k = 0$ (a flat universe), the expansion is eternal.

The ultimate fate of the universe is determined by the ratio of the mean mass density of the universe to a quantity known as the critical density. If the expansion rate of the universe is denoted as $H \equiv \dot{a}/a$, then the critical density is $\rho_C = 3H^2/8\pi G$, and from the Friedmann equation we can see that the sign of k is related to the ratio of the mass density of the universe to the critical density. The present value of this important ratio is usually denoted by Ω_0. If $\Omega_0 > 1$, then $k = 1$ and the evolution will be that of a closed universe, while if $\Omega_0 < 1$, then $k = -1$ and the evolution will be that of an open universe. If Ω_0 is exactly unity, the spatial curvature vanishes. Determinations of the present expansion rate and the present average mass density are not sufficiently accurate to establish whether Ω_0 is larger or smaller than unity. A generous range of uncertainty would be $0.1 \leq \Omega_0 \leq 2$.

The Redshift, Hubble's Law, and the Age of the Universe

One of the implications of the big bang is the redshift of the light from distant sources due to the expansion of the universe. (Light would be shifted toward the blue in a contracting universe.) As follows from the generalization of conservation of en-

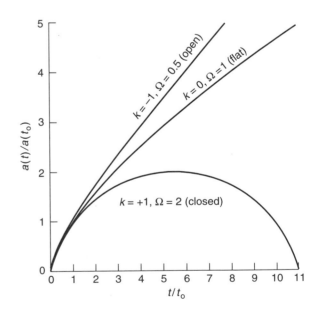

Figure 1 The evolution of the scale factor $a(t)$ as a function of time in closed, flat, and open models. The time is normalized to the present age of the universe in the different models, and the scale factor is normalized to its value today.

Table 1 A Brief History of the Universe

Epoch	Approximate Age of Universe	Temperature	Redshift $1 + z = a(t_0)/a(t)$
Today	14,000,000,000 yr	2.7 K	1.0
Earth forms	9,400,000,000 yr	3.6 K	1.3
Distant quasars form	1,300,000,000 yr	13.6 K	5.0
Formation of atoms	300,000 yr	3,000 K	1,100
BBN ends	3 min	1,000,000,000 K	400,000,000
BBN begins	1 s	10,000,000,000 K	4,000,000,000

ergy, if a photon of wavelength λ_1 is emitted from a source at time t_1, it will be detected at time t_0 with a wavelength λ_0 given by $\lambda_0/\lambda_1 = a(t_0)/a(t_1) \equiv 1 + z$. If a increases in time, then the photon will suffer a redshift as the universe expands. The factor z is known as the redshift. For small redshift ($z < 1$) the relation between redshift and distance is predicted to be linear, given by what is now known as Hubble's law:

$$cz = c(\lambda_0 - \lambda_1)/\lambda_1 = H_0 D, \qquad (3)$$

where D is the distance to the source, and H_0 is a constant known as Hubble's constant. The first prediction of a redshift seems to have been made in a 1917 paper by de Sitter, who unfortunately referred to the redshift as "spurious." In 1923 Hermann Weyl demonstrated that the redshift should increase in proportion to the distance. Hubble's law is only an approximation expected to be valid for $z < 1$. As $z \to 1$ there are a host of effects that lead to a departure from a linear distance–redshift relation.

Although the redshift of the light from distant galaxies was first reported in 1914 by Vesto M. Slipher, not much progress on the issue was made until Hubble was able to establish an extragalactic distance scale. Using the methods he pioneered to establish the distance to galaxies, and using values for the redshifts measured by Slipher and by Hubble and Humason, in 1929 Hubble published observational evidence for the linear dependence of redshift and distance.

The most important cosmological parameter is Hubble's constant, the present value of the expansion rate. The uncertainty in establishing reliable distance indicators to distant galaxies translates into an uncertainty in the Hubble constant, which is usually expressed as $H_0 = 100h$ km·s^{-1}·Mpc^{-1}, where the dimensionless parameter h is usually assumed to be in the range $0.4 < h < 1$, and 1 Mpc (megaparsec) is 3.0856×10^{24} cm. The traditional units of Hubble's constant ($H_0 = 100h$ km·s^{-1}·Mpc^{-1}) are chosen for convenience when working with the redshift in terms of an equivalent Doppler recessional velocity ($v_R = cz$) expressed in km·s^{-1} and the distance in terms of Mpc. Expressed in its "natural" units of inverse time or inverse distance, Hubble's constant is $H_0^{-1} = 9.778 \times 10^9 h^{-1}$ yr, and $cH_0^{-1} = 2998h^{-1}$ Mpc.

Hubble's constant sets the scale for cosmological distances and ages. It is typically the case for very distant objects that the redshift can be determined, but not the distance. Therefore the distance is expressed in terms of the Hubble constant as $D = zH_0^{-1} = 2998h^{-1}z$ Mpc.

The present age of the universe is related to Hubble's constant by $t_0 = H_0^{-1}F(\Omega_0)$, where $F(\Omega_0)$ is a monotonically increasing function of Ω_0, with $F(0) = 1$ and $F(1) = 2/3$ for a matter-dominated universe. For a flat matter-dominated model, the predicted age of the universe, $t_0 = 2H_0^{-1}/3 = 6.52 \times 10^9 h^{-1}$ yr, is consistent with the ages of the oldest stars for H_0 toward the small end of the observational range.

The Microwave Background Radiation and Primordial Nucleosynthesis

The original investigation of Friedmann assumed a matter equation of state. Lemaître also considered the possibility of a radiation-dominated universe. Because of the cosmological redshift, the energy of any photon will decrease in the expansion, and the radiation temperature will decrease as $T(t_0)/T(t_1) = a(t_1)/a(t_0)$. If the early universe was dense enough to establish a thermal distribution of radiation, then there should be a thermal background today with the temperature redshifted from its value in the early universe.

While the prediction of a blackbody cosmic background is an inherent feature of any big bang model, until the 1950s no one was able to predict the present temperature of the radiation. In a series of papers in the early 1950s, Ralph Alpher, Robert Herman, and George Gamow predicted the temperature of the background radiation on the basis of the elements produced in the early universe.

If the universe was ever hotter than about 10^{10}K, then there should have been sufficient thermal energy for nuclear reactions to occur. In nuclear statistical equilibrium at high temperatures the universe consisted only of neutrons and protons in roughly equal number. As the universe cooled, elements were built out of the primordial neutrons and protons. The final abundances of nuclei produced in big bang nucleosynthesis (BBN) depended sensitively upon the ratio of the number density of nucleons to the number density of photons. This ratio is very nearly constant from the time of nucleosynthesis until today, so using the ratio necessary to obtain the correct mix of primordial elements gives the present ratio of the density of nucleons to photons. Coupled with estimates of the mean nucleon density in the universe today, this allows a determination of the present photon density, hence the temperature of the background radiation.

Although the rates for many of the important nuclear reactions were poorly known at the time, in the early 1950s Alpher, Herman, and Gamow predicted a present background temperature of between 1 and 28 K.

In 1964 the background radiation was detected in the famous experiment of Arno Penzias and Robert Wilson with the temperature determined to be around 3 K. Recently the far-infrared radiation absolute spectrophotometer on the cosmic background explorer satellite measured the temperature and spectrum of the background radiation in a frequency interval around the peak of the spectrum that contained about 90 percent of the total energy and found a perfect blackbody spectrum to an accuracy better than 0.03 percent of the peak intensity, and a temperature of $T = 2.726 \pm +0.01$ K.

In the years since the idea of a primordial origin of the light elements was proposed, the predictions of BBN have improved, along with the development of an understanding of how to infer the primordial abundances from present-day observations. It is now believed that hydrogen, deuterium, helium-3, helium-4, and lithium-7 were produced in BBN.

Fundamental Predictions of the Big Bang

The standard big bang model is based upon the principle that on large scales the observable universe is homogeneous and isotropic. The strongest evidence for this is the high degree of isotropy of the cosmic microwave background radiation, where temperature fluctuations have been measured to be less than a few parts in 100,000. Although the distribution of galaxies is not completely uniform, particularly on scales less than about 10 Mpc, where their location in space is correlated with the location of other galaxies, recent large three-dimensional surveys of galaxies have confirmed that the distribution of matter in the universe becomes smoother on larger scales.

The expansion of the universe is another fundamental prediction of the big bang. That galaxies out to z about unity obey Hubble's law is clear. From the observed Hubble expansion, one is confident that the big bang model was a good description back at least as far as $z \sim 1$, or when the universe was about 35 percent of its present age. Less certain is the observational determination of Hubble's constant as well as other cosmological parameters such as Ω_0.

After the Hubble expansion, the strongest evidence for the big bang comes from the existence, isotropy, and spectrum of the cosmic background radiation. Since the universe today is transparent to the microwave radiation, in order to establish a thermal spectrum it must have once been hot enough to ionize the universe. In the standard big bang model this occurred about 300,000 years after the bang.

The fact that the inferred abundances of hydrogen, deuterium, helium-3, helium-4, and lithium-7 seem to agree with predictions of BBN suggests that the big bang picture can be extrapolated back to when the universe was hot enough for nuclear reaction to occur, about one second into the bang.

Whether the big bang model is the final word or not, it seems to be a fundamental description of the universe that will be encompassed by any larger cosmological world view.

See also: COSMIC BACKGROUND EXPLORER SATELLITE; COSMIC MICROWAVE BACKGROUND RADIATION; COSMOLOGICAL CONSTANT; COSMOLOGICAL PRINCIPLE; COS-

MOLOGY; COSMOLOGY, INFLATIONARY; HUBBLE, EDWIN POWELL; HUBBLE CONSTANT; NUCLEOSYNTHESIS; REDSHIFT; UNIVERSE; UNIVERSE, EXPANSION OF

Bibliography

BERNSTEIN, J., and FEINBERG, G. *Cosmological Constants* (Columbia University Press, New York, 1986).

KOLB, E. W., and TURNER, M. S. *The Early Universe* (Addison-Wesley, Reading, MA, 1990).

PEEBLES, P. J. E. *Principles of Physical Cosmology* (Princeton University Press, Princeton, NJ, 1993).

ROWAN-ROBINSON, M. *The Cosmological Distance Ladder* (W. H. Freeman, New York, 1981).

SILK, J. *The Big Bang* (W. H. Freeman, New York, 1989).

WEINBERG, S. *The First Three Minutes* (Basic Books, New York, 1977).

EDWARD W. KOLB

BINARY PULSAR

See PULSAR, BINARY

BINDING ENERGY

See NUCLEAR BINDING ENERGY

BIOPHYSICS

One of the most important phenomena in nature is life. Because of the complexity of life, however, no single branch of science has been able to provide complete understanding of the processes that are involved in living organisms. In an attempt to unify the various branches of science to solve the mysteries of life, the study of biophysics developed.

What is Biophysics?

In 1945 Erwin Schrödinger published *What is Life?*, an important book asserting that not only physics but also biochemistry was essential to answer the question posed by the book's title. Since that time, biophysics has come to be understood as a broad interdisciplinary area encompassing such fields as biology, physics, chemistry, and mathematics. Thus it is a very rich part of modern science and offers tremendous opportunities for basic and applied research.

As biophysics came to be seen as an independent branch of science, the isolated manner of research characterized by models, theories, and techniques to understand biology and life sciences gave way to a more integrated approach. This resulted in new curricula for the training of biophysicists at the undergraduate, graduate, and postdoctoral levels. In addition, the Biophysical Society was formed in the United States in the mid-1950s. Many journals, congresses, and symposiums dedicated to the ever-growing study of biophysics soon followed.

Molecular Biophysics

While biophysics has become a broad area of study covering many kinds of biophysics, this entry will focus on the trend toward and attempted understanding of the molecular basis of life. From this basis, one can eventually explain the behavior of other, larger components, such as cells, tissues, organs, and even entire organisms and ecosystems. While this sounds very promising in theory, it is not easy to understand the molecular basis of life. In fact, such an understanding has been achieved only for a few specialized situations. On the positive side, however, these situations have been so fundamental that they lend further support to the importance of studying and understanding the molecular basis of life.

In the case of genetics and evolution, for instance, the molecular basis was found to be related to the famous DNA molecule. This molecule contains all the ingredients of a chemical language to code for the genetic heritage of viruses and bacteria, as well as plants and mammals. The molecular mechanism for information transfer in nerves was studied. Electric nerve signals (action potentials) were found to be due to ionic transport in membranes and other molecular mechanisms that provide energy and information processing to neural systems. Molecular diseases have been identified and found to be related to specific defects in the molecules themselves. This is the case for sickle-cell anemia, in which a single error in the composition of the hemoglobin molecule leads to a serious systemic disease. Many other molecular diseases have been found, leading to the emergence of the field of molecular medicine.

This molecular approach in biophysics research also led to another revolution in science—biotechnology and genetic engineering. DNA itself has been modified or engineered to perform biological functions, some of which had never before existed.

Emergence of Biophysics

The pioneers of biophysics as a new interdisciplinary area of science looked beyond the established frontiers of physics to find new applications of physics in the study of biology. These scientists included individuals such as Hermann von Helmholtz and Albert von Nagyrapolt Szent-Györgyi. Helmholtz, a major driving force in this area, made contributions through experiments designed to investigate the mechanisms of sight and hearing, as well as in theories of color vision. Szent-Györgyi, who discovered vitamin C, was among the first to use molecular investigation. In particular, he studied how muscles contract. By mixing isolated biomolecules in a test tube and adding calcium ions, he was able to see the actual mechanical action of the fibers. Other important pioneers of molecular biophysics include Max Delbrück, Linus Pauling, James Watson, and Francis Crick.

Initially, the most important equipment used in the application of physics and chemistry to the study of biology was the simple microscope, which had been invented by Antoni van Leeuwenhoek in the seventeenth century and permitted cells and microscopic organisms to be seen and studied. A newer and more profound step was taking place at the end of the nineteenth century when Wilhelm Röntgen discovered the existence of x rays. This discovery permitted submicroscopic structures to be analyzed, including biomolecules and biopolymers (i.e., lipids, proteins, polynucleotides, and polysaccharides). The invention of the electron microscope in the early twentieth century permitted the study of cell membranes, tissues, and cell organelles never before seen in detail.

In addition to new experimental techniques, new areas of physics—nuclear and quantum physics—have been applied to the study of biology and medicine. The application of nuclear physics has led to nuclear and radiation biophysics, while the application of quantum physics has led to the creation of the electron microscope and quantum devices such as lasers and magnetic sensors, including the SQUID (superconducting quantum interference device) and the tunneling microscope.

The field of spectroscopy in physics has led to new areas in biophysics. Raman, electron-spin, and nuclear-spin resonance spectroscopies have been of enormous significance for the study of biology and biochemistry. Nuclear-spin resonance in particular paved the way for magnetic resonance imaging (MRI), an imaging technique that is of great importance to medical physics.

Another fundamental step in biophysics was the invention of the computer and the associated development of computer sciences. For research in molecular biophysics, computers are essential. They facilitate the study of biomolecular structures, and they simplify the extensive calculations necessary for modeling and imaging. They also provide the necessary technology to control most of the modern instrumentation that exists in a biophysics laboratory.

Frontiers of Biophysics

What was the origin of life itself? How does the brain function? Can artificial plants be made? Is it possible to catalog the entire library of genetic data for humans? What are the basic concepts underlying the self-organization of organisms? These are just a few of the questions that are driving researchers in the field of biophysics.

In examining the origins of life itself, scientists are trying to determine what were the first biomolecules to appear and what was the nature of the "molecular evolution" that produced present-day life processes. They want to understand why amino acids, the fundamental building blocks of proteins and other polypeptides, are all left-handed (i.e., their molecular structure is such that they always turn the plane of polarized light to the left). Biophysicists want to know if it is possible to reproduce in the laboratory the original prebiotic (pre-life) or postbiotic conditions.

The key to learning how the brain functions is an understanding of the neurons, cells that make up the circuitry of the brain, and how these cells are connected and communicate with each other. Concepts from information theory, statistical physics, molecular electronics, neural networks, artificial intelligence, and neural biochemistry are introducing new insights for biophysicists working in this area. However, there is more to the study of the brain than its mechanical functioning. Philosophical questions, such as what is the difference between mind and brain, are of great importance.

Biophysicists have been able to learn the basic processes involved in photosynthesis. They know that a light photon of a preferred wavelength is absorbed by a chlorophyll molecule and an electron is transferred to another molecule in a very short time. They also know that the photon energy is transformed into the electrical potential energy of the molecule and that chemical reactions transform this into chemical energy in the plant cell. Again, however, this knowledge is only the beginning. Biophysicists want to know if it is possible to create artificial plant forms. They are trying to identify the common processes between photosynthesis in plants and vision, where the photon energy is transformed into electric signals in the brain. They want to know if it is possible to make use of the sophisticated electronic processes occurring in vision and photosynthesis to build new biocircuits (biochips), initiating a new field of molecular electronics.

Biotechnology and genetic engineering are both important frontiers of biophysics. The topics in this area range from molecular medicine to the human genome project, already in existence as a megaproject with costs estimated in the billions of dollars. This project is of fundamental significance for solving the problems of, in particular, molecular genetics and, in general, molecular biophysics. The goal of the project is no less than obtaining the complete genetic data bank for humans. From these data, predictions can be made about the possible occurrence of and cures for many illnesses and molecular diseases.

Finally, there is the frontier to understand the basic complexity of life. Living systems are made up of an enormous number of atoms, molecules, organelles, and cells. This complexity nevertheless leads to simplicity if it is recognized that, despite the vast number of components, living processes evolve harmonically in time and space. This is certainly due to cooperative effects. A complete human being starts with just one fertilized cell, and once that individual has matured, billions of cells are formed and die each day. The main question in this area for biophysicists is whether it is possible to obtain the basic concepts underlying self-organization from the laws of statistical physics, thermodynamics, and chemistry. These scientists want to know how it is that a large biomolecule, such as a protein, remembers to fold exactly to its original form even when it is denatured by thermal or chemical means. They want to discover what induces bacterial colonies to organize themselves in almost perfect crystalline arrangements. Biophysicists acknowledge that life exists in a delicate equilibrium between order and disorder, but they are still looking for answers to how a living system finds its equilibrium point and how its complex mechanisms maintain this balance once it has been established.

Biophysics as a Career

The study of biophysics, with its vast opportunities for basic and applied research, presents excellent opportunities for ambitious individuals to embark on a career of scientific adventures. To assist in the development of students interested in biophysics and its challenging frontiers, many universities have established undergraduate, graduate, and postdoctoral biophysics programs.

See also: BODY, PHYSICS OF; CAT SCAN; HELMHOLTZ, HERMANN L. F. VON; MAGNETIC RESONANCE IMAGING; MEDICAL PHYSICS; MODELS AND THEORIES; MOLECULAR PHYSICS; NUCLEAR MEDICINE; RÖNTGEN, WILHELM CONRAD; SCANNING TUNNELING MICROSCOPE; SCHRÖDINGER, ERWIN; SPECTROSCOPY; VISION; X RAY, DISCOVERY OF

Bibliography

AUSTIN, R. H., and CHAN, S. *Biophysics for Physicists* (World Scientific, River Edge, NJ, 1992).

BLUMENFELD, L. A. *Problems of Biological Physics* (Springer-Verlag, New York, 1981).

CERDONIO, M. *Introductory Biophysics* (World Scientific, River Edge, NJ, 1986).

SCHRÖDINGER, E. *What is Life?* (Cambridge University Press, Cambridge, Eng., 1945).

SERGIO MASCARENHAS

JOHN R. CAMERON

BIOT–SAVART LAW

Magnetic fields are produced by electric currents. The Biot–Savart law tells us how to calculate the static magnetic field, as a function of position, that is produced by a steady current. Soon after Hans Christian Oersted discovered in 1820 that an electric current will deflect a nearby compass needle, Jean Baptiste Biot and Felix Savart confirmed Oersted's results, and measured the force exerted on a magnet by steady currents. From their data they were able to empirically construct the formula that now bears their name.

Since electric charges come in two types, positive and negative, by convention the direction of electric current is that which is equivalent to the motion of positive charge. Let I denote the positive-charge-flow-equivalent per unit time, the "conventional current." Our objective is to calculate the static magnetic field **B** (or "magnetic induction") as a function of position, given the steady conventional current that produces it. There are three things we must know in order to proceed.

First, what current distribution is the source of the magnetic field that we are to calculate? Let us visualize a steady conventional current I flowing along a thin wire bent into the shape of some path C (see Fig. 1). Second, where do we wish to know the local **B** field that is produced by this current? Let the location of this "field point" relative to the origin of our coordinate system be specified by the vector **r**. Third, what principles of physics enable us to calculate the magnetic field as a function of **r**; that is, how do we calculate **B**(**r**)? When the current I is steady, the required principles are the Biot–Savart law and the superposition principle.

The Biot–Savart law tells how to calculate the infinitesimal magnetic field $d\mathbf{B}(\mathbf{r})$ that is produced by an infinitesimal segment of the steady line current, a segment sufficiently small that it may be considered a point source of the field. Let this infinitesimal

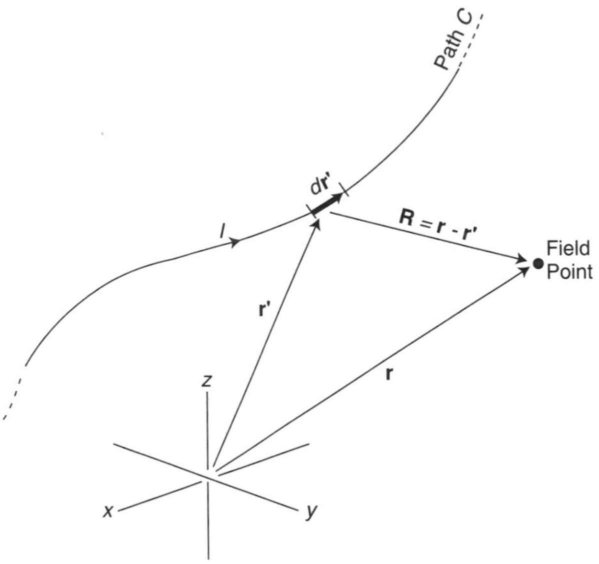

Figure 1 Relative to the coordinate system shown, the vector **r** locates the field point (i.e., the point in space where the field will be calculated). The vector **r**′ locates the source point (i.e., the point in space where the infinitesimal source of the field is located). For the magnetostatic field, that point source is an infinitesimal segment of the current I that flows along the path C. The vector $d\mathbf{r}'$ is tangent to the conventional current I at the source point. The vector $\mathbf{R} = \mathbf{r} - \mathbf{r}'$ is the vector *from* the source point *to* the field point.

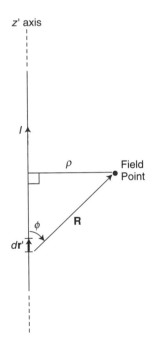

Figure 2 The infinite, straight line of conventional current I flows along the z' axis, from bottom to top in this figure. The magnetostatic field **B** that exists at some distance (ρ) from the line is to be calculated. The vector $d\mathbf{r}'$ is tangent to the infinitesimal point source current; **R** is the vector from the source point to the field point, and ϕ is the angle between $d\mathbf{r}'$ and **R**.

current element be located at the position $\mathbf{r}'$ in our coordinate system. (In this entry the primed coordinates refer to "source points," while unprimed coordinates refer to "field points.") According to Biot and Savart, the magnetic field in vacuum at $\mathbf{r}$, due to the infinitesimal segment of steady current located at $\mathbf{r}'$, is

$$dB(\mathbf{r}) = \frac{k_{\mathrm{m}} I \, (d\mathbf{r}' \times \mathbf{R})}{R^3}, \tag{1}$$

where $\mathbf{R} = \mathbf{r} - \mathbf{r}'$ is the vector from the source point to the field point, $d\mathbf{r}'$ is the infinitesimal displacement of the positive charges that comprise the steady conventional current at $\mathbf{r}'$ (thus $d\mathbf{r}'$ is tangent to C, and points in the direction of conventional current), and k_{m} is a constant whose numerical value depends on the units selected. In SI units, $k_{\mathrm{m}} = 10^{-7}$ N/A^2 *exactly* (chosen so that the ampere is defined in terms of magnetic force), and is often written as $\mu_0/4\pi$, where μ_0 is called the "permeability of free space." In SI units, the magnetic field is measured in tesla (T) or webers per square meter (Wb/m^2). Notice that the Biot–Savart law shares with Coulomb's law of electrostatics and Newton's law of gravitation the feature that the magnitude of the vector field produced by a point source diminishes as $1/R^2$. Unlike Coulomb's or Newton's field laws, however, where the direction of these fields are parallel or antiparallel to $\mathbf{R}$, the magnetic field is perpendicular to $\mathbf{R}$.

We calculate the total magnetic field $\mathbf{B}(\mathbf{r})$ produced by the entire current by applying the superposition principle, because the individual $d\mathbf{B}$ fields combine according to vector addition. Thus

$$\mathbf{B}(\mathbf{r}) = k_{\mathrm{m}} I \int_C \frac{d\mathbf{r}' \times \mathbf{R}}{R^3}, \tag{2}$$

where the integration is over the path C.

To illustrate the use of the Biot–Savart law, let us calculate the magnetostatic field produced by a straight, infinite line of current (see Fig. 2). Let a steady current I run along the z' axis from $-\infty$ to $+\infty$, and let us inquire about the field at an arbitrary point located the distance ρ from the z' axis. The vector $d\mathbf{r}'$ becomes $dz'\,\hat{\mathbf{k}}$, so upon evaluating the cross product, Eq. (1) gives

$$dB = \frac{k_{\mathrm{m}} I(\hat{\boldsymbol{\theta}}\, dz' \sin\phi)}{R^2} \tag{3a}$$

$$dB = \frac{k_{\mathrm{m}} I \hat{\boldsymbol{\theta}}\, dz'\, \rho}{(\rho^2 + z'^2)^{3/2}}, \tag{3b}$$

where $\hat{\boldsymbol{\theta}}$ is the polar unit vector tangent to a circle whose axis of symmmetry is the z' axis. Integrating over all source points, we obtain

$$\mathbf{B}(\rho) = \left(\frac{2k_{\mathrm{m}} I}{\rho}\right)\hat{\boldsymbol{\theta}}. \tag{4}$$

This magnetic field is sketched in Fig. 3.

If the conductor carrying the source current is now allowed to have finite cross section, in place of $I\, d\mathbf{r}'$ we may write the point source element as $\mathbf{j}(\mathbf{r}')\, d^3r'$, where $\mathbf{j}(\mathbf{r}')$ is the current density (i.e., the current per unit area) that exists at $\mathbf{r}'$, and d^3r' is the volume occupied by the current element. We may think of the finite-width conductor as a bundle of thin conductors, and again use the superposition of fields. The Biot–Savart law now takes the form

$$\mathbf{B}(\mathbf{r}) = k_{\mathrm{m}} \iiint_V \frac{d^3r'\, [\,\mathbf{j}(\mathbf{r}') \times \mathbf{R}]}{R^3}, \tag{5}$$

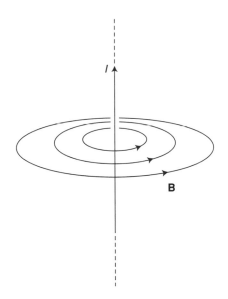

Figure 3 As a result of integrating the Biot–Savart law over an infinite straight line current, it is found that the field's magnitude diminishes as the inverse distance from the line. From the direction of $d\mathbf{B}$ given by the cross product in the Biot–Savart law, and from symmetry, the $\mathbf{B}$ field forms concentric circles around the inifinite current's axis.

where the integration is over the volume V occupied by the entire source current.

The Biot–Savart law is to magnetostatics what Coulomb's law is to electrostatics: the empirical starting point, from which the rest of the theory follows through mathematics and applications of the superposition principle. When we study magnetostatics and electrostatics as stand-alone disciplines, we must postulate the Biot–Savart law quite independent of Coulomb's law. But when we allow the electric and magnetic fields to be time-dependent, we see the Biot–Savart law in a greater context. For instance, if we solve Maxwell's equations for the time-dependent electric and magnetic fields produced by given charge and current distributions, the Biot–Savart law emerges as the only surviving magnetic field in the limiting case of steady currents. Conversely, if we take the Biot–Savart law as given, and require its components to be the space-space components of relativistic field equations, Maxwell's equations result; or if we accept Coulomb's law of electrostatics, then impose the requirements of the special theory of relativity, we anticipate the Biot–Savart law.

See also: AMPÈRE'S LAW; COULOMB'S LAW; ELECTROMAGNETISM; FIELD, ELECTRIC; FIELD, MAGNETIC; MAXWELL'S EQUATIONS; VECTOR

Bibliography

FEYNMAN, R. P.; LEIGHTON, R. B.; and SANDS, M. *The Feynman Lectures on Physics,* Vol. 2 (Addison-Wesley, Reading, MA, 1963).

GRIFFITHS, D. J. *Introduction to Electrodynamics,* 2nd ed. (Prentice Hall, Englewood Cliffs, NJ, 1989).

HALLIDAY, D.; RESNICK, R.; and KRANE, K. S. *Physics,* 4th ed. (Wiley, New York, 1992).

JACKSON, J. D. *Classical Electrodynamics,* 2nd ed. (Wiley, New York, 1975).

LORRAIN, P.; CORSON, D. R.; and LORRAIN, F. *Electromagnetic Fields and Waves* (W. H. Freeman, New York, 1988).

NEUENSCHWANDER, D. E., and TURNER, B. N. "Generalization of the Biot–Savart Law to Maxwell's Equations Using Special Relativity" *Am. J. Phys.* **60,** 35–38 (1992).

PANOFSKY, W. K. H., and PHILLIPS, M. *Classical Electricity and Magnetism,* 2nd ed. (Addison-Wesley, Reading, MA, 1962).

ROJANSKY, V. *Electromagnetic Fields and Waves* (Prentice Hall, Englewood Cliffs, NJ, 1971).

SERWAY, R. A. *Physics for Scientists and Engineers with Modern Physics,* 3rd ed. (Saunders, Fort Worth, TX, 1990).

TIPLER, P. A. *Physics,* 2nd ed. (Worth, New York, 1982).

VANDERLINDE, J. *Classical Electromagnetic Theory* (Wiley, New York, 1993).

DWIGHT E. NEUENSCHWANDER

BIREFRINGENCE

Most optical materials are isotropic; that is, they have a single index of refraction that is the same no matter which direction the light travels and the index is the same for any polarization of the light. Materials that are birefringent have an index of refraction, which depends upon both the direction of travel of light and the polarization of the light. This reflects the building blocks from which the crystal is made. Instead of being little cubes, as in isotropic materials (e.g., table salt), the building blocks are rectangular or hexagonal in shape, for example. Materials like glass and plastic are isotropic because they have no crystal structure and, hence have no special directions.

Let us start with the classic birefringent material, calcite (chemically $CaCO_3$). If we put a piece of calcite on a piece of paper that has a circle drawn on it, we see two circles. Rotate the calcite and notice that one image remains fixed while the other moves around it, as the crystal rotates.

If we now put a linear polarizer on top of the piece of calcite, we see two images, but as we rotate the polarizer, one image disappears. Continuing to rotate the polarizer, we find that the missing image reappears and the other one disappears. Figure 1 shows one dashed circle, but this is only to indicate that we are seeing two different images of the original circle. The two images appear the same except for their position. One finds that the polarizer needs to be rotated 90° to move from one image to the

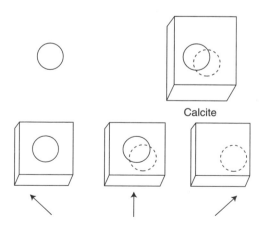

Calcite

Figure 1 Calcite with polarizer oriented as shown by the arrows.

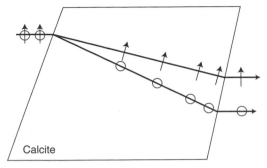

Calcite

○ electric field perpendicular to plane of page

↑ electric field in plane of page

Figure 2 Behavior of an unpolarized ray.

other, indicating that the two images are from the light of perpendicular linear polarizations.

Why are there two images? The last experiment above suggests that the calcite crystal treats the two different polarizations differently. The refraction index for one polarization is different from the refractive index for the other polarization. One index is called n_o, for what is called the ordinary ray and the other index is n_e, for the extraordinary ray. Thus, there are two different Snell's laws for a light ray entering a calcite crystal—one for the ordinary ray, which uses n_o, and one for the extraordinary ray, which uses n_e. Figure 2 shows an unpolarized ray.

Materials that are ordinarily isotropic become birefringent if stressed, a phenomena known as stress birefringence. Engineers use the effect to determine stresses in models. They make a plastic model of the part and then apply a force to the model to stress it while between crossed polarizers. The stress causes the plastic to become birefringent. This changes the polarization of the light passing through the model, which means the light is transmitted by the second polarizer. Try this with a piece of plastic wrap or other transparent plastic between crossed polarizers. The points where the stress is highest will show up as the brightest. For the engineer, these are places where the material is likely to break.

See also: POLARIZATION; REFRACTION; REFRACTION, INDEX OF; SNELL'S LAW; STRESS

Bibliography

JENKINS, F. A., and WHITE, H. E. *Fundamentals of Optics*, 4th ed. (McGraw-Hill, New York, 1976).

RALPH W. ALEXANDER JR.

BLACKBODY RADIATION

See RADIATION, BLACKBODY

BLACK HOLE

As early as 1795, Pierre-Simon Laplace noticed that there were situations in which a star would be invisible. The fundamental principle involves the escape velocity. Throw a rock into the air, and it falls back. Throw it upward faster, and it rises higher and falls back after a longer time. If a rock (or a spacecraft) is thrown from Earth at a speed exceeding 11.1 km/sec it will never fall back; even as it slows, it is moving to a place where gravity is weaker; and even as its speed keeps diminishing, the effect of gravity also decreases, and it will keep going forever. Since Earth is approximately spherical, we can use the simple formula for escape velocity v_{esc} from the surface of a spherical gravitating body of mass M:

$$v_{esc} = \sqrt{\frac{2GM}{R}}, \tag{1}$$

where G is Newton's gravitational constant and R is the radius of the body. What Laplace noted is that as we add more mass to the central body (without changing R), or if we keep M constant while squeezing the object (i.e., reducing R), the escape velocity will increase. This is reasonable because the gravitational field gets stronger in either case. If v_{esc}, which is the minimum speed to escape from the body, exceeds the speed of light (conventionally called c), then, as Laplace noted, it becomes "black" because no light can escape it. This is the description using Newtonian gravity theory of what is more completely described in Albert Einstein's general theory of relativity: a black hole (the name was invented by John Wheeler in 1968).

If we insert the speed c for v_{esc} and rearrange, we obtain the standard expression for the circumference of a black hole:

$$2\pi R = 2\pi \cdot \frac{2GM}{c^2} \tag{2}$$

This expression holds even for the general relativistic spherical black hole. One of the basic tenets of Einstein's relativity is that nothing can travel faster than light. Thus, if light cannot escape, nothing can; and thus this is a hole—things fall in, but they cannot get back out.

For the Sun, the scale defined by Eq. (2) gives $R = 3$ km. Compressing the entire mass of the Sun into such a small region will clearly lead to a very high density configuration; thousands of times denser than nuclear matter, in fact. But black hole formation does not require high densities. More massive black holes form at lower densities, and a black hole of the mass of the Galaxy would have a density comparable to air. The required density to form a black hole scales like M^{-2}, where M is the total mass of the black hole. Very small black holes thus require very large compressions, but large, massive ones can collapse from very ordinary densities.

If the black hole has a small enough mass, about 10^{-5} grams, then its effective density is of the order 10^{99} gm/cm^3. The 10^{-5} g is given a special name, the Planck mass. Such a black hole is dominated by quantum effects and requires for its description a quantum theory of gravity. Such a theory has not been found yet. Although Einstein's equations appear correctly to describe large-scale gravitating objects like planetary systems, large black holes (as occur from the collapse of stars), and even the universe itself, many experts believe there will have to be a modification in the Einstein equations before a successful quantum theory can be written.

In 1974 Stephen Hawking demonstrated that there is a remnant of quantum behavior even for black holes that are much larger than the 10^{-5} gm quantum black hole. This manifests itself by allowing black holes to emit radiation, which eventually carries off their mass. In other words, black holes undergo Hawking evaporation, giving off radiation with a temperature $T = \hbar/8\pi k(GM/c^3)$. Here the presence of the Planck constant $\hbar = 1.05 \times 10^{-27}$ gm cm^2/sec is the signal that we have a quantum process. (Note that k is Boltzman's constant, $k = 1.38 \times 10^{-16}$ergs/K, which is essentially just a conversion factor between energy and temperature.) For the Sun, the Hawking temperature is remarkably low: $T \sim 10^{-7}$ K. The lifetime of a solar mass black hole that undergoes evaporation by this effect is 10^{64} years, which is much greater than the age of the universe. Because the smaller the mass, the higher the temperature of a black hole, small ones evaporate in shorter times. A black hole the mass of an asteroid ($\sim 10^{19}$ gms) would have a lifetime of 10^{10} years, about the same as the age of the universe. A Planck mass black hole, if it could be described by the approximate Hawking method, would live a remarkably short 10^{-43} s. Such a black hole requires the full quantum description.

Large black holes, even though they have very small Hawking evaporation, still can produce very interesting physical effects. In astrophysics, such black holes seem to be the source of energy for bright quasi-stellar-objects, and for many x-ray sources in the sky. In these situations one believes there is an accretion disk of matter orbiting the black hole, held in by its strong gravitational field. The disk itself may be no larger than our solar system. As matter orbits the black hole, some kind of friction or dissipation slows the matter on the innermost orbit, allowing it to fall into the black hole. This dissipation heats the rest of the orbiting matter, causing it to give off energy (including in many cases x rays) that we can detect.

An isolated black hole in space would be a fascinating object. By definition, nothing can escape. A person who falls in can never again communicate with the outside. Fortunately we can understand what goes on inside without experiencing it. As an object, or a person, falls into the black hole, first a tidal force is felt. There is nothing unusual about tidal forces; they are a feature of orbits around any gravitating body. They arise because part of the object is closer to the center and thus more strongly affected by the gravitational field. Fall in feet first, and your feet will pull away from your head. You will be stretched in the direction of your feet. At the same time each of your hands is moving toward the center and must be getting closer together. Thus you are compressed in the transverse direction. These tidal forces affect (very weakly) a skydiver falling freely from a plane, or an astronaut in orbit. Near Earth these tidal forces are never very large. But near a black hole they could be large; for instance, large enough to pull a person apart, long before crossing inside a solar-mass black hole. On the other hand, for a larger, more massive black hole these forces would be milder, and so a human observer would be able to cross inside such a larger black hole.

Once inside a black hole these tidal effects become stronger and stronger. One can prove that so long as the electric charge of the black hole is negligible, there are only two kinds of black hole: a nonrotating, Schwarzschild black hole, and its rotating version, the Kerr black hole. The nonrotating black hole possesses an unavoidable singularity in the future. The rotating kind may provide a way to avoid the singularity. However, there is pretty good evidence suggesting that the disturbance caused by dropping someone, or something, into the rotating hole gets amplified up, to guarantee a singularity there too.

Black holes distort time and space. Einstein's general relativity describes gravity as a distortion of time and space, and these effects predominate near a black hole. For instance, a person outside of the hole would see the diving astronaut "freeze" at the surface of the hole, taking apparently an infinite time to cross, and fading from view as the light emitted by the astronaut reddened. So in a real sense we never see someone fall into the hole. To the diver, however, there is nothing special about the black hole surface. Since the light from exterior objects can follow him in, he sees nothing special at the horizon, though things outside look distorted into a small patch of the sky outside. But no one from outside can watch him after he crosses this event horizon. He sees a definite last image from outside just as the singularity destroys him.

The black hole horizon means that the singularity and the death of the diver are hidden from outside. Every mathematical description of a black hole has the property that the horizon shrouds the singularity. Roger Penrose has thus postulated the cosmic censorship hypothesis, which holds that every gravitational singularity is hidden behind an event horizon.

Computational techniques have been used to study the very complex equations describing interacting black holes. The collision of such black holes would be expected to form very strong gravitational radiation, which may be detectable by terrestrial detectors. Other computational studies focus on the field of black hole critical phenomena, which were discovered by Matthew Choptuik in 1993. He found that in the formation of a black hole there may occur repeated increasingly fine-scale oscillations. This happens just at the threshold of gravitational fields strong enough to form a black hole.

In his book *Geometrodynamics* (1962), John Wheeler proposed a mechanism to make a black hole. It involves imploding a ball of iron, by a surrounding sphere of deuterium bombs (using all the deuterium in the ocean). The exploding deuterium would compress the iron to the point that it collapses to a black hole. Although there is as yet no incontrovertible

evidence, scientists still hope to find definite proof of the existence of black holes without having to resort to such a black hole bomb.

See also: BLACK HOLE, KERR; BLACK HOLE, SCHWARZCHILD; ESCAPE VELOCITY; LAPLACE, PIERRE-SIMON; PLANCK MASS; RELATIVITY, GENERAL THEORY OF

Bibliography

D'INVERNO, R. *Introducing Einstein's Relativity* (Clarendon Press, Oxford, Eng., 1992).

HAWKING, S. W. *A Brief History of Time: From the Big Bang to Black Holes* (Bantam, London, 1988).

LAPLACE, P.-S. "Proof of the Theorem, That the Attractive Force of a Heavenly Body Could Be So Large That Light Would Not Flow Out of It (1799)," in *The Large Scale Structure of Space-Time,* edited by G. F. ELLIS and S. W. HAWKING (Cambridge University Press, Cambridge, Eng., 1973).

SCHUTZ, B. *A First Course in General Relativity* (Cambridge University Press, Cambridge, Eng., 1985).

WHEELER, J. A. *Geometrodynamics* (Academic Press, New York, 1962).

RICHARD A. MATZNER

BLACK HOLE, KERR

A Kerr black hole is a rotating, nonspherical black hole. Black holes are objects so strongly gravitating that their escape velocity exceeds the speed of light, so nothing can escape from them. According to Albert Einstein's theory of general relativity (currently the best description of gravity), a black hole is described as a warping of space, but despite this totaly different description, a spherical nonrotating black hole (i.e., a Schwarzschild black hole) has properties very similar to the Newtonian spherical description. In Newtonian theory the gravitational field is described by only the mass distribution of the source. For instance, one could imagine a rotating body that is "designed" in such a way that it is exactly spherical when rotating at a particular angular rate ω. Such a body would have a spherical gravitational field in the Newtonian description.

General relativity, in contrast to Newtonian gravity, assigns a source strength to the motion (for instance, of rotation) of the mass. Hence a spinning object as described by general relativity has novel elements. The Kerr solution (found by Roy Kerr in 1964) is the black hole for a rotating collapsed object. It depends on the mass M of the black hole, and on its angular momentum per unit mass, but on nothing else. Mathematical theorems have been proven which state that giving the mass, angular momentum, and electrical charge completely specifies the gravitational properties of the black hole. In astrophysical settings the electrical charge is essentially zero. A Kerr black hole with zero angular momentum is the same as a Schwarzschild black hole. Most astrophysicists expect that a large fraction of black holes will be found in binary systems, and in that case matter transfer from the stellar partner to the black hole will be via an accretion disk, so angular momentum will fall on the black hole, meaning that we expect astrophysical black holes to be of the Kerr type.

Kerr black holes differ from spherical Schwarzschild ones because of the strong influence the rotation has both outside and inside the hole. Outside the hole, the phenomenon of frame dragging means that particles, observers, photons of light, and so on are preferentially swept along in the direction of the rotation of the black hole. This phenomenon happens on a much weaker scale even for bodies like Earth, and satellite experiments have been proposed to test it. Close enough to a Kerr black hole (but still outside it), however, there is a region, the ergosphere, where an observer or photon must rotate with the hole; it is impossible to stay still with respect to distant stars. The Penrose process is a way to extract rotational energy from the black hole by splitting particles orbiting in the ergosphere, throwing part into the hole in a way that opposes this rotation, and shooting the other part out to large distances, where it has excess energy due to this process.

The structure inside a Kerr black hole is dramatically different from that of a Schwarzschild black hole. In a Schwarzschild black hole anything (or anyone) falling in gets crushed by the singularity. In Kerr, it is possible to find paths that avoid being crushed by a singularity, and which can reemerge into another universe surrounding the Kerr hole. However, perturbations, such as the gravitational radiation from the object moving through, seem to cause a secondary singularity to appear. There has been a substantial amount of study on whether exotic quantum mechanical effects can be used to overcome this defect. If one could "duck through the wormhole" one might be able to travel to distant locations very quickly, and even produce time travel. The conditions to achieve this would in any case be extreme, but one can consider very advanced civilizations for which this would be a practical question.

See also: BLACK HOLE; BLACK HOLE, SCHWARZSCHILD; NEWTONIAN MECHANICS; PERTURBATION THEORY; RELATIVITY, GENERAL THEORY OF; TIME

Bibliography

D'INVERNO, R. *Introducing Einstein's Relativity* (Clarendon Press, Oxford, Eng., 1992).

SCHUTZ, B. F. *A First Course in General Relativity* (Cambridge University Press, Cambridge, Eng., 1984).

THORNE, K. S. *Black Holes and Time Warps: Einstein's Outrageous Legacy* (W. W. Norton, New York, 1994).

RICHARD A. MATZNER

BLACK HOLE, SCHWARZSCHILD

A black hole is a region from which no light (nor any other object or signal) can escape. Black holes were predicted by Pierre-Simon Laplace, but the first demonstration of their existence within Albert Einstein's description of gravity was by Karl Schwarzschild in 1916. A Schwarzschild black hole is a particular general relativistic description of distorted spacetime that is exactly spherically symmetric. Exact sphericity may seem difficult to achieve in an astrophysical setting, but there are theorems based on the mathematical structure of general relativity stating that if we assume black holes have zero electric charge, stationary black holes are either exactly spherical, and nonrotating, or if they are rotating they have a definite axi-symmetric shape that depends on their angular momentum per unit mass. Thus the two parameters, mass of the black hole and angular momentum per unit mass, completely specify which kind of stationary black hole emerges from a stellar collapse. An object (star) that collapses with precisely zero spin eventually becomes a Schwarzschild black hole. The dynamical features of its formation that might suggest deviations from a spherical shape are lost by being radiated away through gravitational radiation.

Schwarzschild's original solution was found by looking for a static, spherical solution to Einstein's equations describing general relativity. Mathematically this is straightforward because the solution he found required solving only the same equation as in Newtonian theory. But the results have a different interpretation. The full Schwarzschild solution can be written down in terms of its metric. The metric gives the way distances ds are related to changes in coordinates (position labels) in the spacetime:

$$ds^2 = -c^2\left(1 - \frac{2GM}{c^2 r}\right)dt^2 + \frac{dr^2}{1 - 2GM/c^2 r}$$

$$+ \, r^2(d\theta^2 + \sin^2\theta d\phi^2).$$

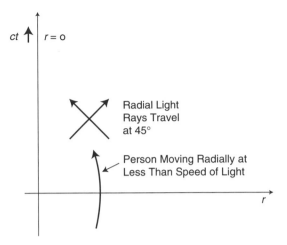

Figure 1 Simple illustration of spacetime.

The Schwarzschild metric involves the physical constants c, the speed of light; and G, the Newtonian gravitational constant, as well as M, the mass of the object. The part involving the angles θ, ϕ gives the rule for measuring distances on the surface of the sphere and is the same as for an ordinary sphere of circumference $2\pi r$. The radial distance is distorted, however, A small change dr in the radial coordinate can become a large proper distance ds when $2GM/c^2 r$ approaches unity because of the denominator associated with the dr^2 term. Similarly, large changes dt in time coordinate t lead to a small proper time change when r approaches the Schwarzschild radius $R_s = 2GM/c^2$. Notice that in general relativity one typically finds the reversal of sign between ds^2 and dt^2 as seen here. The surface $r = 2GM/c^2$ is the location of the black hole surface, also called the horizon because we cannot see beyond it.

The stretching of time associated with the horizon leads to a redshift of events near there, as seen by distant observers. Hence, for instance, it takes on infinite amount of coordinate time as viewed by a distant observer for an object to drop from a finite r down to R_s. This observer says the object never crosses the horizon. But it can be shown that an object or person falling into the black hole will cross the horizon in a perfectly finite amount of local, proper time and will hit the Schwarzschild singularity a short time later.

Only in 1957 did M. D. Kruskal and G. Szekeres provide a way of writing the metric for a Schwarzschild black hole that allows easy extension across $r = R_s$. With the help of their description we are now aware of the full spacetime picture describing a Schwarzschild black hole. In this approach one takes an r–t description of the full spacetime, suppressing the angular coordinates θ, φ. Then if there are no gravitational fields (= flat spacetime) one has a simple figure to describe the whole spacetime (see Fig. 1) The vertical line represents radius = 0, and there is no content to the left of this vertical line. We plot time multiplied by the speed of light (c) vertically because this allows a simple description of light rays—they move at 45° lines in the figure. Because nothing can go faster than light, all physical objects evolve along lines making angles less than 45° to the "vertical" time direction. A nonmoving object is represented by just a straight line.

The corresponding black hole picture is rather more complicated. In Fig. 2 the 45° diagonal rays are the horizon(s) of the black hole. The jagged lines to the future and past are singularities located at $r = 0$. We live outside the black hole, in our universe. Because we cannot move faster than light, we cannot in fact travel from our universe through the black hole to another universe. Instead, we can see that if we fall across the surface of the black holes we must run into the singularity.

The history of such an unfortunate diver is shown in Fig. 2. What the diver last sees from outside the surface of the black hole is determined by the last light ray to follow him across the surface. It is distinctly only part of the full history outside. Furthermore he sees a finite blueshift or redshift of the objects outside. If he has fallen in from infinity, there is no redshift or blueshift from objects at rest outside. For observers outside the black hole, their

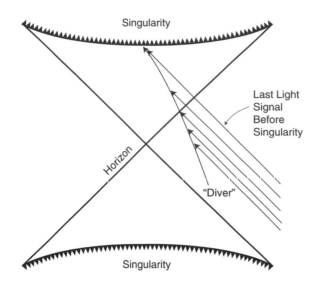

Singularity

Last Light
Signal
Before
Singularity

Horizon

"Diver"

Singularity

Figure 2 Representation of the black hole concept.

last sight of him is as he crosses the horizon, and they see an infinite redshift.

See also: BLACK HOLE; BLACK HOLE, KERR; LAPLACE, PIERRE-SIMON; REDSHIFT; RELATIVITY, GENERAL THEORY OF; SPACETIME

Bibliography

D'INVERNO, R. *Introducing Einstein's Relativity* (Clarendon Press, Oxford, Eng., 1992).

MISNER, C. W.; THORNE, K. S.; and WHEELER, J. A. *Gravitation* (W. H. Freeman, San Francisco, 1973).

THORNE, K. S. *Black Holes and Time Warps: Einstein's Outrageous Legacy* (W. W. Norton, New York, 1994).

RICHARD A. MATZNER

BODY, PHYSICS OF

Many people divide the scientific world into biological sciences and physical sciences. In fact, it was not known until about 1600 that these two broad fields had any elements in common. When, more than 200 years later, Neil Arnott's *Natural Philosophy: General and Medical* was published in London in 1827, it became the first physics book to deal extensively with the physics of the body.

The field of medical physics, which includes the physics of the body as a portion of its scientific territory, grew rapidly in the last half of the twentieth century. However, of the more than 4,000 members of the American Association of Physicists in Medicine (AAPM) in 1995, only a handful worked primarily in the area of physics of the body. Most members of the AAPM work in the area of instrumentation used in medicine—such as radiation therapy, diagnostic radiology, computerized tomography, ultrasound, nuclear medicine, and magnetic resonance imaging.

There are many physical principles involved in the functioning of the body. Some physics principles are obvious, such as the mechanical aspects of walking. However, even walking requires a lot of hidden physics, such as sensory input and feedback mechanisms to maintain balance. Some applications of physics in the body are extremely subtle and many are not yet understood, such as the physical mechanisms for hearing a sound so weak that the motion of the detection mechanism in the ear is about the dimension of the nucleus of an atom.

Force, Energy, and Power

The basic physics concepts of force, energy, and power of the human body are clearly visible in weight lifters, long-distance runners, and sprinters, respectively. It is obvious that the body must convert food energy into energy available to the muscles. The details of this energy conversion are far from trivial. The body does not work like a heat engine or an electric motor. The most analogous energy converting device is the fuel cell.

The details of how food energy is converted to energy useful for our cells required many years of research, primarily by physiological chemists. The body is remarkably efficient at extracting energy from food. Only a few percent of the food energy taken in is excreted in the feces—the remainder is used for body functions or stored as fat.

The average power consumption of a typical adult is about 100 W. The major energy consuming organs in the body in order of their average power consumption are the liver, brain, skeletal muscles, kidneys, and heart. The power consumption of the heart under resting conditions is about 10 W, but during vigorous physical work or exercise, it can exceed 100 W. The efficiency of the skeletal muscles under optimum conditions, such as cycling, is about

20 percent, which exceeds the efficiency of a typical steam engine.

Homeostasis

The heat generated by metabolic processes and by muscular work is dissipated using a number of physical mechanisms. Heat is carried from the core of the body to the skin by the blood. In cold weather, when heat energy must be conserved, the blood returning from the hands and feet avoids the skin and flows near major internal arteries. It uses counter-current heat exchange to conserve body heat.

There are a large number of physical and chemical mechanisms for maintaining the stable internal environment of the body. The medical term for this process is "homeostasis." In physics and engineering, it is called negative feedback. One example is the system used to keep our body at a stable temperature even when the external temperature ranges from below freezing to considerably higher than normal body temperature.

Homeostatic systems often involve sensors we are not aware exist. Basic body mechanisms—digestion, breathing, blood flow, and our five basic senses—function very well with no conscious effort on our part. We can control our breathing rate, but our natural breathing rate optimizes energy consumption. Breathing at rest typically consumes about 2 W.

Muscles

Our muscles are more apparent than most body organs. They produce their force by using the attraction of opposite electric charges. Thus, they can only contract. Muscles always come in pairs; when one is contracting, the other is lengthening. The muscle pairs are not of equal strength. For example, the jaw muscles for chewing are much stronger than the muscles to open the jaws for another bite. Muscular contraction is controlled by electrical stimulation from the brain or sometimes from the spinal cord. A given muscle fiber is a binary device—it is fully contracted or not contracted at all. We produce a stronger force by contracting more muscle fibers.

Blood Circulation

The closed circuit design of blood circulation was not understood until 1616. At that time, William Harvey in England discovered and explained the capillary system. The blood is pumped to the general body circulation by the strong muscle in the left side of the heart. The contraction of the heart produces a relatively high systolic pressure that lasts about one-tenth of a second. About 80 ml of blood is forced into the elastic aorta—the major artery that feeds blood to the general circulation of the body. When the pressure in the aorta drops about in half, the valve between the heart and aorta snaps closed, producing a sound that can be heard with a stethoscope. The pressure in the aorta does not drop to zero because the elastic walls continue to force the blood through the arteries. On the next contraction of the heart, the valve opens again with another sound that is audible through a stethoscope. The valve opens when the pressure exceeds the diastolic pressure in the aorta. In the United States, it is conventional to give blood pressures in terms of the height of a column of mercury (mm Hg) or water (cm H_2O). Much of the world uses the international scientific (SI) unit for pressure, pascal (Pa); 1 Pa = 1 N/m^2. Atmospheric pressure is 101,000 Pa (101 kPa), 760 mm Hg, or 10 m H_2O. Thus, 1 kPa is about 7.5 mm Hg or 10 cm H_2O. The systolic and diastolic blood pressures 120/75 mm Hg in the United States are given as 15/10 kPa in Europe.

As the arteries branch out from the aorta, they become smaller in diameter. The cross-sectional area of all the arteries continues to increase as the blood moves away from the heart. The blood flows slower and finally enters a huge network of fine capillaries. Some of the capillaries are smaller in diameter than a red blood cell. The flexible red blood cells have to double over to get through. If the cross-sectional area of all the capillaries were added up, they would make a tube as large as a sewer pipe, about 30 cm in diameter. The blood that enters the aorta at about 30 cm/s is traveling only about 1 mm/s when it passes through the capillaries. This provides time for the nutrients and oxygen in the blood to diffuse into nearby cells. At the same time, the metabolic products of water and CO_2 diffuse into the blood to be carried to the kidneys and lungs for disposal.

The blood from the capillaries collects into small veins that converge to larger veins. The pressure in the venous system is very low. All the driving pressure from the left side of the heart is dissipated in getting the blood to and through the capillaries. How then does the blood in our feet get back to our heart when we are standing? In fact, several mechanisms are at work. The muscles around the veins are continually contracting to drive the blood toward the heart. Each vein has little valves to keep the blood from flowing back toward the capillaries. Each time the muscles around the veins contract, they drive a little more blood past these valves. The blood creeps back toward the heart. When these little valves break down the condition is called varicose veins.

Once the venous blood approaches the lungs, it gets an unexpected boost. The pressure in our chest cavity is generally negative relative to atmospheric pressure because our lungs would collapse if they were not sealed inside the airtight chest cavity. This negative pressor of about 1 to 2 kPa helps lift the blood toward the entrance to the right side of the heart. The slow movement of the blood in our veins means that at any given time most of our blood—about 80 percent—is in our veins.

After the blood returns to the right side of the heart, it is pumped at rather low pressure (3 to 4 kPa) to the lungs. In the lungs, the blood enters a network of capillaries wrapped around the 300 million small air sacks (alveoli) that have a surface area of about 80 m^2. The blood, which is low on oxygen and high on carbon dioxide, is able to replenish its oxygen supply and get rid of most of its CO_2 in the very brief period it spends in the capillaries.

The physics of diffusion is used extensively in the body. The blood receives and delivers its oxygen and nutrients by diffusion as well as the by-products of metabolism. In the heart and in major muscles, nearly every cell is in contact with a capillary. In fact, there are about 190 km of capillaries in 1 kg of active muscle. Because of the shear numbers involved, the heart is not able to provide blood to all the capillaries in the body at the same time. Under resting conditions, only about 2 percent of the capillaries in your muscles are carrying blood.

Since the heart cannot provide blood to all capillaries at all times, the body must make "decisions" about which organs are to receive blood. In doing so, the brain is always given first priority. If the brain is deprived of blood for even a few minutes, it can be permanently damaged. During vigorous exercise, the blood flow is diverted to major muscles. If substantial blood loss occurs (i.e., more than one-fourth of an individual's supply), the body will shut down blood flow to the kidneys and the skin. All of these bodily reactions are responses to ensure efficient blood flow and, in emergency situations, to save lives.

See also: BIOPHYSICS; CAT SCAN; MAGNETIC RESONANCE IMAGING; MEDICAL PHYSICS; NUCLEAR MEDICINE; ULTRASONICS

Bibliography

CAMERON, J. R.; SKOFRONICK, J. G.; and GRANT, R. M. *Physics of the Body* (Medical Physics Publishing, Madison, WI, 1992).

HOBBIE, R. K. *Intermediate Physics for Medicine and Biology,* 2nd ed. (Wiley, New York, 1988).

LENIHAN, J. *How the Body Works* (Medical Physics Publishing, Madison, WI, 1995).

JOHN R. CAMERON

BOHR, NIELS HENRIK DAVID

b. Copenhagen, Denmark, October 7, 1885; *d.* Copenhagen, Denmark, November 18, 1962; *atomic and nuclear physics.*

Bohr, who ranks among the most prominent figures in the history of science, was a fifth-generation Dane. Academic distinctions have a long tradition in his family. A brother of Peter Georg Bohr, Niels's great-grandfather, was a member of the Royal Norwegian and Swedish Academies of Sciences. His grandfather was rector of the Westenske Institut, a high school in Copenhagen. Christian Bohr, Niels's father, was a distinguished physiologist, rector of the University of Copenhagen in 1905–1906, and was proposed for a Nobel Prize in 1907 and 1908.

Niels's mother, Ellen, hailed from a prominent wealthy Jewish banker's family. David Adler, her father, was a cofounder of Handelsbanken and Privatbanken, Danish banks that are still flourishing, and a member of Parliament. Niels was born in the Adler mansion at Ved Stranden 14 in Copenhagen. He grew up in a harmonious and stimulating family. He had an older sister, Jenny, and a younger brother, Harald, who became a distinguished mathematician.

From 1891 until he completed high school in 1903, Niels attended Gammelholms *Latin og Realskole.* After passing his final exam "with distinction" he enrolled in Copenhagen University, choosing physics as his major, with astronomy, chemistry, and mathematics as minor subjects. In those years he was goalkeeper for Akademisk Boldklub, a well-known soccer club. In 1907 he was awarded a gold medal by the Royal Danish Academy of Sciences and Letters (KDVS) for his answer to a problem on surface vibrations of liquids. In 1911 he received a Ph.D. for his thesis "Studies on the Electron Theory of Metals." In September of that year he left for Cambridge to study with Joseph J. Thomson. From March to July in 1912 he was in Manchester for further work under Ernest Rutherford, the man who was to become the role model for his scientific and personal style.

The timing of Bohr's arrival in Manchester could not have been more propitious. In 1911 Rutherford had published his new model of the atom, consisting of a heavy central body, the nucleus, around which orbit electrons, very lightweight particles. According to the theories of that time, however, such a system should have been unstable; the electrons would fall into the nucleus. It was Bohr's genius to see that new concepts based on quantum theory could serve to stabilize the atom. He made a first rudimentary attempt to do so while still in Manchester, then returned to Denmark, where in August he married Margrethe Nøzlund. They had six sons.

In April 1913 Bohr completed the most important scientific paper of his life, which would make him world famous. It contains his theory of the simplest atom (of hydrogen), and is based on the (for its time) very audacious hypothesis that of all possible electron orbits, then believed to form a continuous set, only a discrete subset of "quantum orbits" actually occur in nature. This hypothesis appeared to defy logic, yet had to be correct, since Bohr was able to make successful predictions that could not have been made on the continuum picture. His success caused his career to make rapid strides, from reader in Copenhagen (1913) to associate professor in Manchester (1914) to full professor in Copenhagen (1916), with an office at the *Polytekniske Læreanstalt* (Institute of Technology)—the university did not have a physics institute or laboratory.

It was Bohr himself who in 1917 took the initiative for creating such an institute. He raised the funds in Denmark and America, oversaw its construction and later extensions, and became its first director, giving guidance to the theoretical as well as experimental programs. In 1921, at the official opening of his *Institut for teoretisk fysik* on Blegdamsvej (renamed the *Niels Bohr Institutet* in 1965), Bohr stated its main function: "To introduce a constantly renewed number of young people into the results and methods of the sciences"—not just Danes but people from all over the world.

And that is what indeed happened, on a spectacular scale. Between 1921 and 1961, 444 physicists from thirty-five countries spent at least a month in Copenhagen. Bohr, one of the most inspiring teachers of his time, was forever ready, in fact eager, to discuss their work with them. Some of these young people became leading physicists, for example Paul Adrien Maurice Dirac, Werner Karl Heisenberg (who wrote his paper on the uncertainty relations at Bohr's institute), and Wolfgang Pauli. Furthermore, beginning in 1929 Bohr initiated a series of conferences to which leading physicists of the world flocked. These meetings made the institute the gathering place for the best physics discussions of that period.

During Bohr's life some 1,200 papers were published from Copenhagen. Of these, about 200 were by Bohr himself. In spite of many other duties, he never ceased doing research. One important result of his early work was his development of the theoretical basis for the periodic table of elements (during the period from 1920 to 1922), which made him the founder of quantum chemistry. It also led to the discovery at his institute of a new chemical element named hafnium (after Copenhagen's Latin name).

Thus Bohr was not only a creative scientist but also the founder, administrator, and intellectual leader of an institute that during its first two decades was the world's most important center of theoretical physics.

Bohr was well aware that his early work was based on illogical premises. However, relief came with the discovery of quantum mechanics, a new theory that wiped out all previous paradoxes. This revolutionary development proceeded in two stages. During the first stage (1925–1926), new equations (matrix equations, wave equations) were formulated that accounted, now in a logical way, for the earlier work of Bohr and others. The second stage began with a paper by Bohr in 1927, which, put in simplest terms, posed the following question: The new equations are clearly good, but what do they mean? Bohr gave the answer in terms of a new logic, a new philosophy, which he called complementarity (and which he considered his most important scientific contribution). In later years Bohr continued to refine these ideas, especially stimulated by objections from Albert Einstein, who could not accept the new logic. However, their intellectual antagonism never diminished their mutual respect and affection.

The next important year in Bohr's career was 1932, when the Bohr family moved into the "Residence of Honor" on the Carlsberg brewery grounds. That year Bohr made an attempt (unsuccessful) to use complementarity in biology and, together with his faithful assistant Léon Rosenfeld, engaged in applying complementarity to the quantum theory of the electromagnetic field. Elsewhere that year, experimental discoveries, notably of the neutron, laid the basis for theoretical nuclear physics.

Then, in 1933, came the rise to power of the Nazis, causing many to flee Germany. Bohr took part in providing help, as board member of the Danish committee for support of refugee intellectuals, and especially by raising funds for providing temporary hospitality at his institute. Among those who came was Georg von Hevesy, who in 1935 began the first effort anywhere of applying artificially radioactive isotopes in biology, thus founding the new science of nuclear medicine. That same year Bohr went after funds for building accelerators for use in physics and biology. The result of his efforts was that in late 1938 a high tension generator went into operation, soon followed by one of Europe's first cyclotrons.

Meanwhile, during 1936 and 1937 Bohr made important contributions to nuclear theory, dealing with neutron physics, the study of what happens when nuclei are bombarded by neutrons. He worked out a model for these processes based on an analogy with a liquid drop that vibrates when disturbed. This theory could explain many phenomena studied in the Los Alamos atomic bomb laboratory.

Early in 1939 it was reported from Berlin that hitting uranium nuclei with neutrons causes fission, their violent rupture. Within weeks the theory of the process was worked out using Bohr's model. The term "fission" was suggested, in analogy with biological cell division, by a coworker of Hevesy. Bohr took the news in person to America, where (at age fifty-three) he also made the crucial discovery that fission by slow neutrons occurs only for the rare isotope uranium 235.

After the German occupation of Denmark in April 1940, Bohr continued to lead his institute until September 1943, when he was forced to flee to England. He then became adviser to the Anglo-American atomic weapons program.

Bohr was less concerned with the war effort, however, than with the changes the new weapons would create in the postwar world. To forestall competition with the Russians it would be necessary, he emphasized, to inform them at once of the Western effort and offer them full future cooperation. In 1944 he met personally with Winston Churchill and with Franklin D. Roosevelt to urge them to do so, but had no success. Nor did his 1948 meeting with Secretary of State George Marshall help. In 1950, and again in 1956, he repeated his proposal in open letters to the United Nations, again without success. Times were not yet ripe for Bohr's far-seeing ideas.

Bohr dutifully accepted social obligations resulting from his stature. From 1939 until his death he was president of the KDVS. Also at one time or other he was president of the Danish Physical Society and of the Society for the Dissemination of Natural Science, chairman of the Danish Atomic Energy Commission, president of the National Association for Combating Cancer, chairman of the governing board of Nordita (the Nordic Institute for theoretical atomic physics), and prime mover in founding the Risø research center, not far from Copenhagen.

On November 18, 1962, Bohr died of heart failure. His ashes rest in the family grave at Assistens Kirkegården in Copenhagen.

The honors bestowed on Bohr during his lifetime were numerous. They include the Nobel Prize (1922) and the first Atoms for Peace Award (1957); many orders of chivalry, including the Elephant Orden (1947), Denmark's highest order; and some thirty honorary degrees. Yet it can be said of him what he himself has said of Rutherford: "In his life all honors imaginable for a man of science came to him, but yet he remained quite simple in all his ways . . . He created around him a spirit of affection wherever he worked."

See also: ATOM, RUTHERFORD–BOHR; BOHR MAGNETON; BOHR'S ATOMIC THEORY; COMPLEMENTARITY PRINCIPLE; FISSION; LIQUID-DROP MODEL; RUTHERFORD, ERNEST; THOMSON, JOSEPH JOHN

Bibliography

PAIS, A. *Niels Bohr's Times* (Oxford University Press, Oxford, Eng., 1991).

ABRAHAM PAIS

BOHR MAGNETON

The Bohr magneton, μ_B, is the basic quantum unit of magnetic moment given by

$$\mu_B = \frac{e\hbar}{2m_e c} = 0.9273 \times 10^{-20} \text{ erg/G}, \qquad (1)$$

where $\hbar$ is the Planck's constant divided by 2π, c is the speed of light, and e and m_e are the electric charge and the mass of an electron, respectively. The Bohr magneton equals the magnetic moment of an electron due to its intrinsic spin $\vec{s}$ ($s = 1/2$) predicted by the one-particle Dirac theory.

The spin and the magnetic moment of the electron were postulated by Samuel A. Goudsmit and George E. Uhlenbeck to interpret the experimental observations, including, for example, the anomalous Zeeman effect and the doublets in alkali atoms. The magnetic moment of an electron due to the orbital angular momentum $\vec{L}$ is given by

$$\vec{\mu_l} = -g_l \mu_B \vec{L}, \qquad (2)$$

where $g_l = 1$ is the orbital g factor. Similarly, the magnetic moment of an electron due to its electron spin is defined by

$$\vec{\mu_s} = -g_s \mu_B \vec{s}, \qquad (3)$$

where $g_s \approx 2$ is the spin g factor. Detailed calculation based on the quantum electrodynamics suggests that $g_s = 2.0023192$.

The magnetic moments of heavier subatomic particles are measured in terms of nuclear magneton

$$\mu_N = \frac{e\hbar}{2m_p c} = 0.5050 \times 10^{-23} \text{ erg/G}, \qquad (4)$$

where m_p is the mass of the proton. The magnetic moment due to the intrinsic spin of a proton or a neutron is given by a similar expression with an opposite sign, that is,

$$\vec{\mu} = g\mu_N \vec{s}. \qquad (5)$$

For a proton, $g = 2.7928$, and for a neutron, $g = -1.9135$. Both differ significantly from what is expected from the one-particle Dirac theory. These deviations can be attributed to the fact that the proton and neutron are not point-like particles, but are particles with internal structure. The nuclear magnetic moments may be measured by microwave and laser (or maser) spectroscopy studies of the hyperfine structure of atomic spectral lines using atomic and molecular beams techniques.

See also: MAGNETIC MOMENT; SPIN, ELECTRON

Bibliography

FANO, U., and FANO, L. *Physics of Atoms and Molecules: An Introduction to the Structure of Matter* (University of Chicago Press, Chicago, 1972).

TU-NAN CHANG

BOHR'S ATOMIC THEORY

In 1913 Niels Henrik David Bohr, a Danish nuclear physicist, proposed a model for the hydrogen atom, the so-called Bohr atom, which, for the first time, permitted a deeper qualitative understanding of atomic structure and atomic spectra. Bohr's atomic theory, which was constructed around a planetary atomic model advocated by Ernest Rutherford and others, may be characterized as a provisional semi-classical quantum-theoretical approach. This theory, however, yields acceptable results solely for a limited number of atomic systems, namely the hydrogen atom and hydrogen-like systems containing only one electron in the outer atomic shell. Although Bohr's theory has been superseded by more sophisticated conceptions, it is still useful for didactic applications.

Bohr's theory of the hydrogen atom is a refined synthesis of the notion of quanta of energy, Hermann Nernst's quantization rule for the kinetic energy or the angular momentum (moment of momentum) of a rotating body or "rotator," and Rutherford's planetary or dynamic atomic model, as well as knowledge available from other areas of physics, especially from spectroscopy. The success of the theory was recognized because of its ability to predict accurately the wavelengths λ of certain lines in the spectrum of atomic hydrogen, the fact that the Rydberg constant (an empirical spectroscopic quantity) could be expressed in terms of known universal constants (i.e., the electronic charge e, the electronic mass m, the vacuum velocity of light c, and Planck's constant h), and that the fine structure splitting of spectral lines could be explained by an extension of the theory.

On the one hand, it was already known that the wavelengths λ of the most prominent lines in the visible spectrum of atomic hydrogen, the Balmer series, could be obtained from Johann Balmer's empirical formula $\lambda = An^2/(n^2 - 4)$, with $n = 3, 4, 5 \ldots$, where A is an experimental constant. The real meaning of this equation was gradually realized only, however, after it had been written in reciprocal form:

$$\frac{1}{\lambda} = \frac{\nu}{c} = R\left(\frac{1}{4} - \frac{1}{n^2}\right) \ (n = 3, 4, 5 \ldots), \quad (1)$$

where ν is the frequency of oscillation belonging to the wavelength λ, and $R = 4/A = 109{,}678 \text{ cm}^{-1}$ is Rydberg's constant. The simple structure of Eq. (1) suggested to Bohr that the frequency of oscillation belonging to one spectral line should be expressible as the beat frequency of two other frequencies of oscillation, $f_n = cR/n^2$ and $f_m = cR/4$, which have to be related somehow to atomic structure:

$$\nu_{mn} = f_m - f_n. \quad (2)$$

Here, the index m refers to the higher frequency. Bohr's problem was the physical interpretation of the two summands f_n and f_m of Eq. (2).

On the other hand, it was known from the theory of Max Planck that the total energy of a harmonic oscillator is given by $E_{\text{total}} = nh\nu$, where $\nu = \omega/2\pi$ is the frequency of oscillation and n is an integral number. This implies that a harmonic oscillator can emit and absorb energy only in small, discrete amounts, each quantum transporting the energy

$$E = h\nu. \quad (3)$$

Combining Planck's formula with Eq. (2) produces the following equation:

$$h\nu_{mn} = h(f_m - f_n) = E_m - E_n, \quad (4)$$

where E_n and E_m are certain energy levels still to be interpreted physically.

Equation (4), which is similar to a quantum relation deduced by Albert Einstein in 1905 to explain photoelectric effect, is generally true. It has become widely known as Bohr's frequency condition. It refers to the emission of a single photon of energy $h\nu_{mn}$; E_m is a higher energy level than E_n.

Another known result of quantum theory available to Bohr was Nernst's quantization condition for the kinetic energy of a rotating particle or body of mass m and moment of inertia $J = ma^2$:

$$E_{\text{rot}} = \left(\frac{J}{2}\right)\Omega^2 = \left(\frac{ma^2}{2}\right)(2\pi f)^2 = \left(\frac{n}{2}\right)hf = \left(\frac{n}{2}\right)\hbar\Omega, \quad (5)$$

where $\Omega = 2\pi f$ is the angular velocity of the rotating body, $\hbar$ is Planck's constant divided by 2π, and n is an integer. (For extended bodies the moment of inertia would have to be replaced by $J = \sum m_f a_f^2$.) A rotating particle or body is a periodic system different from a harmonic oscillator, so that Planck's quantization rule cannot directly be adopted; it has to be modified. Nernst inserted $nhf/2$ instead of nhf on the grounds that a rotator possesses only kinetic energy, whereas a harmonic oscillator also has potential energy available (an equal amount on the average). Since a particle's angular momentum is given by $L = J\Omega = ma^2\Omega$, from Eq. (5) it follows that obviously angular momentum is quantized:

$$L = J\Omega = ma^2\Omega = n\hbar. \quad (6)$$

These two quantization prescriptions were thoroughly applied in Bohr's atomic theory, Eq. (3) in order to quantize the radiation field with which the atom interacts and Eq. (5) to quantize the mechanical motion of the electron rotating around the atomic nucleus. To appreciate the fundamental difference of the two quantization instructions, note that the total energy of the rotator, which for the rotator equals its kinetic energy, can be written with the help of Eq. (6) as $E_{\text{rot}} = J\Omega^2/2 = JL^2/(2J^2) = n^2\hbar^2/(2J)$, which indicates that it increases quadratically with quantum number n and not linearly, as in the case of the harmonic oscillator.

Bohr's work had been motivated by progress achieved in contemporary atomic physics. From the results of scattering experiments performed with the help of beams of electrons, and later with beams of α particles, it could be concluded that nearly all of the atomic mass was confined to the center of the atom and that most of the physical space occupied by an atom is actually empty. These experimental results raise the question of how in such an atomic structure (i.e., a structure that has a radius of about 10^{-8} cm and is composed of Z electrons [each of charge $-e$] orbiting about a nucleus with a positive charge Ze and a radius of approximately $3 \cdot 10^{-12}$ cm) can the oppositely charged particles permanently keep apart in the presence of the attracting Coulomb force (Ze^2/a^2, where a is the radial distance of an electron from the nucleus). In the Rutherford–Bohr planetary atomic model this stability problem is settled by assuming that the electrons rotate around the nucleus in such a manner that the electrostatic force is counterbalanced by the centrifugal force.

Consider, as Bohr did, the simplest atom, the hydrogen atom, which accommodates only one electron ($Z = 1$) in the atomic shell. As long as there are no restrictions inflicted by quantum conditions, the electron can rotate on circular and elliptical orbits with an angular velocity $\Omega = v/a$ in accordance with Kepler's laws, v being the orbital velocity. For the sake of simplicity, a circular electron orbit shall be considered. If the electron were at rest, then under the action of the Coulomb force

$$F_1 = \frac{e^2}{a^2},$$ (7)

it tends to fall toward the nucleus. This attracting force is counterbalanced by the centrifugal force

$$F_2 = ma\Omega^2.$$ (8)

From the balance condition $F_1 = F_2$, it follows that

$$\frac{e^2}{a^2} = ma\Omega^2.$$ (9)

This is Kepler's third law: The squares of the period $T(2\pi/\Omega)$ of revolution of different planets around the Sun are proportional to the cubes of the major semi-axes a of their orbits.

For later considerations the equation describing the total energy E of this system is needed. The energy E is the sum of the kinetic energy E_k and the potential energy E_p. The kinetic energy is given by $E_k = mv^2/2 = ma^2\Omega^2/2$. Combining the latter relation with Eq. (9), the result is

$$E_k = \frac{(ma^2\Omega^2)}{2} = \frac{e^2}{2a}.$$ (10)

Since the potential energy in the electrostatic field is

$$E_p = -\frac{e^2}{a},$$ (11)

the total energy of the system becomes

$$E = E_k + E_p = -\frac{e^2}{2a},$$ (12)

which turns out to be a negative quantity, since work must be invested to move the electron away from the nucleus to a distance $a \to \infty$. With the help of Eqs. (9) and (12), the radius of the circular electron orbit, a, and the electron's frequency of revolution, f, may be expressed in terms of the total energy E:

$$a = -\frac{e^2}{2E}$$

$$f = \frac{\Omega}{2\pi} = \left(\frac{1}{\pi e^2}\right)\left(\frac{2}{m}\right)^{1/2}(-E)^{3/2}.$$ (13)

From the classical viewpoint, a system described by Eq. (13) would not be stable since the electron is under permanent acceleration and should, therefore, continuously emit electromagnetic energy in the form of waves of gradually growing frequency f until it ultimately collapses into the nucleus. During this radiative process the negative quantity E increases, and the electron's radial distance from the nucleus, a, decreases.

Since in a real hydrogen atom the electron does not collapse into the nucleus, Bohr was convinced that there must be exceptions to this classical picture. He introduced the following two postulates:

1. Among the classical electron orbits there exist certain "stationary states," in which the circulating electron does not lose energy by radiating.
2. There can occur radiative transitions between two stationary states, and the frequency of the radiation ν of a single event of emission, involving a transition of the atomic system from its initial energy level E_m to the final energy level E_n is [in accordance with Eq. (4)] given by

$$\nu_{mn} = \frac{(E_m - E_n)}{h} \qquad (14)$$

In particular, among the set of possible stationary states of the atom, there is a ground state, in which the electron's distance from the nucleus is lowest. This ground state must be radiationless, because there is no energy available to be emitted. Bohr's first postulate thus clearly represents a departure from classical physics, although one might argue that in classical electromagnetic theory the possibility of self-interaction of a circulating electron with its own radiation field had previously never been seriously taken into account.

If Bohr's first postulate is accepted, then the second one, stating that there can occur transitions between the stationary states, becomes plausible, if the remainder of the system, especially the mechanism bringing about the energy exchange with the surroundings, shall be described exclusively in classical terms. Indeed, in the regime of classical physics, transitions of the electron spiralling between two adjacent stationary states are the only mechanisms thinkable that could be made responsible for the energy exchange.

From the tactical plan to modify classical physics as little as possible, Bohr's correspondence principle has emerged. The principle, which is still consulted in modern physics, has served for general orientation in the early development as well as in later extensions of the theory.

To get hold of the stationary states of the hydrogen atom, Bohr applied Nernst's quantization rule for the rotator because he was aware that the circulating electron represented a physical realization of a rotator. Bohr was convinced that there was a link between classical and quantum theory when he inspected the process of creation of a hydrogen atom. This latter process involves both an atomic nucleus at rest, which carries a positive charge e, and a free electron at infinite distance, where $f_\infty = 0$. Because of the influence of the Coulomb field, the electron is then captured by the nucleus, around which it continuously spirals until it finally circulates in the ground state of the newly born H atom, where its frequency of revolution is f_0.

Now, during the very process of creation, the circulating electron emits, in accordance with classical electrodynamic theory, a continuous wave train of a frequency of oscillation, which equals the electron's spontaneous frequency of revolution, and its kinetic energy increases while the total energy of the complete system decreases. The emitted electromagnetic wave train (photon) is a frequency chirp, the center frequency and energy of which should comply with Bohr's frequency condition [Eq. (4)]. The center frequency of the chirp can be estimated, however, from the edge frequencies of its spectrum: It should be on the order of the arithmetic mean between f_∞ and f_0. Therefore, one can tentatively write

$$h\nu_{\infty 0} = \frac{h(f_\infty + f_0)}{2} = \frac{h(0 + f_0)}{2} = E_\infty - E_0, \qquad (15)$$

so as to obtain for the energy of the ground state the expression

$$-E_0 = \left(\frac{h}{2}\right)f_0. \qquad (16)$$

This outcome, however, is fully compatible with Nernst's quantum condition [Eq. (5)], from which it follows for $n = 1$ (with the negative sign resulting because the zero of energy has been adjusted at the radius $a \to \infty$).

Thus, there apparently exists a concrete link or correspondence between classical and quantum theory, which concerns a pair of frequencies of revolution f_m and f_n on the one hand and the atomic emission frequency ν_{mn} on the other hand. Since the discrepancy in frequencies is largest for the extreme case just considered, this difference in frequency should decrease with increasing quantum number n. Therefore, it can be expected that for increasing quantum numbers n, Bohr's atomic theory approaches asymptotically classical theory. This does not imply, however, that at some high quantum number a metamorphose between atomic and classical theory actually takes place.

The compatibility of Eq. (16) with Eq. (5) suggests that the model considerations seem to be sound. In the present context the correspondence principle does not need to be consulted. By combining either Eq. (5) with Eq. (12) or Eq. (6) with Eq. (9), it is possible to pick out from the classical infinite manifold of Keplerian electron orbits a discrete set of stationary states, each characterized by another radius a. If Ω is eliminated between Eq. (6) and Eq. (9), the following set of orbital radii is obtained:

$$a = \frac{n^2\hbar^2}{me^2}, \qquad (17)$$

where n is referred to as the quantum number, which specifies the stationary state. For $n = 1$ the radius of the ground state, a_0, results:

$$a_0 = \frac{\hbar^2}{me^2} = 0.529 \cdot 10^{-8} \text{ cm}. \qquad (18)$$

The outer diameter of the atom, when in the ground state, is therefore about 1 Å, which compares well with the expected order of magnitude; the gas-kinetic interaction cross sections are of this order of magnitude.

If instead the radius a is eliminated, an expression for the various frequencies of revolution f in the stationary states is gained:

$$f = \frac{\Omega}{2\pi} = \frac{e^4 m}{2\pi n^3 \hbar^3}. \qquad (19)$$

Inserting $n = 1$ yields for the ground state:

$$f_0 = \frac{\Omega}{2\pi} = \frac{e^4 m}{2\pi\hbar^3} = 6.58 \cdot 10^{15} \text{ rev}. \qquad (20)$$

Since the rotation of the electron around the nucleus is fastest in the ground state, it follows from Eq. (4) that the lines observed in the optical spectrum of hydrogen should correspond to optical frequencies $\nu < f_0$, a condition which is fulfilled by this numerical result because the frequencies of visible light are on the order of $0.6 \cdot 10^{15}$ Hz.

The combination of Eq. (12) with Eq. (17) gives the energy levels of the stationary states:

$$E = -\frac{e^4 m}{2n^2\hbar^2}. \qquad (21)$$

This expression generates the Balmer terms belonging to the quantum numbers. They define the energy levels of the stationary states.

Letting $n = 1$, the energy of the ground state is found to be

$$E_0 = -\frac{e^4 m}{2\hbar^2} = -13.6 \text{ eV}. \qquad (22)$$

That is, an energy of 13.6 eV has to be invested, or, what amounts to the same thing, an ionization voltage of 13.6 V has to be applied externally to the hydrogen atom, if its electron is to be removed from the ground state and completely extracted from the atom.

By manipulating the latter equations, one can obtain for arbitrary quantum numbers n the following relations for the radii, frequencies of revolution, and energies of the stationary states:

$$a = a_0 n^2$$

$$f = \frac{f_0}{n^3} \qquad (23)$$

$$E = -\frac{E_0}{n^2}.$$

It is seen that the atom's radius increases quadratically with quantum number n and that an ensemble of discrete electron orbits with radii a_0, $4a_0$, $9a_0$ and so on is obtained. Since the radii soon become excessively great, one expects that in dense matter only the lowest quantum numbers will play a role. Hydrogen atoms having large quantum numbers are observable in outer space, and they may be produced artificially in the laboratory.

Figure 1 shows a diagram of the location of the various energy levels computed for the hydrogen atom. The energy values are negative because there exists attraction between the electron and the nucleus. During the process of creation of a hydrogen atom, the captured electron moves in the direction of the stationary states of lower energy. For $n = 1$ the lowest energy state (the stable ground state) is obtained. The separation of energy levels shrinks in inverse proportion with the quantum number, so that the neighboring levels above the ground state have 1/4, 1/9, and so on, of its energy, and the levels in the vicinity of the level of zero energy are extremely dense.

Once the Balmer terms [Eq. (21)] are known, it is possible by means of Eq. (14) to derive the formula for the entire system. If, for convenience of comparison with the original Balmer formula [Eq. (1)], the meaning of the indices n and m is exchanged so that the index n refers to the higher energy state, then the general formula reads

$$\nu_{nm} = \frac{E_n - E_m}{h} = \left(\frac{cR}{h}\right)\left[\left(\frac{1}{m^2}\right) - \left(\frac{1}{n^2}\right)\right], \qquad (24)$$

where the factor R is the Rydberg constant:

$$R = \frac{2\pi^2 me^4}{h^3 c}.$$ (25)

Balmer's formula [Eq. (1)] is obtained for the special case $m = 1$. The theoretical value of the Rydberg constant is in satisfactory agreement with the experiment.

If other constant values of m are inserted in Eq. (24), further line series are predicted that have also been spectroscopically observed. In particular, for $m \leq 5$ the following line series named after their discoverers have become known:

$m = 1$: Lyman series (ultraviolet),

$m = 2$: Balmer series (visible),

$m = 3$: Paschen series (infrared),

$m = 4$: Brackett series (infrared),

$m = 5$: Pfund series (infrared).

In all cases the experimentally observed line centers agree very well with the theoretical predictions made with the help of Eq. (24). Figure 2 indicates the electron orbits within the hydrogen atom and the transitions among them, which are involved in the individual series.

Bohr's atomic theory so far described lends itself to several generalizations. Thus, it is possible to correct the theory in a simple manner for the finite mass of the nucleus, M, which in the previous considerations was set equal to infinity. To achieve this, the electronic mass m in the various expressions for the energies and frequencies must be replaced everywhere by the reduced mass

$$\mu = \frac{mM}{M + m}.$$ (26)

Otherwise, these formulas remain the same. The Rydberg constant [Eq. (25)] must be multiplied by a factor of $M/(M + m)$. Experimental evidence for the correctness of this correction factor can be found by inspecting the frequency shift exhibited by a hydrogen isotope, such as deuterium.

Another generalization of the theory concerns hydrogen-like atomic structures that have a single electron in the outer shell, such as the He$^+$ ion ($Z = 2$) or the Li^{++} ion ($Z = 3$). Equation (24) is then to be replaced by

$$\nu_{nm} = \left(\frac{2\pi^2 \mu Z^2 e^4}{h^3}\right)\left(\frac{1}{m} - \frac{1}{n}\right).$$ (27)

The line spectra of a number of hydrogen-like atomic systems, especially those of the alkali atoms and earth alkali ions, can be understood if one takes

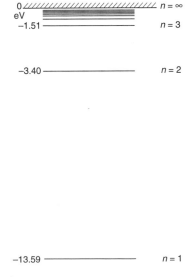

Figure 1 Separation of energy levels in a hydrogen atom. Ionization of the atom requires an energy of 13.6 eV.

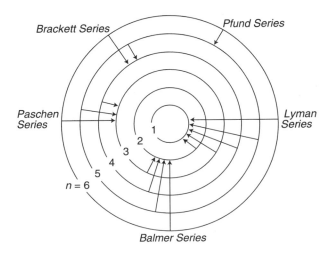

Figure 2 Identification scheme for the various spectral series and the corresponding transitions between the lowest stationary electron orbits of the hydrogen atom. The circle with $n = 1$ represents the ground state.

130

into consideration that the outer electron revolves fairly far away from the residual electrons, so that the combined effect of the latter boils down to a reduction of the electrostatic force exerted on the outer electron. The combined influence of the nucleus and the inner electrons can then be accounted for by introducing a corresponding effective nuclear charge.

A further generalization involves the application of the Bohr model in the area of semiconductor technology, where it is used to make rough estimations of the energies and radii of impurity levels in semiconductor crystals, such as doped germanium or silicon, the materials employed to manufacture transistors and other electronic components. In this application, the balance equation [Eq. (9)] is replaced by the equation

$$\frac{e}{a^2} = \frac{m^* v^2}{a},$$

(28)

where m^* is a parameter known as the effective mass. The parameter takes all periodic crystal forces into account, to which a conduction electron made available by the implanted impurity is subjected. Starting from Eq. (28) and resorting to Bohr's line of attack, it is possible to construct a straightforward theory that permits the prediction of the approximate ionization potential of an impurity.

Furthermore, Arnold Sommerfeld and others generalized the Bohr theory in the respect that they replaced Nernst's quantization condition by a more sophisticated one that is applicable to more complicated systems having many degrees of freedom. If p and q denote, for example, momentum and location, or angular momentum and angle, or other conjugate coordinates, then the quantization prescription is expressed by the phase integral

$$I = \oint p \, dq = nh,$$

(29)

which has to be taken along all infinitesimal elements of the particle's path. The circle shall emphasize that the integral has to be computed for a full period of revolution. The Sommerfeld–Wilson quantization integral allows it to quantize electron motion on elliptical orbits, and it can also deal with the relativistic variation of mass. In the latter case the energy formula contains an additional Balmer term:

$$E_{n,k} = -\left(\frac{RZ^2}{n^2}\right)\left[1 + \left(\frac{\alpha^2 Z^2}{n}\right)\left(\frac{1}{k} - \frac{3}{4n}\right) + \dots\right]$$

$$(n = 1, 2, 3\dots; \, k = 1, 2 \dots n).$$

(30)

Here the integral numbers n and k are respectively referred to as the principal quantum number and the azimuthal quantum number. The dimensionless quantity

$$\alpha = \frac{2\pi e^2}{hc} \approx \frac{1}{137}$$

(31)

is Sommerfeld's fine-structure constant. The close agreement of this formula with spectroscopic observation has been one of the greater triumphs celebrated by the Bohr–Sommerfeld theory.

Despite its usefulness in a number of interesting applications and its appealing inherent simplicity, Bohr's theory of the hydrogen atom has never been developed into a powerful simple theory that could also be applied to atomic structures not resembling the hydrogen atom. Unfortunately, even for the helium atom, which has two electrons in the atomic shell and is the next atom in the periodic table of elements, Bohr's theory yields unacceptable results. Ideas originally derived during the course of the development of the theory, such as the correspondence principle or the conception of the stationary state, have survived, however, and found their way into the more powerful and refined modern theory known as quantum mechanics.

See also: ATOM; ATOM, RUTHERFORD-BOHR; ATOMIC PHYSICS; ATOMIC THEORY, ORIGINS OF; CORRESPONDENCE PRINCIPLE; COULOMB'S LAW; ELECTROMAGNETIC RADIATION; ENERGY LEVELS; FINE-STRUCTURE CONSTANT; GROUND STATE; KEPLER'S LAWS; PHOTOELECTRIC EFFECT; PLANCK, MAX KARL ERNST LUDWIG; QUANTIZATION; QUANTUM MECHANICS; QUANTUM NUMBER; QUANTUM THEORY, ORIGINS OF; RUTHERFORD, ERNEST; SPECTRAL SERIES; STATIONARY STATE

Bibliography

ADLER, R. B.; SMITH, A. C.; and LONGINI, R. L. *Introduction to Semiconductor Physics,* (Wiley, New York, 1964).

BOHR, N. *Collected Works* (North-Holland, Amsterdam, 1981).

BORN, M. *Atomic Physics,* 7th ed. (Blackie, London, 1962).

HERZBERG, G. *Atomic Spectra and Atomic Structure,* 2nd ed. (Dover, New York, 1944).

SOMMERFELD, A. *Atomic Structure and Spectral Lines* (Methuen, London, 1934).

WILFRED KRAUSE

BOLTZMANN, LUDWIG

b. Vienna, Austria, February 20, 1844; *d.* Duino, near Trieste, Italy, September 5, 1906; *kinetic theory, statistical mechanics, radiation theory, philosophy of physics.*

"Loschmidt's body is now disintegrated into atoms. How many, we can calculate from the principles he established." We learn much about the nineteenth-century Austrian physicist Ludwig Boltzmann in this excerpt from his 1895 obituary for his friend and colleague Josef Loschmidt. Boltzmann was one of the most creative and productive theoretical physicists of the nineteenth century. This quotation shows as well his commitment to mechanical and atomic models, his robust and uninhibited sense of humor, and his striking use of language.

Boltzmann was born in 1844. His father was an Austrian tax official, and the family appears to have been firmly middle class; but there are few biographies of Boltzmann, and very little has been written about his upbringing and early education. He studied physics at the University of Vienna and received his doctorate in 1866. Among his most influential teachers was Josef Stefan, to whose experimental work on the T^4 law describing the dependence of radiation on temperature Boltzmann later gave a theoretical justification. Stefan also introduced Boltzmann to the work of the British physicist James Clerk Maxwell when he handed Boltzmann (still a university student) a copy of Maxwell's book on electricity and magnetism, along with an English grammar: At the time, as Boltzmann tells us, he "understood not a word of English." In later years, Boltzmann became a strong and influential supporter of Maxwell's theory.

Boltzmann is best known for his many contributions to statistical mechanics and kinetic theory—the study of how the large-scale properties of matter follow from the behavior of microscopic atoms and molecules that obey the laws of mechanics. Along with Maxwell, Rudolf Clausius, and the American physicist Josiah Willard Gibbs, he helped establish these subjects as central to our understanding of physics. Among his chief concerns was the demonstration that the second law of thermodynamics, and particularly the concept of entropy, could be derived from the laws of mechanics and an atomic theory of matter.

Perhaps his best-known contribution to this subject grew out of a hotly disputed puzzle in nineteenth-century physics. The laws of mechanics are "reversible"—they do not give a direction to time. If we were to take a movie of Earth circling the Sun, we could run the film backwards and not be able to tell anything was wrong. But the laws governing heat (thermodynamics) are *not* reversible, and do show a direction in time. A glass of boiling water spontaneously cools, but the same water at room temperature does not spontaneously absorb heat from its surroundings and begin to boil! How then can one derive the laws governing heat from the laws of mechanics? This paradox implicit in Boltzmann's work was clearly pointed out by his friend Loschmidt in 1876.

In response, in 1877, Boltzmann considered a large collection of atoms that had a constant total energy. He then calculated the number of possible combinations of individual atomic energies consistent with that total energy, and showed that one could assign a probability to each combination. (For example, the probability that one single atom carries all of the energy of the system is very small.) A careful analysis of these probabilities shows that for large systems, one can identify the entropy with the logarithm of the probability of a state. Furthermore, it is overwhelmingly likely that the system will be found in its most probable state. Thus, the large-scale thermodynamic behavior of the system depends not only on the laws of mechanics but on statistical laws as well; hence, the "statistical" in statistical mechanics. Boltzmann's specific approach is still used in many textbooks. It is also the source of the (often vague) statement in introductory texts that entropy is a measure of molecular "disorder." Historically, Boltzmann's approach was central to the work of Max Planck and Albert Einstein in formulating the concept of the "quantum" around the turn of the century. It is hard to imagine that quantum physics would have developed as it did without Boltzmann's contributions. In a touch that would doubtless have appealed to Boltzmann's sense of humor, the modern formulation of his relation between entropy and probability, $S = k \ln W$, was inscribed on his tombstone in Vienna. The constant of proportionality k that we call Boltzmann's constant was in fact introduced into physics by Planck and Einstein in their applications of Boltzmann's method.

Atomic theories of matter were by no means universally accepted around the turn of the century, and their status was vigorously and publicly debated. The influential physical chemist Wilhelm Ostwald, for example, sharply attacked atomic theories and strongly supported a theory of "energetics" that, he thought, would make atomic hypotheses superfluous. Boltzmann's enthusiastic support of atomic theories led him to play a forceful and prominent role in these disputes during the 1890s.

Boltzmann was an enthusiastic and successful teacher who injected the full force of his personality into his teaching. His students included Paul Ehrenfest and Lise Meitner (the second woman to obtain a Ph.D. in physics in Vienna). He also wrote two influential textbooks, *Lectures on Mechanics* and *Lectures on Maxwell's Theory of Electricity and Light*. A third more advanced treatise, *Lectures on Gas Theory*, deeply influenced the young Albert Einstein, who may have read Boltzmann's other texts as well. In these books, Boltzmann presented not only the physics but also his own ideas on how physics should be interpreted. He speaks at length, for example, on the central position of mechanics and Newton's laws of motion, and the status of atomic and molecular models; and he makes extensive use of elaborate mechanical models in developing electromagnetic theory. This personal approach, not unusual at that time, makes for lively and interesting reading. It makes one realize that physics is a human science, and that different physicists can give surprisingly different interpretations and outlooks on the same physical laws.

Boltzmann had a deep love of music and was an enthusiastic pianist. He also enjoyed traveling and visited the United States on several occasions. In his essay "Journey of a German Professor to Eldorado," he gives an amusing account of his travels and experiences in California. He was not only one of the most creative and influential physicists of the nineteenth century but also one of its most interesting personalities as well.

See also: BOLTZMANN CONSTANT; BOLTZMANN DISTRIBUTION; MEITNER, LISE; STATISTICAL MECHANICS; STEFAN–BOLTZMANN LAW; THERMODYNAMICS, HISTORY OF

Bibliography

BLACKMORE, John, ed. *Ludwig Boltzmann: His Later Life and Philosophy, 1900–1906* (Kluwer, Norwell, MA, 1995).
BOLTZMANN, L. *Theoretical Physics and Philosophical Problems*, edited by B. McGuinness (Reidel, Boston, 1974).
BOLTZMANN, L. "Journey of a German Professor to Eldorado." *Phys. Today* **45**, 44–51 (1992).
BRODA, E. *Ludwig Boltzmann: Man—Physicist—Philosopher* (Oxbow, Woodbridge, CT, 1983).
HIEBERT, E. N. "The Energetics Controversy and the New Thermodynamics" in *Perspectives in the History of Science and Technology*, edited by D. H. D. Roller (University of Oklahoma Press, Norman, 1971).
KLEIN, M. J. *Paul Ehrenfest, Volume I: The Making of a Theoretical Physicist* (North-Holland, Amsterdam, 1970).
KLEIN, M. J. "Mechanical Explanation at the End of the Nineteenth Century." *Centaurus* **17**, 58–72 (1972).
KUHN, T. S. *Black-Body Theory and the Quantum Discontinuity, 1894–1912* (Oxford, New York, 1978).

CLAYTON A. GEARHART

BOLTZMANN CONSTANT

The Boltzmann constant (k_B), named for the Austrian physicist Ludwig E. Boltzmann, defines the proportionality between the absolute temperature and energy. The quantity $k_B T$ has the unit of energy and at temperature T = 300 K is about 1/40 eV. More specifically, the dimensionless ratio of the kelvin (K) and the joule (J) is by definition Boltzmann's constant $k_B = 1.3807 \times 10^{-23}$ J/K.

Determination of Value

The value of k_B can be determined from the ideal gas law $PV = N k_B T = nRT$, where P is the pressure, V is the volume, N is the number of molecules, n is the number of moles, and R is the gas constant $R = 8.3145$ J·mol^{-1}·K^{-1} so $k_B = R/(N/n)$ where N/n is Avogadro's number (6.0221×10^{23} mol^{-1}). For this reason, Boltzmann's constant is sometimes also called the gas constant per molecule.

Max Planck, not Boltzmann, first obtained the value of k_B. Planck derived a law that correctly gives the spectral energy distribution of the radiation emitted by a blackbody. A blackbody is a body that emits the maximum amount of thermal radiation.

From Planck's law one can derive Wien's displacement law (which states that the wavelength at which the blackbody radiation is a maximum times the corresponding temperature is a constant) and the Stefan–Boltzmann law (which says the total emitted radiation of a blackbody is proportional to the fourth power of the temperature). From these two laws, k_B and h (Planck's constant) can be determined.

Curie's law is $M = CH/T$, where M is the magnetization and H is the magnetic field. It is the magnetic analog of the ideal gas law. Using Curie's law one can obtain a value for Boltzmann's constant by fitting the magnetization (M) to the ratio of magnetic field (H) to temperature for appropriate magnetic systems in which Curie's law is satisfied. The fit determines Curie's constant C, which involves k_B.

There are other ways to determine the Boltzmann constant, not necessarily with high accuracy. For example, in Brownian motion (the motion of dust particles in air) the mean square displacement of a particle depends on time and is proportional to $k_B T$. Boltzmann's constant has also been obtained from high-altitude Rayleigh scattering measurements. The Rayleigh scattering involves scattering of electromagnetic radiation (e.g., light) from randomly positioned particles whose size is much smaller than the wavelength of the scattered radiation.

Role in Thermodynamics

The most common occurrence of Boltzmann's constant in thermodynamics is in the already mentioned ideal gas law. It should also be noted that the Dulong–Petit law states that the heat capacity per mole of a solid at high temperature is $3R$ (one mole of an ideal gas at high temperature has one half of this heat capacity). Since 1 cal is 4.186 J, $3R$ is about 6 cal/mol. In general, by the law of equipartition for a classical system, each degree of freedom that contributes to the kinetic energy plus each degree of freedom that contributes quadratically to the potential energy adds $0.5\ k_B T$ to the internal energy (or $0.5\ RT$ per mole).

Role in Statistical Mechanics

Statistical mechanics is the branch of physics that describes the behavior of systems with many identical constituents. Let S be the entropy of the system, a measure of its disorder, and W the number of microstates of the isolated system consistent with its en-

ergy and number of constituents. Then $S = k_B \ln W$. This result is carved on Boltzmann's memorial in the Central Cemetery in Vienna and is the fundamental link between thermodynamics and statistical mechanics. Thus, k_B plays the role of the proportionality constant linking a statistical concept and entropy. For S to have its equilibrium value, W must be a maximum.

See also: AVOGADRO NUMBER; BOLTZMANN, LUDWIG; CURIE TEMPERATURE; ENTROPY; GAS CONSTANT; IDEAL GAS LAW; MOTION, BROWNIAN; PLANCK, MAX KARL ERNST LUDWIG; RADIATION, BLACKBODY; RADIATION, THERMAL; STATISTICAL MECHANICS; STEFAN–BOLTZMANN LAW; THERMODYNAMICS

Bibliography

CALLEN, H. B. *Thermodynamics and an Introduction to Thermostatistics,* 2nd ed. (Wiley, New York, 1985).

HAAR, D. TER. *Elements of Statistical Mechanics* (Rinehart, New York, 1954).

HUANG, K. *Statistical Mechanics,* 2nd ed. (Wiley, New York, 1987).

KITTEL, C., and KROEMER, H. *Thermal Physics,* 2nd ed. (W. H. Freeman, San Francisco, 1980).

TOLMAN, R. C. *The Principles of Statistical Mechanics* (Oxford University Press, Oxford, Eng., 1938).

J. D. PATTERSON

BOLTZMANN DISTRIBUTION

The Boltzmann distribution describes the probability that a system comprised of a large number of atoms or molecules has a certain energy at a fixed temperature. If the system is at a temperature T and the possible values of the energy are $E_0, E_1, E_2, \ldots, E_i, \ldots$, then the probability $P(E_i)$ that the system has particular energy E_i is proportional to the Boltzmann factor $e^{-E_i/k_B T}$:

$$P(E_i) \propto e^{-E_i/k_B T}. \tag{1}$$

The constant k_B is Boltzmann's constant with the value 1.38×10^{-23} J·K^{-1}, and $e = 2.71828\ldots$ is the base of the natural logarithm. The combination

$1/k_BT$ frequently appears as β, so the Boltzmann factor is also written $e^{-\beta E_i}$. The term "Boltzmann distribution" refers to the set of all possible Boltzmann factors. The Boltzmann factor decreases rapidly as energy increases, so Eq. (1) states mathematically the simple observation that a system is more likely to be in a state of low energy than in a state of high energy.

If there are two energy states of interest, say E_0 and E_1, then the probability that the system is in the state with energy E_1 relative to the probability that it is in the state with energy E_0, is

$$\frac{P(E_1)}{P(E_0)} = \frac{e^{-E_1/k_BT}}{e^{-E_0/k_BT}} = e^{-(E_1 - E_0)/k_BT}. \qquad (2)$$

In words, the relative probability depends on the ratio of the energy difference between the states to the temperature (expressed in energy units since it is multiplied by Boltzmann's constant). The proportionality constant between the probability of occupation of a particular state and the Boltzmann factor is usually labeled Z^{-1}, so that

$$P(E_i) = \frac{e^{-E_i/k_BT}}{Z}. \qquad (3)$$

The constant Z is called the canonical partition function. Since the system must be in one of the states accessible to it, the sum of the occupation probability over all the states i is unity. This identifies Z as

$$Z = \sum_{i=1}^{n} e^{-E_i/k_BT}, \qquad (4)$$

or the sum of the Boltzmann factors for each state.

The average values, and their temperature dependence, of thermodynamic properties like energy, entropy, and pressure of a system are calculated using the Boltzmann factors. The average of any quantity is obtained by multiplying the value of the quantity by the probability of its occurrence, and adding such products for each possible value. For instance, the average energy is

$$U = \sum_{i=1}^{n} E_i P(E_i) = \frac{\sum_{i=1}^{n} E_i e^{-E_i/k_BT}}{Z}. \qquad (5)$$

A simple example is a system that can be in one of two states (usually called a two-level system). The energy difference between the states is ε, and for simplicity, suppose the energy of the lower state is zero and the energy of the upper is ε. An example of a system that approximates this ideal is an atom with two closely spaced energy levels that are very different in energy from any others. The average energy of the system is

$$U = \frac{E_1 e^{-E_1/k_BT} + E_2 e^{-2E_2/k_BT}}{e^{-E_1/k_BT} + e^{-E_2/k_BT}}$$

$$= \frac{0e^{-0/k_BT} + \varepsilon e^{-\varepsilon/k_BT}}{e^{-0/k_BT} + e^{-\varepsilon/k_BT}}$$

$$= \varepsilon \frac{e^{-\varepsilon/k_BT}}{1 + e^{-\varepsilon/k_BT}}. \qquad (6)$$

Equation (6) shows that as the temperature tends to zero, the exponentials become essentially zero so that average energy goes to zero. This implies that only the lowest (ground) state, with zero cnergy, is occupied at low temperature. "Low temperature" means that the thermal energy k_BT is much smaller than the energy difference between the states, so there is not enough energy to excite the system into the higher energy state. At high temperature, the energy difference between the states is insignificant compared to the thermal energy

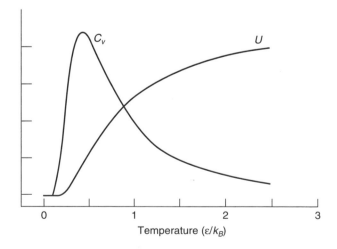

Figure 1 Average energy U and specific heat C_V of a two-level system, as a function of temperature. The temperature is expressed in units of ε/k_B.

$k_B T$, so both states are populated about equally. This is confirmed by Eq. (6), since in the high temperature limit the exponential factors are approximately unity, which gives an average energy of $\varepsilon/2$. Figure 1 shows the temperature dependence of the average energy.

From the average energy, it is simple to deduce an easily measured property, the specific heat of the system. The specific heat is the derivative of the average energy with respect to temperature with the volume held constant:

$$C_V = \left(\frac{\partial U}{\partial T}\right)_V. \qquad (7)$$

Substitution of Eq. (6) into Eq. (7) gives

$$C_V = \frac{\varepsilon^2}{k_B T^2} \frac{e^{-\varepsilon/k_B T}}{(1 + e^{-\varepsilon/k_B T})^2}. \qquad (8)$$

Figure 1 also shows the specific heat of the two-level system as a function of temperature. The specific heat goes to zero as the temperature decreases to zero and also as the temperature increases to infinity. The maximum at $k_B T \approx 0.4\varepsilon$ is called the Schottky anomaly.

Sometimes confused with the Boltzmann factor is a related quantity, the Gibbs factor. The Boltzmann factor is proportional to the probability of finding a system with a fixed number of particles in a state with a particular energy if that system is in contact with a heat reservoir (a much larger system at a fixed temperature) with which it can exchange energy. The Gibbs factor $e^{(\mu N_i - E_i)/k_B T}$ is the analogous expression for a system in contact with a reservoir with which it can exchange energy and particles. N_i is the number of particles in the state with energy E_i, and μ is the chemical potential. Just as the partition function is the sum of Boltzmann factors, so the grand partition function is the sum of Gibbs factors.

See also: BOLTZMANN, LUDWIG; BOLTZMANN CONSTANT; ENERGY LEVELS; GROUND STATE; SPECIFIC HEAT; STATE, EQUATION OF; THERMODYNAMICS

Bibliography

KITTEL, C., and KROEMER, H. *Thermal Physics*, 2nd ed. (W. H. Freeman, New York, 1980).

REIF, F. *Berkeley Physics Course*, Vol. 5: *Statistical Physics* (McGraw-Hill, New York, 1967).

WALDRAM, J. R. *The Theory of Thermodynamics* (Cambridge University Press, Cambridge, Eng., 1985).

JANET TATE

BOLTZMANN STATISTICS

See MAXWELL–BOLTZMANN STATISTICS

BOND

See COVALENT BOND; HYDROGEN BOND; IONIC BONDING

BORN, MAX

b. Breslau, Germany (now Wrocław, Poland), December 11, 1882; *d.* Göttingen, Germany, January 5, 1970; *quantum theory.*

Born was the elder of Gustav and Margarethe Born's two children. When Born was only four years of age, his mother died. His father, an anatomy professor at the University of Breslau, later married Bertha Lipstein, and they had one son.

Following his early years of education at the Kaiser Wilhelm Gymnasium in Breslau, Born entered the University of Breslau in 1901. Although he took a wide variety of courses, Born concentrated mainly on mathematics and physics. In 1904 Born moved on to the University of Göttingen, where he was greatly influenced by the works of mathematicians David Hilbert, Felix Klein, and Hermann Minkowski. Born received his Ph.D. in 1907; his dissertation was on the theory of elastic stability. Following a brief six-month period of study at Cambridge, where he attended lectures by Joseph J. Thomson, Born returned to Germany.

Born's research during this period was related to Albert Einstein's special theory of relativity. Using

Einstein's theory and Minkowski's mathematical approach, Born was able to obtain a simplified method for calculating the mass of an electron. His work at the University of Göttingen, where he had returned as a lecturer, was centered on optics and the study of crystals. He was able to show that the properties of crystals depended on the arrangement of their atoms, and he created an exact theory of the heat capacity of a crystal as a function of temperature.

In 1913 Born married Hedwig Ehrenberg (they had one son and two daughters), and two years later he accepted an assistant professorship under Max Planck at the University of Berlin. During this period Born met and developed a personal relationship with Einstein. Born's work in Berlin continued to center on crystal theory.

It was not until 1921, when Born was appointed director of Göttingen's Physical Institute, that his work turned to the development of a new theory of "quantum mechanics." This new quantum theory, created by Werner Heisenberg and developed with Born, Pascual Jordan, and Wolfgang Pauli, was to have a highly mathematical basis. In his studies on the phenomenon known as atomic scattering, Born developed a method known as the Born approximation and argued that quantum mechanics could only provide a statistical description of a particle's location based on probability; it could not provide a definite description of a particle's location. Soon after, Heisenberg published his uncertainty principle, which states that it is impossible to determine a particle's location *and* its momentum simultaneously.

With Adolf Hitler's rise to power during the early 1930s and the promulgation of anti-Semitic civil service laws, Born, who was of Jewish origin, lost his professorship. Born then left Germany for England and became Stokes lecturer at Cambridge, a position he held from 1933 until 1936, when he was appointed Tait professor of natural philosophy at the University of Edinburgh.

Born remained at Edinburgh until his retirement in 1953. He then returned to Germany, where he took an active stand regarding scientists' ethical and social responsibilities. He was especially active in attempts to stop further development of nuclear weapons and limit access to existing weapons.

In 1954, the year after his retirement, Born was awarded the Nobel Prize for physics for "his fundamental research in quantum mechanics, especially for his statistical interpretation of the wave function."

See also: ATOMIC PHYSICS; EINSTEIN, ALBERT; PAULI, WOLFGANG; PLANCK, MAX KARL ERNST LUDWIG; QUANTUM MECHANICS, CREATION OF; RELATIVITY, SPECIAL THEORY OF; SCATTERING; THOMSON, JOSEPH JOHN; UNCERTAINTY PRINCIPLE; WAVE FUNCTION

Bibliography

BORN, M. *My Life and My Views* (Scribners, New York, 1968).

BORN, M. *My Life: Recollections of a Nobel Laureate* (Taylor & Francis, Bristol, PA, 1978).

BORN, M. *Atomic Physics,* 8th ed. (Dover, New York, 1989).

BORN, M. *Physics in My Generation,* 2nd rev. ed. (Springer-Verlag, New York, 1989).

DALE KINSEY

BOSE–EINSTEIN CONDENSATION

See CONDENSATION, BOSE–EINSTEIN

BOSON

See FERMIONS AND BOSONS

BOSON, GAUGE

"Gauge boson" is the generic name for a particle that mediates the interactions between particles in a quantum gauge theory. Four such spin-1 bosons are known to exist: the massless photon that mediates electromagnetic interactions between charged particles, the massive W and Z bosons that mediate weak interactions, and the gluon which is responsible for the strong interaction between quarks. A fifth massless spin-2 gauge boson, the graviton, is assumed to mediate gravitational interactions, although there is no direct experimental evidence for its existence.

A gauge theory is one in which there exists a symmetry under which the fields in the theory transform nontrivially but, under which, physical phenomena are invariant. The simplest theory of this type is the Maxwell theory of electromagnetism. The (electromagnetic) gauge potential $[A_\mu(x)]$ can be modified $[A_\mu(x) \rightarrow A_\mu(x) + \partial_\mu\phi(x)$, where ϕ is a scalar field] in such a way that the electric and magnetic field strengths derived from A_μ are unchanged. This property is known as a local gauge invariance. If $\phi(x)$ is independent of the space time point x, then the invariance is said to be a global invariance. Such symmetries are associated with conserved charges. In the Maxwell theory, the conservation of electric charge is a direct consequence of the gauge invariance. When the theory is quantized, each classical field corresponds to a particle; the particle corresponding to the gauge potential is the gauge boson. In the quantum theory of electromagnetism (quantum electrodynamics or QED), there exists a particle, the photon, with spin equal to one, that is the quantum field corresponding to A_μ. The gauge symmetry ensures that the photon is massless. The interaction between charged particles, such as electrons, is mediated in the quantum theory by the exchange of photons. Under the gauge symmetry the phase of the electron field $[\psi(x)]$ changes thus: $\psi(x) \rightarrow exp(-i\phi(x)/e)\psi(x)$, where e is the electric charge of the electron. The interaction between two charges can be viewed as the emission of a photon by one of them and its absorbtion by the other. The massless nature of the photon results in a force of infinite range between the charges called the Coulomb interaction.

In the case of QED, the gauge particle (photon) does not carry the charge (electric charge) of the gauge symmetry. This is a consequence of the simple nature of the gauge invariance of QED. The weak and strong interactions are also described by gauge theories. In the latter case the gauge charge is called color. The quarks carry one of three color indices, i. In this case the gauge transformation on the quark fields rotates the colors among themselves, rather than simply changing the phase: $q_i(x) \rightarrow \sum_{j=1,3} exp(-i\phi(x)_{ij}/g) q(x)_j$. Hence, the gauge parameter $\phi(x)_{ij}$ and the gauge field of the theory, the gluon, itself carries the gauge charge. Unlike photons, the gluons can interact with each other by virtue of this charge. Unlike photons, the gluons

can interact with each other by virtue of this charge. The symmetry is exact and the gluons, like the photon, are massless. However, the force is not of infinite range. The self interactions of the gluons result in the strength of the gluon coupling to quarks becoming large at low energy or long distance. The resulting strong coupling results in quark confinement.

In the case of weak interactions, pairs of particles having the interactions (quarks or leptons) are transformed into one another by the gauge symmetry. For example a neutrino is transformed to an electron. The gauge boson involved in this case (the W boson) carries electric charge. As in the case of the gluons, the W bosons can interact with each other since they also carry the gauge charge. However, weak interactions are short ranged and weakly coupled. In order to explain this feature, the bosons must be massive. The mass breaks the gauge symmetry and arises because, although the full theory possesses the symmetry, its ground state does not. An analogous situation arises in the case of a ferromagnet. The underlying theory of magnetism (Maxwell theory) is rotationally invariant; no special orientation is preferred. However in a ferromagnet this symmetry is broken and a preferred direction (the direction of polarity of the magnet) is selected. The breaking of the symmetry of weak gauge invariance results in the generation of a mass for the W (and Z) gauge bosons. This spontaneously broken gauge symmetry is the only known way to consistently describe massive spin-1 particles in a quantum theory and this type of theory is therefore uniquely selected as the theory of weak interactions.

See also: BOSON, W; BOSON, Z; CHARGE, ELECTRONIC; ELEMENTARY PARTICLES; GAUGE INVARIANCE; GROUND STATE; INTERACTION, ELECTROMAGNETIC; INTERACTION, STRONG; INTERACTION, WEAK; PHOTON; QUANTUM ELECTRODYNAMICS; QUANTUM FIELD THEORY; SYMMETRY

Bibliography

LENG, T. P., and LI, L.-F. *Gauge Theory of Elementary Particles* (Oxford University Press, Oxford, Eng., 1984).
QUIGG, C. *Gauge Theories of Strong, Weak, and Electromagnetic Interactions* (Benjamin-Cummings, Redwood City, CA, 1983).

IAN HINCHLIFFE

BOSON, HIGGS

The discovery of the weak nuclear force, and the development of a comprehensive theory to explain it, has been one of the profound achievements of twentieth-century physics. Two consequences of the weak force are beta decay (a form of radioactivity) and hydrogen fusion (which ultimately is responsible for the energy we receive from the Sun). The theory of the weak force has been tested to great accuracy at high-energy particle accelerators. Nevertheless, one aspect of the theory has yet to be verified—the existence of a hypothetical elementary particle called the Higgs boson.

The "invention" by theoretical physicists of an elementary particle that is subsequently discovered in the laboratory has played a key role in the development of theories of fundamental particles and forces of nature. In 1930, Wolfgang Pauli invented the neutrino in order to explain certain anomalies in beta-decay radioactivity. Twenty-six years after his bold prediction, the neutrino was discovered in the laboratory. By 1961, a theory of the weak force had been formulated by Sheldon Glashow (and others) that invoked the existence of another new set of fundamental particles, called W and Z bosons. The term "boson" describes a class of particles whose interactions with ordinary matter transmit a force of attraction or repulsion. For example, the electromagnetic force between charged particles is transmitted through interactions with the photon (the quantum of light). Likewise, the interactions of the W and Z bosons with matter transmit the weak force. However, unlike the photon (which has no mass), the W and Z bosons were predicted to be very massive (nearly 100 times heavier than the proton). As a result, very high energy colliding particle beams are required to produce the W and Z bosons in the laboratory. In the collision process, energy is converted to mass (as predicted by Einstein's relativity theory, which implies the equivalence of mass and energy). With sufficient energy, the heavy W and Z bosons can be created. Indeed, the W and Z bosons were detected for the first time in high-energy particle collisions in 1983, and their predicted properties were verified.

Yet, Glashow's theory of the weak force cannot be complete; a careful analysis reveals that the theory fails to describe correctly the properties of elementary particle collisions at extremely high energies. Enter another inventor of elementary particles, Peter Higgs, who in 1964 postulated the existence of a new boson that now bears his name. What was the theoretical motivation that led to the invention of the Higgs boson? The answer is connected to the origin of mass. All elementary particles that comprise matter possess mass. In contrast, consider the bosons that are involved in the transmission of the fundamental forces. As noted above, the photon is masless. This is not accidental but rather is a consequence of a deep theoretical principle (called gauge invariance) that underlies the theory of electromagnetism and light. The same theoretical principles were used to construct the theory of the weak force; hence, the theory at first seemed to require that the W and Z bosons (in analogy with the photon) should also be massless. But the experimental properties of the weak force require the W and Z bosons to be very massive (a fact confirmed by their discovery in 1983). Calculations in Glashow's theory, however, show that certain high-energy scattering processes are predicted to occur with a probability greater than 100 percent, clearly a nonsensical outcome. The origin of this unacceptable result can be traced back to the fact that the W and Z bosons possess mass.

Remarkably, adding the Higgs boson to Glashow's theory can cure its flaws and yield sensible theoretical predictions in all cases. Through the interaction energy of the W and Z bosons with the Higgs boson, masses for the W and Z bosons are generated without violating any fundamental principles. (Mass as a consequence of interaction energy is in accord with Einstein's mass-energy equivalence.) In the period 1967 to 1968, Steven Weinberg and Abdus Salam took the invention of Peter Higgs and discovered how to incorporate the Higgs boson into a more complete theory of the weak force. In doing so, they combined the theory of the electromagnetic and weak forces into a unified description, called the electroweak force. The theory also implies that the electron mass is a consequence of the interactions of the Higgs boson with electrons. Apparently, the origin of mass is connected in some deep way with the existence of the Higgs boson. Although this may be true for the W and Z bosons and the electron, not all mass can be attributed to the existence of the Higgs boson. For example, most of the mass of the proton is due to the interaction energy of the strong force among its constituent quarks. Can one generate mass for the W and Z bosons and the electron in a similar manner, say, by inventing a new strong subnuclear force (not yet discovered), and dispense with the Higgs boson entirely? Many theorists have tried to do this, but so far the results have been unsatisfactory. So, the question remains undecided—is the mass of the W and Z bosons and the electron due to the Higgs boson or is it due to the existence of a new strong force between particles? Ultimately, experiment will decide which path nature has chosen.

If the Higgs boson exists, many of its properties are already known from the Weinberg and Salam theory of the electroweak force. Unfortunately, the theory does not predict a definite value for the Higgs boson mass; at best, it asserts that this mass cannot be larger than about 10 times the mass of the Z boson. Nevertheless, such information is useful since it indicates the minimum collider energy necessary to be sure that one can produce the Higgs boson and discover it. The most intensive search is presently being undertaken at the large electron-positron (LEP) collider at the *Consell Européen pour la Recherche Nucléaire* (CERN), the main European particle physics laboratory in Geneva, Switzerland. As of 1996, LEP had not succeeded in finding evidence for the Higgs boson. But, perhaps this is not too surprising since the LEP collider does not have sufficient energy to create and detect the Higgs boson if it is heavier than the Z boson. CERN is now developing and preparing to construct the world's highest energy particle collider, called the large hadron collider (LHC), which will collide two proton beams with a collision energy equivalent to 14 trillion volts. The LHC is scheduled to produce its first proton collisions in the year 2005. If the Higgs boson exists, it will be prolifically produced in such collisions—perhaps 1 million Higgs bosons per year. Yet, its discovery will not be easy. Although produced in great numbers, each Higgs boson is extremely unstable and decays almost immediately into other elementary particles. To prove that the Higgs boson has been produced, one must reconstruct its presence from the debris it has left behind. This is not an impossible task; nevertheless, it requires particle detectors of a very specialized nature and extremely sophisticated data analysis. Much work has already been devoted to developing the tools and techniques necessary for this task. Based on this work, particle physicists are confident that if the Higgs boson exists, it will be detected either in the near future at the LEP collider or later at the LHC.

Does the Higgs boson exist? Or does the existence of mass require fundamentally new phenomena that await discovery by future experiments? The answers, although not known today, will be found in the physics textbooks of the twenty-first century.

See also: BOSON, GAUGE; BOSON, NAMBU–GOLDSTONE; BOSON, W; BOSON, Z; DECAY, BETA; NEUTRINO; NUCLEAR FORCE; PARTICLE PHYSICS, DETECTORS FOR

Bibliography

GUNION, J. F.; HABER, H. E.; KANE, G.; and DAWSON, S. *The Higgs Hunter's Guide* (Addison-Wesley, Reading, MA, 1990).

VELTMAN, M. J. G. "The Higgs Boson." *Sci. Am.* **255** (5) 76–84 (1986).

WEINBERG, S. *Dreams of a Final Theory* (Vintage Books, New York, 1994).

HOWARD E. HABER

BOSON, NAMBU–GOLDSTONE

Nambu–Goldstone bosons are massless excitations that arise as a consequence of the spontaneous breakdown of a global symmetry. For each independent symmetry that is spontaneously broken, there is a Nambu–Goldstone boson. The origin of these particles, and their connection to spontaneous symmetry breaking, is perhaps best illustrated by a simple example. Consider two quantum fields σ and π associated with two different particles interacting through the potential

$$V = \lambda(\sigma^2 + \pi^2 + f)^2.$$

This potential is invariant under an arbitrary rotation (by an angle θ) of the σ and π fields into each other:

$$\sigma = \cos\theta\,\sigma' + \sin\theta\,\pi'$$

$$\pi = -\sin\theta\,\sigma' + \cos\theta\,\pi'.$$

This symmetry, if it is preserved by the dynamics, leads to equal masses for the σ and π particles. If it is spontaneously broken, it will lead to the appearance of a Nambu–Goldstone boson.

For the case at hand, these two alternatives are determined by the sign of the parameter f in the potential. If $f > 0$ the symmetry is preserved. The quadratic term in the potential informs us then that, as expected from the symmetry, σ and π have equal mass $m = 2\sqrt{\lambda f}$. If, on the other hand, $f < 0$ the potential has a maximum when $\sigma = \pi = 0$, with the minimum occurring at $\sigma^2 + \pi^2 = -f$. The fact that this minimum is not unique but is characterized by a curve in the $\sigma - \pi$ plane forces a breakdown of the symmetry. The spectrum of excitations in this case can be obtained by picking some possible minimum of the potential and again looking for the quadratic curvature at this point. This is simply done by writing, for example, $\sigma = \rho + \sqrt{-f}$ and expanding the potential about $\rho = \pi = 0$. One has

$$V = \lambda(\rho^2 + 2\sqrt{-f}\,\rho + \pi^2)^2.$$

Now the excitation associated with the ρ field has a mass $m_\rho = 2\sqrt{-2\lambda f}$, while the π field is massless. Clearly, the spectrum no longer has any symmetry at all and, as a result of the symmetry breakdown, a Nambu–Goldstone boson appears in the spectrum.

The best example in nature of Nambu–Goldstone bosons is provided by the π mesons. In reality, these particles have a small mass and so cannot be real Nambu–Goldstone bosons. What is true is that their mass is much smaller than that of any of the other strongly interacting particles (hadrons). This argument, however, does not really suffice to make this identification. Fortunately, one can adduce good theoretical reasons for supposing that the π mesons are approximate Nambu–Goldstone bosons. Quantum Chromodynamics (QCD), the theory that describes the strong interactions and determines the spectrum of hadrons as bound states of quarks and gluons, is well established. This theory, in the limit of vanishing quark masses, possesses a large global symmetry. Although not all quarks are light, the approximation of neglecting the mass of the u and d quarks is very good. So one would expect that the experimentally observed spectrum of hadrons containing these quarks should reflect this approximate symmetry. Although this spectrum indeed shows traces of symmetry—for example, the proton and the neutron have almost the same mass—the total symmetry present in QCD is not apparent in the spectrum. For instance, the full symmetry predicts that the proton and neutron should each have partners of the same mass and this is not seen experimentally. To reconcile this with the otherwise excellent understanding of the theory of the strong interactions, it is neces-

sary to suppose that part of the global symmetry of QCD, present in the limit in which one neglects the u and d quark masses, is spontaneously broken. The (approximate) Nambu–Goldstone bosons associated with this broken symmetry are identified with the π mesons. Indeed, one can show that the square of the mass of these particles is directly proportional to the sum of the u and d quark masses. Therefore, π mesons really would become Nambu–Goldstone bosons in the limit of massless quarks.

See also: BOSON, GAUGE; BOSON, HIGGS; BOSON, *W*; BOSON, *Z*; EXCITED STATE; FERMIONS AND BOSONS; HADRON; INTERACTION, STRONG; QUANTUM CHROMODYNAMICS; SYMMETRY BREAKING, SPONTANEOUS

Bibliography

WEINBERG, S. *Dreams of a Final Theory* (Vintage Books, New York, 1994).

ROBERTO D. PECCEI

BOSON, *W*

The *W* boson, a spin-1, electrically charged gauge boson of mass 80.33 ± 0.15 GeV/c^2, is responsible for mediating decays that transform quark or lepton types.

The *W* boson was discovered at the European Center for Nuclear Research (CERN) in Geneva in 1983. The UA1 (Underground Area) experiment observed events produced by collisions of protons and antiprotons each of energy 270 GeV in a storage ring. The *W* bosons were produced by quark-antiquark annihilation and detected via their decay into an electron and neutrino or muon and neutrino. The 1985 Nobel Prize in physics was awarded to Carlo Rubbia and Simon Van Der Meer for this discovery. The largest sample of *W* bosons, and the most precise determination of its mass, is now provided by the proton-antiproton collider located at Fermilab near Chicago.

Parity is maximally violated in the coupling of quarks and leptons to the *W* boson. Left-handed quarks and leptons (i.e., ones with their spin aligned against their direction of motion) couple while right handed ones do not. The coupling strength (g) is related to the W mass (M_W) and Fermi constant (G_F) of nuclear beta decay by $G_F = g^2/8M_W^2$; g is related to the electric charge of the electron in the Glashow-Weinberg-Salam model of weak and electromagnetic interactions. The mass of the W boson was predicted prior to its discovery by using the known values of these quantities and results from experiments involving neutrino scattering.

Lepton flavors (corresponding to the electron, the muon, the tau, and their associated neutrinos) are conserved by the W couplings, hence, for example, the decay of a muon produces an electron, a muon-type neutrino, and an electron-type antineutrino; $\mu^- \rightarrow e^- \nu_\mu \bar{\nu}_e$. Quark flavors are not conserved in the couplings to a W which can be written in the form $\bar{U}_i K_{i,j} D_j$. Here U_i refers to one of the three flavors of charge $\frac{2}{3}$ quarks (up, charm, and top) and D to one of the three flavors of charge $-\frac{1}{3}$ quarks (down, strange, and bottom). The (Kobayashi–Maskawa) matrix $K_{i,j}$ parameterizes the strength of the flavor violation. The simplest quark transition ($i = j = 1$) produces $d \rightarrow ue^- \bar{\nu}_e$, which is the quark transition responsible for nuclear beta decay or the decay of the neutron to proton, electron, and antineutrino. As a consequence of these couplings, only the lightest baryon, the proton, is stable with respect to decay by weak interactions. (Isolated neutrons are not stable, but many atomic nuclei are stable because neutrons inside these nuclei would result in two protons in identical quantum states, which is forbidden by the Pauli exclusion principle.)

See also: ANTIMATTER; ATOM; ATOMIC PHYSICS; BARYON NUMBER; CHARGE; COLLISION; DECAY, BETA; ELECTROMAGNET; ELECTROMAGNETIC FORCE; ELECTROMAGNETISM; ELECTRON; ENERGY; FERMI, ENRICO; FERMIONS AND BOSONS; INTERACTION; INTERACTION, ELECTROMAGNETIC; INTERACTION, STRONG; INTERACTION, WEAK; LEPTON; MASS; MASS-ENERGY; MUON; NEUTRINO; NEUTRON; PARITY; PAULI'S EXCLUSION PRINCIPLE; PROTON; QUARK; SPIN; SPIN, ELECTRON; SPIN AND STATISTICS

Bibliography

ARNISON, G., "Experimental Observation of Isolated Large Transverse Energy Electrons with Associated Missing Energy." *Phys. Lett.* **122B**, 103 (1983).

GLASHOW, S. "Partial Symmetries of Weak Interactions." *Nucl. Phys.* **22**, 579 (1961).

SALAM, A. "Weak and Electromagnetic Interactions" in *Elementary Particle Theory*, edited by N. Svartholm (Almqvist and Wiksell, Stockholm, 1968).

WEINBERG, S. "A Model of Leptons." *Phys. Rev. Lett.* **19**, 1264 (1967).

IAN HINCHLIFFE

BOSON, Z

The Z boson is a spin-1, electrically neutral gauge boson of mass 91.187 ± 0.007 GeV/c^2. It decays into quark-antiquark, lepton-antilepton, or neutrino-antineutrino pairs. The existence of the boson was first inferred from experiments on neutrino scattering. The dominant neutrino scattering process involves the exchange of a charged W boson and the transmutation of the neutrino into a charged lepton. In the early 1970s experiments at the European Center for Nuclear Research (CERN) in Geneva, Switzerland, and at Fermilab, located near Chicago, Illinois, found evidence for the interactions of neutrinos that did not involve the production of charged leptons. Additional evidence arose from the observation that the scattering rate of polarized electrons from an unpolarized nucleus showed a slight asymmetry; the cross section for polarization parallel and antiparallel to electron's momentum is not equal. This violation of parity can be accounted for if, in addition to photon exchange which dominates the cross section and conserves parity, there is a contribution from Z exchange. Additional evidence is provided from the observation of (very small) parity violating effects in atomic physics. The mass of the Z boson was inferred from these measurements by using the couplings predicted by the Glashow–Weinberg–Salam model of weak and electromagnetic interactions. The UA1 (underground area) experiment observed events produced by collisions of protons and antiprotons each of energy 270 GeV in a storage ring located at CERN. The Z bosons were produced by quark-antiquark annihilation and detected via their decay into an electron and positron or muon and antimuon. The 1985 Nobel Prize in physics was awarded to Carlo Rubbia and Simon Van Der Meer for this discovery.

Approximately 10 million Z bosons have been produced in e^+e^- annihilation at the LEP (large electron positron) collider located at CERN. This large sample enables the couplings of the Z to quarks and leptons to be measured with great accuracy. A smaller number of Z's have been produced at the Stanford Linear Accelerator Center in California. The latter has a polarized electron beam which enables more detailed studies of the Z-boson couplings to be made. The total decay width of the Z is determined from the behavior of the process $e^+e^- \rightarrow Z \rightarrow \mu^+\mu^-$ measured as a function of the energy of the e^+e^- system. This can then be combined with the measured decay rate into quarks and leptons to infer the decay rate into neutrinos even though the decay $Z \rightarrow \nu\bar{\nu}$ is not observable directly since the neutrinos do not interact within the detector. The data indicate the number of massless neutrinos is 2.988 ± 0.023. Since evidence for the muon, electron, and tau type neutrinos exists, these data exclude a fourth light neutrino and provide indirect evidence that there are only three generations of quarks and leptons.

Unlike the W boson, the Z boson does not play a direct role in the decay of quarks and leptons. This result follows from the fact that there is no evidence for a coupling a Z to lepton-antilepton or quark-antiquark where the quarks or leptons are of different flavor. The decay of a neutral K meson [a bound state of a strange quark (s) and an antidown quark ($\bar{d}$)] to e^+e^- has not been seen, and hence the coupling Z to $s\bar{d}$ is very small. In the Glashow–Weinberg–Salam model of weak and electromagnetic interactions all such flavor changing couplings of the Z boson are zero.

See also: BOSON, GAUGE; BOSON, W; DECAY, BETA; FERMIONS AND BOSONS; INTERACTION; INTERACTION, STRONG; INTERACTION, WEAK; LEPTON; MUON; NEUTRINO; NEUTRON; PARITY; QUARK; QUARK, DOWN; QUARK, STRANGE

Bibliography

ARNISON, G., et al. "Experimental Observation of Lepton Pairs of Invariant Mass Around 95 heV/c^2 at the CERN SPS Collider." *Phys. Lett.* **126B,** 398 (1983).

AUBERT, B., et al. "Measurement of Rates for Muonless Inelastic Neutrino and Anti-Neutrino Scattering." *Phys. Rev. Lett.* **32,** 1454 (1974).

GLASHOW, S. "Partial Symmetries of Weak Interactions." *Nucl. Phys.* **22,** 579 (1961).

HASERT, F. J., et al. "Observation of Neutrino-Like Interactions Without Muon or Electron in the Gargamelle Neutrino Experiment." *Phys. Lett.* **46B,** 138 (1973).

PRESCOTT, C. Y., et al. "Parity Non-Conservation in Inelastic Electron Scattering." *Phys. Lett.* **77B,** 347 (1978).

SALAM, A. "Weak and Electromagnetic Interactions" in *Elementary Particle Theory,* edited by N. Svartholm (Almqvist and Wiksell, Stockholm, 1968).

SCHAILE, D. "Tests of the Electroweak Theory at LEP." *Fortschr. Phys.* **42,** 429 (1994).

WEINBERG, S. "A Model of Leptons." *Phys. Rev. Lett.* **19,** 1264 (1967).

IAN HINCHLIFFE

BOYLE'S LAW

Robert Boyle was born in 1627 in Ireland, but he spent much of his life in Oxford and London. He was a renowned experimenter on gases, and he was one of the founders of the Royal Society of London, which was established to support scientific experiments and discussion in a time when science was not well organized.

Boyle's experimental apparatus was a uniform glass U-tube, open to the atmosphere at one end and closed at the other end. Mercury was poured into the open end, trapping a sample of air for the experiment at the closed end. Boyle recorded the height of the column of air and calculated its volume. The pressure of the air could be measured and corrected to take into account atmospheric pressure. As more mercury was poured into the open end, the absolute pressure of the trapped air increased with a corresponding decrease in the volume of the air. A graph of representative data is shown in Fig. 1, using both the English system of units and SI metric units.

In 1662 Boyle announced his discovery that the pressures and volumes of a closed sample of air are "in reciprocal proportions." In other words,

$$P \propto \frac{1}{V},$$

where P is the pressure and V is the volume; $PV =$ constant, at constant temperature. It was well-known to Boyle that air expands when heated, so the restriction "at constant temperature" was obviously necessary for this simple law. Boyle maintained the

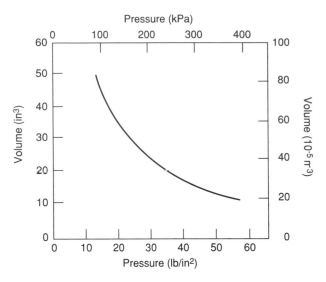

Figure 1 Graphic representation of the relationship between the volume of the air and the absolute pressure of the trapped air when mercury is poured into a tube.

temperature of the air at a fixed value during the experiment by using a wet cloth to cool the trapped air as it was being compressed.

Later experiments showed that all simple gases obey Boyle's law at temperatures much above their boiling points, except that the various products of pressure and volume yield different constants for different temperatures of the same sample of gas. Boyle's law is one of the experimental laws needed to establish empirically the ideal gas law, which is the basis of the kinetic theory of gases.

An application of Boyle's law is the filling of a medicine dropper with a liquid by means of a process known as "suction." The bulb of the dropper is squeezed, which decreases the volume of air in the bulb. As the dropper is inserted into the liquid, there is a smaller volume of air but the pressure is still at atmospheric pressure inside the bulb. When the bulb is released, the volume of the trapped air increases while the pressure inside the bulb decreases to a value *less* than the outside atmospheric pressure, according to Boyle's law. Thus, for a tube of uniform cross-sectional area, there is less force exerted on the liquid by the air in the bulb and more force exerted on the liquid by the atmosphere. The resulting net force on the liquid causes the liquid to move into the bulb, according to Newton's second law of motion.

A bicycle pump also uses Boyle's law to explain its pumping action. Unlike the medicine dropper, the air trapped inside a bicycle pump experiences an increase of pressure as the volume decreases. In this case, Newton's second law pumps in the direction opposite to the medicine dropper.

See also: GAS; IDEAL GAS LAW; KINETIC THEORY; PRESSURE, ATMOSPHERIC; THERMOMETRY

Bibliography

SCHOOLEY, J. F. *Thermometry* (CRC Press, Boca Raton, FL, 1986).

RICHARD H. DITTMAN

BRAGG'S LAW

The first observation of the diffraction of x rays by crystals was made by Max von Laue and his students Walter Friedrich and Paul Knipping in Munich in 1912. When a crystal was placed in a fine beam of x rays, diffracted rays came out in various directions, making spots on a photographic plate. However, the interpretation of the results was by no means straightforward. It was known at that time that crystals are made by the repetition of atoms or groups of atoms at regular intervals in each of three dimensions, that is, placed at the points of a three-dimensional lattice, but the repeat distances and the nature of the repeated atom groups were not known. It was known that x rays coming from the metal target of an x ray tube had a continuous distribution of wavelengths, with high intensity for some particular wavelengths, but the exact values of these wavelengths were unknown. Von Laue gave a theory of x-ray diffraction in 1912 but, although the theory was essentially correct, it was complicated and not very useful for the many people who, within a year, were confirming the original experimental results.

One of these people was William Lawrence Bragg (later Sir Lawrence Bragg), then aged twenty-two and a student at Cambridge, who had been involved with the work of his father, William Henry Bragg, professor of physics in Manchester, on studies with x rays. W. L. Bragg introduced the simplifying concept of the reflection of x rays from planes of atoms and derived the equation known as Bragg's law.

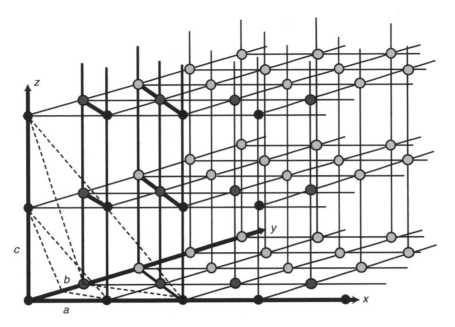

Figure 1 Part of a space lattice formed by translations a, b, and c along the x, y, and z axes, respectively. Planes with Miller indices 110 (dark lines) and 121 (dashed lines) are indicated.

It was known at the time, particularly by mineralogists, that the regular arrangement of atoms (or groups of atoms) in a crystal implied that the atoms lay on sets of regularly spaced planes, resulting in the planar faces of crystals and the breaking of crystals along definite cleavage planes. Such sets of parallel planes were denoted by the Miller indices, $hk\ell$ (see Fig. 1). If the repeat distances of the atom arrays were denoted by a, b, and c along the x, y, and z axes, and one of a set of parallel planes passed through the origin, then the next plane of the set would intersect the axes at a/h, b/k, and c/ℓ. The distance between planes of a parallel set is $d_{hk\ell}$. For rectangular axes, simple geometry gives

$$1/d_{hk\ell}^2 = h^2/a^2 + k^2/b^2 + \ell^2/c^2. \qquad (1)$$

For such a set of planes, Bragg's law is

$$2d \sin \theta = n\lambda, \qquad (2)$$

where λ is the wavelength of the x rays, d is the interplanar distance, θ is the angle made by the incoming and reflected rays with the planes, and n is an integer. The derivation follows from the geometry of Fig. 2.

If rays are incident on the planes at an angle θ and reflected at the same angle, then the path difference between rays reflected by successive planes is $2d \sin \theta$. If this path difference is $n\lambda$, an integral multiple of the wavelength, then reflected rays from all planes are in phase and add up to give a strong reflected beam. Thus, Eq. (2) gives immediately the geometry of the diffracted beams from a crystal. It was shown later that the intensities of the diffracted beams depend on the nature and relative positions of the atoms in the group associated with each crystal lattice point. For a single atom, the scattered am-

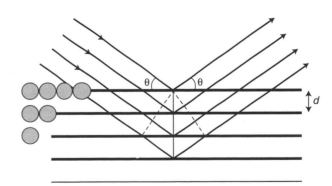

Figure 2 The geometry of Bragg's law. Incident and reflected rays make an angle θ with planes of spacing d representing planar arrays of atoms in a crystal.

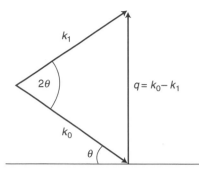

Figure 3 Vector diagram equivalent to Fig. 2. The incident and reflected beams are represented by vectors $\mathbf{k}_0$ and $\mathbf{k}_1$ of magnitude $1/\lambda$. The diffraction vector $q = \mathbf{k}_1 - \mathbf{k}_0$ has magnitude $2\lambda^{-1} \sin \theta$.

plitude is proportional to the atomic scattering factor $f(\theta)$. For a group of atoms at positions relative to the lattice point denoted by vectors $\mathbf{r}_i$, or coordinates x_i, y_i, z_i, the amplitude is

$$F_{hk\ell} = \Sigma_i f_i \exp\{2\pi i(hX_i + kY_i + \ell Z_i)\}, \qquad (3)$$

where $X_i = x_i/a$, $Y_i = y_i/b$, and $Z_i = z_i/c$. Then the diffracted beam intensity for diffraction by the $hk\ell$ planes is proportional to $|F_{hk\ell}|^2$.

Bragg's law is widely used for the interpretation of not only x-ray diffraction experiments but also for neutron diffraction and electron diffraction. It is applied routinely for the identification of crystalline phases by powder diffraction methods. In the case of fine powders with small crystals in random orientations, diffracted beams come out of the specimen at all the angles $2\theta_{hk\ell}$ associated with all sets of planes h, k, ℓ of the crystal lattice, that is, in a set of cones of half angle $2\theta_{hk\ell}$ with the incident beam as axis and with relative intensities given by Eq. (3). The cones of radiation, detected on a photographic film or with an electronic counter, give patterns that are characteristic of the crystalline phase and so can be used for "fingerprinting." Computer data bases now available contain the d values and relative intensities for several hundred thousand substances and allow rapid identification of unknown materials or mixtures of phases.

Bragg's law, however, is much more powerful than is implied by the relationship between d values and diffraction angles given in Eq. (2) and used for powder patterns. The concept of reflection from planes of atoms, as in the reflection of light from a flat glass surface, implies a vector relationship, defining the directions of the diffracted beams. Thus if, as in Fig. 3, a vector $\mathbf{k}_0$ of magnitude $1/\lambda$ is drawn in the incident beam direction, and a vector, $\mathbf{k}_1$ of length $1/\lambda$ is drawn in the diffracted beam direction, the difference $\mathbf{k}_1 - \mathbf{k}_0$, the diffraction vector is a vector perpendicular to the diffracting planes of length $|\mathbf{q}| = (2 \sin \theta)/\lambda = 1/d$.

The concept of a reciprocal lattice was developed in the nineteenth century. If a vector is drawn from an origin perpendicular to the $hk\ell$ set of planes and of length $1/d_{hk\ell}$, the vector defines the h, k, ℓ reciprocal lattice point. The reciprocal lattice is a regularly spaced set of points defined by (for rectangular axes) distances of $1/a$, $1/b$, $1/c$ along the axes, so that the distance of the h, k, ℓ reciprocal lattice point from the origin is given by Eq. (1).

Thus Bragg's law implies that a strong diffracted beam will be produced if the diffraction vector extends from the reciprocal lattice origin to a reciprocal lattice point. Or, for a given incident beam vector $\mathbf{k}_0$, drawn to the reciprocal lattice origin, a diffracted beam will be given if a sphere of radius $1/\lambda$, drawn around the starting point P of the vector $\mathbf{k}_0$, passes through a reciprocal lattice point as in Fig. 4. Hence Bragg's law in vector form is identical with the concept of the sphere of reflection, or

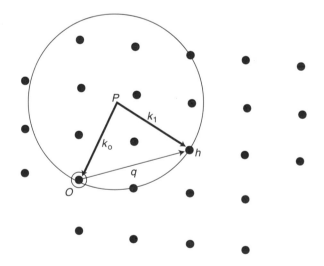

Figure 4 The Ewald sphere construction in reciprocal space. If a sphere of radius λ^{-1}, drawn through the origin O, with center at P, the starting point of vector $\mathbf{k}_0$, passes through the $hk\ell$ reciprocal lattice point (denoted by $\mathbf{h}$), a diffracted beam will be produced in the direction of $\mathbf{k}_1$.

Ewald sphere, proposed by P. P. Ewald in his doctoral thesis in 1913 and universally used as a basis for determining the geometry of diffraction patterns given by x rays, neutrons, or electrons from single-crystal specimens.

The simple concept of Bragg's law, applied to reflection from crystal planes, became an important basis for the development of the analysis of crystal structures by x-ray diffraction methods by W. L. Bragg and his many followers, and for the development of x-ray spectroscopy by W. H. Bragg and his many followers. For their pioneering work, the Braggs, father and son, received the Nobel Prize in physics for 1915.

See also: CRYSTALLOGRAPHY; CRYSTAL STRUCTURE; DIFFRACTION; SCATTERING; SPECTROSCOPY, X-RAY; X RAY

Bibliography

BRAGG, W. L. *The Crystalline State,* Vol. 1: *A General Survey* (Bell & Sons, London, 1949).

BUNN, C. *Crystals: Their Role in Nature and in Science* (Academic Press, New York, 1964).

EWALD, P. P., ed. *Fifty Years of X-Ray Diffraction* (N. V. A. Oosthoek's Uitgeversmaatschappij, Utrecht, 1962).

JCPDS–INTERNATIONAL CENTER FOR DIFFRACTION DATA. *JCPDS–ICDD Powder Diffraction File* (JCPDS–ICDD, Swathmore, PA, 1990).

JOHN M. COWLEY

BREEDER REACTOR

See REACTOR, BREEDER

BREMSSTRAHLUNG

Bremsstrahlung is a German compound word that translates as "braking radiation." It refers to the electromagnetic radiation produced when an electric charge is accelerated. This radiation can be derived from Maxwell's equations as a classical effect. In the quantum theory of electrodynamics (QED) bremsstrahlung occurs through the production of photons. The name reflects the fact that by radiating this energy the particle is slowed somewhat.

The phenomenon of bremsstrahlung is the major way in which high-energy electrons traveling through matter are slowed and stopped. When such electrons traverse strong electric field regions inside atoms, they are accelerated (i.e., deflected). This acceleration causes the electron to emit bremsstrahlung photons. Each photon can produce an additional electron, together with its antiparticle positron. Each of these particles again produces bremsstrahlung radiation in the atomic field regions. Thus one gets a shower of pair-produced electrons and positrons of successively lower energies.

When any two energetic charged particles interact, they accelerate one another. Hence bremsstrahlung effects must be taken into account in understanding high-energy collisions in particle physics experiments. These effects are referred to as radiative corrections and are calculated from QED.

Bremsstrahlung also plays an important role in accelerator design, because its effects must be considered whenever a particle is accelerated in any way. In accelerators a major bremsstrahlung effect is the production of radiation whenever the beam of charged particles is deflected by a magnetic field. This is referred to as magnetic bremsstrahlung. In a circular accelerator, or synchrotron, magnets steer the beam around the ring, providing an acceleration toward the center of the circle. Bremsstrahlung in a synchrotron due to this acceleration is also called synchrotron radiation. The energy lost from the beam by synchrotron radiation must be restored by accelerating regions in the ring to keep a beam circulating at a constant energy. This effect becomes a major limiting factor on the design of storage rings for high-energy electrons. The radius of the storage ring must grow with the desired energy to keep the energy loss due to synchrotron radiation at an acceptable level. The synchrotron radiation itself turns out to be a useful tool to study matter on the atomic scale since it is an intense source of x rays. A number of electron synchrotrons have now been built, specifically designed to maximize the synchrotron radiation production for this purpose. For protons, which have much larger mass than electrons, the amount of synchrotron radiation at a given beam energy is much less, and so it is not as important a factor in the design of proton synchrotrons.

A peculiar modern variant of the word bremsstrahlung is the English-German hybrid term "beamstrahlung." This term is used by accelerator physicists to describe the intense radiation produced by a small but high-density bunch of charged particles as it passes through a similar but oppositely moving (and perhaps oppositely charged) bunch. This effect can cause a high-energy electron to radiate a substantial fraction of its energy. Accelerator designers work hard to find designs for the bunch shapes at the collision point that can minimize such effects while maintaining high collision rates.

See also: ACCELERATOR; ACCELERATOR, HISTORY OF; ELECTROMAGNETIC RADIATION; MAXWELL'S EQUATIONS; PHOTON; QUANTUM ELECTRODYNAMICS; RADIATION, SYNCHROTRON; SYNCHROTRON

Bibliography

PURCELL, E. M. *Berkeley Physics Course*, Vol. 2: *Electricity and Magnetism* (McGraw-Hill, New York, 1965).

HELEN R. QUINN

BREWSTER'S LAW

Polarized light may be obtained by several methods, one of which is by reflection. An incident unpolarized light beam reflected from a surface is polarized to a certain extent depending on the angle of incidence. For a particular angle of incidence, the reflected light is found to be completely polarized (i.e., the electric field always oscillates in the same direction). This angle of incidence is called Brewster's angle, after David Brewster, who also invented the kaleidoscope.

When the angle between the reflected and the refracted rays is a right angle, as shown in Fig. 1, then the reflected beam is completely polarized with the electric field vector parallel to the surface. This angle of incidence is the polarizing angle, or Brewster's angle. The refracted ray is described by Snell's law,

$$n_1 \sin \theta_1 = n_2 \sin \theta_2,$$

where n is the index of refraction of the medium and θ is the angle between the ray and a normal to

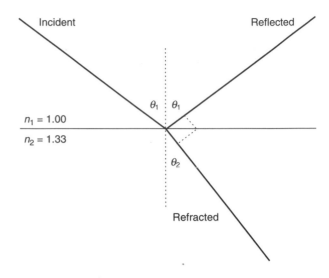

Figure 1 Reflection and refraction of an incident light beam.

the surface. Since the angle of reflection is equal to the angle of incidence, and at the polarizing angle

$$\theta_1 + \theta_2 + 90° = 180°,$$

then

$$\frac{n_2}{n_1} = \frac{\sin \theta_1}{\sin \theta_2} = \frac{\sin \theta_p}{\cos \theta_p} = \tan \theta_p,$$

which is Brewster's law. For example, the polarizing angle for light incident from air (index of refraction 1.00) onto water (index of refraction 1.33) is

$$\theta_p = \tan^{-1} 1.33 = 53.1°.$$

The polarizing angle depends somewhat on the wavelength of the light since the index of refraction is a function of wavelength. For ordinary glass the dispersion is such that the polarizing angle does not change much over the visible spectrum.

The incident light sets the electrons in the atoms or molecules of the reflector/refractor into oscillation. The reflected beam is the reradiation from these atoms. If the reflected and refracted rays are perpendicular to each other, only the vibrations perpendicular to the plane of incidence can contribute to the reflected beam. Those atoms in the plane of incidence have no radiation component perpendicular to the 90° direction. Thus the reflected ray is polarized. Reflected light from water, pavement, and other similar surfaces is therefore nearly linearly polarized, with the electric field vector vibrating preferentially in the horizontal plane. Polarizing sunglasses with their planes of transmission vertical will reduce such reflected glare.

Polarization by a stack of glass plates is possible if the plates are oriented such that the reflected ray is completely polarized. Since most of the light is transmitted through the plates, however, the reflected polarized beam is relatively weak. The transmitted refracted ray is partially polarized, and the degree of polarization can be made larger by including a larger number of plates. Two such piles of plates, one as a polarizer and the other as an analyzer, may be used in tandem as a polariscope.

Lasers produce polarized light when the end plates have been tilted or beveled to the polarizing angle (Brewster's windows). Then there is complete transmission for light with the electric field vector parallel to the plane of incidence. The normal component of the electric field is partially reflected at each interface with each traversal of the laser discharge tube, and the resulting laser beam is thereby polarized. Brewster windows will let pass only one direction of polarization, but without reflection losses.

See also: LASER; POLARIZED LIGHT; REFLECTION; REFRACTION; REFRACTION, INDEX OF; SNELL'S LAW

Bibliography

JENKINS, F. A., and WHITE, H. A. *Fundamentals of Optics,* 4th ed. (McGraw-Hill, New York, 1976).
MOLLER, K. D. *Optics* (University Science Books, Mill Valley, CA, 1988).
SERWAY, R. A. *Physics for Scientists and Engineers,* 3rd ed. (Saunders, Philadelphia, 1990).

JULIETTE W. IOUP

BRILLOUIN ZONE

Brillouin zones are geometrical constructions that have great utility in theoretical solid state physics for describing electrons, and excitations such as phonons, in crystals. Underlying any crystal structure is a lattice, defined as an infinite set of regularly repeating points in space with the arrangement and orientation about each point being identical. A Brillouin zone can be associated with any real-space lattice describing a given crystal structure.

For any real-space lattice a reciprocal-space or *k*-space lattice can be defined. Position vectors in real space have dimensions of length, (e.g., A or m); those in reciprocal space have dimensions of reciprocal length (A^{-1} or m^{-1}). It is on this *k*-space lattice that Brillouin zones are defined. Specifically, the first Brillouin zone is the region in *k* space that is closer to any given reciprocal lattice point than to any other.

Three-dimensional real-space lattices can be constructed from three noncoplanar fundamental lattice vectors, **a, b,** and **c,** using vector addition:

$$\mathbf{R} = h\mathbf{a} + k\mathbf{b} + l\mathbf{c},$$

where the **R** are lattice points and *h, k,* and *l* have all integer values. Fundamental reciprocal-lattice vectors, **a*, b*,** and **c*** are then defined as

$$\mathbf{a}^* \equiv 2\pi \frac{\mathbf{b} \times \mathbf{c}}{\mathbf{a} \cdot \mathbf{b} \times \mathbf{c}},$$

$$\mathbf{b}^* \equiv 2\pi \frac{\mathbf{c} \times \mathbf{a}}{\mathbf{a} \cdot \mathbf{b} \times \mathbf{c}},$$

$$\mathbf{c}^* \equiv 2\pi \frac{\mathbf{a} \times \mathbf{b}}{\mathbf{a} \cdot \mathbf{b} \times \mathbf{c}},$$

and the points, **K,** of the reciprocal lattice are integer combinations of these:

$$\mathbf{K} = h\mathbf{a}^* + k\mathbf{b}^* + l\mathbf{c}^*.$$

See Fig. 1 for examples in one and two dimensions.

To construct the first Brillouin zone, pick any reciprocal lattice point as the origin, *O.* Planes that are perpendicular bisectors to **K** vectors are called Bragg planes, which correspond to x-ray and electron diffraction lines. Consider the **K** vectors pointing from *O* to the first several lattice points nearest *O* and construct their Bragg planes. The smallest region about *O* enclosed by these Bragg planes is the

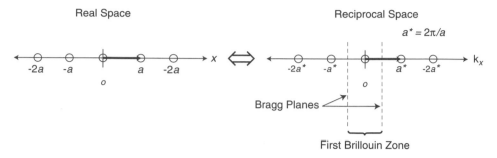

(a) One-Dimensional

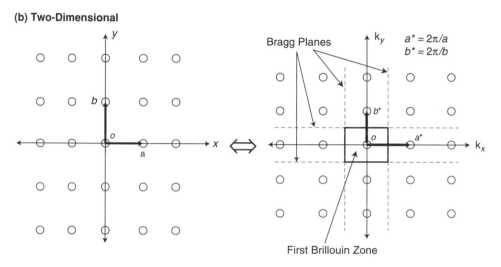

(b) Two-Dimensional

Figure 1 Brillouin zones in one and two dimensions.

first Brillouin zone (see Fig. 1). Similarly, the second Brillouin zone is the region outside the first that can be reached without crossing any Bragg planes.

For two-dimensional lattices, the first Brillouin zone is either rectangular or hexagonal. The simplest three-dimensional case is the simple cubic lattice: $\mathbf{a} = a\hat{x}$, $\mathbf{b} = a\hat{y}$, $\mathbf{c} = a\hat{z}$. It has a simple cubic reciprocal lattice and a cubic first Brillouin zone with sides of dimensions $2\pi/a$. The first two Brillouin zones of the three-dimensional face-centered cubic lattice of copper, diamond, and other important solids is shown in Fig. 2.

According to quantum mechanics (or wave mechanics), particles such as electrons and phonons in a crystal behave like waves or functions of the form $\sin(k_x x + k_y y + k_z z)$ and $\cos(k_x x + k_y y + k_z z)$, where the $k_i = 2\pi/\lambda_i$ are wave numbers; λ_i being wavelengths. It is natural to describe these solutions in terms of a three-dimensional k space. For finite crystals with N lattice sites, only N discrete k values are allowed, corresponding to the different wavelengths

that fit within the crystal. The longest such wavelength has the dimensions of the crystal (Na in one dimension); the shortest has the dimensions of a single lattice constant. These correspond to k values at the Brillouin zone center ($k = 2\pi/Na \approx 0$ for

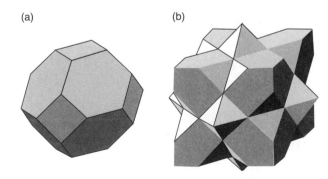

Figure 2 Brillouin zones of the face-centered cubic lattice: (a) first Brillouin zone and (b) second Brillouin zone.

150

large N) and edge ($2\pi/a$), respectively. Thus, only one Brillouin zone in reciprocal space is needed to describe fully all of the allowed k values, or solutions, to these fundamentally important problems in solid state physics.

See also: CRYSTAL; CRYSTAL STRUCTURE; ELECTRON; PHONON; WAVE MECHANICS

Bibliography

ASHCROFT, N. W., and MERMIN, N. D. *Solid-State Physics* (Saunders, Philadelphia, 1976).

BLAKEMORE, J. S. *Solid-State Physics,* 2nd ed. (Saunders, Philadelphia, 1974).

BURNS, G. *Solid-State Physics* (Academic Press, San Diego, CA, 1990).

KITTEL, C. *Introduction to Solid-State Physics,* 6th ed. (Wiley, New York, 1986).

WILLIAM F. OLIVER III

BROGLIE, LOUIS-VICTOR-PIERRE-RAYMOND DE

b. Dieppe, France, August 15, 1892; *d.* Paris, France, March 19, 1987; *quantum theory, relativity.*

De Broglie suggested in 1923 that atoms of matter have an inescapable wave-like property, one that controls their movement in microspace. His ideas were heavily influenced by those of Albert Einstein of a decade and a half earlier that wave-like light must be ascribed an inescapable particulate property. De Broglie won the 1929 Nobel Prize in physics for completing this permutation of wave/particle with matter/light, showing ultimately that the two perceivable manifestations of nature—namely radiation and substance—are actually both reflections of a more fundamental entity beneath human ability to perceive. In its fully developed form, this "wave-particle dualism" rejects the Platonic dictum that the microscopic realm recapitulates what people perceive in the macroscopic.

Born to highly aristocratic parents whose lives were deeply imbedded in French diplomacy (and following the unexpected death of the youngest of four children), de Broglie had all of the benefits of wealth and social connection in early twentieth-century France. The family's prior generations included several marshals of France and members of the most illustrious academies of the nation. De Broglie was raised to study history and politics, like his father, grandfather, and great-grandfather had before him. However, in 1911 de Broglie's course of study changed when his older brother (Maurice), an accomplished experimental physicist recently emancipated from his tradition-bound family, introduced de Broglie to the excitement and furor of the discussion by some intellectuals who had taken up a mathematical interpretation of the new radiations discovered in the field of physics. De Broglie was transfixed; he abandoned the solemn traditions of his family and threw his lot in with the interests of his elder brother as guide and support.

The subjects being investigated by these students included, among other novelties, the x-ray photoelectric effect: the release of negatively charged electrons from metals struck by x rays. The velocity of the emitted electrons could easily be measured, as could the velocity of the electrons that produced the x rays in the first place. The problem was that these two velocities were of about the same magnitude. If the x rays spread from their source in all directions, the velocity granted to the released photoelectron should have had much lower magnitude than the electrons producing the original radiation.

De Broglie's contribution to the discussion of this problem was the realization that while Einstein's theory of relativity predicted that the observed frequency of a moving particle had to seem less to a stationary observer, quantum theory predicted that the frequency had to be greater. In order to help reconcile this conflict, de Broglie showed that, were a guiding plane wave to precede an electron about its atomic orbit, the location of the electron could be fixed at the moving point where the vibration of the plane wave and that of the relativistic vibration constructively interfere: where their vibrations remain permanently in phase. He showed that the wavelength of this planar vibration had to be inversely proportional to the classical momentum of the particle. In other words, the more accurately one knows the particulate property, the less one knows of the wave property. On this assumption, de Broglie derived what had eluded Niels Bohr, the originator of the modern theory of the atom, namely a physical reason that only specific electron orbits should exist at all. They correspond to the resonant standing waves of collar-like bands able to vibrate only at defined frequencies, not in between. "This beautiful result is the best justification we can give for our way of addressing the problem of quanta" (Broglie, 1924b).

De Broglie was not finished. If there is a wave associated with electrons (or indeed any particle), then were electrons to pass through a small enough hole, they would demonstrate diffraction, just as light does. De Broglie tried to get his brother's Parisian experimentalists to do the tests, but failed. The successful results to this novel prediction came from America and Britain in 1927. Before the awarding of de Broglie's Nobel Prize, his theory was developed mathematically by Erwin Schrödinger into what is now called wave mechanics.

De Broglie was elected to the French Academy of Sciences in 1933 and became its perpetual secretary (director) in 1942. He held this position until failing health forced him to step down in 1975.

The main contribution that de Broglie made to physics after the 1930s was in the dissemination of an intellectual understanding of modern physics to a newly interested (mostly French) audience. With his older brother, he helped to encourage French industry to improve productivity by incorporating research laboratories for application of the new sciences (particularly electronics) to the development of their products. For many years he accepted the nonmechanistic interpretation placed on his main accomplishment by other physicists, and tried to further his successes in ways using the solely statistical realities of quantum mechanics. Later in his life, de Broglie rejected the appeal to statistics, as had Einstein, and instead appealed to a theory akin to classical determinism hidden from general perception as a means to avoiding what Einstein had called "God playing dice."

De Broglie was an essential player in drawing French physics into the modern period. A reticent, self-effacing aristocrat, he applied his considerable analytical skills to a fundamental problem at the forefront of his newly adopted field and achieved a symmetrical and viable resolution to the central problem in his brother's experimental research program. Both brothers helped raise modern physics in France to a level equivalent to that practiced in Germany, Britain, and the United States, particularly after World War II.

See also: BOHR, NIELS HENRIK DAVID; BROGLIE WAVELENGTH, DE; QUANTUM MECHANICS; WAVE MECHANICS; WAVE–PARTICLE DUALITY

Bibliography

BROGLIE, L. DE. "A Tentative Theory of Lightquanta." *Philos. Mag.* **47,** 446–458 (1924a).

BROGLIE, L. DE. *Recherches sur la théorie des quanta* (Masson, Paris, 1924b).

WEATON, B. R. "Louis de Broglie and the Origins of Wave Mechanics." *Phys. Teach.* **22,** 297–301 (1984).

WHEATON, B. R. *The Tiger and the Shark: Empirical Roots of Wave–Particle Dualism* (Cambridge University Press, Cambridge, Eng., 1991).

WHEATON, B. R. "The Laboratory of Maurice de Broglie and the Empirical Foundations of Matter-Waves." *La Découverte des Ondes de Matière* (Technique and Documentation, Paris), 25–39 (1994).

BRUCE R. WHEATON

BROGLIE WAVELENGTH, DE

Classical physics, which is based primarily on observations we make with our own senses, divides all physical objects into two completely distinct classes: particles and waves. While particles are objects with a definite location, a wave extends over all of space. A particle is described in terms of its mass and energy, while waves are characterized by their wavelength, frequency, and amplitude. Particles travel in straight lines unless a force deflects them, while a wave passing by an obstacle will undergo bending, or diffraction, of the wave crest even though no force has acted on the wave. Two particles will only affect each other's motion if they exert forces on each other, while waves with the same frequency emanating from different locations will interfere with each other to produce locations where no wave amplitude is found.

When Louis de Broglie presented his thesis in 1924 before the faculty of the University of Paris, they found his hypothesis that particles could act like waves under appropriate conditions so radical that they refused to accept the thesis. In fact, de Broglie was completely correct. By the time he finished his doctoral studies, there was a lot of evidence that light had a dual wave–particle nature. Light acted as a wave when it showed interference in Thomas Young's 1801 double-slit experiment and when it underwent Bragg diffraction from the atoms in a crystal. However light had to act as a particle in

Max Planck's 1900 theory of blackbody radiation, Albert Einstein's 1905 theory of the photoelectric effect, and Arthur Holly Compton's 1923 observation of the Compton shift in the wavelength of light scattered by an electron. It was de Broglie's suggestion that if light could act this way, so could ordinary particles. His work laid the basis for the modern theory of quantum mechanics, in which all physical objects are capable of behaving either as a particle or as a wave depending on the experimental conditions. So objects we always think of as particles, such as planets, rocks, atoms, electrons, and protons, can show wave behavior like interference and diffraction under the appropriate conditions.

When a particle reveals its wave-like nature, we call it a de Broglie wave. The de Broglie wavelength λ of the particle is given by the formula

$$\lambda = \frac{h}{mv},$$

where m is the mass of the particle, v is its speed, and h is Planck's constant.

This equation, known as de Broglie's relation, provides a link between its particle properties (mass and speed) and one of its wave properties. The frequency f of the de Broglie wave is related to the energy E of the particle by $E = hf$. De Broglie's theory does not relate the amplitude of the wave to any particle property. Later work by Max Born established that the square of the wave amplitude at a point in space determined the probability that a particle would be observed there in a given experiment.

To understand how de Broglie waves can be observed, consider an experiment where particles are shot straight through a circular hole of diameter D to strike a screen placed behind it. If the objects have no wave properties, the particles will hit the screen inside a circle of the same diameter D. However, a wave passing through the same hole would diffract and so the part of the screen the wave struck would fill a circle of diameter larger than D. The angle θ through which the wave is diffracted is given by the Fraunhofer formula $\sin \theta = 1.22\lambda/D$. According to de Broglie, the wave-like nature of the particles will cause some of the particles shot through the hole to travel on bent paths like the diffracted wave crests, even though no force was applied to the particles.

For most common objects, the amount of deflection that occurs is so small that it is practically impossible to detect. Suppose the diameter D of the hole is 1/10 in. and the objects have a speed of 10 mph. An object with a mass of 1 kg has a de Broglie wavelength of 1.5×10^{-34} m and would deflect by less than 4×10^{-30} degrees. A common but much less massive object like a mosquito will deflect by at most 4×10^{-21} degrees. In contrast, an electron under the same conditions has a wavelength of 0.0064 in. and will deflect by up to 4.5 degrees. So objects like baseballs, planets, and mosquitoes have unobservable wave effects and act like perfect particles. It is particles with very little mass, such as electrons and protons, that can show their wave-like side easily.

It is important to realize that de Broglie waves, which are an intrinsic property of any particle, are very different from the waves usually encountered in daily life, such as sound waves or waves on the surface of the ocean. These kinds of waves arise from the organized motion of a collection of many particles and exist only because of the forces the particles exert on each other. In contrast, the de Broglie wave nature of particle is an inherent property of a single particle and does not appear as the result of some force.

After being rejected, de Broglie's thesis languished on a shelf for three years. His ideas were finally confirmed by experiments done in 1927 by Clinton J. Davisson and Lester H. Germer at Bell Laboratories and George Paget Thomson at the University of Aberdeen. Davisson and Germer were studying the reflection of electrons from the surface of a crystal of nickel. In addition to the specular reflection they expected from a particle-like electron bouncing back from the surface, they also observed that electrons were reflected from the surface in other specific directions. The set of directions they observed coincided with the Bragg reflection of waves from the atoms in the crystal, which had only been observed before when x rays were used. Davisson wrote to Born, who pointed out how their data justified de Broglie's theory. The pattern they found was unequivocal evidence the electrons were acting as waves just as de Broglie had predicted, and the wavelength of the waves agreed with de Broglie's formula. Davisson and Germer's work and Thomson's observations of the diffraction of electrons passing through thin metal films finally led to the acceptance of de Broglie's thesis. For his work, de Broglie was awarded the Nobel Prize in physics in 1929.

De Broglie's work on the wave nature of particles has resulted in several important technological developments. Davisson and Germer's technique of using the wave nature of the electron to make a diffraction pattern of a solid has been extended to become a useful diagnostic tool for studying the arrangement of atoms and molecules in liquids and solids. To produce a useful Bragg diffraction pattern, the wavelength should be in the range of 0.5 to 3 Å, so that it is about the same as the distance between the atoms being studied. For electrons, this requires their kinetic energy to be in the range of 10 to 150 eV. While they are inside the solid, the electrons feel strong electric forces from the electrons and ions of the solid. These strong forces will scatter the electrons and prevent them from escaping from the solid. Only those electrons that undergo Bragg reflection right at the surface ever escape, so their diffraction pattern indicates the structure of the surface. This technique of Low Energy Electron Diffraction (LEED) is now a standard method for studying the structure of solid surfaces. Neutrons are used to study the interior of a solid or liquid. Since neutrons have no electric charge and only interact weakly with the nuclei by the nuclear force, neutrons that undergo Bragg reflection deep in the material have a good chance of escaping to be detected. A neutron with a de Broglie wavelength of 1.8 Å spacing has a kinetic energy of about 1/40 eV.

One of the most useful practical developments from de Broglie's theory is the electron microscope. In an ordinary optical microscope, light reflected from some object passes through lenses to produce an enlarged image that is observed by the eye or photographed. To resolve two objects close to each other, the light from each object must be focused as sharply as possible. However, as the light passes through the lenses, its wave nature causes some diffraction that leads to unavoidable blurring of the focused image. For two objects separated by a distance less than the wavelength of the light (4,000 Å), their blurred images overlap too much to be distinguished. This limits the usefulness of optical microscopes to magnifications of 1,200X or less.

In the electron microscope, the light is replaced by electrons. A source of electrons replaces the lightbulb that illuminates the sample. After being reflected from the sample, the electrons pass through lenses consisting of magnetic fields that deflect the electrons just as glass lenses deflect light. The image is produced in a manner similar to a television screen by having the electrons strike a phosphorescent screen causing it to glow. The diffraction ef-

fects are limited by the de Broglie wavelength of the electrons, which for electrons with an energy of 100,000 eV is 0.04 Å. This results allows a magnification of 1,000,000X before blurring by diffraction is significant.

See also: BROGLIE, LOUIS-VICTOR-PIERRE-RAYMOND DE; DAVISSON–GERMER EXPERIMENT; DIFFRACTION; DOUBLE-SLIT EXPERIMENT; ELECTRON MICROSCOPE; INTERFERENCE; PHOTOELECTRIC EFFECT; QUANTUM MECHANICS; RADIATION, BLACKBODY; WAVE MOTION; WAVE–PARTICLE DUALITY; WAVE-PARTICLE DUALITY, HISTORY OF

Bibliography

BINNIG, G., and ROHRER, H. "The Scanning Tunneling Microscope," *Sci. Am.* **253**, 50 (1985).
BOORSE, H. A.; MOTZ, L.; and WEAVER, J. H. *The Atomic Scientists: A Biographical History* (Wiley, New York, 1989).
FURTH, R. "The Limits of Measurement." *Sci. Am.* **183**, 48 (1950).
GAMOW, G. *Mr. Tompkins in Paperback* (Cambridge University Press, Cambridge, Eng., 1965).
MEDICUS, H. A. "Fifty Years of Matter Waves." *Phys. Today* **27**, 39 (1974).

DANA A. BROWNE

BROWNIAN MOTION

See MOTION, BROWNIAN

BUBBLE CHAMBER

The bubble chamber played an important role in the development and understanding of particle physics in what is sometimes termed the "golden age" of the field from the 1950s to the 1970s as large accelerators were developed to produce beams of protons and electrons at ever-increasing energies and intensities. The invention of the bubble chamber is attributed to Donald A. Glaser, who in 1950 began work on a better detector than the cloud chamber to register rare events. Glaser recognized that passage of a charged particle through a superheated liquid would produce a trail of bubbles that could be photographed. (It is often told that he got his idea from watching bubbles form in beer; however, in a review article discussing his early work, the only role beer seemed to play was that once he did try to heat bottles of beer or soda and open them with and without a gamma source present. The only apparent result was a very messy laboratory!) He reported his original success with diethyl ether in Pyrex glass containers holding a few cubic centimeters of liquid and with the expansion and recompression done manually by the use of a crank-and-piston arrangement. The expansion and recompression cycle is used to achieve a super-heated condition and to refresh the chamber by removing old bubbles. After the expansion cycle, bubbles will form in the superheated liquid along the path of ionization left by a charged particle passing through the medium. In diethyl ether, the bubbles grow to more than a millimeter in size in 300 μsec. Glaser predicted that liquid hydrogen could be used as well, and John Wood at Berkeley was the first to succeed in making such a chamber.

It is amusing that Glaser was first refused beam time at the Brookhaven Cosmotron, an early workhorse accelerator, since they considered it a "misuse of public funds" to put such a frivolous device in a beam line. He did manage to get his first device set up at a crack in the shielding wall and with his first 36-exposure roll of film, he generated enough excitement that there was then no problem convincing his colleagues that he had invented an important new particle detector.

The beauty of the hydrogen bubble chamber is that the liquid serves as both the interaction target and the detector. Thus, one can record a particle interacting with a proton as long as charged particles are eventually produced somewhere in the interaction. The beam particles can be charged particles, such as protons and pions, but they can also be neutrinos, which interact rarely with other matter. The particle interactions can be easily visualized and a single event can verify the existence of a new particle, as was the case for one of the most famous bubble chamber results, the discovery of the Ω^-.

In liquid hydrogen, and most other materials used in bubble chambers, the expansion of the bubble to photographable size is leisurely after the initial explosive expansion of the bubble that occurs in a fraction of a nanosecond. (Glaser's ether is one of the exceptions.) The bubbles expand to 0.1 mm size in about 1 msec and then are photographed. If one replaces the hydrogen with deuterium, one has the closest thing possible to a neutron target by comparing results seen when only protons are present; pure neutron targets are not possible since the neutron is unstable. Chambers using hydrogen, deuterium, and helium operate at very low temperatures and are known as cryogenic chambers. It is also possible to use heavy liquids that can operate at temperatures at or above room temperature. By placing the chamber in a magnetic field, the momenta of the charged particles can be found by measuring the curvature of the particle trajectories.

The major cause of the rapid decrease in the use of the bubble chamber was its inability to be triggered rapidly enough to respond to a particle passing through it. Since the areas of nucleation only persist on the order of nanoseconds, there is no method by which the mechanical decompression can occur fast enough to use a particle passing through the chamber to trigger the decompression cycle. Thus, extremely rare interactions require searching through an enormous number of photographs. Even though various automatic scanners for bubble chamber photographs were developed, they do not compete well with electronic detectors where no photographs are involved and computers can easily process many millions of recorded particle interactions. Another major disadvantage of the hydrogen bubble chamber is its inability to detect photons, which are the decay products of many particles of interest such as the neutral π meson.

See also: ACCELERATOR; ACCELERATOR, HISTORY OF; CLOUD CHAMBER; PARTICLE PHYSICS; PARTICLE PHYSICS, DETECTORS FOR

Bibliography

GLASER, D. "The Bubble Chamber." *Sci. Am.* **192** (2), 46–50 (1955).

HARIGEL, G. G.; COLLEY, D. C.; and CUNDY, D. C., eds. "Bubbles 40: Proceedings of the Conference on the Bubble Chamber and Its Contributions to Particle Physics." *Nucl. Phys. B (Proc. Suppl.)* **36**, 3–80 (1994).

HENDERSON, C. *Cloud and Bubble Chambers* (Methuen, London, 1970).

L. DONALD ISENHOWER

BUOYANT FORCE

An object partially or completely submerged in a fluid is subjected to an upward force due to an effect best known as Archimedes' principle. Archimedes' principle is the consequence of fluid statics. The fluid exerts pressure to every boundary surface between the object and the fluid in the direction perpendicular to the boundary surface. Since the hydrostatic pressure on the boundary increases with depth, the total effect is a net upward force on the submerged object. This is the buoyant force or buoyancy.

Since the hydrostatic pressure distribution near and at the surface of an object of irregular shape submerged in a fluid does not depend on what kind of material the object is made of, we can replace the object with the same fluid of the same shape. In other words, the inside of the closed imaginary boundary formerly occupied by the object is now filled with the surrounding fluid. We should still have the same pressure distribution at the imaginary boundary even though the volume is now filled with the fluid like the rest of the volume within the larger container. Now the body of the fluid inside the boundary surface is in mechanical equilibrium; that is, it does not rise or fall. Since all objects in the gravitational field are pulled downward by gravity, there must be an upward force equal to the weight of this body of fluid that replaced the original object in order to maintain the equilibrium. Thus the buoyant force must be equal in magnitude to the weight of the body of water within the imaginary boundary formerly occupied by the object. In other words, the buoyant force that acts on a body partially or completely submerged in a fluid is equal in magnitude to the weight of the fluid the body displaces and is directed upward. This is Archimedes' principle.

It is possible to estimate by Archimedes' principle the fraction f of the volume of a partially submerged iceberg. The weight of the iceberg W is the product of its volume V, its density ρ, and the acceleration due to gravity g:

$$W = \rho V g.$$

The buoyant force is given by the weight W' of the sea water displaced by the submerged portion of the iceberg. The volume of the submerged portion is fV, and the weight of this sea water is given by the product of its volume fV, density of sea water ρ', and g:

$$W' = \rho' f V g.$$

According to Archimedes' principle, $W = W'$. (Note that these two forces have opposite directions and, if we are to use a vector notation, it will be $\mathbf{W} + \mathbf{W}' = \mathbf{0}$.) Thus we have

$$\rho V g = \rho' f V g$$

or

$$f = \frac{\rho}{\rho'}.$$

The density of ice ρ at $0°C$ is approximately 0.92×10^3 kg/m^3 and the density of sea water ρ' is about 1.03×10^3 kg/m^3. Thus the fraction f of the iceberg that is submerged turns out to be about 0.89. This means that only about 10 percent of an iceberg is visible above the ocean level.

See also: ARCHIMEDES' PRINCIPLE; FLUID DYNAMICS; FLUID STATICS

Bibliography

RESNICK, R.; HALLIDAY, D.; and KRANE, K. J. *Physics,* 4th ed. (Wiley, New York, 1992).

CARL T. TOMIZUKA

C

CALORIC THEORY

See HEAT, CALORIC THEORY OF

CALORIMETRY

Calorimetry is a measure of the heat flow to or from a sample as its properties evolve. Calorimetry provides insight into the energies associated with changes in the temperature, phase, pressure, volume, and composition, and many other properties of materials. Calorimetry provides insight into the course of chemical reactions (heats of reactions), the energy content of materials (heat capacity and entropy), and the physics of phenomena such as superconductivity and magnetism.

During calorimetric investigations of phenomena, materials are generally isolated in containers known as calorimeters (see Fig. 1). Generally, the measurement and control of temperature and heat flow necessitates a thermal link between calorimeter and thermal bath. During operation of the calorimeter, the heat flow between the sample and the container is measured before, during, and after a

thermal event such as an explosion or an ordering of magnetic spins in the sample (see Fig. 2).

Calorimetry provides a measurement of the heat released during a chemical reaction. For example, if

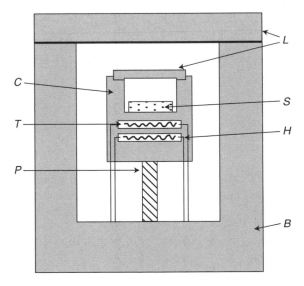

Figure 1 A sketch of a calorimeter. In that this is a schematic form, this could correspond to a number of different calorimeters, including oxygen combustion or quasi-adiabatic heat capacity. The components include *C*, calorimeter; *T*, thermometer; *P*, supporting post, part of the thermal link; *B*, thermal bath; *H*, heater; *S*, sample; and *L*, lids.

oatmeal is oxidized completely (i.e., burned in oxygen) in a bomb calorimeter, the temperature rise ΔT of the calorimeter vessel can be used to calculate the heat ΔQ of this reaction. In the ideal case of adiabatic isolation, the calorimeter heat capacity C would be used to find ΔQ from $\Delta Q = C \cdot \Delta T$. In practice, some heat leaks out to the thermal bath during the reaction and compensations for this effect are made. In the present example, the heat of oxidation of oatmeal is reported to the consumer in units of calories per serving. The actual units are kilocalories per serving, but it is clear that the supermarket is a repository of calorimetric data.

Determination by calorimetry of the energies of formation of different phases allows the optimization of processes for fabricating new materials. A change in energy of a system (such as in a calorimeter vessel during the reaction of gases to form thin films) depends solely on its initial and final physical states, not on the path followed in the course of the change of the system. Changes in the internal energy, ΔU, of a system can be determined at constant volume in a calorimeter. Given the first law of thermodynamics ($\Delta U = \Delta Q + \Delta W$, where ΔQ is the heat added to the system and ΔW is the work done on the system), constant volume ($\Delta V = 0$) calorimetric measurements of the heat of formation of thin film reaction products correspond directly to the energy of the new phases ($\Delta W = -P\Delta V = 0$, so $\Delta U = \Delta Q$). Constant pressure measurements are more common in calorimetry, in which case work is done due to volume change. In this case, the change in the enthalpy state function ($H = U + PV$) can be determined from the heat flow, $\Delta H = \Delta Q$. Measurements of the heat capacities and entropies of the new phases by means of calorimetry allow the free energy of formation of the material to be determined from ΔH.

See also: ENTHALPY; ENTROPY; HEAT CAPACITY; THERMODYNAMICS; THIN FILM

Bibliography

KITTEL, C., and KROEMER, H. *Thermal Physics* (W. H. Freeman, San Francisco, 1980).

PITZER, K. S., and BREWER, L. *Thermodynamics*, 2nd ed. (McGraw-Hill, New York, 1961).

ZEMANSKY, M. W. *Heat and Thermodynamics*, 4th ed. (McGraw-Hill, New York, 1957).

ERIC J. COTTS

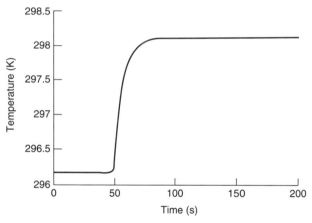

Figure 2 A plot of temperature versus time during a typical thermal event in a calorimeter, such as the oxygen combustion of oatmeal. The heat flow between the sample and the container is measured before (0 to 50 s), during (50 to approximately 51 s), and after the thermal event. The slope of the curve is finite before the thermal event because there is a finite heat leak between calorimeter and thermal bath. The thermal event was of relatively short duration (less than or equal to 1 s), but the response time of the calorimeter was much longer (approximately 50 s, that is, a time constant of approximately 10 s). The heat leak and the finite response time of the calorimeter are compensated for in the computation of ΔQ.

CAPACITANCE

When electric charge is transferred from one initially neutral conductor to another—if those conductors are separated by a dielectric insulator—an electric potential difference develops between the conductors. Capacitance is a measure of the ratio of charge transferred to the electric potential developed.

Since electric charge is neither created nor destroyed in the process, the charge gained by one of the conductors will always equal the charge lost by the other one, and is often referred to as the charge, q stored by the system.

All physical objects exhibit capacitance. If you shuffle your feet on the carpet and touch a door knob, the resulting spark is evidence that your body had some electric charge stored on it; it has capacitance. The mathematical relation that defines capacitance is

$$C \equiv \frac{q}{V}.$$

In the MKS system, q is the number of coulombs of electric charge transferred, V is the voltage that exists between the conductors, and C is the resulting capacitance. The unit used for capacitance is the farad, F, named in honor of the English physicist, Michael Faraday. Thus if 1 C of electric charge is transferred when a potential difference of 1 V is applied between two conductors, they have a capacitance of 1 F.

A farad is a very large amount of capacitance. The more common unit is the microfarad, μF, which is one-millionth of a farad, or 10^{-6} F. The capacitance of a short piece of wire might be a few picofarads (pF) or 10^{-12} F.

The capacitance of a spherical conductor of radius r is given by

$$C = 4\pi\varepsilon r,$$

where ε is the dielectric constant of the medium surrounding the sphere.

Passive electric circuit components designed to have a specified amount of capacitance are called capacitors. The capacitance of a parallel plate capacitor is given by

$$C = \varepsilon\frac{A}{d},$$

where A is the area of the plates, d is their separation, and ε is the dielectric constant of the insulator separating the plates.

Varactors are special purpose semiconductor diodes whose capacitance varies with the amount of voltage applied to them.

All of the components in an electric circuit have capacitance. This tends to limit the high frequency response of any circuit. Thus high-frequency circuits such as radio or television circuits have to be designed so as to minimize this distributed capacitance.

Capacitance measurement in the past was rather difficult, requiring the user to balance a bridge circuit. Modern capacitance testers provide a digital read-out, often under microprocessor control, and with virtually no operator intervention.

See also: CAPACITANCE; CHARGE; CONDUCTOR; DIELECTRIC CONSTANT

Bibliography

BOYLSTAD, R. L. *Introductory Circuit Analysis,* 7th ed. (Merrill, New York, 1994).
HALLIDAY, D.; RESNICK, R.; and KRANE, K. *Physics,* 4th ed. (Wiley, New York, 1992).

ROBERT R. LUDEMAN

CAPACITOR

A capacitor is an electric circuit component that is designed to have a specified amount of capacitance. The standard symbols used to represent fixed value and variable capacitors in electric circuit diagrams are shown in Fig. 1. The symbol reflects the idea that a capacitor can be thought of as consisting of two parallel metal plates separated by an insulating material known as a dielectric. The capacitance, C of such a capacitor is given by

$$C = \varepsilon\frac{A}{d}, \tag{1}$$

where A is the area of the plates, d is their separation, and ε (epsilon) is the dielectric constant of the insulator separating the plates.

Perhaps the practical capacitor that comes closest to the above conceptualization is the multiplate variable tuning capacitor that is still used in some small radios. The capacitance of such capacitors can be

(a)　　　(b)

Figure 1 (a) Symbol for a fixed value capacitor. (b) Symbol for a variable capacitor.

varied from a minimum value of around 50 pF to a maximum capacitance of a few hundred picofarads.

As its name implies, ceramic disk capacitors are constructed by depositing thin metal films on each side of a thin ceramic disk. Hookup wires are attached to the metal films, which are the capacitor plates. The ceramic disk is the dielectric. The entire package is then coated with a protective layer of ceramic material. Typical values for such capacitors range from the order of 10 pF to near 0.01 μF.

To obtain capacitances in the range from about 0.001 μF to around 1 μF, a tubular design is often used. A tubular capacitor consists of two long strips of metal foil, usually aluminum, separated by an insulating dielectric such as paper or a plastic such as mylar. Hookup wires are attached to the metal foils, and the whole structure is rolled together to form a tubular configuration.

In order to achieve larger capacitances in packages of manageable size, the surface of the above aluminum strips is often roughened or texturized in order to increase their surface area. The dielectric is then replaced by blotting paper soaked in a chemical electrolyte. The components are then rolled together and packaged. The result is known as an aluminum electrolytic capacitor.

As part of the manufacturing process, a voltage is applied to the newly formed electrolytic capacitor. This results in an electrochemical reaction that oxidizes the aluminum strip that is to become the positive plate of the capacitor. The aluminum oxide that is thus formed is a very good insulator and becomes the capacitor's dielectric. Since the dielectric is chemically deposited, it is very thin, and it takes the shape of the texturized aluminum plate, effectively reducing d in Eq. (1) to molecular dimensions. Such electrolytic capacitors can have capacitances of as much as 100,000 μF or more.

Sintered tantalum has been used to replace the texturized aluminum in some electrolytic capacitors. This further improves the capacitance-to-size ratio. Perhaps the ultimate step in this direction is the use of activated carbon as the capacitor electrodes to form what is sometimes called a super cap. These capacitors can achieve a capacitance of as much as 1 F in a volume of about 1 in.[3]

Since it is the dielectric that insulates the plates of a capacitor, its thickness is directly related to the voltage difference that can be safely placed across a capacitor. That voltage rating can be under 10 V for some capacitors. Both the capacitance and the maximum voltage rating are usually indicated on the

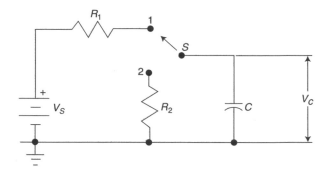

Figure 2 Capacitor charging and discharging circuit.

capacitor. Further, the required polarity is indicated on electrolytic capacitors. Since the electrochemical reaction that creates the dielectric is usually reversible, reversing its polarity of an electrolytic capacitor will often destroy it.

Most if not all capacitor applications depend on one simple qualitative statement: The voltage across a capacitor cannot change suddenly. This observation stems from the definition of capacitance, which is

$$C \equiv \frac{q}{V}. \tag{2}$$

The above definition says that an instantaneous change in voltage would require an instantaneous change in charge and that would require an infinite current flow. Thus a relatively slow voltage change is required.

Many capacitor applications involving dc circuits can be illustrated by the circuit of Fig. 2. These include applications as diverse as power supply filters and timer control circuits. When switch S is thrown to position 1, current flows from supply V_S through resistor R_1, charging capacitor C. When S is thrown to position 2, C is discharged through resistor R_2.

If S is flipped to position 1 at $t = 0$, Eq. (2) can be integrated over the time interval from zero to t. The result is the following capacitor charging equation:

$$-\frac{t}{R_1 C} = \ln \frac{V_S - V_t}{V_S - V_0}$$

or

$$e^{-t/R_1 C} = \frac{V_S - V_t}{V_S - V_0}, \tag{3}$$

(a)   (b)

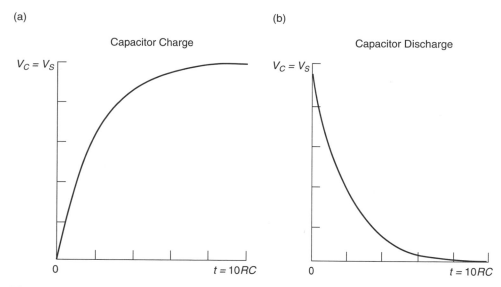

Figure 3   Response to the circuit in Fig. 2. (a) Capacitor charging graph (switch in position 1). (b) Capacitor discharging graph (switch in position 2).

where V_0 is the initial voltage across C when S is flipped to position 1 and V_t is the voltage across C after t seconds. The resulting capacitor charging graph is shown in Fig. 3a.

When S is flipped to position 2, C is discharged through resistor R_2. Equation (3) then becomes the capacitor discharge equation:

$$-\frac{t}{R_2 C} = \ln \frac{V_t}{V_0}$$

or   (4)

$$e^{-t/R_2 C} = \frac{V_t}{V_0}.$$

A graph of Eq. (4) solved for V_t is shown in Fig. 3b. Both of the graphs in Fig. 3 support the assertion that the voltage across a capacitor cannot change suddenly.

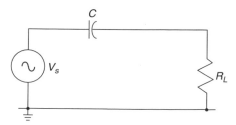

Figure 4   A capacitor used as a signal coupling device.

A common capacitor application involving ac circuits is the signal coupling circuit shown in Fig. 4. Capacitor C couples the ac signal from the source v_s to the load R_L while at the same time blocking the passage of any dc which may be present.

Since C and R_L in Fig. 4 form an ac voltage divider, the amount of v_s that reaches R_L depends on the capacitive reactance of C, which is given by

$$X_C = \frac{1}{2\pi f C}.$$ (5)

Equation (5) indicates that if the frequency f is relatively high, good signal coupling will occur. The cutoff frequency f_c is arbitrarily defined to be that frequency at which $R_L = X_C$ or

$$f_c = \frac{1}{2\pi X_C R_L}.$$ (6)

At frequencies below f_c, the signal is significantly attenuated. Thus the circuit of Fig. 4 is often called a high-pass circuit or a differentiating circuit. Analogously, if the resistor and the capacitor exchange places, the circuit becomes a low-pass circuit or an integrator.

See also: CAPACITANCE; CIRCUIT, AC; CIRCUIT, DC; ELECTRICAL RESISTANCE; ELECTRICAL RESISTIVITY

Bibliography

BOYLSTAD, R. L. *Introductory Circuit Analysis,* 7th ed. (Merrill, New York, 1994).

HALLIDAY, D.; RESNICK, R.; and KRANE, K. *Physics,* 4th ed. (Wiley, New York, 1992).

ROBERT R. LUDEMAN

CARNOT, NICOLAS-LÉONARD-SADI

b. Paris, France, June 1, 1796; *d.* Paris, France, August 24, 1832; *thermodynamics.*

Carnot was the elder of two sons in a distinguished Burgundian family. His father (Lazare Carnot) was a leading military figure who had served with distinction in the early years of the revolutionary wars of France and wielded great political power as a member of the five-man Directory that governed the country from November 1795, until Napoleon Bonaparte's coup d'état in September 1797. Carnot's education was governed until 1812 by his father, who was also a notable mathematician. In that year, after a brief period of preparation at the Lycée Charlemagne in Paris, Carnot entered the Ecole Polytechnique. Despite the disruption of his studies during the defense of Paris in the spring of 1814, he graduated and proceeded in January 1815 to the Ecole de l'Artillerie et du Génie at Metz. He was commissioned as a military engineer in April 1817, but his army career was undistinguished and unhappy, and in January 1819 he went on the reserve list, where he remained until he resigned his commission in 1828.

From 1819 until his death during the cholera epidemic of 1832, Carnot lived a quiet, studious life in Paris. In these years, his only significant academic contact was with Nicolas Clément, professor of industrial chemistry at the Conservatoire Royal des Arts et Métiers. Nevertheless, supporting himself modestly by inherited family money and his reduced military pay, Carnot wrote his 118-page *Réflexions sur la Puissance Motrice du Feu (Reflexions on the Motive Power of Fire),* a work that the Scottish physicist William Thomson (later Lord Kelvin) was one of the first to recognize as a masterpiece, though long after its publication in 1824. "Nothing in the whole range of Natural Philosophy," Thomson wrote of the book in 1849, "is more remarkable than the establishment of general laws by such a process of reasoning." In the *Réflexions,* Carnot laid some of the most important foundations of modern thermodynamics, using an argument that possessed rigor and elegance, despite being constructed almost entirely without the use of mathematics.

Although the results were above all of theoretical significance, the problem that Carnot tackled in the *Réflexions* had its origins in the practical world of power technology. Like many engineers and physicists of his day, Carnot was intrigued by the economy of a generation of medium- and high-pressure steam engines that had been developed in Britain during the Napoleonic Wars and only became known in France after 1814. The most notable of these engines was that of the Cornish engineer Arthur Woolf, which applied a technique, patented by James Watt but not seriously exploited by him, for enhancing the amount of work that could be obtained from steam. The technique consisted in avoiding unnecessary waste by allowing the steam to expand in the cylinder of an engine until its pressure was no longer sufficient to move the piston.

Attempts to explain the economy of the "expansive" engine had yielded little agreement by the time Carnot turned to the question, probably about 1820. His treatment, as presented in the *Réflexions,* was a highly abstract one. It was based on the caloric theory, which treated heat as an invisible, weightless fluid whose accumulation in a body was manifested as a rise in temperature. In Carnot's ideal engine, heat entered the working substance (usually steam, although the theory could be applied equally well to air or any other elastic fluid) at the high temperature of the boiler. Then, after the working substance had yielded its work, expanding and cooling in the process, the heat would pass from it to the condenser as the steam condensed. Essentially, therefore, heat had "fallen" from the higher temperature of the boiler to the lower temperature of the condenser, and it was this "fall" that, in Carnot's theory, yielded work or, as he called it, motive power.

There was a close analogy between Carnot's conception of the source of work in a heat engine and the operation of a water wheel. In both cases, a substance (heat or water) passed from a higher to a lower "level," producing work without being consumed. In this conception of an ideal engine, the influence of Lazare's theoretical investigation of the work produced by a water-powered engine is unmistakable, and there can be little doubt that both the general approach and certain key details of the *Réflexions* owe much to this earlier work. Apart from the fundamental notion of the "fall" of heat, Carnot was able to draw from his father's analysis the core of what he developed in the *Réflexions* as the concepts of reversibility and the closed cycle of operations.

Using these concepts, Carnot presented what can be regarded as a form of the second law of thermodynamics. By insisting that the production of work required the presence of both a hot and a cold source and the passage of heat between these sources, he was breaking totally new ground. Also novel was his insistence that, in calculating the work produced by an ideal heat engine, the working substance had to be restored to its original conditions; in other words, the cycle of operations—the "Carnot cycle" of modern thermodynamics—had to be completed. Carnot's cycle was reversible, and by an ingenious and subtle argument be proved that the "Carnot engine" was the most efficient of all engines that could operate between the same two temperature intervals. The unfamiliarity of such ideas probably mystified the few contemporaries who read the *Réflexions*. At all events, the book was largely ignored, with only the engineer Emile Clapeyron (in 1834) making any significant use of its argument until it was finally recovered, independently and almost simultaneously, by Rudolf Clausius and William Thomson a quarter of a century after its publication.

Perceiving the fundamental error of Carnot's assumption that work could be produced without some heat being consumed, Clausius and Thomson showed how the essentials of his argument could be reconciled with the newly discovered principle of the conservation of energy (or first law of thermodynamics). In their version of Carnot's theory, the amount of heat leaving the system at the lower temperature was less than the amount entering at the higher temperature by an amount proportional to the quantity of work produced. Although this reconciliation between the first and second laws of thermodynamics moved Carnot's theory definitively into the main stream of modern physics, it was only in 1878 that the full extent of Carnot's achievement became apparent. It was then that his brother Hippolyte presented to the *Académie des Sciences* a set of manuscript notes that had remained in his possession since Carnot's death. These notes showed that at some time between the publication of the *Réflexions* in 1824 and his death eight years later, Carnot himself had reached the conclusion that work could

only be produced if a corresponding amount of heat was consumed; he had even calculated a remarkably accurate value for the mechanical equivalent of heat. The posthumous recognition was generous, and Carnot's reputation has remained undimmed ever since.

See also: CARNOT CYCLE; CLAUSIUS, RUDOLF JULIUS EMMANUEL; ENGINE, EFFICIENCY OF; HEAT, CALORIC THEORY OF; HEAT ENGINE; KELVIN, LORD; THERMODYNAMICS; THERMODYNAMICS, HISTORY OF

Bibliography

CARDWELL, D. S. L. *From Watt to Clausius: The Rise of Thermodynamics in the Early Industrial Age* (Heinemann, London, 1971).

CARNOT, S. *Reflexions on the Motive Power of Fire, A Critical Edition with the Surviving Manuscripts*, translated and edited by R. Fox (Manchester University Press, Manchester, Eng., 1986).

WILSON, S. S. "Sadi Carnot." *Sci. Am.* **240** (2), 102–114 (1981).

ROBERT FOX

CARNOT CYCLE

Converting work into thermal energy is easy—rubbing sticks may eventually make fire. The reverse process, converting heat into useful work, is more difficult. Practical, if inefficient, steam engines first appeared about 1700. But how does one define efficiency? The usual definition is

$$\text{efficiency} = \varepsilon = \frac{\text{what is obtained}}{\text{what is expended}}.$$

What is obtained from an engine is its work (W). Ultimately, what is expended is the energy contained in fuel; it is fuel that provides the necessary heat. Let Q_{in} represent heat entering an engine. The efficiency is

$$\varepsilon = \frac{\text{what is obtained}}{\text{what is expended}} = \frac{W}{Q_{in}}.$$

Clever inventors such as Thomas Newcomen and James Watt improved early engines. But were there limits to the possible improvements, or could increasingly ingenious designs create an engine of 100 percent efficiency? The physics of the eighteenth century was not up to the task of predicting the maximum attainable efficiencies. Even the concept of heat was not understood; many thought heat to be an invisible fluid called caloric (a vestige of this idea remains today when the unfortunate term "heat flow" is used). Today we know that heat is energy transferred due to a temperature difference.

Sadi Carnot

Nicholas-Léonard-Sadi Carnot was a retired military officer and engineer. In his short life he wrote only one scientific article, but it would ensure his enduring fame. Carnot's results have withstood the test of time and are of fundamental importance to thermodynamics.

Carnot made two important observations. Early engines vented their steam after doing work (and our automobiles still spew out hot exhaust). The condensed steam could have been reused again and again in an endless cycle; Carnot realized that all heat engines can be regarded as a cyclic process. Although unknown to Carnot, the first law of thermodynamics states that: $Q = \Delta U + W$. Heat (Q) can change the internal energy of a system (ΔU) and do work (W). In a cyclic process the system returns to its original state, so $\Delta U = 0$ for the cycle; thus heat equals the work performed, $Q = W$.

Carnot also grasped the fact that heat engines must eject heat (whether it be into steam condensers or into the tailpipe emissions of today's automobiles). All heat engines involve both the absorption of heat (Q_{in}) and the ejection of heat (Q_{out}); thus the net heat for a cycle is $Q = Q_{in} - Q_{out}$ and the efficiency becomes

$$\varepsilon = \frac{W}{Q_{in}} = \frac{Q_{in} - Q_{out}}{Q_{in}} = 1 - \frac{Q_{out}}{Q_{in}}.$$

The Carnot Cycle

A real steam engine is a complex device of valves, boilers, pistons, and steam undergoing rapid temperature changes—too complex to determine ε. Carnot was forced to invent a simpler, idealized engine that he was capable of analyzing; this has become known as the Carnot cycle. Carnot already knew that heat must be both absorbed and rejected; the simplest situation is to restrict the engine to have heat exchange at two fixed temperatures only, T_{hot} and T_{cold}. The constant temperature requirement means that this engine must be in continuous thermal equilibrium and thus can only move infinitely slowly; today his engine is said to be *strictly reversible*. In addition to the constant temperature phases, there are two other steps during which the engine is insulated and not allowed to interact with the rest of the world except by performing mechanical work. Today the Carnot engine is said to undergo isothermal (no temperature change) and adiabatic (zero heat) processes.

A further word about reversible processes follows: Generally, a process is reversible if a small change results in a small effect. Consider a small compression of the gas inside an insulated cylinder; the result is a small increase in temperature and pressure. Undo the compression and the temperature and pressure drop to their precompression values. Next consider a free expansion: a cylinder has a partition with gas trapped on only one side of the partition. To create a small change, poke a small hole into the partition. Gas spews into the empty part of the cylinder. Reversing the small change (plugging the hole) does not bring the system back to its original state; this is an irreversible situation. Imagine a video of these two examples. View the scenes played forward and in reverse. If the video played in reverse appears normal, then that process is reversible; such is the case for the cylinder compression. For the free expansion, gas returning by itself to one side of the partition will appear totally unnatural.

The Second Law of Thermodynamics

Carnot was able to analyze his idealized engine. Using simple arguments he derived properties fundamental to thermodynamics—they are consequences of what is today called the second law of thermodynamics:

1. All Carnot engines operating between the same T_{hot} and T_{cold} will have the same efficiency no matter how or of what they are constructed.
2. Since every Carnot engine operating between the same T_{hot} and T_{cold} is equivalent, the efficiency can only depend on T_{hot} and T_{cold}. Later, Rudolf

Clausius would show the temperature dependence is

$$\varepsilon_{\text{carnot}} = 1 - \frac{Q_{\text{out}}}{Q_{\text{in}}} = 1 - \frac{T_{\text{cold}}}{T_{\text{hot}}}.$$

3. No real engine (operating between the same temperatures) can have an efficiency that exceeds $\varepsilon_{\text{carnot}}$.

Suppose an engine did exist whose efficiency exceeded $\varepsilon_{\text{carnot}}$, contradicting Carnot's findings. To see the consequences, run this superefficient engine together with a Carnot engine. Remember that the Carnot cycle is reversible. If operated in reverse it becomes a refrigerator: the work input causes heat to be absorbed in a cold area and heat to be ejected into a hot area.

Carefully design the superefficient engine so that all its work output is used to run a Carnot refrigerator; thus there is no net work from the system. The superengine absorbs heat from a hot area while the refrigerator ejects heat into the hot area; the superengine ejects heat into a cold area while the refrigerator absorbs heat out of the cold area. It is possible to prove that combinations of superefficient engines and Carnot refrigerators can result in a net transfer of heat from cold to hot. This transfer seems to be spontaneous because no net work was done. Such spontaneous processes have never been observed. Clausius showed that the absence of spontaneous thermal energy transfer from cold to hot is another, equivalent, statement of the second law of thermodynamics.

Will it ever be possible for ingenious designers to devise an $\varepsilon = 100$ percent engine? For this to happen Q_{out} or T_{cold} must be zero. Yet there must always be some waste heat and the temperature of the exhaust is never at absolute zero. The impossibility of $\varepsilon = 100$ percent is usually referred to as the Kelvin–Planck statement of the second law. But what are realistic values of ε, say for a typical automobile? The maximum efficiency is set by the operating temperatures T_{hot} and T_{cold}. Our society usually dumps waste heat directly into the environment, thus T_{cold} is perhaps 300 K. Steel begins to lose its strength near 900 K, setting an upper limit for T_{hot}. $\varepsilon_{\text{carnot}}$ is equal to $1 - \frac{300}{900}$, which is 0.67, or 67 percent. Real, irreversible engines suffering friction and other losses will have efficiencies much less than this.

Another form of the second law can be obtained using a result due to Clausius, the Clausius inequality. A thermodynamic quantity exists called the entropy; roughly speaking, entropy is a measure of the disorder of a system. Consider any process involving a thermodynamic system and its surroundings: the Clausius inequality can be used to prove that the total entropy change must be greater than zero for any real (irreversible) process. This is the most general form of the second law.

See also: CARNOT, NICOLAS-LÉONARD-SADI; CLAUSIUS, RUDOLF JULIUS EMMANUEL; ENTROPY; THERMODYNAMICS; THERMODYNAMICS, HISTORY OF

Bibliography

FENN, J. B. *Engines, Energy, and Entropy* (W. H. Freeman, San Francisco, 1982).

FEYNMAN, R. P.; LEIGHTON, R. B.; and SANDS, M. *The Feynman Lectures on Physics,* Vol. 1 (Addison-Wesley, Reading, MA, 1966).

KITTEL, C. *Thermal Physics* (Wiley, New York, 1969).

MENDOZA, E., ed. *Reflections on the Motive Power of Fire by Sadi Carnot and Other Papers on the Second Law of Thermodynamics* (Dover, New York, 1960).280

TIPLER, P. A. *Physics for Scientists and Engineers,* 3rd ed. (Worth, New York, 1991).

ROBERT H. DICKERSON

JOHN MOTTMANN

CATHODE RAY

Cathode rays were discovered gradually during the second half of the nineteenth century, after the invention of excellent vacuum pumps by Heinrich Geissler. Until then it was known that high voltage discharge through poor vacuums could produce glows (as in neon signs). But with high vacuums the discharge glows became fainter and fainter, and the glass opposite the negative terminal (the cathode) began to glow with a greenish fluorescence. By using cathodes that were slightly concave the greenish fluorescence could be focused on a spot at the other end of the tube, and made brighter with the help of a special coating. The cathode rays could be deflected by passing them through electric and magnetic fields.

Considerable controversy, especially between German and British physicists, ensued as to the nature of cathode rays. Were they electromagnetic waves or subatomic particles of some sort? Finally, in 1897, the great English physicist Joseph J. Thomson showed conclusively that cathode rays were indeed subatomic particles having a charge-to-mass ratio of 1.750019×10^{11} C/kg.

Modern cathode-ray tubes use heated filaments to provide electrons for their beams, and two sets of parallel plates, perpendicular to each other, provided vertical and horizontal deflections of the electron beam. Many cathode-ray oscilloscopes have been made with two cathodes and sets of deflecting plates. Thus they can show two sets of traces that can be compared simultaneously.

Usually, the vertical deflection of the beams is determined by the voltage sources being investigated, while the horizontal deflections are produced by a sawtooth voltage of appropriate amplitude and timing. The horizontal time scale can be set anywhere from a few microseconds to 10 s per division, while the vertical sensitivity can range from 2 mV to 10 V per division.

Cathode-ray oscilloscopes are useful in many kinds of scientific research—to see and measure, for instance, electric impulses transmitted by nerves, or generated by light signals from a distant pulsar, or to check the characteristics of some experimental research device. They are, of course, also used in industrial production lines, and in every television repair shop.

See also: CHARGE; ELECTRON, DISCOVERY OF; OSCILLOSCOPE; THOMSON, JOSEPH JOHN; VACUUM TECHNOLOGY

Bibliography

ANDERSON, D. L. *The Discovery of the Electron* (Van Nostrand, New York, 1964).

DAVID L. ANDERSON

CAT SCAN

Computerized Axial Tomography (CAT) is perhaps the most revolutionary x-ray imaging technique that has been developed since the discovery of x rays at the end of the nineteenth century. Godfrey N. Hounsfield and Allan M. Cormack both received the 1979 Nobel Prize in physiology and medicine for their discovery and development of the CAT technique. The basic principle involved in CAT is the reconstruction of an axial slice of a patient's body from a series of x-ray transmission projections or profiles. The resulting images are unique because they retain the three-dimensional (3-D) nature of the distribution of organs and vessels in our bodies. There is no overlapping of tissues (e.g., the heart superimposed on the spine) such as occurs in conventional film radiography. Also, in CAT, the x-ray beam is collimated, to be very narrow (e.g., 10 mm) and the x-ray detectors are often collimated resulting in the detection of very few scattered x rays emanating from the patient. Scattered x rays are those whose directions change due to interactions with electrons in objects (e.g., tissues), and their rejection results in images that have a much higher contrast. Indeed, CAT images typically display about 10 times more contrast than conventional film radiographs. Soft tissues and body fluids that have similar x-ray attenuation properties and cannot be distinguished in radiographs are often easily distinguished in CAT images.

Data Acquisition and Image Reconstruction

For each slice that is imaged, the CAT scanner acquires a large number of projections (e.g., approximately 1,000) around the patient, where each projection contains a large number of data points (Fig. 1). Each data point represents the integrated x-ray attenuation along a given line. The purpose of CAT reconstruction is to obtain the x-ray attenuation coefficient of each point in the slice from the projection data. The reconstruction is performed using a computer that is interfaced to the scanner, and the computed attenuation coefficients are stored in an image matrix in the computer's memory. The most commonly used reconstruction technique is filtered backprojection. Backprojection refers to the smearing of the collected data back onto the image matrix (Fig. 2) along the same direction in which it was acquired. If backprojection alone is employed for each data point to reconstruct an image, then the sharpness of the image is reduced by blurring. With appropriate filtering of the data before backprojection, this blur-

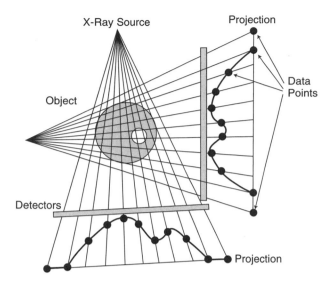

Figure 1 Data Acquisition. Two projections, and eleven data points in each projection are shown. Each data point represents the integrated attenuation along the line connecting the detector to the x-ray source.

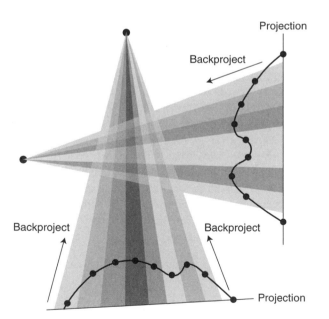

Figure 2 Backprojection. Data points are smeared back along the same direction as they were acquired. Each data point leaves a "trail" on the image. The higher the value of the data point, the darker the trail. If backprojection is performed for all data points and all projections, a blurred image is obtained.

ring is eliminated, and a faithful reconstruction is obtained (Fig. 3).

CAT Scanner Geometries and Characteristics

The first CAT scanners became commercially available in 1973. They acquired the x-ray transmission projections by a slow process involving a pencil beam of x rays, two detectors, and a translate-rotate scanning geometry. Present-day scanners use three scanning geometries. "Third-generation" scanners employ a fan beam of x rays and a detector array, both of which rotate about the patient. "Fourth-generation" scanners employ a stationary ring of detectors and a rotating fan beam of x rays. "Fifth-generation" or ciné-CAT scanners employ an enormous x-ray tube target that surrounds the patient. That target is scanned electronically to produce a moving fan beam of x rays that scans a single slice of a patient in about 50 ms. The scan time for slices with the third- and fourth-generation scanners are typically 1 or 2 s. For all generations, slice thicknesses are variable and range from about 1 to 10 mm. From 500 to 1,000 detectors are employed in third- and fifth-generation scanners, and up to 4,800 in fourth-generation scanners. These detectors are either high pressure xenon gas ionization chambers or scintillation crystals backed by photodiode arrays. Another form of scanning termed "helical" CAT has also been developed. Helical CAT is achieved on both third- and fourth-generation systems by continuous x-ray beam rotation about the patient, who is simultaneously continuously transported on a movable table through the gantry. If one were to trace the intersection of the incident x-ray beam on the patient's body it would form a helix or a spiral. The projection data are interpolated to create parallel adjacent slices similar to those in conventional CAT. Advantages of helical CAT include decreased patient motion, arbitrary selection of slice position, and finer slice incrementation for better detailed 3-D reconstruction. Helical CAT scanning is characterized by a pitch factor, defined as the table increment distance per x-ray tube revolution divided by the slice thickness. Typically, a pitch factor of 1 is employed, although pitch factors of 1.5 and more are sometimes used to scan greater volumes per unit time.

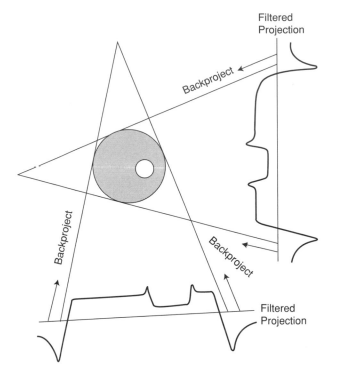

Filtered
Projection

Backproject

Backproject

Backproject

Filtered
Projection

Figure 3 Filtered Backprojection. When projection data are first filtered and then backprojected, a faithful reconstruction is obtained.

CAT Number and Display

Each picture element (pixel) of the CAT image has a CAT number associated with it. That CAT number represents the relative x-ray attenuation of the tissues within the volume element or voxel at that location. A voxel is a parallelepiped having a face corresponding to the pixel (e.g., a 0.5 mm × 0.5 mm square) and a depth corresponding to the slice thickness (e.g., 10 mm). The CAT number is defined by the equation:

$$\text{CAT\#} = 1000\,\frac{\mu_v - \mu_w}{\mu_w},$$

where μ_v and μ_w are the linear x-ray attenuation coefficients of the tissue(s) in the voxel and water, respectively. Most CAT scanners employ polyenergetic x-ray beams having maximum energies of about 120 keV and effective (equivalent monoenergetic) energies of about 70 keV. The predominant mechanism of x-ray attenuation at this effective energy is Compton scatter, the probability of which is proportional

to the mass density of the tissue. Hence, the CAT numbers in the pixels are representative of the mass densities. Typical CAT numbers are as follows: −1,000 for air, −100 for fat, 0 for water, 50 for muscle, and 2,000 for bone. The CAT numbers are digitized to 12 bits (4,096 levels) and are displayed on monitors capable of showing 256 gray levels. CAT scanners employ a technique termed "window level and width adjustment" to enhance image contrast within selected ranges of CAT numbers. The "level" is the central CAT number and the "width" is the range of CAT numbers that are converted to 256 gray levels for display. For example, when imaging the abdomen, one might use a window level of 20 and width of 450. Tissues having CAT numbers of −205 to +245 would be displayed via a linear transformation as 256 gray levels in the image. Adjustments of the window level and width permit optimized viewing of different tissues (e.g., lung and bone) within the same image without necessitating a rescanning of the patient.

Spatial Resolution

Spatial resolution refers to the ability of a system to distinguish two closely spaced objects. The spatial resolution of CAT images is typically 10 times worse than that of conventional radiographic images. Thus, it is the excellent contrast resolution rather than spatial resolution that makes CAT an indispensable diagnostic tool. CAT spatial resolution depends on many factors, including focal spot size (the width of the x-ray beam at its source), detector aperture (the effective detector size), number of projections, number of data points in each projection, reconstruction matrix size, and the reconstruction filter. Typical pixel sizes are 0.9 mm for scans of the body and 0.5 mm for scans of the head.

CAT Artifacts

No imaging system is perfect, and CAT scanners are no exception. CAT images may suffer from a number of artifacts, including beam hardening, partial volume, streak, motion, and ring artifacts. The x-ray attenuation and CAT number of a tissue depend on the incident x-ray energy. For the polyenergetic beams used in CAT, low-energy x rays are preferentially attenuated as the x-ray beam penetrates the object. This effect is called beam hardening, and it is an important cause of CAT artifacts. It may manifest itself as web-like streaks or areas of decreased intensity near the bones. Partial volume artifacts are encountered when two or more tissue types are contained within a voxel. The resulting reconstructed pixel value represents a CAT number that is an average of the CAT numbers of the tissues involved. When the imaged slice contains a very attenuating object, such as a surgical clip or a dense bone, streak artifacts may be observed in the reconstructed image. Another important cause of artifacts is patient motion, which may blur the reconstructed image. In third-generation CAT scanners, ring artifacts may appear in the image if one or more of the detectors are out of calibration.

See also: BIOPHYSICS; BODY, PHYSICS OF; MAGNETIC RESONANCE IMAGING; MEDICAL PHYSICS; NUCLEAR MEDICINE; X RAY

Bibliography

BROOKS, R. A., and DI CHIRO, G. "Principles of Computer Assisted Tomography (CAT) in Radiographic and Radioisotopic Imaging." *Phys. Med. Bio.* **21,** 689–732 (1976).

CORMACK, A. M. "Representation of a Function by Its Line Integrals, with Some Radiological Applications." *J. Appl. Phys.* **34,** 2722–2727 (1963).

GORDON, R.; HERMAN, G. T.; and JOHNSON, S. A. "Image Reconstruction from Projections." *Sci. Am.* **233,** 56–68 (1975).

HOUNSFIELD, G. N. "Computed Medical Imaging (Nobel Prize Lecture)." *Science* **210,** 22–28 (1980).

KAK, A. C., and SLANEY, M. *Principles of Computerized Tomographic Imaging* (IEEE, New York, 1988).

KALENDER, W. A.; SEISSLER, W.; KLOTZ, E.; and VOCK, P. "Spiral Volumetric CT with Single-Breath-Hold Technique, Continuous Transport, and Continuous Scanner Rotation." *Radiology* **176,** 181–183 (1990).

LEDLEY, R. S.; DI CIRO, G.; LUESSENHOP, A. J.; and TWIGG, H. L. "Computerized Transaxial X-Ray Tomography of the Human Body." *Science* **186,** 207–212 (1974).

MITCHELL M. GOODSITT
BERKMAN SAHINER

CAUSALITY

Causality pertains to the cause-effect relationship between particular states or events. The experience of striking a nail with a hammer and its sinking into a board is a common one. But what is the nature of the relationship between the striking of the nail with the hammer and its sinking into the board? An account of causality would explain what this and all other circumstances in which the concept is applied have in common.

Natural philosophers have sought this common core concept of causality over the ages, abstracting various general accounts from what is distinctive about particular examples and testing them against other examples. But there has been no agreement on a single account. Some philosophers suspect that the multitude of senses of "cause" that coexist is an indication that the concept applies to our theories of nature rather than to nature itself, and reflects the complexity of the inferential links in our theories and the variety of our interests in seeking causes. In any case, the various proposals suggested by philosophers provide the conceptual options for discussions of causality.

The search for a general analysis of causality is usually traced back to Aristotle, who included it in his *Physics*. Aristotle's concepts of material, formal, efficient, and final causes still are part of our thinking about causes, as is his concern with the possibility of chance or spontaneity.

The eighteenth-century philosopher David Hume presented the analysis of causality that has motivated much of the discussion since. Hume dedicated himself to the elimination of metaphysics from the study of nature. Thus he had no patience with Aristotle's "efficacy," the idea that there is in objects a "causal power" to bring about other states of affairs. Hume claimed that if we are scrupulous in analyzing the world as it is presented to our senses, we find in purported causal relationships only three circumstances: (1) whenever one state of affairs occurs some particular other state does as well; (2) the first state precedes the second; and (3) the two states are contiguous in space. We never experience some "cause" as a separate empirical component. Much of the systematic discussion of causality over the last 200 years has been related to Hume's three conditions.

The causality implicit in Newton's law of gravity, for instance, involves instantaneous action-at-a-distance, thus violating Hume's conditions 2 and 3. Concerns about such an idea of causality were important in the origins of the special theory of relativity, which features a limit on the speed of propagation of causal influences. On another tack, many natural correlations satisfying Hume's condition 1 are not regarded as causal. Something more than condition 1 seems necessary for a causal relationship.

One approach to articulating what more is necessary recognizes in Humean regularities instances of "uniformities of nature," and seeks to analyze such uniformities in terms of "natural laws." Two states are to be regarded as causally related when the occurrence of the effect state follows from some law of nature, instantiated by the cause state and other initial conditions.

The main problem with this approach is again related to noncausal correlations; there are laws that are clearly not causal laws. Such "empirical" laws—such as Bode's law or Kepler's laws in astronomy—merely summarize observed regularities. How does one in a noncircular way characterize a law as causal? One answer to this question appeals to the idea of counterfactuality. If a second state would not have occurred without some first state, then the states may be causally related. Causal laws are those that pick out this kind of relation between states. Appealing to counterfactuality in a general account of the causal relationship, however, runs into difficulties where multiple causes—each sufficient to produce an effect—occur simultaneously, as sometimes happens in cases of deaths from multiple injuries.

Such difficulties involving the complicated states related by natural laws led to a careful analysis of the initial and consequent conditions. Initial conditions, for instance, are generally individually necessary and collectively sufficient for the effect. A counterfactual condition is clearly a necessary condition. Sophisticated contemporary accounts of causality in terms of conditions ignore laws and counterfactuality altogether. But such accounts have problems formulating nontemporal criteria distinguishing causes from effects.

Other counterexamples to a necessary-and-sufficient-condition analysis of causality are the many instances in which effects are only predisposed by particular causes, not guaranteed by them. A consideration of such counterexamples has given rise to probabilistic accounts of causality, in which a cause is considered to increase the probability of an effect. The Humean constant conjunction of states (condition 1) yields to correlations in probabilistic accounts of causality. But even correlations may not be a sign of direct causal links between events. Rather, two correlated effects may be linked to a single common cause, as the correlation between lowering barometers and stormy conditions is linked to the onset of low pressure weather systems. Arguably, this probabilistic, common-cause link is the weakest of all causal links.

Probabilistic analysis of causation encompasses multiple necessary conditions and admits that no combined sufficient condition need exist in causal relationships. This last feature has given rise to much contemporary debate, particularly in discussions of the foundations of quantum mechanics and the quantum theory of measurement. Some are willing to introduce sheer chance into the notion of cause, while others have resisted, notably Albert Einstein. Explicating the notion of probability in probabilistic causal accounts is far from trivial.

Causality in quantum theory has been a major component of discussions of the Einstein–Podolsky–Rosen thought experiment and Bell's theorem, with its experimental tests. These far-reaching discussions of the correlations of measured properties of spatially separated composite quantum systems have included the hypothesis that such systems exhibit nonlocal causation (locality), perhaps violating the special relativistic limits. If not, and if the probabilistic common cause link is the weakest of all causal links, then the systematic correlations exhibited by these quantum systems are not underlain by "causes" recognizable in any classical sense.

See also: ACTION-AT-A-DISTANCE; BELL'S THEOREM; EINSTEIN, ALBERT; EINSTEIN–PODOLSKY–ROSEN EXPERIMENT; KEPLER'S LAWS; LOCALITY; NEWTON'S LAWS; PROBABILITY; QUANTUM MECHANICS; QUANTUM THEORY OF MEASUREMENT; RELATIVITY, SPECIAL THEORY OF, ORIGINS OF

Bibliography

ARMSTRONG, D. *What is a Law of Nature?* (Cambridge University Press, Cambridge, Eng., 1983).

CUSHING, J. T., and MCMULLIN, E., eds. *Philosophical Consequences of Quantum Theory: Reflections on Bell's Theorem* (University of Notre Dame Press, Notre Dame, IN, 1989).

LEWIS, D. "Causation." *Journal of Philosophy* **70**, 556–567 (1973).

MACKIE, J. L. *The Cement of the Universe* (Clarendon, Oxford, Eng., 1974).

SALMON, W. *Scientific Explanation and the Causal Structure of the World* (Princeton University Press, Princeton, NJ, 1984).

E. ROGER JONES

CAVITATION

On the surface of some objects, such as a ship propeller moving at high speed in water, local lowering of pressure takes place according to Bernoulli's law. When this pressure becomes lower than that of the saturated vapor pressure, tiny pre-existing voids are expanded, collecting water vapor and dissolved gases to form bubbles. This phenomenon is called (hydraulic) cavitation. A wealth of physical, chemical, and biological effects accompany cavitation. The main areas of application and research are hydraulic and acoustic cavitation. Hydraulic cavitation is associated with all kinds of hydraulic systems, for example, ship propellers, turbine blades, hydrofoils, and pumps. The occurrence of cavitation lowers the efficiency, leads to a quick destruction of the parts being exposed to cavitation, and is the source of intense noise radiated into the liquid. Acoustic cavitation appears in a liquid when a sound field is set up at a sufficiently high sound pressure amplitude. Then a tension is set up in the liquid in the lower-pressure phase of the sound wave that expands the always existing tiny voids to cavitation bubbles. Acoustic cavitation is unwanted in underwater acoustics, where signals are to be transmitted over a long range (SONAR), and in ultrasonic diagnostics, where, for example, the interior of the human body is scanned with the aid of sound waves. In industry, acoustic cavitation has found some application in cleaning complex parts, in emulsification, and disintegration. It appears around kidney stones in extracorporeal shock-wave lithotripsy and in the brain after shock loading of the head in accidents.

Cavitation is a very violent physical phenomenon. The cavities formed upon rupture of the liquid in a high tension region tend to get spherical due to the action of the surface tension of the liquid. When they subsequently collapse in a high pressure region, energy is concentrated into a small volume of the cavity by compression of the gas and vapor content it acquired during growth. As a result, strong shock waves are radiated into the liquid, and a light flash is emitted into the surroundings. The light emission caused by collapsing cavitation bubbles is best seen in acoustic cavitation. Therefore, it has come to be known as sonoluminescence. The light flashes are faint, but they can be seen with the dark-adapted eye in water when a strong sound wave is set up leading to cavitation. The duration of the light flashes is very short, shorter than 10^{-9} s. The dynamics of cavitation bubbles thus belongs to the fastest mechanical processes. Since extreme physical conditions are set up in a cavitation bubble, conditions that are difficult to reach otherwise, chemists are exploring the possibilities of using cavitation in producing exotic substances. That area is called sonochemistry, where once again acoustic cavitation is used as a simple means to produce cavitation. It has even been speculated that cavitation may be used for nuclear fusion, but the temperatures and pressures reached so far have not been encouraging. The main obstacle seems to be instabilities of the spherical form during cavity collapse, which gives rise to another spectacular effect: high speed liquid jet formation with velocities reaching at least 200 m/s. Liquid jet formation is most pronounced when the cavity collapses in the vicinity of a solid wall.

Since the end of the nineteenth century, when cavitation was first observed on ship propellers, it has become of major concern in ship building, in particular in the military sector where the more important issues are ship speed and silence. Cavitation limits the speed of conventionally designed ships (which have propellers and hydrofoils) and gives away a ship's location via the noise produced from the oscillating and collapsing bubbles.

See also: LIQUID; PRESSURE, VAPOR; WATER

Bibliography

LEIGHTON, T. G. *The Acoustic Bubble* (Academic Press, London, 1994).
YOUNG, F. R. *Cavitation* (McGraw-Hill, London, 1989).

W. LAUTERBORN

CELESTIAL MECHANICS

Celestial mechanics is the field of astronomy that deals with the motions of the planets (major and minor), moons, and comets in the solar system. It has been extended to include the motions of stars in pairs, triples, and so forth, and to clusters of stars in larger aggregations. Celestial mechanics stands as the first discipline in which the circle of science closed: from observations of the planets, to induction and the formation of hypotheses, to development of a complete theory, and then back to observations with quantitative predictions. Celestial mechanics is also special for another reason—it was the first part of science to demonstrate that the events discerned in the heavens could be related to occurrences on Earth. Thus, the possibility of understanding the origins of the solar system and the universe could become a reality by utilizing the physics of our immediate vicinity and applying it across the cosmos.

The most ancient astronomers made observations of the "fixed stars" and the "wandering stars." The latter consisted of seven objects (i.e., the Sun, the Moon, Mercury, Venus, Mars, Jupiter, and Saturn). "Wandering stars," by and large, became what we call planets. Although without a physical theory to account for the motions of planets and their mutual phenomena, which is especially difficult during opposition (when they are opposite the Sun on the sky as seen from the earth), mathematical constructs to predict these events were devised. The most notable was Claudius Ptolemy's theory of epicycles. This kind of induction culminated in Johannes Kepler's synthesis of Tycho Brahe's observations of the solar system and led to Kepler's three laws of planetary motion. They state that (1) every planet in the solar system moves about the Sun in an ellipse with the Sun at one focus, (2) that in doing so the area swept out in a unit of time is equal to the area swept out in the same unit of time anywhere else in the orbit, and (3) that the square of the period of revolution is proportional to the cube of the size of the orbit (technically the semimajor axis).

Real physics began with Galileo Galilei (the inventor of the telescope) and made its first quantitative strides with Isaac Newton. It was Newton who formulated the laws of motion and it was Newton who generalized from the gravitational attraction of commonplace bodies toward Earth (the proverbial apple that allegedly fell on his head) to the gravitational attraction of the Moon toward Earth. (Along the way he had to invent integral calculus to solve the general gravitational attraction problem for Earth; much of mathematics has been generated by celestial mechanics.) He recognized that the underlying phenomena were the same in both cases, only the rate of acceleration changed with the distance between the two bodies. Deducing from the actual motion of the Moon that the strength of the attraction was inversely proportional to the square of the distance, he went one step further yet and postu-

lated that the same force, with the same characteristics, was the foundation for the mechanics of the solar system. Next he combined his three laws of motion and his law of gravitation and deduced Kepler's laws of motion. That is, he predicted, from the mathematical formulation of the physics, the statements comprising Kepler's laws. Other predictions followed and Newtonian gravitation became accepted as the binding force of the universe.

To make this extrapolation believable required one more convincing piece of evidence—that the stars themselves moved, in a quantitatively predictable fashion, in a way that could be perfectly explained by the theory. This step had to await the discovery of double stars (or binary stars); a couplet of stars revolving around each other just as the Moon revolves around Earth. It was William Herschel's analysis of many years of observations of several double stars (which he thought were merely optical doubles, i.e., two stars along the same line of sight) that made it clear the gravitation was, in fact, responsible for their motions and hence universal. Astronomers and astrophysicists followed with the other laws of physics, thereby providing the basis for the general framework in which modern astrophysics works. Today binary pulsars provide an opportunity to test the theory of general relativity in much the same way.

One of the most acclaimed triumphs of classical celestial mechanics was the prediction of the existence of Neptune from observations of Uranus. Uranus was discovered in 1780 and it soon became clear that its motion through the sky did not follow a forecast from Newton's laws. One possible explanation could be yet another body in the solar system. John Couch Adams and Urbain Jean Joseph Leverrier separately performed the computations necessary to deduce the location and mass of the perturbing body. When a German astronomer directed his telescope to the appropriate place (in September 1846), there was Neptune. In a similar fashion came Pluto in 1930. Some astronomers still use this tool to search for Planet X.

A second major event in the development of celestial mechanics occurred on January 1, 1801. Another new planet (actually an asteroid or minor planet) was discovered as it was approaching the Sun (we now call it Ceres). Hence, it soon was lost in conjunction. The problem was how to find it again after it passed behind the Sun. Karl Friedrich Gauss took this quest up, laying the foundations for the computation of orbits from very little observational data. This method became of renewed importance in the space age. Now there were hundreds and then thousands of new objects to compute orbits for, there was computational capacity undreamt of by Gauss (i.e., the electronic digital computer), and there were new observing techniques such as radar and computer-controlled optical telescopes with which to track the satellites and their associated rocket bodies.

These computers were soon applied to more complicated celestial mechanics problems that had long been analytically unsolved (although leading to many discoveries in pure mathematics as astronomers continued to devise new techniques to employ on them). In particular, the motions of stars in clusters of stars, i.e., objects consisting of hundreds or hundreds of thousands of stars, were studied numerically. Larger-scale simulations dealt with billions of stars (i.e., galaxies). Gradually some understanding of the evolution and stability of these objects has been reached.

See also: COPERNICAN REVOLUTION; COPERNICUS, NICOLAUS; GALILEI, GALILEO; KEPLER, JOHANNES; KEPLER'S LAWS; NEWTONIAN MECHANICS; PLANETARY SYSTEMS; SOLAR SYSTEM

Bibliography

BROUWER, D. *Methods of Celestial Mechanics* (Academic Press, New York, 1961).
TAFF, L. G. *Celestial Mechanics* (Wiley, New York, 1985).

LAURENCE G. TAFF

CELSIUS

See TEMPERATURE SCALE, CELSIUS

CENTER OF GRAVITY

Consider a rigid body of mass M. Any rigid body can be thought of as being composed of a large number

n of smaller masses m_i such that the sum of all m_i is the total mass

$$\sum m_i = M. \tag{1}$$

Note that $\sum$ indicates a sum over all the mass elements numbered $1 - n$. The center of gravity of the rigid body is the point within the rigid body through which an applied force $\mathbf{F}$ balances all linear and angular accelerations due to a gravitational field $\mathbf{g}(\mathbf{r}_i)$, where $\mathbf{r}_i$ is the position of the ith particle relative to an arbitrary origin such as the center of mass of the body. The center of gravity is the equilibrium point through which the sum of all external forces $\mathbf{F}_{ext}$ and torques τ_{ext} is zero. If $\mathbf{F}$ is applied at the center of mass, then

$$\mathbf{F}_{ext} = \mathbf{F} + \sum m_i \mathbf{g}(\mathbf{r}_i) = 0, \tag{2}$$

$$\tau_{ext} = (\sum m_i \mathbf{r}_i) \times \mathbf{g}(\mathbf{r}_i) = 0. \tag{3}$$

Note that $\times$ denotes a vector cross product.

Equations (2) and (3) are complicated by the generality that the force of gravity on a rigid body changes according to the distance of each mass element from the center of the earth. Thus the gravitational field is written as $\mathbf{g}(\mathbf{r}_i)$. The problem can be simplified if we assume that the gravitational field is uniform throughout the rigid body such that $\mathbf{g}(\mathbf{r}_i) = \mathbf{g}$. In this case, the equilibrium condition becomes

$$\mathbf{F}_{ext} = \mathbf{F} + M\mathbf{g} = 0, \tag{4}$$

$$\tau_{ext} = (\sum m_i \mathbf{r}_i) \times \mathbf{g} = 0. \tag{5}$$

Thus, in Eq. (4) we find that the applied force is $\mathbf{F} = -M\mathbf{g}$. As for the torque equation, recall that center of mass $\mathbf{r}_{cm}$ is defined as

$$\mathbf{r}_{cm} = \frac{1}{M} \sum m_i \mathbf{r}_i \tag{6}$$

and is independent of any external fields, including gravity. Thus, in a uniform gravitational field, the equilibrium condition for torque becomes

$$\mathbf{r}_{cm} \times \mathbf{g} = 0. \tag{7}$$

Together, Eqs. (4) and (7) show that for a rigid body in a *uniform* gravitational field, the gravitational forces and torques are equilibrated (such that there is no linear or angular acceleration) by a force $\mathbf{F} = -M\mathbf{g}$ applied at the center of mass. In this case, the center of gravity is the center of mass, the point through which the gravitational force acts.

If we do *not* assume that the gravitational field is uniform, then determining the center of gravity is more complicated. The uniform field $\mathbf{g}$ becomes position dependent such that $\mathbf{g} = \mathbf{g}(r_i)$, where r_i is the distance of the *i*th mass element from the center of mass of the body to which it is gravitationally attracted, in this case the center of the earth. For a rigid body whose position is not rotationally symmetric with respect to the earth, the center of gravity will not coincide with the center of mass. Furthermore, if the rigid body is in rotational motion, the center of gravity will move in a corresponding trajectory around the center of mass of the rigid body.

To illustrate how determining the center of gravity depends on whether one assumes gravitational uniformity, consider a twirling baton. First, let the rubber knobs on each end of the baton have the same mass. If we assume a uniform gravitational field, then as the baton twirls the changing distance of the knobs from the center of the earth will not affect the baton's motion; that is, there will be no differential gravitational torque on the two ends of the baton and the center of gravity will correspond to the center of mass. However, if we do take into account the nonuniformity of the *g*-field, then we have to calculate the gravitational force and corresponding torque on each end of the baton as a function of its position as it rotates. Since the gravitational force depends on the inverse square of the distance between attracting bodies (in this case, each mass element of the baton and the earth), mass elements closer to the earth will experience a greater force and torque than those farther away.

Is it reasonable to assume that the earth's gravitational field is uniform, at least near the earth's surface? Going back to the twirling baton, let the baton's length be 1 m. Consider the baton to be at the top point of a 2-m throw and oriented vertically such that the top knob is 2.5 m above the earth's surface (at the equator) and the bottom knob is 1.5 m above the earth's surface. Ignoring the mass of the baton's bar, take the difference between the gravitational force on the bottom knob and the gravitational force on the top knob using Newton's

gravitational force law. Using the radius of the earth $r_e = 6.37 \times 10^6$ m, M_e = mass of the earth, and the mass of each knob to be *m*, we find that

$$F_{\text{bottom}} - F_{\text{top}} = 3.14 \times 10^{-7}(GM_e m/r_e^2), \qquad (8)$$

where G is the gravitational constant. The difference in the gravitational force between the two knobs is negligible compared to the gravitational force on each knob, thus showing that at or near the earth's surface it is reasonable to assume that the gravitational field is uniform.

Finally, a simple experiment can be performed to determine the center of gravity of a rigid body. The experiment is done most easily with a flat object, but is generalizable to three-dimensional bodies. First, suspend the body from any point on its edge. From the suspension point, draw a vertical line through the body. Next, pick another point on the body from which to suspend it and draw another vertical line. Any one line will pass through the center of gravity; the intersection of two such lines is the center of gravity. This experiment illustrates the equilibrium conditions that define center of gravity [Eqs. (4) and (7)]. The tension in the string used to suspend the body is simply the applied external force, $\mathbf{F} = -M\mathbf{g}$. Suspending the body by a point causes it to rotate into an orientation that minimizes the local gravitational potential energy, thus equilibrating the gravitational torque on the body. The vertical lines drawn from two or more such suspension points will thus pass through the center of gravity.

In summary, the center of gravity of a rigid body is the equilibrium point at which the sum of all external forces and torques equals zero. In general, the center of gravity is a function of the distance of each mass element in the rigid body from the center of the earth, but near the earth's surface it is reasonable to assume that the earth's gravitational field is uniform. In this case the center of gravity equals the center of mass.

See also: CENTER OF MASS; CENTER-OF-MASS SYSTEM; GRAVITATIONAL FORCE LAW; INVERSE SQUARE LAW; RIGID BODY

Bibliography

BLATT, F. J. *Principles of Physics* (Allyn & Bacon, Newton, MA, 1983).

HALLIDAY, D., and RESNICK, R. *Physics,* 3rd ed. (Wiley, New York, 1978).

P. W. "BO" HAMMER

CENTER OF MASS

Every physical system (i.e., an object or set of objects chosen for study), has a center of mass, a point that moves as if all the mass of the system were concentrated there. Using the center of mass often simplifies the analysis of motion, such as by separating motion *of* the center of mass from motion *about* the center of mass, or in calculating rotational inertia using the parallel axis theorem, or in calculating angular momentum using the remarkable theorem that the rate of change of angular momentum equals the net torque about the center of mass, even if the center of mass is accelerating.

It does not matter whether the system is a rigid body, a nonrigid solid, a liquid, a living thing, a galaxy made up of many stars, or even a beam of radiation, such as light or x rays. Every system has a center of mass. This point may or may not be inside the matter of the system. For example, a circular hoop with all its mass on the circumference will have center of mass at the center of the circle, where there is no mass.

The center-of-mass point is the weighted average position of all the mass in the system. It can be calculated for nonrelativistic systems ($v \ll c$) as

$$\mathbf{R}_{CM} = \frac{\sum m_k \mathbf{r}_k}{\sum m_k},$$

where the summations are over all mass particles making up the system, with the kth particle having mass m_k and located at a position $\mathbf{r}_k$ in some chosen coordinate system. Here $\sum m_k = M$ is the total mass of the system. For a continuous solid body, the summations are ordinarily replaced by integrations over the mass distribution of the body, using infinitesimal masses dm in place of the individual particle masses m_k.

With $\mathbf{R}_{CM}$ defined as above, we can differentiate with respect to time to obtain $M\mathbf{V}_{CM} = \sum m_k \boldsymbol{v}_k$, which is just $\mathbf{P}_{CM} = \sum \mathbf{p}_k = \mathbf{P}$, the total system momentum.

So the center of mass is the point whose momentum is the total momentum of the system.

It is somewhat simpler and more readily applicable to relativistic systems to approach the center of mass via the total system momentum, as we show in the next paragraph. In addition, this approach has the advantage of emphasizing center-of-mass velocity, which is what principally makes center of mass an important concept. The fundamental idea is that there exists a point associated with any physical system that moves with the total system momentum as if the total mass of the system were concentrated at that point. That point is the center of mass.

The equation for total system momentum that is valid even for speeds that are a significant fraction of the speed of light is

$$\mathbf{P} = \sum \mathbf{p}_k = \gamma_{CM} M \mathbf{V}_{CM},$$

where $\gamma_{CM} M = \sum \gamma_k m_k$. This equation gives

$$\mathbf{V}_{CM} = \frac{\mathbf{P}}{\sum \gamma_k m_k}.$$

Here γ is the Lorentz factor,

$$\gamma = \frac{1}{\sqrt{1 - (v^2/c^2)}},$$

where v is the speed of the object and c is the speed of light. This equation for $\mathbf{V}_{CM}$ can be integrated to find $\mathbf{R}_{CM}$. Nevertheless, we often will not need more than $\mathbf{V}_{CM}$, since this allows the definition of center-of-mass coordinates, a set of coordinates moving with the center of mass of the system being studied. In these coordinates, the total momentum of the system is zero, and we are able to focus entirely on internal system motions (relative motions) and rotations about the center of mass. So identifying the center of mass allows one to separate a problem into two parts: motion of the center *of* mass and motion *about* the center of mass.

Motion of the Center of Mass

The motion of individual particles in classical physics is governed by Newton's laws. Any physical system or object is composed of very many tiny particles whose individual motion is determined by

$$\frac{d\mathbf{p}_k}{dt} = \mathbf{F}_k,$$

where $\mathbf{F}_k$ is the net force on particle k. The total force on the system is $\mathbf{F} = \Sigma \mathbf{F}_k$. This total force is just the external force because the internal forces all cancel in this vector sum. Newton's third law tells us that for every internal force there is an equal and opposite force. So the external force is

$$\mathbf{F} = \frac{\Sigma d\mathbf{p}_k}{dt} = \frac{d(\Sigma \mathbf{p}_k)}{dt} = \frac{d\mathbf{P}}{dt} = \frac{M\,d\mathbf{V}_{CM}}{dt} = M\mathbf{A}_{CM}.$$

This says the center of mass moves according to Newton's second law just as if it were a point particle of mass M. When we can approximate a body or system as a point, then the motion will just be the motion of the center of the mass.

Center of Mass of a Rigid Body

A rigid body is able to move along a trajectory (translational motion) and rotate (rotational motion), but it cannot deform; the particles making up the body remain in the same relative positions as the body moves. The center of mass of a rigid body is determined by the geometry of the body and always stays in the same position with respect to the body. Translational or rotational motion of the body does not change the position of the center-of-mass relative to the body itself. In the real world there is no truly rigid body; however, the idealization that solids subject to forces small enough to cause no significant deformation can be treated as rigid bodies is a good approximation.

The motion of the center-of-mass with respect to a stationary observer outside the rigid body can be described by $\mathbf{V}_{CM}$. In center-of-mass coordinates, that is, coordinates moving with the center of mass of the body, the rigid body has no translational motion; its total momentum is zero. However, it still has rotational motion, and the rotational motion of the body has been separated from the translational motion, which simplifies our problem. In this reference frame, the rotational motion we study is rotation about the center of mass. By this means, we can describe the most general motion of our rigid body as translational motion of the center of mass plus rotation about the center of mass.

Symmetry and Center of Mass

If the mass particles making up a system are arranged with some definite symmetry that is maintained as the system moves, then the center of mass will lie on the symmetry plane or axis. Since the center of mass is the weighted average position of the mass of the system, that average will reflect the geometric symmetry of the system. For example, a uniform thin rectangular plate will have center of mass at the geometric center of the plate. A uniform circular ring of mass will have center of mass at the center of the circle. Taking account of the symmetry simplifies the calculation of the center of mass.

See also: MOMENTUM; NEWTONIAN MECHANICS; NEWTON'S LAWS; RIGID BODY

Bibliography

ASIMOV, I. *Understanding Physics* (George Allen & Unwin, London, Eng., 1966).

FEYNMAN, R. P.; LEIGHTON, R. B.; and SANDS, M. *The Feynman Lectures on Physics*, Vol. 1 (Addison-Wesley, Reading, MA, 1963).

HALLIDAY, D., and RESNICK, R. *Fundamentals of Physics*, 3rd ed. (Wiley, New York, 1988).

WILLIAM E. EVENSON

CENTER-OF-MASS SYSTEM

When dealing with a physical system consisting of more than one moving and colliding particle or object it is often convenient to separate the motion into that of the center of mass of the system and the motion of the system's constituents relative to the center of mass. The center-of-mass coordinate system is then the frame of reference in which the center of mass is stationary. One can visualize this by attaching the reference frame to the center-of-mass point. This center-of-mass frame is to be distinguished from an external, fixed laboratory frame against which we commonly suppose motion to take place.

There are two good reasons for using the center-of-mass system as opposed to the laboratory system. First, in the center-of-mass system the equations that

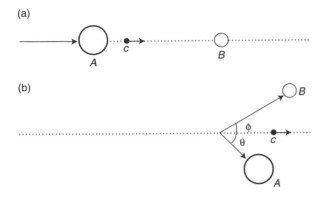

Figure 1 Laboratory reference frame (a) before collision and (b) after collision.

describe motion can be simplified. By fixing the reference frame to the center of mass one can focus on the interactions between the constituents of the system and reduce the number of parameters used to describe the motions of the system. The center-of-mass system is especially useful when describing collisions of bodies since the total momentum is zero in this frame of reference before and after the collision. The second reason is that the motion of the center of mass is often not important since it is only the motions and hence the energies of the system's constituents relative to each other that is available to change the internal states of the system. For example, the stars in a galaxy may pull and collide with each other, but the fact that the galaxy as a whole is rushing through space does not affect these interactions. Consequently it is not necessary to consider the motion of the center of mass of the galaxy when discussing the internal motions of the stars.

As an example of the use of the center-of-mass system consider the collision of two objects such as balls on a pool or billiard table. Figure 1 shows the situation as viewed in the laboratory reference frame. Initially (Fig. 1a), object A of mass m_A moves towards B, which is at rest, with velocity v_A and momentum $p_A = m_A v_A$. The center of mass, indicated by the point c, moves with velocity $v_c = m_A v_A / (m_A + m_B)$. After the collision (Fig. 1b) the center of mass continues to move in a straight line at the same constant velocity since no external forces have been applied. The objects A and B move off with momenta p_A' and p_B' at angles θ and ϕ, respectively. The values of the angles cannot be predicted from energy and momentum considerations alone; they depend on the internal details of the collision (for example, did the balls collide head-on or at a glancing angle?).

The picture is simplified if we now change to a coordinate system in which the center of mass is always at rest. Now A and B initially move toward c with velocities $m_B v_A / (m_A + m_B)$ and $-m_A v_A / (m_A + m_B)$ and equal and opposite momenta (Fig. 2a). After the collision (Fig. 2b) A and B also move with equal and opposite momenta $q_A' = -q_B'$ (since momentum is conserved) and at an angle γ to the initial direction of A. The system is now determined in terms of q_A' and γ, two parameters as opposed to four in the laboratory frame of reference. We may note that if the collision is elastic so that there is no change in the internal energy, then collision in the center-of-mass system simply serves to rotate the velocity vectors. That is, the balls move off with the same initial speeds but in different directions. The relationship between the angles in the center-of-mass and laboratory systems can be found by translating the laboratory system at the same velocity as the center-of-mass so that the center of mass now appears at rest.

The center-of-mass system is frequently used in particle physics when analyzing the collisions between elementary particles in accelerators. Since it is only the energy of one particle relative to another that is available to excite the internal structure or create new particles, the uniform motion of the center of mass of the system is of no consequence. The energies and momenta in the center-of-mass system are thus the important ones. It can be shown that at very high energies (where relativistic effects become important) that the energy available in the center-of-mass system only increases as the square root of the laboratory energy. For this reason modern parti-

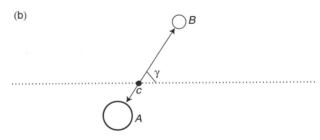

Figure 2 Center-of-mass reference frame (a) before collision and (b) after collision.

cle physics is carried out using two colliding beams of subatomic particles rather than one beam and a fixed target.

See also: ACCELERATOR; CENTER OF MASS; COLLISION; ENERGY, INTERNAL; FRAME OF REFERENCE, INERTIAL; FRAME OF REFERENCE, ROTATING

Bibliography

FRAUENFELDER, H., and HENLEY, E. M. *Subatomic Physics,* 2nd ed. (Prentice Hall, Englewood Cliffs, NJ, 1991).

LANDAU, L. D., and LIFSHITZ, E. M. *Mechanics,* 2nd ed. (Pergamon, New York, 1969).

SCHIFF, L. I. *Quantum Mechanics,* 3rd ed. (McGraw-Hill, New York, 1968).

JOHN P. SHARPE

CENTRIFUGAL FORCE

Centri*fugal* forces fall into a class of forces called pseudo-forces, in contrast to centri*petal* forces, which all have a clear physical origin or mechanism. In the case of motion along a curved path, pseudo-forces originate from a desire to write an equation of the form of Newton's second law in a frame of reference where they are not valid (any frame where the motion is not in a straight line at a constant speed). The term mv^2/r, the scalar form of centrifugal force, is added to maintain the form of the equation.

However, just because centrifugal force is a pseudo-force does not mean that people do not "experience" it. A detailed everyday example is the passenger who feels a force or push against the door of the car when making a left turn at the corner of an unbanked road. This force is labeled the centrifugal force (from the Latin words for center-fleeing). The passenger in the car perceives and interprets these effects in an accelerated frame of reference. Contrast this with the point of view of an observer standing on the curb watching the car turning. The driver makes a left turn and the frictional force between the car's tires and the road provides the centripetal (center-seeking) force that initiates the change in direction. The seat-of-the-pants frictional centripetal forces in turn help divert the passenger's direction to follow the car. However, the external observer

notes that only the "seat of the pants" experiences this force. The passenger's body and head continue to travel in a straight line, and so the curved path of the car door intersects the straight path of the passenger's body, producing a collision. As this collision continues, the car door pushes on the passenger's right side, forcing the body to follow the car's path.

For every action force there is an equal and opposite reaction force. The external observer in the inertial frame sees the action force as the car door pushing on the passenger, but the passenger in the noninertial frame perceives the push against the door as an action force. It is this personal experience that colors the passenger's judgment as to the respective origin of the forces.

The applications of centrifugal forces include the human centrifuge to subject pilots to the forces experienced in flying or space travel, the blood centrifuge to separate blood cells from blood plasma due to their differing densities, and the gas centrifuge to separate isotopes of uranium with differing masses. Large space stations could simulate gravity by rotation. Fair or theme park rides that spin the occupant in a small car or in a large cylinder with people along the outside wall, motorcyclists traversing the interior circumference of a metal cage vertically, or roller coasters negotiating vertical loops are all common applications of centrifugal forces.

See also: CENTRIPETAL FORCE; CORIOLIS FORCE; FRAME OF REFERENCE, INERTIAL; FRAME OF REFERENCE, ROTATING; NEWTON'S LAWS

Bibliography

GRIFFITH, W. T. *The Physics of Everyday Phenomena* (Wm. C. Brown, Dubuque, IA, 1992).

HEWITT, P. *Conceptual Physics* (Little, Brown, New York, 1994).

GEORGE BISSINGER

CENTRIFUGE

One of the most common instruments in use is the centrifuge. The uses of this instrument in its various forms range from jobs as mundane as separating cream from raw milk or spinning excess water from clothes in a washing machine to determining the size and shape of molecules in research laboratories. The fundamental concept of sedimentation, the motion of a body in a centrifugal field, is simple.

Before the concept can be clearly understood, however, two basic "forces" must be investigated. Since bodies in motion tend to travel in a straight line, a body constrained to travel along a curved path is constantly trying to continue along the straight path that is tangent to its current path. This results in a tendency for the body to appear to pull away from the center of curvature of its path. This is generally referred to as the centrifugal (center-fleeing) force. The true force involved, however, is centripetal (center-seeking), which acts to keep the body moving along the circular path. For example, when a ball is twirled at the end of a rubber string, the elasticity of the rubber string acts as a centripetal force to keep the ball going in a circle, preventing it from flying away from the center of the circle. If the string is released, centripetal force will be lost and the ball will fly off in a straight line tangent to the curved path. The tendency by which the ball tries to leave the circular path is dependent upon the angular speed at which it is traveling (revolutions per unit of time). If the ball is twirled faster, the elastic will stretch until it can apply a larger centripetal force equal to the tendency to take a straight course. If a person were sitting on the twirling ball, the force would feel just like gravity pulling the individual away from the center of rotation. Therefore, centrifugal force is usually measured in proportion to Earth's gravity.

A modern scientific centrifuge usually consists of a high speed motor that rotates a spindle upon which is mounted a rotor. The rotor can be a solid unit containing holes into which tubes are inserted. The rotor can also have buckets that are hinged, allowing them to swing out when the centrifuge is started. Centrifuges can spin the rotor at speeds up to 90,000 rpm, producing forces approaching 700,000 times the force of gravity. That means a 4 oz. pencil would appear to weigh more than 3 tons! The pencil would rapidly sink through a solution under these conditions. This apparent sinking through the solution is really movement away from the center of the centrifuge and is called sedimentation.

Particles are usually sedimented in tubes containing liquid solutions. When the solution is spinning in the centrifuge, a number of parameters in addi-

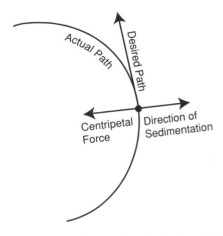

Figure 1 The circular path of a ball twirled at the end of a rubber string.

tion to the rotation rate affect the velocity at which the particle sediments. (1) The more massive the particle, the faster it sediments. (2) The denser the solution, the slower the particle will sediment because when objects displace fluid they create an upward force (buoyancy) whose magnitude is proportional to the mass of solvent displaced. When a greater mass of solvent is displaced per unit volume of the particle, the buoyant effect increases. (3) Denser particles move faster because they displace a smaller volume of solvent relative to their own mass, so less buoyancy is created. (4) The greater the friction between the particle and the solvent, the slower it moves. Friction is dependent upon both the viscosity of the solution and the shape of the particle; round particles create less friction, and oblong particles create more friction. By carrying out experiments under controlled conditions, scientists can determine the mass and shape of particles, even molecules. In addition, centrifugation can be used to separate from each other particles that differ in any of these properties (size, density, or shape). The velocity of sedimentation is defined by the following equation: $v = \omega^2 rm(1 - \bar{v}\rho)/f$, where v is the velocity, ω is the angular velocity in radians per second (one revolution equals 2π radians), r is the distance from the center of rotation, m is the mass of the particle, $\bar{v}$ is the partial specific volume of the particle (the reciprocal of its density), ρ is the density of the solution, and f is the frictional coefficient of the sedimenting particle.

See also: CENTRIFUGAL FORCE; CENTRIPETAL FORCE

Bibliography

GRIFFITH, W. T. *The Physics of Everyday Phenomena* (Wm. C. Brown, Dubuque, IA, 1992).
HEWITT, P. *Conceptual Physics* (Little, Brown, New York, 1994).

MARVIN PAULE

CENTRIPETAL FORCE

Early science dealt with circular motion a great deal. Circular motion was used to describe the motions of the stars overhead. Unfortunately early models of star motion stated that this circular motion was "natural" for heavenly bodies and so did not require any forces to produce or maintain. Now our understanding of mechanical motions presumes that the only way an object travels at a constant speed in a straight line is if the net force acting on it is zero. If an object moves in a curved path with a center such as a circle, it is subject to centripetal forces (the Latin root word *centripetus* means center-seeking). Centripetal forces are not any particular type of force. Any force whose direction is toward the center of a circle along a radius, even if only briefly, can be a centripetal force. This includes the following forces: gravitational between objects with mass; electrostatic between oppositely charged particles; mechanical between a string and the object attached to its end when swung over the head; and even frictional as in the contact between a car's tires and the road when the car makes a turn.

Our concepts of force were pioneered by such seventeenth-century physicists as Isaac Newton. Newton's laws are used almost exclusively in the description of force-induced movement such as the gravitational motions of stars, planets, satellites, and baseballs. It was the curved orbit of the Moon that led Newton to isolate one of the first clear examples of centripetal force. A rather common mistake is to confuse centripetal force with its pseudo-force cousin, centrifugal (from the Latin for center-fleeing) force. One of the first recorded examples of this confusion occurred when Newton, during his initial analysis of the gravitational force, made the mistake of thinking it was centrifugal. Letters between him and his contemporary Robert Hooke clearly show that Newton's confusion over this matter was cleared up with Hooke's assistance.

Centripetal forces commonly appear during everyday activities in rotating systems such as a merry-go-round. The seat-of-the-pants friction from the horse directed toward the axis of rotation is the centripetal force that keeps the rider from sliding free of the platform and off in a straight line. The fact that any object moving along a path that is not a straight line can have a centripetal force acting on it even if only briefly is an indication of how commonly the centripetal force can be observed. Sometimes multiple centripetal forces act on an object. The curveball in baseball actually has two centripetal forces acting on it, vertically there is gravity and horizontally the pressure difference between the ball faces (assuming the spin axis is vertical), which exerts a force approximately perpendicular to the baseball's direction.

See also: CENTRIFUGAL FORCE; CENTRIFUGE; MOTION, CIRCULAR

Bibliography

COHEN, I. B. "Newton's Discovery of Gravity." *Sci. Am.* **244** (March), 166–176 (1981).
GRIFFITH, W. T. *The Physics of Everyday Phenomena* (Wm. C. Brown, Dubuque, IA, 1992).

GEORGE BISSINGER

CHADWICK, JAMES

b. Bollington near Macclesfield, England, October 20, 1891; *d.* Cambridge, England, July 24, 1974; *radioactivity, nuclear physics.*

Chadwick was the first child of John Joseph Chadwick, a cotton spinner, and his wife Ann Mary (*née* Knowles). He was brought up by his grandparents in the country village of Bollington, where he received his early schooling. Later, he moved to join his parents (who, by this time, had produced two other sons and a daughter) in Manchester. He was educated at the Manchester Municipal Secondary School, where he developed strong interests in physics and mathematics.

In 1908 he entered Manchester University, where he gained First Class Honours in Physics in 1911. Under the direction of Ernest Rutherford, Chadwick undertook research on the technique of gamma-ray measurement, for which he was awarded an M.Sc. degree in 1912. The following year, he was awarded a prestigious 1851 Exhibition Research Studentship, enabling him to work with Hans Geiger (a former pupil and coworker of Rutherford) at the Physikalisch-Technische Reichsanstalt in Berlin-Charlottenburg, where he carried out important work on the beta-ray spectra of radio-elements. Trapped by the outbreak of war in 1914, he was interned at Ruhleben, near Spandau, for the duration. This experience had a significant effect on his physical constitution and later character and temperament; Chadwick was often seen as a rather dour and forbidding person, though those who knew him well note that his apparent severity masked an essentially sympathetic, but shy character.

Returning to Manchester in 1918, Chadwick began work again with Rutherford, and subsequently moved with his mentor to the Cavendish Laboratory, Cambridge, where he earned his Ph.D. in 1921 with a thesis on the field of force surrounding the atomic nucleus. In 1923 he was appointed Rutherford's Deputy Director of Research, in which capacity he played a central role in training young research students and in organizing the work of the Cavendish Laboratory.

During the 1920s Chadwick collaborated with Rutherford and others in a series of researches designed to elucidate the structure of the atomic nucleus. Chief among these were experiments on the artificial disintegration of the nuclei of light elements. A series of puzzling results from such experiments in Berlin and Paris led Chadwick, in February 1932, to postulate the existence of the neutron, an uncharged nuclear constituent. For this work, which transformed experimental and theoretical nuclear physics, he was awarded the Nobel Prize for physics in 1935. During the same period, he also carried out important work on the scattering of alpha particles in helium (which verified predictions from the new wave mechanics) and on the photodisintegration of the deuteron.

In 1935 Chadwick left the Cavendish Laboratory to become Lyon Jones Professor of Physics at the University of Liverpool, a post which he held until 1948. At Liverpool, Chadwick revitalized a moribund physics department, and embarked upon the construction of a cyclotron. During World War II, he coordinated British experimental work on the development of uranium fission for military purposes. When the work was integrated into the United States' Manhattan Project in 1943, he headed the British contingent at Los Alamos. For this work, he was knighted in 1945. From 1948 to 1958, he was Master of Gonville and Caius College, Cambridge, during which time he continued to be influential in science policy-making in postwar Britain.

See also: NEUTRON; NEUTRON, DISCOVERY OF; NUCLEAR PHYSICS; RADIOACTIVITY; RUTHERFORD, ERNEST; WAVE MECHANICS

Major Publications

CHADWICK, J. "The Charge on the Atomic Nucleus and the Law of Force." *Philos. Mag.* **27,** 112–125 (1920).

CHADWICK, J. *Radioactivity and Radioactive Substances* (Pitman, London, 1921).

CHADWICK, J. "Observations Concerning the Artificial Disintegration of Elements." *Philos. Mag.* **2,** 1056–1075 (1926).

CHADWICK, J. "The Scattering of α-Particles in Helium." *Proc. R. Soc. London, Ser. A* **128,** 114–122 (1930).

CHADWICK, J. "The Existence of a Neutron." *Proc. R. Soc. London, Ser. A* **136,** 692–708 (1932).

RUTHERFORD, E., and CHADWICK, J. "A Balance Method for Comparison of Quantities of Radium and Some of Its Applications." *Proc. Phys. Soc. London* **24,** 141–151 (1912).

RUTHERFORD, E.; CHADWICK, J.; and ELLIS, C. D. *Radiations from Radioactive Substances* (Cambridge University Press, Cambridge, Eng., 1930).

Bibliography

GOWING, M. *Britain and Atomic Energy, 1939–1945* (Macmillan, London, 1964).

GOWING, M., and ARNOLD, L. *Independence and Deterrence* (Macmillan, London, 1974).

HENDRY, J., ed. *Cambridge Physics in the Thirties* (Hilger, Bristol, Eng., 1984).

MASSEY, H., and FEATHER, N. "James Chadwick." *Biographical Memoirs of the Fellows of the Royal Society* **22,** 11–70 (1976).

JEFF HUGHES

CHAIN REACTION

In brief, a chain reaction is a self-perpetuating series of nuclear reactions.

In 1934 the Italian physicist Enrico Fermi and his collaborators began a ground-breaking series of experiments in Rome that eventually led to the discovery of nuclear fission. They bombarded various naturally occurring elements with neutrons and found that in many cases radioactivities were produced. When uranium was bombarded with neutrons, the activities that were produced were interpreted by Fermi's group as being from elements with higher atomic number (nuclear charge) than uranium. However, the experiments were repeated by the German chemists Otto Hahn and Fritz Strassmann, whose careful chemical studies on the reaction products indicated without a doubt that the elements produced were lower in atomic number than uranium. They also determined that radioactivities from many different elements were produced. In 1939 Hahn and Strassmann were led to conclude that the uranium had been "split" into two heavy fragments by a neutron-induced "explosion" of the nucleus. Lise Meitner and Otto Frisch provided a physical explanation for this process and named it nuclear fission.

The fission of a nucleus like ^{235}U, the isotope of uranium having a mass of 235 (with 92 protons and 143 neutrons), is readily initiated by the capture of a slow neutron. A slow or thermal neutron has a kinetic energy that is typically a small fraction of an electron volt (eV). Although the kinetic energy brought in by the neutron is negligible, the increased binding energy of the compound system consisting of neutron plus uranium nucleus is significant. This energy increase can trigger a distortion of the combined system, pushing it into a shape somewhat along the lines of an elongated liquid drop. If the distortion is sufficiently large, it will disturb the delicate balance between internal charge repulsion trying to disrupt the nucleus and the surface tension which tends to hold it together. As a result, the nucleus can split into two roughly equal fragments. Since there is a difference in total binding energy of a uranium nucleus and two elements near the middle of the periodic table, the fission of one uranium nucleus results in the release of about 200 MeV of energy (1 MeV = 10^6 eV). Most of this energy appears in the form of kinetic energy of the fragments, and it represents the equivalent of about 0.09 percent of the mass energy (mc^2) possessed by the composite system before fission. When the fragments stop in the surrounding material, their kinetic energy is converted to heat. It was recognized early on that this large energy release was orders of magnitude greater than the few eV that are produced by chemical reactions.

Along with the two fragments, an average of about 2.5 fast neutrons are emitted during each fission event. The energies of these fast neutrons extend up to 10 or 15 MeV, with a most probable value of 1 MeV. The presence of the additional neutrons produced during fission introduces the possibility that, once initiated, a self-sustaining sequence of fission reactions could take place. This is known as a chain reaction. A chain reaction will be possible if at least one of the additional neutrons produced by a fission event can itself be made to induce the fission of another uranium nucleus.

For a chain reaction to operate most efficiently, the fast neutrons produced in fission must first be slowed down, or moderated, since the capture probability follows the "$1/v$" law, where v is the neutron velocity. This law implies that the probability that a neutron will be captured and thus be able to induce fission of a uranium nucleus is proportional to the time it spends in the vicinity of the nucleus. The best way to slow down or thermalize the neutrons is to allow them to have many collisions with a light nucleus such as carbon, which will not absorb or react with the neutrons. Because it takes about 100 such collisions to thermalize the neutrons, typical slowing-down times are in the millisecond range.

It has already been mentioned that at least one fission-produced neutron must induce another fission event if a chain reaction is to take place. This is not a trivial condition, because the neutrons can be lost by absorption, escape, and competing reactions. If realized, a self-replicating reaction system can be seen to be a source of heat energy if it is allowed to proceed under controlled conditions. This is the situation found in the core of a nuclear power reactor, where the neutrons are first slowed by interacting with a graphite or other moderator, and then allowed to fission the ^{235}U fuel. In a reactor the neutron economy, or the number of neutrons available at a given time for inducing fission, is tightly controlled by inserting or removing neutron absorbers, typically made of the element cadmium. The absorbers remove neutrons without themselves undergoing fission.

One additional property that influences whether a chain reaction may take place in an assembly of fissionable material is the physical size of the reacting system. For a neutron, the probability of escaping depends on the surface area, while the probability of initiating another fission event is proportional to the volume. There will therefore be a minimum critical size that the reactor must have in order to sustain a chain reaction. For a spherical graphite-uranium reactor, this critical size is somewhat larger than 100 m^3, making for quite a bulky object.

If the number of neutrons produced by chain reactions in fissioning material is allowed to increase without limit, the production of a huge energy release in a short time becomes possible. Such uncontrolled fission is the source of energy in an atomic bomb, where the time scale for the energy release is on the order of microseconds. Since the material is dispersed by the explosion, the fission must be induced by fast neutrons, because there is no time to wait for thermalization. It is therefore necessary to use a material that can be fissioned by neutrons of all energies and that does not offer the possibility of competing reactions that reduce the number of neutrons available for fission. Such a material is pure ^{235}U or ^{239}Pu (plutonium, element number 94), each of which was used in atomic bombs that were responsible for bringing an end to World War II.

The critical size for these devices is estimated to be less than 0.005 m^3.

See also: ENERGY, NUCLEAR; FERMI, ENRICO; FISSION; FISSION BOMB; RADIOACTIVITY; REACTOR, BREEDER; REACTOR, FAST; REACTOR, NUCLEAR

Bibliography

KRANE, K. S. *Introductory Nuclear Physics* (Wiley, New York, 1988).

CARY N. DAVIDS

CHAMBER

See BUBBLE CHAMBER; CLOUD CHAMBER; IONIZATION CHAMBER

CHAOS

Chaos is the complicated nonperiodic evolution in time of deterministic nonlinear dynamical systems that in the long term is unpredictable and equivalent to a random process because of exponential sensitivity to the initial conditions.

Although the term "chaos" was first applied to the complicated nonperiodic behavior of nonlinear dynamical systems by Tien-Yien Li and James A. Yorke in 1975, the essence of this phenomenon was understood by Henri Poincaré through his attempts (first published in 1892) to solve the three-body problem of planetary motion.

That a system governed by deterministic laws can exhibit effectively random behavior runs directly counter to our normal view of determinism, stated forcefully in the early nineteenth century by Pierre-Simon Laplace:

An intellect which at any given moment knew all the forces that animate Nature and the mutual positions of the beings that comprise it . . . could condense into a single formula the movement of the greatest bodies of the universe and that of the lightest atom; for such an intellect nothing could be uncertain; and the future just like the past would be present before its eyes.

A century later, based on his study of the three-body problem, Poincaré recognized the hidden assumption in Laplace's "clockwork universe" worldview and formulated the proper intuition for understanding chaos:

[E]ven if it were the case that the natural laws had no longer any secret for us, we could still only know the initial situation *approximately*. If that enabled us to predict the succeeding situation *with the same approximation*, that is all we require, and we should say that the phenomenon had been predicted. . . . But it is not always so; it may happen that small differences in the initial conditions produce very great ones in the final phenomena. . . . Prediction becomes impossible, and we have the fortuitous phenomenon.

Despite Poincaré's remarkable insight and the work of many mathematicians in the intervening years, chaos remained generally unappreciated by the scientific community until the pioneering computational studies of Yoshisuke Ueda and Edward N. Lorenz in the early 1960s revealed in a striking manner that chaos could occur in (simplified models of) common physical phenomena, such as nonlinear electric circuits or the weather.

To understand chaos one must appreciate several concepts about dynamical systems. Our solar system, with the nine planets orbiting around the Sun (and interacting with each other) is a familiar dynamical system; abstractly, a dynamical system consists of a state—more precisely, a vector in state space, which for the Sun and nine planets is a 60-dimensional space describing the locations and momenta of each of the ten bodies—and a dynamical rule describing how the state evolves in time (for the solar system, the rule is Newton's universal law of gravitation). For a deterministic dynamical system the law of evolution contains no explicit stochastic (i.e., random) terms. Typically, one uses "dynamical system" to refer to both the physical system and the (idealized, approximate) mathematical equations modeling it. Flows are dynamical systems in which time changes continuously, whereas for maps time changes by discrete increments. Possible non-chaotic behaviors of dynamical systems include (1) fixed points, equilibria for which no motion occurs; (2) periodic motions, in which states recur at a single fixed frequency in time; and (3) quasi-periodic motions, in which many different, isolated frequencies appear in the evolution. In contrast to these regular motions, chaos is non-periodic, showing no isolated frequencies.

Remarkably, even very simple dynamical systems—flows involving as few as three first-order differential equations, or maps of a single variable—can exhibit chaos. This result contradicts our understanding of random processes such as "noise." As in the Brownian motion of a particle buffeted by a sea of molecules, noise typically arises from the effects of a very large number of independent, unobserved agents and is modeled by adding explicit stochastic terms to the dynamical equations.

The exponential sensitivity to initial conditions that characterizes chaos is quantified by the Lyapunov exponent, λ: For an M-dimensional flow or map, there are M Lyapunov exponents, and chaos occurs if any of these is positive. The physical dimension of λ is inverse time, and in essence $1/\lambda$ measures the rate at which predictability is lost as the chaotic systems evolves. One system exhibiting exponential sensitivity is an inverted broom, balanced vertically on its handle: If the broom is perfectly vertical, it is in an unstable equilibrium. The direction in which it will eventually fall depends on "a very small cause which escapes our notice," just as Poincaré observed. Just as exponential sensitivity to initial conditions is the analytic essence of chaos, its geometric essence is stretching (the exponential separation) and folding (due to the nonlinearity of the system).

Most dynamical systems in the real world are dissipative, meaning that they do not conserve mechanical energy or volumes in state space. As a consequence, dissipative dynamical systems can contain attractors, which are defined as subsets of the state space to which initial conditions evolve as time increases without limit ($t \rightarrow \infty$) (i.e., after transients have died out). Attractors corresponding to the regular motions of dynamical systems have integer dimensions—zero for a fixed point, one for a periodic orbit, n for an quasi-periodic orbit with n frequencies—whereas those corresponding to chaotic motions are fractals (i.e., have a fractional dimension and are known as strange attractors.

An important example of a dissipative dynamical system exhibiting chaos arises in the study of the motion of the earth's atmosphere for purposes of weather prediction. The atmosphere is a strongly driven (heated by the sun, cooled by the oceans) and damped (frictional losses to the earth's surface) dissipative system. Poincaré himself gave weather forecasting as a qualitative example of sensitivity to initial conditions, but it was the work of Lorenz in the 1960s that provided the critical quantitative modeling. As a very simplified model of atmospheric thermal convection, Lorenz formulated a dynamical system of three first-order, nonlinear differential equations. When he simulated these equations on a digital computer, he found that the solutions were extremely sensitive to the initial conditions; to emphasize this point, he entitled a public lecture describing his results "Predictability: Does the Flap of a Butterfly's Wings in Brazil set off a Tornado in Texas?" He thereby coined the term "the butterfly effect" to describe the sensitivity to initial conditions.

Underlying this chaotic behavior is the celebrated Lorenz attractor, which is shown in Fig. 1. From the three perspectives we see a two-lobed, nonplanar geometrical object that extends over a finite region of the state space. Looking at the attractor in soft focus, it appears much like a "figure 8," but there is a crucial difference when one focuses sharply; namely, the orbit never closes on itself (otherwise it would be periodic) so it must snake forever through the phase space, avoiding itself continually. Further, because of chaos, motions starting from nearby initial points on the attractor will separate exponentially in time. The two-lobed structure of the attractor suggests a useful analogy to emphasize this sensitivity. Choose two very nearby initial points and follow their evolution in time. Call each loop around the right lobe "heads" and around the left

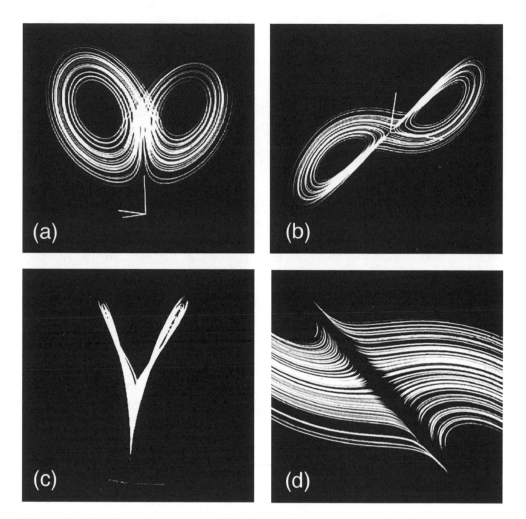

Figure 1   The Lorenz attractor. The three perspective views (a, b, c) reveal the two-lobed, nonplanar structure and thickness of the attractor. The close-up (d) illustrates the fact that although confined to a finite region of the three-dimensional state space, the orbit never closes, even as time approaches infinity, so that the motion is nonperiodic, that is, never repeats itself. The attractor has a fractal dimension of approximately 2.04, which is between that of an area and a volume.

lobe "tails." Then the asymptotic sequences of heads and tails corresponding to the two initial points will be completely different and totally uncorrelated to each other, exactly as if they had come from two sequences of tosses of a coin. Of course, the nearer the initial points, the longer their motions will remain similar. But for any initial separation, there is a finite time beyond which the motions appear totally different.

The geometric structure of the Lorenz attractor is a consequence of the stretching and folding characteristic of chaos: The stretching is as just described and the folding is reflected in the finite size of the attractor, which means that the orbits must bend back on themselves, rather than diverging to infinity. When the attractor folds back on itself, different orbits are brought near to each other, but they can never intersect. The resulting structure is like the *mille feuille* pastry made by a baker by a similar stretching and folding process, except that in this mathematical model it continues on all scales. As a consequence, the Lorenz attractor is "more" than a planar object but "less" than a solid one; it is a fractal strange attractor, with a dimension of approximately 2.04, that is, between that of an area and a volume.

Not all interesting dynamical systems are dissipative; conservative or Hamiltonian systems are isolated systems without friction, so that energy is conserved and the volumes in state space (called phase space for Hamiltonian systems) are preserved by the time evolution (Liouville's theorem). The solar system provides a natural and historically important physical realization of a (nearly) conservative dynamical system. Although the solar system has long been viewed as the paragon of Newtonian orderliness, and Laplace and Joseph-Louis Lagrange proved (but only to first-order in the planetary masses) that its motions were regular, Poincaré's realization that chaos could lurk in the solar system has received full confirmation in numerical studies made during the 1980s and 1990s. For instance, the Kirkwood gaps in asteroids, the changing rotational period of Saturn's moon Hyperion, and the orbit of Pluto are all examples of chaotic motion. Results based on numerical studies of 200 million years of evolution of equations describing the orbits of the eight main planets have shown that the solar system is overall chaotic, not quasi-periodic, with a surprisingly large maximum Lyapunov exponent.

Chaos in the solar system does *not* mean that Earth could at any second fly off into interstellar space. The solar system as we know it appears to have been stable for 4.5 billion years, and the example of the Lorenz attractor shows that chaotic motions can occur in confined regions of state space that have fairly regular shapes. But this chaos does mean that no matter how accurately present planetary motions are known, motions over cosmic time scales are unpredictable.

The logistic map is a remarkably simple nonlinear dynamical system that exemplifies many of the central concepts of chaos. This discrete-time, dissipative dynamical system is described by the equation

$$x_{n+1} = rx_n(1 - x_n) \qquad (1)$$

and can be interpreted, for instance, as modeling a population of insects that has discrete generations. In this interpretation, x_n is the size of the population of the nth generation, scaled such that the maximum population that can be supported has the value 1, and r represents the fecundity of the population (larger r implies more successful reproduction). Importantly, equations of this form arise in many circumstances; for instance, in certain cases

the successive maxima of one of the variables in the Lorenz equations, when plotted as x_{n+1} versus x_n, give a logistic-like curve.

The logistic map exhibits an amazing range of dynamical behaviors as the parameter r is varied over the range $0 < r < 4$. These behaviors are illustrated in Fig. 2, which shows the attractor of the logistic map as a function of r. For $0 < r < 1$, the attractor is a fixed point (call it x^*) at $x_n = x^* = 0$, so the population dies out. At $r = 1$, there is a sudden change in the solution (this is known technically as a bifurcation), and for $1 < r < 3$, the attractor is a nonzero fixed point—$x^* = (1 - 1/r)$—so that there is a stable population. At $r = 3$, there is a bifurcation to a period-two attractor, so that the population oscillates between two stable values. As r is further increased, the further bifurcations lead to attractors of periods 4, 8, 16, and in fact *all* powers of 2, all of this occurring before the limiting value $r_c \sim 3.5699$. Above $r_c \sim 3.5699$, the map exhibits deterministic chaos interspersed with "windows" of stable periodic orbits.

The sequence of period-doubling bifurcations illustrates one way in which systems can change from regular to chaotic motion. In addition to the period doubling transition to chaos, several other types of transitions have been identified and studied. Intermittency describes a transition in which a periodic cycle begins to show irregular, "chaotic" bursts at intermittent intervals, which increase in duration, eventually leading to full chaos. This transition is actually observed in the logistic map by decreasing r from just above the point at which period-three orbit first becomes stable, $r = r_3 \sim 3.84$. Crises are sudden transitions to chaos caused by the interaction of two different attractors, at least one of which is chaotic; in the logistic map, a crisis transition occurs as r is decreased to 4 from above. Although not exemplified by the logistic map, a quasi-periodic transition to chaos has also been identified. Despite this considerable progress, the general problem of classifying all means by which dynamical systems can transform from regular to chaotic behavior remains unsolved.

Far more important than their occurrence in the simple logistic map is the universality associated with some of these transitions to chaos. In his analysis of the period-doubling transition, Mitchell Feigenbaum argued that certain quantitative features of this transition (for instance, a quantity related to the rate at which the successive transitions occur as r is

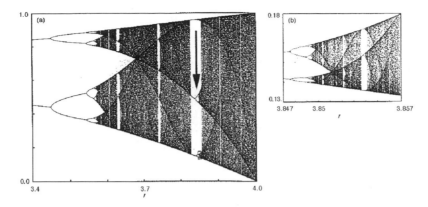

Figure 2 (a) The attracting set for the logistic map generated by plotting 300 values of the iterated function (after transients have died out) for each of 1,150 values of the control parameter r. For $r < 1$, the attractor is a fixed point at $x = 0$, while for $1 < r < 3$, it is a fixed point at $1 - 1/r$. At $r = 3$, the fixed point bifurcates to a period-2 cycle and then successively to cycles of periods 4, 8, 16, and so forth as r increases, generating the period-doubling cascade. Above $r_c \sim 3.5699 \ldots$, the map exhibits deterministic chaos interspersed with "windows" of stable periodic orbits. For example, cycles of periods 6, 5, and 3 can be seen in the three larger gaps to the right. (b) A magnified region [shown as a small rectangle in (a)] illustrates the self-similar nature that occurs on a smaller scale, revealing that this attracting set is a fractal.

changed) should be universal, so that they are independent of the simple model and in fact should have the same numerical values in any system in which the transition is observed. This prediction has been strikingly confirmed in a wide variety of experiments on real world systems, ranging from fluid flows through acoustics to electrical circuits.

In the parameter region $r_c < r < 4$, chaos and periodicity are intermingled in a very complex manner. The existence of periodic windows for certain ranges of r is evident from Fig. 2. In the periodic windows the Lyapunov exponent is negative, whereas in the chaotic regions it is positive. In addition to the stable periodic cycles reflected in the attracting set, there are (infinitely) many unstable periodic cycles in the chaotic region. For instance, in the period doubling cascade, one can show that at each bifurcation, the cycle that was stable for smaller r remains a possible motion of the system but becomes unstable, like the inverted broom. The set of unstable cycles forms a sort of "skeleton" of the chaotic motion and its existence is very important for the concept of "controlling chaos."

In view of the complexity of the dynamics above r_c, it is not at all surprising that the logistic map, like the

typical problem in chaotic dynamics, defies direct analytic approaches. For $r = 4$, however, there is an elegant trick that permits analytic confirmation of many of our earlier statements about chaos. If we let $x_n \equiv \sin^2 \theta_n$, then the logistic map can be rewritten as

$$\sin^2 \theta_{n+1} = 4 \sin^2 \theta_n \cos^2 \theta_n$$

$$= (2 \sin \theta_n \cos \theta_n)^2 = (\sin 2\theta_n)^2. \quad (2)$$

Hence for $r = 4$ the map is simply the square of the doubling formula for the sine function, and we see that the solution is $\theta_{n+1} = 2\theta_n$. In terms of the initial value, θ_0, this gives

$$\theta_n = (2^n)\theta_0 \equiv (e^{n \ln 2})\theta_0. \quad (3)$$

In Eq. (3) we see explicitly the exponential sensitivity to initial conditions—the "stretching" of chaos—and also that the Lyapunov exponent for $r = 4$ is $\lambda = \ln 2 \sim 0.693$, as found by numerical studies. To recover the value of x_n from θ_n, we must use the $\sin^2$

function, which provides the (nonlinear) "folding" aspect of chaos. Equation (3) also makes manifest the difference between determinism and predictability. By writing θ_n as a binary number with a finite number of bits—as one would in any digital computer—we see that for $r = 4$ the logistic map amounts to a simple bit shift operation, plus a nonlinear "mask." (Parenthetically, this is exactly the way the pseudo-random number generators used in computers are constructed.) When this process is carried out on a real computer, round-off errors replace the right-most bit with garbage after each operation, and each time the map is iterated, one bit of information is lost. If the initial condition is known to 48 bits of precision, then after only 48 iterations of the map no information about the initial condition remains, and predictability is lost.

In the end, it is fair to ask how the "chaos revolution" of the 1980s and 1990s has changed our understanding of, and our ability to model or control, nature. Is chaos really a "new science?" Leaving aside the historical question of who knew what when, the recent widespread appreciation of chaos has certainly changed our perspectives on problems ranging from philosophy through basic science to engineering. Philosophically, it has finally severed the link between determinism and predictability and even quantified (by the Lyapunov exponent) the rate at which predictability is lost in a chaotic system. Chaos has also clarified the fundamental tenets of statistical mechanics, making clear how the slightest bit of "coarse graining" in a chaotic system leads from microscopic reversibility to macroscopic irreversibility. Although much further work is needed, chaos has provided a firmer foundation for our understanding of turbulence, especially near the onset. The impact of chaos on quantum mechanics, although initially controversial and still oft-debated, has been very important; chaos theory has shown how best to calculate the spectra of quantum systems which have irregular motions in their classical limits, and experiments on a number of atomic and solid-state systems have shown unmistakable traces of chaotic classical motions in quantum observables.

In terms of applications, chaos is becoming increasingly relevant, as the following three examples illustrate. First, time series of data ranging from sunspots to stock market prices, which previously were often characterized as "just noise," have been reanalyzed using new nonlinear methods motivated by chaos. In essence, these methods rely on the ability to reconstruct the dynamical system from the time series alone, a procedure which in some cases can be justified rigorously. If the reconstructed system is chaotic (and if the leading Lyapunov exponent is not too large), then the short-term predictability that chaos permits (but noise does not) can provide valuable insight into the data. Second, nearly all industrial processes, from chemical refining through composite materials processing to water reclamation, involve mixing different substances, and chaos offers deep insights into how to do this quickly and thoroughly. Third, and perhaps the most exciting, are applications of "controlling chaos" or "chaotic control." Although many different schemes have been proposed, an idea central to most of them is easily understood by combining the example of the inverted broom with our earlier discussion of the logistic map. Suppose the system we are trying to control is in the region of chaotic behavior of the logistic map; recall that this region is replete with unstable periodic orbits of many different periods (in fact, for $r = 4$, of all possible periods). Notice that a period-n orbit is a fixed point for the nth iterate of the map. But stabilizing an unstable fixed point is precisely the problem of stabilizing the inverted broom; we can all do this by (carefully) moving our hand so that the location of the balance point changes. This intuition suggests, correctly, that a system undergoing chaotic motion can be stabilized into a wide range of periodic motions, in fact with relatively small stabilizing forces. Such chaotic control has already been demonstrated in electronics, laser systems, fluid flows, chemical reactions, and physiological processes. Indeed, chaotic control may underlie the technology of the next generation of cardiac pacemakers and defibrillators.

See also: DETERMINISM; FRACTAL; MOTION, PERIODIC; PHASE SPACE; PROBABILITY; TURBULENCE

Bibliography

CRUTCHFIELD, J. P.; FARMER, J. D.; PACKARD, N. H.; and SHAW, R. S. "Chaos." *Sci. Am.* **255** (Dec.), 46–57 (1986).

GLEICK, J. *Chaos: Making a New Science* (Viking Press, New York, 1987).

HOLTE, J., ed. *Chaos: The New Science* (University Press of America, Lanham, MD, 1993).

KERR, R. A. "Does Chaos Permeate the Solar System?" *Science* **244**, 144–145 (1989).

LORENZ, E. N. *The Essence of Chaos* (University of Washington Press, Seattle, 1993).

MANDELBROT, B. *The Fractal Geometry of Nature* (W. H. Freeman, New York, 1977).

POINCARÉ, H. *Science and Method,* trans. by F. Maitland (Dover, New York, 1952).

SHINBROT, T. "Chaos: Unpredictable Yet Controllable?" *Nonlinear Science Today* **3**, 2–8 (1993).

DAVID K. CAMPBELL

CHARGE

Electric change is a property of some elementary particles, such as the electron and the proton. A particle that carries charge acts as a source for the electromagnetic field. Static charged particles interact as described by Coulomb's law. Charges can be "positive" or "negative." Charges of like sign (both positive or both negative) repel one another, and charges of opposite sign (one positive and the other negative) attract one another. The charge of the electron is said to be negative; the charge of the proton is positive. The designations positive and negative are arbitrary and are used for historical reasons. (This terminology was introduced by Benjamin Franklin in 1747.)

The unit of charge (in the CGS system of units) is the electrostatic unit (esu). This unit is defined so that if two pointlike bodies each carry 1 esu of charge and are separated by 1 cm, the electrostatic force acting between them is 1 dyne. (In the SI system of units, the unit of charge is the "coulomb"; 1 coulomb is 2.99792458×10^9 esu.) The charge of the proton, denoted by e, is approximately 4.803207×10^{-10} esu. The charge of the electron is $-e$. The charge of a macroscopic body is the algebraic sum of the charges of all of the particles that the body contains—in ordinary matter (where the only charged particles contained in the body are protons and electrons) this charge is the number of protons minus the number of electrons, multiplied by e.

Charge Conservation

Charge cannot be created or destroyed. Thus, the total electric charge in an isolated system never changes. This statement is called the law of charge conservation. The total charge in a finite region of space can change only if charged particles cross the boundary of the region. Mathematically, if Q_V is the total electric charge enclosed inside a volume V, then the rate of change of the charge per unit time can be expressed as

$$\frac{d}{dt} Q_V(t) = -I_{\partial V}(t), \tag{1}$$

where $I_{\partial V}$ is the (outward-directed) electric current passing through the boundary ∂V of the volume V. The total charge can be expressed as the integral of a charge density per unit volume ρ, and the current can be expressed as the integral of a current density $\mathbf{J} = \rho \mathbf{v}$ (where $\mathbf{v}$ is the locally measured average velocity of the charge carriers),

$$Q_V(t) = \int_V \rho(\mathbf{x},t), \qquad I_{\partial V}(t) = \int_{\partial V} d\mathbf{a} \cdot \mathbf{J}(\mathbf{x},t). \tag{2}$$

Using Stokes's theorem, the relation in Eq. (1) can be rewritten as

$$\frac{\partial}{\partial t} \rho(\mathbf{x},t) = \mathbf{\nabla} \cdot \mathbf{J}(\mathbf{x},t), \tag{3}$$

which is the differential form of the charge conservation law.

The charge conservation law Eq. (3) is required for the consistency of electrodynamics—it can be derived from Maxwell's equations. By taking the divergence of both sides of the Maxwell equation

$$\mathbf{\nabla} \times \mathbf{B} = \frac{1}{c} \frac{\partial \mathbf{E}}{\partial t} + \frac{4\pi}{c} \mathbf{J}, \tag{4}$$

using the vector identity $\mathbf{\nabla} \cdot (\mathbf{\nabla} \times \mathbf{B}) = 0$, and invoking Gauss's law

$$\mathbf{\nabla} \cdot \mathbf{E} = 4\pi\rho, \tag{5}$$

we obtain Eq. (3). Heuristically, we can see why charge conservation follows from Gauss's law as follows: Gauss's law says that the electric field lines can never end except at the position of an electrically charged particle. (The number of field lines emanating from a particle is proportional to the magnitude of its charge—field lines are directed outward from a positively charged particle, and inward toward a negatively charged particle.) Furthermore, the dynamical laws that govern the evolution of the electric field are continuous in time and local in space. The total charge contained in a region is the number of electric field lines that pierce the boundary of the region. If electric charge were to spontaneously appear (or disappear) deep inside a region of space, the number of electric field lines passing through the boundary (which might be very far away) would have to change instantaneously. Thus, if the charge were to change, a nonlocal and discontinuous rearrangement of the field lines would have to occur, and this is not allowed to happen.

At a still deeper level, why must Gauss's law be satisfied? To seek an answer, one must look beyond classical electromagnetic theory and consider the structure of the underlying quantum theory. From this viewpoint, Maxwell's electrodynamics is the theory of the interactions of a massless and chargeless spin-1 particle (the photon) with matter. The form of the quantum theory of a massless spin-1 particle coupled to matter is highly constrained by fundamental principles such as relativistic invariance and the conservation of probability. It seems that the only self-consistent theory of this kind is one in which the photon couples to a conserved current—thus, in the classical theory, the charge that acts as a source for the electromagnetic field must be absolutely conserved.

Because of the above considerations, the law of conservation of charge has a special status among the conservation laws that are known to experimental physics. For example, the law of baryon number conservation is respected in all processes that have ever been studied experimentally. Yet baryon number conservation is not required for the consistency of physical law, and many physicists believe that processes that change baryon number are not forbidden; rather, they are just extremely rare. In contrast, it would be truly shocking if violations of the law of charge conservation were ever detected.

Note that the law of charge conservation does not disallow the production of charged particles in physical processes—it only requires that whenever a positively charged particle is produced, a negatively charged particle must be produced simultaneously so that the total charge remains unchanged. In particular, for every charged particle there is a corresponding antiparticle that has exactly the same mass but the opposite value of the charge. (The

existence of such antiparticles follows from very general principles of relativity and quantum mechanics.) Corresponding to the negatively charged electron is the positively charged positron, and corresponding to the positively charged proton is the negatively charged antiproton. Electron-positron pairs, or proton-antiproton pairs, are frequently produced in energetic particle collisions that are observed in experiments at particle accelerators. Similarly, the law of charge conservation does not prevent a particle-antiparticle pair (with total charge zero) from annihilating into uncharged particles (such as photons).

Conservation of charge and of energy are responsible for the absolute stability of the electron. The electron is the very lightest of the particles that carry electric charge, so there is no process where the electron decays to other particles that is allowed by the conservation laws. In contrast, the decay of the proton is not forbidden by these laws, and many physicists expect that proton decay can occur, though it is so rare that it has never been seen.

Charge Quantization

As far as is known, the total charge of an isolated system must always be an integer multiple of the fundamental charge e, a phenomenon called charge quantization. One consequence of charge quantization (and the most compelling experimental evidence in favor of it) is that the electric charge of the proton and the electron are exactly equal and opposite—this is verified experimentally to one part in 10^{21}.

There is overwhelming evidence that protons are composite objects that contain elementary constituents called quarks, and that the quarks carry the electric charge $\frac{2}{3}e$ or $-\frac{1}{3}e$. However, it is believed that quarks are always permanently confined inside particles (such as the proton) whose total charge is an integer multiple of e. It is not possible to separate a fractionally charged quark from other quarks or antiquarks by a distance much exceeding 10^{-13} cm.

From the charge quantum e and the fundamental constants c (the speed of light) and $\hbar$ (Planck's constant divided by 2π), a dimensionless number can be constructed. This number, $e^2/\hbar c$, is denoted by α and called the fine-structure constant. Its numerical value (determined most accurately in experimental studies of the quantum Hall effect) is approximately $\alpha^{-1} = 137.03599$. The fine-structure constant is of central importance in fundamental physics, and its numerical value has never been explained by theory.

Why is charge quantized? Theoretical physicists have offered two explanations (that turn out to be closely related). In 1931 P. A. M. Dirac proposed the existence of a new type of elementary particle, called the magnetic monopole, that carries magnetic charge. These magnetic monopoles would act as sources for the magnetic field just as electrically charged particles are sources for the electric field. Dirac emphasized that the existence of magnetic monopoles could be compatible with the principles of quantum mechanics only if electric charge is quantized. In 1974 Howard Georgi and Sheldon Glashow proposed that the theories of the strong, weak, and electromagnetic interactions could all be combined together in one grand unified theory (GUT), and they noted that grand unification would require electric charge to be quantized. In fact, grand unified theories also predict the existence of magnetic monopoles, so in a sense the proposals of Dirac and of Georgi and Glashow are not really logically independent. Neither the existence of monopoles nor the validity of grand unification have been confirmed experimentally—they are purely hypothetical. However, that these hypotheses provide a natural explanation for charge quantization is one of the strong reasons why many physicists take the proposals seriously. If magnetic monopoles really do exist, then magnetic charge, like electric charge, must be exactly conserved, and the lightest magnetic monopole, like the electron, must be an absolutely stable particle. Furthermore, magnetic charge, if it exists at all, is required to be quantized.

Gravitational "Charge"

The idea that Gauss's law requires electric charge to be exactly conserved has a close gravitational analog. As charge acts as a source for the electromagnetic field in Maxwell's electrodynamics, so energy acts as a source for the gravitational field in Einstein's general relativity. Thus, exact conservation of energy is required for the mathematical consistency of general relativity. However, energy in general relativity differs from charge in electrodynamics in a subtle way: the energy cannot be expressed as the integral of a local density as in Eq. (2). The essential difference between energy and charge is that the gravitational field can carry energy (and so can be a source for itself), while the electric field does not

carry charge. As a result, the total energy, unlike the total electric charge, is not merely the algebraic sum of its parts.

See also: CONSERVATION LAWS; FIELD, GRAVITATIONAL; FINE-STRUCTURE CONSTANT; GAUSS'S LAW; GRAND UNIFIED THEORY; QUANTIZATION; QUARK

JOHN PRESKILL

CHARGE, ELECTRONIC

The electronic charge is the electric charge carried by the electron. Precise knowledge of the electron's charge and of its magnetic moment played an important role in the development of the physics of the twentieth century. Accurate measurement of these quantities has long challenged the ingenuity of experimentalists, some of whom won the Nobel Prize for their achievements.

The Unit of Charge

The first precise definition of the magnitude of charge was developed after Charles Augustin Coulomb measured the electric force F_{el} between small charged metallic spheres, A and B, of variable separation r and charges Q_A and Q_B. Coulomb was able to vary the Q values systematically by assuming that charge is shared equally when spheres of the same size come into contact. His result, called Coulomb's law, was $F_{\text{el}} = kQ_AQ_B/r^2$, where k is a constant of proportionality. With $k = 1$, this allows a definition of the so-called electrostatic unit (esu) of charge as being that charge which, placed equally on two spheres 1 cm apart, produces between them a force of 1 dyne (1 dyne = 1 gm·cm/s^2); thus (1 esu)^2 = (1 dyne)·(1 cm^2). Later, a more convenient unit of charge, the coulomb (C), was defined in terms of the standard unit of electric current, the ampere (A), by 1 A = 1 C/s. These units are related by the equation 1 esu $\approx 3.34 \times 10^{-10}$ C.

The Charge of the Electron

Beginning in 1907, a series of experiments undertaken by Robert Andrews Millikan and his students provided the first definitive proof of the particulate nature of charge and gave the first accurate value of the charge Q_e of the electron, discovered a decade earlier by John Joseph Thomson. The basic idea of Millikan's famous oil-drop experiment was to observe a change in the force exerted on a small oil drop by a constant electric field when the charge on the drop changed. The finding that there was a minimum change would establish the existence of a minimum charge.

The experiment involved two horizontal brass plates, between which a variable voltage could be applied. The falling of an oil drop, obtained from a spray just above a tiny hole in the top plate, was viewed through a microscope. The drop's terminal velocity, achieved when the force of gravity on the drop was exactly balanced by the viscous force arising from collisions with the air molecules, could then be measured; from this the mass m of the drop could be calculated with the help of a formula for the viscous force called Stokes's Law. Before the drop hit the bottom plate, a vertical electric field E was applied and varied until the drop, with (unknown) charge Q, was levitated; at this stage the gravitational force $F_g = mg$ and the electric force $F_e = |Q|E$ were equal so that $|Q| = mg/E$ could be calculated. The charge on the drop could also be varied by the use of ultraviolet light. Millikan found that the values of $|Q|$ were all integral multiples of a minimum value e, which he identified, apart from a minus sign, as the charge carried by an electron: $Q_e = -e$. His final result was

$$e \approx 4.80 \times 10^{-10} \text{ esu} = 1.60 \times 10^{-19} \text{ C},$$

with an accuracy of a few parts per thousand. This is in good agreement with the current value, which is uncertain to only 0.3 parts per million,

$$e = 1.60217733(49) \times 10^{-19} \text{ C}.$$

In describing his experiment in 1937, Millikan stated that the number of electrons on a droplet could be counted "with quite the same certainty with which we can count our fingers and toes" and that "anyone who has seen the forgoing experiment . . . has proved for himself the existence of the electron with as much certainty as if he had seen a visible object."

Electron Magnetic Moment

An electron carries an intrinsic angular momentum or spin $\mathbf{s}$, with which there is associated a magnetic moment μ_e, proportional to $\mathbf{s}$ in magnitude but opposite in direction. In the presence of a magnetic field $\mathbf{B}$, the electron's energy changes by an amount $\mu_e B$ if $\mathbf{s}$ is parallel to $\mathbf{B}$ and by $-\mu_e B$ if antiparallel to $\mathbf{B}$. A natural unit for such moments is the Bohr magneton, $\mu_B = e\hbar/2m_e c$ ($\hbar$ = Planck's constant divided by 2π), in terms of which one writes

$$\mu_e = g_e \mu_B,$$

where g_e is a dimensionless number called the electron g-value.

Until 1947 it was believed that $g_e = 2$, in agreement with the prediction of a relativistic wave equation for the electron discovered by Paul Adrien Maurice Dirac. Then Polykarp Kusch and Henry M. Foley measured with great precision the change in the energy levels of alkali atoms in a magnetic field (Zeeman effect) and found a small but definite departure from the value $g_e = 2$. The difference $\mu_e^{an} \equiv \mu_e - 2\mu_B$ is called the anomalous magnetic moment of the electron and is usually written in the form $\mu_e^{an} = 2a_e\mu_B$, with $a_e \equiv (g_e - 2)/2$ called the electron magnetic moment anomaly. Julian Schwinger soon explained that this arose from the interaction of the electron with its own radiation field and calculated that $a_e \approx \alpha/2\pi$, where $\alpha \equiv e^2/\hbar c \approx 1/137$ is the fine-structure constant; this gave $a_e^{th} \approx 1.16 \times 10^{-3}$, in good agreement with the Kusch–Foley result, $a_e^{exp} \approx 1.19 \pm 0.05 \times 10^{-3}$.

Since that time, a_e has been studied with ever-increasing accuracy. The current experimental value is

$$a_e^{exp} = 1,159,652,188.4(4.3) \times 10^{-12}.$$

The most recent theoretical value is

$$a_e^{th} = 1,159,652,201.4(30) \times 10^{-12},$$

with most of the uncertainty arising from uncertainty in the precise value of α. Thus a_e is one of the most precisely measured and calculated numbers in all of physics and a touchstone for the validity of quantum electrodynamics, the quantum theory of the interaction of photons and charged particles.

Proton Charge

The proton is the positively charged nucleus of the hydrogen atom, which is known to be electrically neutral to high accuracy. Since this atom contains only one electron, the proton charge Q_P must be very closely equal in magnitude but opposite in sign to that of the electron. Experiments involving attempts to deflect a beam of neutral atoms by a static electric field indicate that the difference between Q_P and e is less than $10^{-20}e$.

See also: BOHR MAGNETON; CHARGE; COULOMB'S LAW; ELECTRON; ELECTRON, DISCOVERY OF; FINE-STRUCTURE CONSTANT; MAGNETIC MOMENT; PROTON; QUANTUM ELECTRODYNAMICS; ZEEMAN EFFECT

Bibliography

COHEN, E. R., and TAYLOR, B. N. "The 1986 Adjustment of the Fundamental Physical Constants." *Rev. Mod. Phys.* **57,** 1121 (1987).

HUGHES, V. W.; FRAZER, I. J.; and CARLSON, E. R. "The Electric Neutrality of Atoms." *Z. Phys. D: Atoms, Molecules, and Clusters* **10,** 145 (1988).

KINOSHITA, T., ed. *Quantum Electrodynamics* (World Scientific, Singapore, 1990).

KINOSHITA, T. "New Value of the α^3 Electron Anomalous Magnetic Moment." *Phys. Rev. Lett.* **75,** 4728 (1995).

MILLIKAN, R. A. *The Electron* (University of Chicago, Chicago, 1924).

MILLIKAN, R. A. "Electron, The." *Encyclopedia Britannica* (Encyclopedia Britannica, New York, 1937).

JOSEPH SUCHER

CHARGE CONJUGATION

Associated with every known elementary particle in nature is an antiparticle. The antiparticle of the electron e is the positron $\bar{e}$, a particle with the opposite charge and the same mass as the electron. Similarly, there is an antiproton $\bar{p}$ and an antineutron $\bar{n}$. The neutrinos ν also have antiparticles $\bar{\nu}$. Particles

and antiparticles, when they meet, can annihilate, producing other forms of energy.

Because particles and antiparticles are so intimately related, it is natural to ask whether there is a symmetry that connects them. For example, since an electron and a positron each are surrounded by an electric field of the same magnitude and opposite sign, it is natural to speculate that the laws of nature are symmetric if one replaces the electrons everywhere by positrons (and vice versa), and one changes the sign of the electric and magnetic fields. This symmetry is called "charge conjugation" and is usually denoted by the letter C. C is an exact symmetry of quantum electrodynamics (QED), the theory of electrons interacting with photons that describes extremely accurately the phenomena of atomic physics. In this theory, we can explain a bit more precisely what is meant by C. If attention is confined to electrons, there are six basic states in this theory: the two spin states of the electron, the two of the positron (the antiparticle of the electron), and the two polarization states of the photon (light). Charge conjugation is the statement that the laws of QED are invariant under replacement everywhere of electrons by positrons of the same spin. Because this exchange changes the sign of the electromagnetic current, and because the electromagnetic field couples to the current, it is also necessary to reverse the sign of the vector and scalar potentials. This just generalizes our earlier statement that the electric field surrounding an electron has an opposite sign to that surrounding a positron. This symmetry is an exact property of the equations of QED. In quantum chromodynamics (QCD), the theory of the strong interactions, such a symmetry also holds.

Charge conjugation is useful in classifying the states of systems in QED and in strong interactions, and provides important selection rules. One example is provided by positronium, a bound state of an electron and a positron, whose energy levels are similar to those of the hydrogen atom. Because the positron and electron can annihilate, positronium is unstable, decaying to photons. In positronium the levels can be labeled by the angular momentum L and the spin S; C is related to these by $C = (-1)^{L+S}$. The photon has $C = -1$. The statement that C is a symmetry means that if one starts with a state of definite C, one ends with a state of the same C. For example, the positronium state with $S = 1$ and $L = 0$ is observed to decay only to states with three photons, not with two. This can be understood in terms of the fact that $C = -1$ for this state, and that pairs of photons always have $C = 1$, so the decay is forbidden. Similar classifications are useful for the strongly interacting particles, the hadrons. The lightest spin-zero mesons, for example (the π's, K's, and η) are observed experimentally to have $C = +1$. On the other hand, in the quark model these particles are bound states of quarks and antiquarks with $L = S = 0$. As in QED, $C = (-1)^{L+S}$, so the picture is consistent.

Within the framework of QED, it turns out that there is nothing one can do to spoil this symmetry. However, in the weak interactions, this symmetry is violated. This violation is closely related to the famed violation of parity invariance, discovered in 1956 by Chien-Shiung Wu, following the suggestions of Tsung-Dao Lee and Chen Ning Yang. One way to understand this violation is to consider the behavior of neutrinos. Experimentally, neutrinos are always observed to have their spins aligned parallel to their momenta; their antiparticles have spin aligned opposite to their momenta. This situation violates parity. Momenta change sign under parity, but angular momenta (spins) do not, so it does not make sense to even speak of parity in such a circumstance. But this situation is also C-violating. C, by definition, relates the behavior of particles to that of antiparticles with the same spin. Yet for the neutrino there is no such particle.

Even though it is well established that C is violated in nature, there are still relations between the properties of particles and antiparticles. For some time after the discovery of parity violation, it was thought that if one performs C followed by P, this might be a good symmetry. This is possible, since CP would take a neutrino with spin along its momentum into an antineutrino with spin antiparallel to its momentum, which is consistent with the experimental observations described above. However, studies of the behavior of neutral K mesons in the 1960s revealed that even this symmetry does not hold. Slight differences were observed in the behavior of the neutral K meson and its antiparticle. These experiments also demonstrated that T (time reversal) is violated. To date, this is the only system in which CP and T violation have been observed. This is because CP-violating phenomena typically occur at rates at least 1,000 times smaller than ordinary parity-violating weak interactions. Future experiments, in particular experiments with the so-called B mesons (mesons containing b quarks), are expected to provide further insight into these phenomena.

There are certain relations between particles and antiparticles that may be exact. These follow from a symmetry in which one first performs C, then P, then T, or CPT. This symmetry is believed to be exact, both because there is no experimental evidence for its violation and because CPT is conserved by all quantum field theories, theories of the type that so successfully describe elementary particle interactions. This result is known as the "CPT theorem." This theorem ensures, for example, that the masses of particles are exactly equal to those of antiparticles, as are their lifetimes.

There have been many speculations about the role of the symmetries C, P, and T in a more fundamental theory. For example, while in the theory of weak interactions C is not a symmetry, it is possible that C and P or combinations of these might be exact properties of the laws of nature, which are "spontaneously broken" much the way that rotational invariance is broken in a magnet. Such theories typically predict the existence of new types of particles that might be detected experimentally. Alternatively, it is quite possible that at a fundamental level, there is no hint of symmetries such as C and P, and that they emerge only "accidentally," as approximate symmetries of the interactions that are most important at relatively low energies.

See also: CHARGE; *CPT* THEOREM; PARITY; PHOTON; QUANTUM CHROMODYNAMICS; QUANTUM ELECTRODYNAMICS; QUARK

Bibliography

KANE, G. *The Particle Garden, Our Universe as Understood by Particle Physicists* (Addison-Wesley, Reading, MA, 1995).

PERKINS, D. H. *Introduction to High-Energy Physics* (Addison-Wesley, Reading, MA, 1995).

MICHAEL DINE

CHARLES'S LAW

Jacques Charles, born in 1746 in France, was an experimental physicist famous for experiments with gases, including the construction of the first hydrogen gas balloon. Charles never published the law that bears his name, but Joseph Gay-Lussac (who carried out similar experiments in 1801 and 1802) acknowledged that Charles discovered the law about 1787.

Charles and Gay-Lussac found that if a gas is held at constant pressure, then the change of volume of the gas ΔV is directly proportional to the change of temperature ΔT; that is,

$$\Delta V \propto \Delta T$$

at constant pressure. It is surprising that the fractional change of volume per degree change in temperature is almost exactly the same for all simple real gases, regardless of chemical composition. If it is assumed that Charles's law is exactly the same for all gases, then Charles's law is one of the experimental laws needed to establish empirically the ideal gas law, which is the basis of the kinetic theory of gases.

Consider an experiment in which a gas is in a flexible container, such as a thin, easily stretched balloon. The pressure inside and outside the balloon is always constant at 1 atm. Raise and lower the temperature and measure the volume of the balloon at several different temperatures. A hypothetical graph of the data is shown in Fig. 1, where the temperatures are measured in degrees Celsius (°C). The graph is seen to be linear over the ordinary range of temperature. The extrapolation of the straight line shows that the volume of the gas shrinks to zero at $-273°C$ (more accurately, $-273.15°C$), which would be the case for an ideal gas, whereas a real gas would condense into a liquid at a higher temperature.

Temperatures lower than $-273.15°C$ are meaningless for gases (as well as liquids and solids), so $-273.15°C$ is the lowest possible temperature, called the absolute zero of temperature. If Charles's law is simplified by writing it without considering changes in state or phase, then

$$V \propto T$$

at constant pressure, where the absolute volume V requires that the absolute temperature T must be measured on a new absolute scale of temperature, namely, the Kelvin scale.

The obvious application of Charles's law is the hot-air balloon. At all times the air inside the balloon is at constant pressure, namely, equal to the ex-

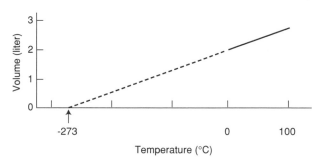

Figure 1 Graphical representation of the relationship between temperature and the volume of a balloon.

ternal atmospheric pressure. As the portable gas furnace raises the temperature of the air inside the balloon, the volume of the trapped air increases, in accordance with Charles's law. As the volume of the trapped air increases, Archimedes' principle states that the buoyancy (upward force) of the air increases to lift the balloon as well as the gondola and occupants.

See also: ABSOLUTE ZERO; ARCHIMEDES' PRINCIPLE; GAS; IDEAL GAS LAW; PRESSURE; TEMPERATURE SCALE, KELVIN; THERMOMETER; THERMOMETRY

Bibliography

MIDDLETON, W. E. K. *A History of the Thermometer and Its Use in Meteorology* (Johns Hopkins University Press, Baltimore, MD, 1966).

RICHARD H. DITTMAN

CHEMICAL PHYSICS

Chemical physics is the field in which the techniques of theoretical and experimental physics as well as mathematics are applied to chemical systems. The goal is to describe chemical and physical systems and their transformations in terms of the detailed properties and dynamics of the atoms, molecules, nuclei, and electrons of which they are composed. Implicit in the concept of a chemical system is a regime of energy that characterizes the interactions and disturbances to which the system is subjected. In general the chemical and physical phenomena are low energy, and while this is to some degree relative, it typically ranges at most up to some tens to hundreds of electron volts (eV). Thus, there will not be sufficient energy available to cause changes in the numbers of electrons and nuclei present in the system. Much of what is of interest occurs at energies below 1 eV, and another common unit of energy used is the kilocalorie (kcal, where 23 kcal $\equiv$ 1 eV) and fractions thereof. The related field of atomic physics, by comparison, generally concerns phenomena characterized by energies in the range of tens to thousands of eV.

Chemical physics arose as a distinct discipline during the 1920s and 1930s, as witnessed by the inception of the field's leading journal, the *Journal of Chemical Physics*. A list of the principle architects of the field includes, among others, such luminaries as Niels Bohr, Max Born, Peter Debeye, Paul Dirac, Albert Einstein, Vladimir Fock, Werner Heisenberg, Egil Hylleras, J. Robert Oppenheimer, Wolfgang Pauli, Erwin Schrödinger, John Slater, and Eugene Wigner from the physics community and Charles Coulson, Henry Eyring, Willard Gibbs, Joseph Hirschfelder, Robert Muliken, Linus Pauling, Michael Polanyi, and E. Bright Wilson from the chemistry community. In the 1930s, the first chemical physicists were trained as such, with the classic example being Hirschfelder, who actually did two Ph.D. theses, one with Wigner and one with Eyring, at Princeton.

The fundamental theoretical tools that the chemical physicist uses to develop a rigorous understanding and description of the systems of interest are quantum mechanics, classical mechanics, statistical mechanics, and classical electrodynamics. Statistical mechanics, the classical form of which is based on the Liouville equation, and the quantal form of which is based on the von Neumann equation, is used to describe the behavior of macroscopic systems in terms of the properties and behavior of their constituent particles. Often, it is also necessary to employ other areas of theoretical physics not in the above list, for example, the special theory of relativity (to describe the lower lying electrons in atoms with large nuclear charge or to treat various effects associated with purely relativistic "spin" degrees of freedom), as well as quantum electrodynamics and quantum field theory (e.g., to treat intense radiation fields interacting with matter). In particular, whether the particles composing the system of interest are fermions or bosons plays a

fundamental role due to the necessity of imposing appropriate identical particle exchange symmetry on the wave function associated with the system. Indeed, one can reasonably state that the entire subject of chemistry is a reflection of the Pauli exclusion principle and the existence of both the ordinary spatial coordinates and spin. Thus, effects that are normally associated with high energy processes, or the most fundamental aspects of matter can still persist even when one is interested in the low-energy regime.

Whether classical equations are adequate for describing a specific chemical physics phenomenon, or whether one needs quantal equations depends on the details of interest as well as particle masses, their momenta, and the time scale for the phenomena. The fundamental difference between quantum and classical physics hinges on the fact that dynamically related variables such as the x position and the x component of momentum cannot be measured simultaneously with arbitrary accuracy. This is reflected in the fact that the quantum operators representing these variables do not commute in the quantum theory. Instead, they satisfy the commutation rule

$$xp_x - p_x x = i\hbar, \tag{1}$$

where $\hbar$ is the fundamental quantum of action. The transition from quantum to classical mechanics can be explicated by a careful analysis of the limit as $\hbar$ tends to zero. The extraction of this limit is mathematically delicate, but many quantal phenomena can be described, at least semiquantitatively, by using an appropriate combination of classically calculated actions and quantum expressions.

A general program of chemical physics is the description of equilibrium and nonequilibrium bulk phenomena in terms of the properties and interactions of the nuclei and electrons composing the system. This is done by determining the average or expected behavior of a macroscopic system (containing an enormous number of atoms and molecules). We will concentrate on the quantum mechanical approach to how this is achieved, and then indicate how the corresponding classical approximation is carried out. The equation describing the bulk system is the von Neumann equation for the density operator, ρ, given by

$$i\hbar \frac{\partial}{\partial t}\rho = H\rho - \rho H, \tag{2}$$

or

$$i\hbar\frac{\partial}{\partial t}\rho = [H, \rho] = L\rho, \qquad (3)$$

where ρ describes the probabilistic behavior of the system. $[H,]$ or L is termed the Liouvillian and is the commutator of the Hamiltonian H, which in Eq. (3) is applied to the density operator. The Liouvillian generates time derivatives and is the quantum analog of the Poisson bracket of the classical Hamiltonian. Equilibrium states are described by special steady state $(\partial\rho/\partial t = 0)$ solutions of Eq. (3); they commute with H, implying they are functions only of H and perhaps any conserved quantities such as total angular momentum. The specific solution is somewhat arbitrary from the point of view of equilibrium thermodynamics and depends on the choice of fundamental thermodynamic variables. For example, the canonical ensemble (constant temperature and number of particles) solution is

$$\rho = \frac{e^{-H/kT}}{Q}, \qquad (4)$$

where k is the Boltzmann constant and Q fixes the normalization of ρ (it is called the partition function), while the microcanonical ensemble (constant total energy and number of particles) solution is

$$\rho = \delta(E - H). \qquad (5)$$

The expressions for the classical probability function (distribution function) are formally the same but with the important difference that $\exp(-H/kT)$ can then be factored as $\exp(-K/kT)\exp(-V/kT)$, where K is the total kinetic energy of the system and V is the total system potential, and T is the finite temperature of the system. Quantum mechanically, such a product expression is only approximate because K and V do *not* commute. Again, this reflects the fact that position does not commute with its corresponding momentum in quantum mechanics.

We emphasize that Eq. (3) holds for *all* systems in quantum mechanics, but it is used primarily to treat macroscopic systems. In that case, K and V are the operators for all N particles (where N is of the order of Avogadro's number). In order to reduce such expressions to manageable proportions, one

must be able to describe the dynamics in terms of the separate motions of each individual atom, molecule, or electron perturbed by the interactions with the other atoms, molecules, or electrons. One approach to doing this is to develop density expansions for ρ, with the lowest contribution coming from a density operator describing noninteracting particles. Higher terms involve pairs of particles, plus their interactions, then triples of particles, plus their interactions, and so on. Generally, this is achieved quantally by performing partial traces of ρ (represented as a matrix) in the von Neumann equation. Classically, one reduces the Liouville equation satisfied by the classical phase space density. The density describing single particles (the singlet density matrix) is obtained by tracing ρ over all but the quantum numbers of the particle of interest. The fact that the labeling of the same kinds of particles is arbitrary is taken into account by normalizing the singlet (and higher) densities. This normalization is distinct from the effects of fermions or bosons and also occurs in the classical theory. If the potential is expressed as a sum of single-particle potentials, pair potentials, three-particle potentials, and so on, then the tracing operation generates coupled integro-partial differential equations in which the singlet density is coupled to the pair densities, triplet densities, and so on. The terms involving only the singlet density describe its change in time due to free dynamics (the streaming terms), and the terms coupling to the pair density build in the interactions of pairs of particles (binary collisions); the terms coupling to the triplet density build in the interactions of triples of particles (three-body collisions). Such hierarchies are known generally as the BBGKY hierarchy (standing for Born–Bogoliubov–Green–Kirkwood–Yvon), and it continues up to the N-particle density itself. In order to produce something tractable, this hierarchy must be truncated. This is done essentially on the basis of density arguments. At relatively low densities, the importance of three-body collisions (which require three particles to be close to one another simultaneously) should be low. This truncation must be done carefully, however. For example, one must include in a binary collision approximation terms describing the collision of an atom with a bound diatomic molecule, even though such a collision involves three atoms simultaneously close together.

Once the equation is truncated, it is usual to introduce some sort of factorization of the higher densities in terms of the lower ones. This procedure is based on the molecular chaos assumption. Essentially, this assumption states that *prior* to a collision, the collision partners are *uncorrelated*. It is this assumption which builds into the resulting equations the irreversibility that underlies the second law of thermodynamics. Mathematically, it results in factoring pair densities into products of single particle densities, thereby yielding closed sets of equations (which now are nonlinear, integro-partial differential equations) to be solved. At this stage, the singlet densities are written as a reference (equilibrium) density plus a (small) perturbation. The perturbation is expressed in terms of the effect of the binary collisions, the equations are linearized, and solved. The simplest level of this approximation ultimately requires the solution of the Schrödinger equation for the binary scattering amplitude, in terms of which the physical processes resulting from the collisions are expressed. These include viscosity, heat conductivity, diffusion, and line broadening. If the system permits chemical reactions, thermal rate constants are obtained by including reactive binary collisions.

The above discussion has been from the point of view of kinetic theory. There are alternative treatments that focus on the formal solution of Eq. (3), easily achieved by the method of integrating factors:

$$\rho = e^{-iLt/\hbar}\rho(0). \qquad (6)$$

This expression is then evaluated by dividing the time t into M segments, $\tau = t/M$, which are sufficiently short that the neglect of commutators (above second order) is justified. The insertion of resolutions of the identity between the factors representing each time step τ leads to the path integral of Richard Feynman:

$$\rho(t) = \prod_{j=1}^{M}\int dx_j e^{-iV(x_j)\tau/2\hbar}$$

$$\langle x_j | e^{-iK\tau/\hbar} | x_{j-1}\rangle$$

$$e^{-iV(x_{j-1})\tau/2\hbar}\rho(t=0|x_0). \qquad (7)$$

Here, x_j is the entire set of coordinates describing the configuration of the system at time interval $j\tau$. The principal difficulty in evaluating this expression is that the free propagator, $\langle x_j | e^{-iK\tau/\hbar} | x_{j-1}\rangle$, with

real τ, is extremely oscillatory. The equilibrium statistical mechanical case results, however, from a purely imaginary time step and leads to computationally powerful techniques. An extremely active area of research deals with approaches that divide the N-body system into a small number of particles of interest (the subsystem) and the rest of the particles (the bath). The coupling to the bath is usually taken as linear, and the bath degrees of freedom are taken to behave harmonically, enabling them to be treated analytically. The subsystem contains sufficiently few particles so that it can be treated quite accurately. Subsystems that are evolving in real time, but are coupled to a large enough bath to remain at equilibrium, can be treated by use of complex times. Very powerful approaches to the theory of thermal reactive rates have been derived based on this type of formulation (e.g., the flux-flux autocorrelation approach). Path-integral-based approaches are extremely popular for treating processes occurring in condensed phases (at higher densities).

Another approach that is popular employs projection operators to partition Eq. (3) into coupled equations for a subsystem density operator and a bath density operator. This yields coupled first-order, linear differential equations that are then solved formally (again by the method of integrating factors). The equation for the subsystem density can be interpreted as a driven, first-order differential equation in time. In addition to the dynamics associated with the subsystem Hamiltonian (analogous to the streaming terms in the kinetic theory approach), one also has an integral (nonlocal) operator that is interpreted as the memory kernel associated with coupling to the bath. There is another term which can be interpreted as a friction. These two terms are related through the fluctuation-dissipation theorem. This approach has been widely used for treating subsystems immersed in a bath, such as a molecule embedded in a solid or liquid, a small portion of a very large molecule (e.g., a functional group in a protein), or a molecule or atom interacting with a solid surface. The formulation lends itself especially well to mixed classical-quantal computational techniques that are able to treat large numbers of atoms.

In these approaches, the quantum dynamics reduces ultimately to solving a Schrödinger-type equation for a few particles. In the case of quantum equilibrium statistical mechanics, one typically needs to calculate the energy eigenvalues that contribute to the sum in the partition function. If one is interested in various kinds of time-dependent processes, then the calculations will typically involve solving equations describing collisions resulting in transitions between eigenstates of the subsystems. In the case of atoms and molecules, the energy eigenstates are generally calculated within the framework of the Born–Oppenheimer separation of the nuclear and electronic degrees of freedom. This separation is based on the large time scale difference for electronic and nuclear motions, which reflects the large difference in the particle masses. During the time required for the nuclei to move significantly, the electrons will have already traversed the region of configuration space available to them, thereby establishing the quantal interferences that produce an eigenstate. Thus, one may first solve an electronic Schrödinger equation in which the nuclei are fixed:

$$H_{el}(q, Q)\Psi_{el,s}(q, Q) = E_s(Q)\Psi_{el,s}(q, Q), \quad (8)$$

where q is the set of electronic coordinates (including spin, in general), Q is the set of nuclear coordinates, and the energy $E_s(Q)$ is a function of Q and includes the nuclear-nuclear repulsion. The fact that the quantized electronic energy adjusts smoothly to changes in the nuclear positions implies that this is an adiabatic approximation. The energy $E_s(Q)$ acts as the potential energy for the motion of the nuclei, giving rise to the nuclear Schrödinger equation,

$$H_{nucl}(Q)\psi(Q) = E\psi(Q). \quad (9)$$

This equation is solved subject to appropriate boundary conditions (reflecting whether bound vibrational-rotational states or scattering states are desired). One requires that the *total* wave function (electronic times nuclear) satisfy the correct symmetry, including antisymmetry under exchange of any pair of fermions. This necessitates that the electron and nuclear spins also be taken into account in forming the total wave function.

Enormous amounts of effort have been expended in the decades since the inception of quantum mechanics toward the development and application of efficient methods of doing all aspects of quantum dynamics. Highly developed computer codes are available now for electronic state calculations. Current research is focused on methods in which the electronic energy is formulated as a functional of the density, rather than trying to solve the wave equation. In principle there exists a functional dependence which is exact (but unknown); the formulation is known as the density functional approach. Very promising results are obtained using a functional consisting of terms arising from an average field description, plus appropriate exchange and correlation terms to correct approximately the average field terms.

The nuclear quantum dynamics problem is less automated, but great progress has been made also. Particularly promising approaches include direct calculation of the microcanonical rate constant, wave-packet-based approaches, and variational methods. Difficulties associated with changes in coordinates occurring during chemical rearrangements have been addressed using a single hyper-radius that can describe all possible products, including complete dissociation. Wave packets have also been used to obtain a more general form of the time-independent Schrödinger equation, termed the time-independent wave packet Schrödinger equation, which has extremely nice properties for obtaining scattering information for a range of collision energies, and for bound state calculations. Negative imaginary absorbing potentials have been introduced to eliminate the necessity of taking detailed account of every possible molecular arrangement. By placing them in regions where there is essentially zero probability of reflection back into the strong interaction region, one can obtain quantitatively accurate results for scattering in the remaining coupled arrangements. They have also been used computationally and as a formal tool to develop an exact quantum transition state approach to reactive scattering. (Transition state theory is one of the earliest methods suggested for calculating reaction rates. It assumes that there is a point of no return in the nuclear motion, so that if this point or configuration is reached and the momentum along a reaction coordinate points in the product direction, reaction must occur. It neglects the possibility of reflection back to the reagents.) Numerically exact calculations have been carried out for the simplest reactions; namely $H + H_2$ and its isotopic variants and the most detailed state-to-state experimental measurements have been simulated. Totally quantitative agreement requires the inclusion of the most subtle of particle symmetries.

For more complex systems, approximations are still required for full theoretical simulation of experiment. For inelastic scattering, the most successful are the sudden approximations, which depend on the translational scattering motion being much faster than the rotational degrees of freedom. Best described are collisions dominated by a short-range, repulsive anisotropy. For reactive scattering at or near thresholds for reaction, the opposite condition is satisfied and the translational motion is slow compared to the bending and vibrational degrees of freedom. One then expects an adiabatic treatment to be better (the faster internal degrees of freedom access their available configuration space sufficiently to create the interferences that produce quantization). Variational transition state theory is another popular approach to treat more complex systems. This is based on the fact that the assumption that *all* collisions reaching the transition are reactive produces an upper bound to the reactive probability. Finally, an enormous number of simulations of collisions and intramolecular dynamics are done by solving Newton's equations of motion, with appropriate averaging over initial conditions (the quasiclassical trajectory method). Classical trajectories are also widely used to simulate the dynamics of extremely large systems, such as proteins. Classical statistical mechanical results are inferred by invoking the ergodic hypothesis equating time and ensemble averages.

Finally, we comment that the number of chemical physicists working on these areas (as well as others) is huge and space does not permit citation of all the names associated with the modern developments discussed above. Recent Nobel Prizes have been awarded to chemical physicists Robert Mulliken, Dudley Herschbach, Yuan Lee, John Polanyi, and Rudolph Marcus.

See also: ATOMIC PHYSICS; BOHR, NIELS HENRIK DAVID; BORN, MAX; COLLISION; DEGREE OF FREEDOM; DIRAC, PAUL ADRIEN MAURICE; EINSTEIN, ALBERT; ELECTROCHEMISTRY; EQUILIBRIUM; FERMIONS and BOSONS; GIBBS, JOSIAH WILLARD; HEISENBERG, WERNER KARL; KINETIC THEORY; OPPENHEIMER, J. ROBERT; PAULI, WOLFGANG; PAULI'S EXCLUSION PRINCIPLE; QUANTUM MECHANICS; SCHRÖDINGER, ERWIN; STATISTICAL MECHANICS; THEORETICAL PHYSICS; WAVE FUNCTION

Bibliography

BERNSTEIN, R. B., ed. *Atom-Molecule Collision Theory* (Plenum, New York, 1979).

BROECKHOVE, J., and LATHOUWERS, L. *Time-Dependent Quantum Molecular Dynamics* (Plenum, New York, 1992).

BOWMAN, J. M., ed. *Advances in Molecular Vibrations and Collision Dynamics* (JAI Press, Greenwich, CT, 1994).

CHANDLER, D. *Introduction to Modern Statistical Mechanics* (Oxford University Press, Oxford, Eng., 1987).

FEYNMAN, R. P., and HIBBS, A. R. *Quantum Mechanics and Path Integrals* (McGraw-Hill, New York, 1965).

HIRSCHFELDER, J. O.; CURTISS, C. F.; and BIRD, R. B. *Molecular Theory of Gases and Liquids* (Wiley, New York, 1964).

LEVINE, R. D. *Quantum Mechanics of Molecular Rate Processes* (Oxford University Press, Oxford, Eng., 1968).

McQUARRIE, D. *Statistical Mechanics* (Harper & Row, New York, 1976).

ZHANG, J. Z. H., and WYATT, R. E., eds. *Dynamics of Molecules and Chemical Reactions* (Marcel Dekker, New York, 1995).

DONALD J. KOURI

CHEMICAL POTENTIAL

The chemical potential, introduced by Josiah Willard Gibbs, is the central energy function of chemical thermodynamics. The chemical potential of a molecular species in a solution, in a gas phase, or in any other chemical environment, is the amount of work which could be obtained, per mole, if a small amount of the species were transferred to a standard reference environment by some hypothetical reversible process. The choice of a standard reference state for a chemical compound or element is arbitrary, usually a matter of convenience, so long as the same reference states are used consistently in comparing chemical potentials of reactants and products.

The preferred units of chemical potential are joules per mole, although other units such as kilocalories per mole or joules per kilogram may be encountered. The customary symbol for a chemical potential is the Greek lowercase *mu* labeled with the following species subscripts: μ_{H_2O}, μ_{N_2}, μ_{Ag}.

What makes chemical potentials important is that the work available from any chemical reaction can be calculated by subtracting the sum of the chemical potentials of the reaction products from the sum of the chemical potentials of the reactants. Thus, for the reaction $2H_2 + O_2 \rightarrow 2H_2O$, the work available per mole of O_2 consumed is given by $2\mu_{H_2} + \mu_{O_2} - 2\mu_{H_2O}$. (Note that the stoichiometric coefficients that are needed to make the reaction balance are also the coefficients used in summing the chemical potentials.) When this available work is positive the reaction will proceed spontaneously and irreversibly, at least in the presence of suitable catalysts or initiators. When the available work is negative, it is the reverse reaction which proceeds spontaneously. When the reaction reaches equilibrium, the sums of the chemical potentials are the same for both products and reactants. At that point no further chemical work is available from the reaction.

Whenever the concentration or partial pressure of a chemical species is low enough that none of its molecules are appreciably affected by the presence of other molecules of the same kind, then the chemical potential has a particularly simple and general linear dependence on the natural logarithm of the species concentration:

$$\mu_i = \mu_i^0(T,P) + RT \ln C_i.$$

Here R is the universal gas constant in joules per mole per degrees kelvin, T is the Kelvin temperature, and C_i is the concentration, or partial pressure P_i in convenient units. At the very low concentration limit all measures of concentration differ from one another merely by constant factors that become additive constants when the logarithm is taken. The quantity $\mu_i^0(T,P)$ depends on temperature and total pressure, but not on concentration, and is usually derivable from standard thermodynamic tables. It is evidently the intercept of the straight line at $\ln C_i = 0$, that is, at unit concentration, in a plot of μ_i versus $\ln C_i$. It often happens that this unit concentration is outside the dilute range where the ideal, strictly linear, dependence is found, but the data in the thermodynamic tables customarily refer to this ideal intercept extrapolated from the low concentration range, not to the actual value of μ_i at the unit concentration.

It is from this linear dependence of μ_i on $\ln C_i$ that the law of mass action for reaction equilibrium, or for equilibrium across phase boundaries, comes about. For example, at equilibrium the chemical potential of dissolved Cl_2 in water must equal the chemical potential of Cl_2 gas in the overlying vapor phase. Thus

$$\mu_{Cl_2}^0(\text{water}) + RT \ln C_{Cl_2} = \mu_{Cl_2}^0(\text{vapor}) + RT \ln P_{Cl_2}.$$

This relation leads directly to the proportionality

$$P_{Cl_2}(\text{vapor}) = K_H\, C_{Cl_2}(\text{water}),$$

where K_H, the so-called Henry's law constant, is a function of temperature and total pressure obtainable from the relation

$$RT \ln K_H = \mu_{Cl_2}^0(\text{water}) - \mu_{Cl_2}^0(\text{vapor}).$$

In a similar manner one can obtain the equilibrium condition on the partial pressures, P_i, for the gas phase reaction $3H_2 + N_2 \rightleftarrows 2NH_3$ such that

$$\frac{P_{NH_3}^2}{P_{H_2}^3 \cdot P_{N_2}} = K_P,$$

and the equilibrium constant K_P is given by

$$RT \ln K_P = 3\,\mu_{H_2}^0(\text{gas}) + \mu_{N_2}^0(\text{gas}) - 2\,\mu_{NH_3}^0(\text{gas}).$$

The negative gradient of any potential defines a force directed from the higher potential to the lower potential. This is true also of chemical potential: The tendency for a species to diffuse away from a region of high concentration to a region of lower concentration can be thought of, and described mathematically, as a chemical flux, $\mathbf{J}_i$, driven by a chemical force equal to the negative gradient of the chemical potential, and proportional to concentration C_i and to a mobility factor M_i, which would be the same for *any* applied external force (as in sedimentation).

Thus

$$\mathbf{J}_i = -M_i C_i \nabla \mu_i.$$

At the low concentration limit we have

$$\nabla \mu_i = RT \nabla \ln C_i = \frac{RT \nabla C_i}{C_i},$$

where

$$\mathbf{J}_i = -RT M_i \nabla C_i,$$

from which the product RTM_i can be identified as none other than the diffusion constant D appearing in Fick's first law of diffusion.

When the number of moles of species i, N_i, in a mixed chemical system at equilibrium, is increased infinitesimally by the reversible transfer of an amount dN_i from a standard reference state, the amount of chemical work done on the system is given by $\mu_i dN_i$. Allowing for possible variations of all the chemical species, the total amount of chemical work will be given by the sum of such terms over all the components:

$$\sum_i \mu_i dN_i.$$

This sum can be introduced directly into the usual differential expressions for the four energy state functions (internal energy U, Helmholtz free energy A or E, enthalpy H, and Gibbs free energy G) that derive from the combination of the first and second laws of thermodynamics.

Thus, for the Helmholtz free energy A (also denoted E), where $A = U - TS$, we obtain

$$dA = -SdT - PdV + \sum_i \mu_i dN_i,$$

and for the Gibbs free energy G, where $G = U - TS + PV$, we obtain

$$dG = -SdT + VdP + \sum_i \mu_i dN_i.$$

The differential expression for Helmholtz free energy implies that the chemical potential is the same as the partial derivative of A with respect to the mole number N_i, holding temperature, volume, and all the other mole numbers constant:

$$\mu_i = \left(\frac{\partial A}{\partial N_i}\right)_{T,V,\text{ all other } N}.$$

This equation may be used to calculate the chemical potential from a statistical mechanical model since A itself depends simply and directly on the partition function Z: $A = -kT \ln Z$.

Evidently the chemical potential can also be identified as the partial molar Gibbs free energy, that is, the partial derivative of G with respect to the mole number N_i, holding temperature, pressure, and all the other mole numbers constant:

$$\mu_i = \left(\frac{\partial G}{\partial N_i}\right)_{T,P,\text{ all other } N}.$$

The Gibbs free energy G and moles numbers N are all extensive variables, all increasing proportionately to the size of the system as the system is multiplied without changing the composition. However, temperature and pressure are intensive variables, independent of the size of the system. Therefore, a form of Euler's theorem for homogeneous functions (i.e., $U = TS - PV + \sum \mu_i N_i$) can be used to prove that

$$G = \sum_i \mu_i N_i.$$

This result would be trivial if the various chemical potentials were constants independent of the mixture composition, but they usual depend strongly on composition. Allowing for these changes one finds for the differential form

$$dG = \sum_i \mu_i dN_i + \sum_i N_i d\mu_i.$$

Comparing this with the previous expression for dG reveals that

$$SdT - VdP + \sum_i N_i d\mu_i = 0,$$

which is known as the Gibbs–Duhem relation. For changes in composition at constant temperature and pressure this reduces to $\sum N_i d\mu_i = 0$, an equation with remarkable consequences for simple solutions in which the solvent is almost pure, and any solute is dilute enough that its chemical potential has the ideal linear dependence on the logarithm of concentration. Then we have $N_\text{solvent} d\mu_\text{solvent} = -N_\text{solute} d\mu_\text{solute}$, but in the ideal dilute solution,

$$d\mu_\text{solute} = RTd\,(\ln N_\text{solute}) = \frac{RT\,dN_\text{solute}}{N_\text{solute}}$$

so that

$$d\mu_\text{solute} = -\frac{RT\,dN_\text{solute}}{N_\text{solute}}.$$

Thus the chemical potential of the solvent declines linearly as the amount of solute increases, by the same amount per mole for any and all solutes so long as they are sufficiently dilute to behave ideally. This decline in chemical potential lowers the freezing point of the solvent, lowers the vapor pressure, raises the boiling point, and accounts for the phenomenon of osmotic pressure: the hydrostatic pressure needed to bring the solvent chemical potential back up to equilibrium with that of pure solvent. Any of these effects can be used to determine the molecular weight of an unknown solute.

See also: CHEMICAL PHYSICS; DIFFUSION; ENERGY, FREE; ENERGY, INTERNAL; ENTHALPY; EQUILIBRIUM; GIBBS, JOSIAH WILLARD; THERMODYNAMICS

Bibliography

ATKINS, P. *Physical Chemistry,* 5th ed. (W. H. Freeman, San Francisco, 1994).

KLOTZ, I. M., and ROSENBERG, R. M. *Chemical Thermodynamics,* 5th ed. (Wiley, New York, 1994).

WALL, F. T. *Chemical Thermodynamics,* 3rd ed. (W. H. Freeman, San Francisco, 1974).

PAUL C. MANGELSDORF JR.

CHROMATIC ABERRATION

See ABERRATION, CHROMATIC

CHROMODYNAMICS

See QUANTUM CHROMODYNAMICS

CIRCUIT, AC

When an alternating voltage is applied to a circuit that contains resistors, capacitors, and inductors, the response is more complex than in the case of direct current (dc) circuits. This is because capacitors and inductors alter the phase of the applied voltage. Kirchhoff's laws are still valid since they represent charge conservation and energy conservation. Their application to alternating current (ac) circuits, however, requires some modification.

The mathematical analysis of basic ac circuits yields differential equations directly analogous to those describing simple mechanical systems. A series L–R–C circuit is diagrammed in Fig. 1. If the potential difference between the terminals of the ac source is given by

$$V(t) = V_0 \sin(\omega t),$$

the current through each component of the circuit is given by

$$I(t) = I_0 \sin(\omega t - \varphi),$$

where the amplitude of the current is related to the amplitude of the applied potential difference by a formula similar to Ohm's law for dc circuits,

$$I_0 = \frac{V_0}{Z}.$$

Z, called the "impedance" of the circuit, is the ac analog of resistance and for this circuit is given by

$$Z = \sqrt{R^2 + \left(\omega L - \frac{1}{\omega C}\right)^2}.$$

The phase shift between the current in the circuit and the applied voltage is

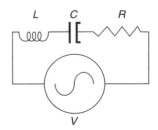

Figure 1

$$\varphi = \tan^{-1}\left(\frac{\omega L - \dfrac{1}{\omega C}}{R}\right).$$

In general, $\omega L = X_L$ is called the "inductive reactance" and $1/\omega C = X_c$ is the "capacitive reactance." Each reactance measures the propensity of the inductor or capacitor to carry a current for a given applied potential difference, just as the impedance Z measures the same propensity for the entire circuit.

Although the current is the same in all of the series components, the amplitude and phase of the voltage across each component differs. The voltage across the resistor is in phase with the current in the circuit and has an amplitude $V_R = I_0 R$; the voltage across the capacitor is $(\pi/2)$ radians behind the current and has an amplitude $V_C = I_0 X_C$; the voltage across the inductor is $(\pi/2)$ radians ahead of the current and has an amplitude $V_L = I_0 X_L$.

Because the impedance (and hence the current amplitude) depends on the applied frequency, a condition of "resonance" is said to exist when the frequency is such that the amplitude of the current has a maximum $I_0 = V_0/R$. The resonance frequency is given by

$$\omega_0 = \frac{1}{\sqrt{LC}}\left(1 - \frac{CR^2}{4L}\right).$$

A parallel ac circuit is shown in Fig. 2. If the driving voltage is again given by

$$V(t) = V_0 \sin(\omega t),$$

we find

$$I(t) = I_0 \sin(\omega t - \varphi),$$

where, as in the series circuit,

$$I_0 = \frac{V_0}{Z},$$

but the impedance now is given by

$$Z = \frac{1}{\sqrt{\dfrac{1}{R^2} - \left(\dfrac{1}{\omega L} - \omega C\right)^2}},$$

and the phase difference is now

$$\varphi = \tan^{-1}\left(\frac{\dfrac{1}{\omega L} - \omega C}{\dfrac{1}{R}}\right).$$

The voltage across all of the parallel components will be the same but the currents through each will be different. The phase differences between the currents through and the voltages across L, R, and C are the same as in the series case. The amplitudes of the

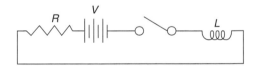

Figure 3

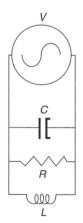

Figure 2

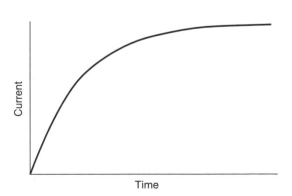

Figure 4

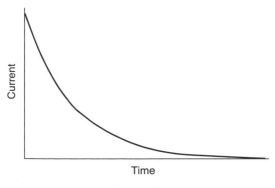

Figure 5

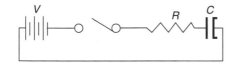

Figure 6

currents are given by $I_L = V_0/X_L$, $I_R = V_0/R$, and $I_C = V_0/X_C$.

By using complex numbers, the method of analyzing ac circuits is a straightforward extension of Kirchhoff's laws as applied to dc circuits. If the source of ac in a parallel L–R–C circuit is provided by the input from a radio or television receiver, the remainder of the parallel circuit is called a "tank" and it can serve as a "tuner" since it will respond with a resonance to a "tuned-in" frequency ω_0.

In general, capacitors and inductors contribute to a "slowing down" of the response of a circuit. For example, consider a simple series L–R circuit shown in Fig. 3. Figure 4 shows the "response" of this circuit when the switch is closed. If the inductance were zero the rise of the current to its final (asymptotic in Fig. 4) value would be instantaneous. The switching on has been slowed by the nonzero L. The rise of the current after the switch is closed is given by

$$I(t) = \frac{V}{R}(1 - e^{-(Rt/L)}).$$

When the switch in Fig. 3 is opened the current does not drop instantly to zero as it would with no inductance present; instead the current decays as shown in Fig. 5. This decay is described by

$$I(t) = \frac{V}{R}e^{-(Rt/L)}.$$

In both opening and closing the switch in Fig. 3 the response of the circuit is governed by an exponential function. The "characteristic time" or "decay constant" or "growth constant" or, more commonly, the "time constant" for the circuit is (L/R). The time constant measures the "sluggishness" of the response of the circuit to a change in the voltage applied. Quantitatively, the time constant is the amount of time required for the exponential factor to reach e^{-1}.

Figure 6 shows a series R–C series circuit. Figure 5 shows the current through the capacitor when the switch is closed (this is the same curve that reflected the decay of the current in the L–R circuit that was turned off). In this case the current is given by

$$I(t) = \frac{V}{R}e^{-(t/RC)}.$$

Figure 7 represents the voltage across the plates of the capacitor as charge builds up (this is similar to the figure used to illustrate the growth of current in an L–R circuit that is turned on). The voltage across the capacitor plates is described by

$$V(t) = V(1 - e^{-(t/RC)}).$$

If the charged capacitor is then discharged through the resistor as shown in Fig. 8, the current

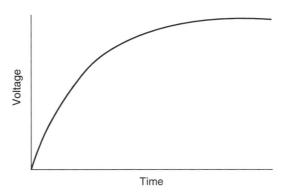

Figure 7

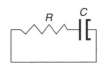

Figure 8

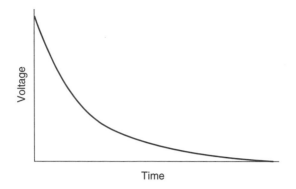

Figure 9

in the capacitor is shown by the graph in Fig. 5; the voltage across the discharging capacitor is shown by the similar curve in Fig. 9. These curves are described by

$$V(t) = V_0 e^{-(t/RC)}$$

and

$$I(t) = I_0 e^{-(t/RC)}.$$

The time constant of the R–C circuit is seen to be the product RC.

If a capacitor bears an initial charge Q_0 on its plates and then is connected to an inductor L as shown in Fig. 10, the voltage across the capacitor is shown by the curve in Fig. 11. The equation describing this voltage is

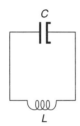

Figure 10

$$V(t) = \frac{Q_0}{C}\cos(\omega_0 t).$$

Similarly, the current through the circuit is given by

$$I(t) = -\frac{Q_0}{\sqrt{LC}}\sin(\omega_0 t),$$

where the frequency of oscillation of this circuit is nearly the same as the resonance frequency of the corresponding L–R–C circuit,

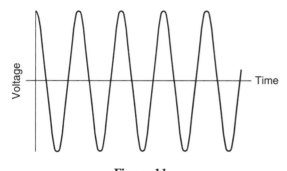

Figure 11

$$\omega_0 = \frac{1}{\sqrt{LC}}.$$

If a resistance is now introduced into the series circuit shown in Fig. 10, with the capacitor initially charged to Q_0, the second-order differential equation governing the circuit is identical to that for a damped, simple harmonic, mechanical oscillator, and depending on whether the quantity $R^2C/4L$ is greater than, equal to, or less than unity, the voltage (and current) will be overdamped, critically damped, or underdamped as shown in Fig. 12.

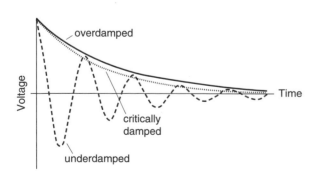

Figure 12

Bibliography

BOAS, M. L. *Mathematical Methods In The Physical Sciences,* 2nd ed. (Wiley, New York, 1983).

MARION, J. B., and HORNYAK, W. F. *Physics for Science and Engineering* (Saunders College Publishing, New York, 1981).

OHANIAN, H. C. *Physics,* 2nd ed. (Wiley, New York, 1989).

REITZ, J. R., MILFORD, F. J., and CHRISTY, R. W. *Foundations of Electromagnetic Theory* (Addison-Wesley, Reading, MA, 1979).

DELO E. MOOK II

CIRCUIT, DC

Although the abbreviation "dc" probably originally stood for "direct current," today it is commonly used to refer to any of the parameters relevant to circuits in which the current flows in only one direction. Thus we commonly refer to dc voltage, dc current, or even dc circuits.

An electric circuit consists of any number of components that are electrically connected. A complete electric circuit consists of components that are connected in such a way as to form one or more closed loops around which it is possible for an electric current to flow. When the terminals of two or more components are connected so that there is no voltage difference between them, that connection is called a node. A branch is a single current path consisting of one component or several components in series which are connected between two specified nodes. In practical circuits there is often one common node to which many of the circuit components are connected. That common node is usually referred to as ground because it is frequently connected to the chassis of the device containing the circuit. For convenience, unless specified otherwise, node voltages are usually referenced to the ground node.

A circuit which illustrates the above features is shown in Fig. 1a. Note that node C is a rather extended node since it is the junction of resistors R_2, R_3, R_4, and R_5. The ground node G also covers a fair amount of territory.

Several branches exist between node C and the ground node. One consists of resistors R_1, R_2, and source V_{S1}. A second branch consists of resistor R_3 alone. Resistor R_4 is a third branch and the last branch is formed by resistors R_5 and R_6 and source V_{S2}.

When circuits such as that in Fig. 1a are analyzed, the value of all of the components is usually known. The circuit is said to be solved when the value of all node voltages and all component currents are known. The solution of such circuits is based on two conservation ideas. They are known as Kirchhoff's current law (KCL) and Kirchhoff's voltage law (KVL). KCL states that the algebraic sum of

(a)

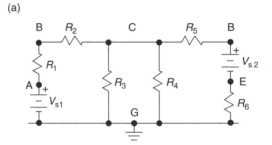

(b)

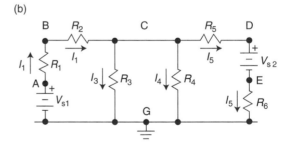

(c)

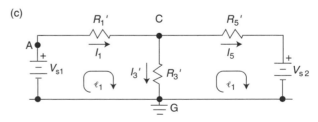

Figure 1 (a) A simple dc circuit. (b) The circuit with currents arbitrarily assigned. (c) The simplified circuit ready for analysis.

the currents entering a node is equal to zero amps; that is,

$$\Sigma I_n = 0. \tag{1}$$

KVL states that the algebraic sum of the voltages around a closed loop is equal to zero volts; that is,

$$\Sigma V_\ell - 0. \tag{2}$$

The first step in the solution of circuits such as that in Fig. 1a is to arbitrarily assign current directions for each resistor. Although the direction chosen is completely arbitrary, it must not be changed during the course of the analysis. If the current actually flows opposite to the direction chosen, its calculated value will be negative. Currents have been assigned in Fig. 1b.

Notice that I_2 is not indicated in Fig. 1b. The reason is that R_1 and R_2 are in series so $I_2 = I_1$. The same situation applies to I_5 and I_6.

The second step in the analysis is to simplify all series and all parallel circuits of Fig. 1b. Although this step is not mandatory, it simplifies the analysis. Note that a "prime notation" is used in the simplified circuit of Fig. 1c. Thus,

$$R_1' = R_1 + R_2,$$

$$R_3' = R_3 \parallel R_4 \left(\text{i.e.,} \ \frac{1}{R_3'} = \frac{1}{R_3} + \frac{1}{R_4} \right),$$

$$R_5' = R_5 + R_6,$$

$$I_3' = I_3 + I_4.$$

The simplified circuit of Fig. 1c can be analyzed using any one of several techniques. The most straightforward method is known as the "branch current method." In the case of the circuit of Fig. 1c, since there are three unknown currents, I_1, I_3', and I_5, three independent circuit equations are required. One equation can be obtained by applying KVL to the loop identified as ℓ_1 in Fig. 1c. Beginning at the lower left-hand corner of the loop and recording the voltage gains and losses as the loop is traversed in a clockwise direction, we have

$$V_{S1} - I_1 R_1' - I_3' R_3' = 0. \tag{3}$$

A second independent equation can be obtained by applying KVL to the loop identified as ℓ_2 in Fig. 1c. Again beginning in the lower left-hand corner of the loop and recording the voltage gains and losses encountered as the loop is traversed in a clockwise direction yields:

$$I_3' R_3' - I_5 R_5' - V_{S2} = 0. \tag{4}$$

A third independent equation can be obtained by applying KCL to the node identified as node C in Fig. 1c. The result is

$$I_1 - I_3' - I_5 = 0. \tag{5}$$

If the circuit component values are known, Eqs. (3), (4), and (5) can be solved simultaneously to obtain currents I_1, I_3', and I_5. These current values can then be substituted back into the original circuit to solve for the required information.

Using the method of loop currents or node voltages often reduces the number of equations that must be solved simultaneously but usually at the expense of greater complexity.

See also: CIRCUIT, AC; CURRENT, DIRECT; KIRCHHOFF'S LAWS

Bibliography

BOYLSTAD, R. L. *Introductory Circuit Analysis,* 7th ed. (Merrill, New York, 1994).

HALLIDAY, D.; RESNICK, R.; and KRANE, K. *Physics,* 4th ed. (Wiley, New York, 1992).

ROBERT R. LUDEMAN

CIRCUIT, INTEGRATED

An integrated circuit (IC) is an electronic circuit in which all of the components are made on a single semiconductor substrate. Components such as transistors, diodes, resistors, and even small capacitors can be included. The substrate usually used is silicon. These chips, smaller than a fingernail, may contain as many as several million circuit components. ICs are the heart of most modern devices such as wrist watches, cameras, automobile-engine monitoring and control modules, home entertainment systems, and of course calculators and computers, to name a few.

It is difficult to identify the beginnings of the IC since many individuals and laboratories were working in the field simultaneously. Further, each technological advance grew out of previous work. Most of those involved agree that in making their contribution, they stood on the shoulders of giants. A good example of this interdependence is the development of the planar transistor. Previously a vertical structure had been employed in transistor construction in order to facilitate making the necessary electrical connections to the various transistor regions. This made the integration of many transistors on a single chip difficult. Planar-transistor technology provided a flat surface which greatly simplified their integration. Jean Hoerni developed the planar transistor in 1958. Using this technology, the first IC was constructed later that same year. What followed was an almost explosive development of technology for both linear and digital ICs.

Early linear ICs were simply bipolar transistor implementations of the vacuum-tube operational amplifier (op amp) that had been developed several years earlier. They were rather temperamental devices that required external compensation to prevent them from breaking into uncontrollable oscillation. The device that really brought IC op amps to the forefront of linear amplifier technology was Fairchild's internally compensated μA741 introduced in 1968. It is still used today.

Early digital ICs consisting of only a few transistors used what is now known as small scale integration (SSI). The next step in IC evolution was the development of medium scale integration (MSI), consisting of circuits containing up to about one-hundred transistors. Next came large scale integration (LSI) with up to around 1,000 transistors, and finally very large scale integration (VLSI), containing millions of transistors. To achieve such component densities, components in some ICs have dimensions of less than one-hundredth the diameter of a human hair.

The construction of ICs involves the use of a photolithographic process to print the required circuit layout on the silicon substrate. The procedure is much the same as that used to print a message or a picture on a tee shirt. The circuit pattern is then chemically etched into the silicon. VLSI chips may require several layers of such patterns and as many as 250 separate steps for their manufacture. Due to the close tolerances which must be maintained, many of those steps must be carried out in a "clean room" where the temperature and humidity are controlled to within less than 1 percent and where there is less than one dust particle per cubic foot of air. That is over 100 times cleaner than the cleanest of hospital operating rooms.

There are several reasons why it is desirable to pack as much circuitry into each IC as possible, most of them having to do with economics. If more circuitry can be put on each chip, the finished product will contain fewer parts and so it will be less expensive to manufacture. Further, with more powerful ICs, more features can be built into the product containing the chip, giving it a competitive edge in the marketplace. Another advantage of high density chips is the speed at which they can operate. The less distance the signal has to travel between components, the less time it will take for data to be processed. Faster computers sell better than slow ones.

Probably the greatest stimulus for the development of ICs is the synergistic relationship that exists between computers and ICs. ICs need a big market in order to spread their high development costs over many units. Computers supply that market. Computers have an insatiable demand for more memory, more speed, and smaller size, all characteristics ICs can supply.

It is interesting to speculate about how much farther the trend toward higher and higher circuit density will go. In 1965 Gordon Moore, one of the cofounders of Intel Corporation, made the rather startling prediction that has come to be known as Moore's law. After having participated in IC development since its beginning, Moore observed that chip density doubles every year. Obviously exponential growth such as this cannot be sustained indefinitely but for the first twenty years of IC development, Moore's law was amazingly accurate. It now appears that the pace is beginning to slacken with a doubling time of more like two years, but since the limitations imposed by the laws of physics have not yet been approached, the trend toward ICs with greater circuit density and more speed is not likely to end any time soon.

See also: CAPACITOR; DIODE; RESISTOR; SEMICONDUCTOR; TRANSISTOR

Bibliography

CARR, J. J. *Integrated Electronics: Operational Amplifiers and Linear ICs with Applications* (Harcourt Brace, New York, 1990).

GARROD, A. R., and BORNS, R. J. *Digital Logic: Analysis, Applications and Design* (Saunders, Philadelphia, 1991).

MALIK, N. R. *Electronic Circuits, Analysis, Simulation, and Design* (Prentice Hall, Englewood Cliffs, NJ, 1995).

ROBERT R. LUDEMAN

CIRCUIT, PARALLEL

Two electrical components can be connected in a circuit such that each one acts as if it is connected directly to the energy source. Such a circuit, called a parallel circuit, is illustrated in Fig. 1. In this figure each of the two resistances is connected by wires directly to the battery. Figure 2 shows a slight variation on that circuit, which requires less wire. However, the effect is the same. One can trace a path from the energy source directly to each of the devices that are converting the electrical energy without passing through any other resistor. For simple parallel circuits, this arrangement is a fundamental property of the circuit.

Because each device is connected, in effect, directly to the energy source, each behaves in the same manner that it would if it were the only component in the circuit. Thus, the current through each of the components is identical to the current that would move through that component if it were alone in a circuit and connected to the same energy source. Thus the number of electrons passing through each component per second can be determined by knowing the current for the device in a simple circuit. Because each device requires a fixed number of electrons, the total number of electrons per second supplied by the energy source must be the sum of those required by each device. Thus, the total current for the circuit in Fig. 2 is

$$I_1 + I_2 = I_{\text{total}}.$$

The labels for current in Fig. 2 indicate where each of these magnitudes of currents exists.

Adding a third component to the circuit would give us the arrangement in Fig. 3 and would add the current for the third device to the total current in the circuit. In this case the total current splits at point A so that the current I_1 goes through R_1 and the current $I_2 + I_3$ continues down the wire to point B, where the current splits again to its respective devices.

In general, as more components are added in parallel to the circuit, the current of the circuit increases. Ohm's law indicates that as the current in the circuit increases, the resistance decreases if the voltage remains the same. This behavior of having the resistance decrease as components are added to the circuit is typical of a parallel circuit. Combining the addition of currents with Ohm's law provides a way to obtain a relation for the total resistance in a parallel circuit. The voltage across each component will be equal to that across any other component. We know that the total current is equal to the sum of the individual currents:

$$I_{\text{total}} = I_1 + I_2 + I_3 + \cdots + I_n.$$

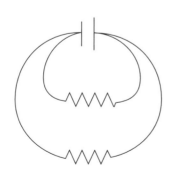

Figure 1 Basic parallel circuit.

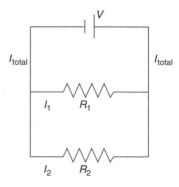

Figure 2 Variation of a parallel circuit.

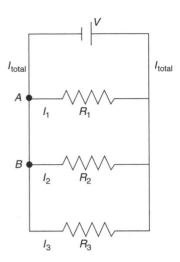

Figure 3 Parallel circuit with three components.

Dividing by the voltage V,

$$\frac{I_{total}}{V} = \frac{I_1}{V} + \frac{I_2}{V} + \frac{I_3}{V} + \cdots + \frac{I_n}{V}.$$

Then noting that I_n/V equals $1/R_n$, we can write this equation as

$$\frac{1}{R_{total}} = \frac{1}{R_1} + \frac{1}{R_2} + \frac{1}{R_3} + \cdots + \frac{1}{R_n}.$$

This equation provides a way to calculate the resistance of any set of devices for which the individual resistances are known and which are connected in parallel. It displays the property that the number of devices increases and the resistance decreases. Moreover, the total resistance of a parallel circuit is always less than the resistance of any individual component in the circuit.

Most devices that convert electrical energy to other forms in a house are connected in parallel with each other. Thus, they operate independently of what else might be operating in the same circuit. If a large number of devices operate in the same parallel circuit, the current could increase beyond the capacity of the circuit. For example, a typical hair dryer requires a current greater than 10 A. If two people decide to dry their hair simultaneously, and plug their hair dryers into the same circuit, they will require a total current in excess of 20 A. Many household circuits become unsafe if the current in their wires exceeds 20 A. Therefore, a fuse or circuit

breaker is placed in series with the rest of the devices. This circuit breaking device turns off when the current exceeds its rated value. Thus the two hair dryers are likely to cause the circuit breaker to "trip" or the fuse to "blow" and shut down all of the electrical current in the entire circuit.

Alternating current circuits may contain inductors and capacitors in addition to resistors. In this case one must consider the impedance of the components in determining the total impedance of the circuit.

See also: BATTERY; CAPACITOR; CIRCUIT, AC; CIRCUIT, DC; CIRCUIT, SERIES; CURRENT, ALTERNATING; CURRENT, DIRECT; ELECTRICAL RESISTANCE; ELECTRICAL RESISTIVITY; INDUCTOR; RESISTOR

Bibliography

NILSSON, J. W. *Electric Circuits* (Addison-Wesley, Reading, MA, 1990).
RIDSDALE, R. E. *Electric Circuits for Engineering Technology* (McGraw-Hill, New York, 1984).

DEAN ZOLLMAN

CIRCUIT, SERIES

The simplest electrical circuit that one can create consists of one energy source, such as a battery, and one device that converts electrical energy to other forms, such as a resistor, with wires connecting the battery to the resistor (see Fig. 1). Typically, however, a circuit will have more than one resistor in it. When two resistors are connected in a circuit, they can be connected such that each electron passes through one of the resistors and then passes through the second resistor. Such a circuit is shown in Fig. 2. Here, each electron deposits some of its energy in the resistor R_1 and some of its energy in R_2. This diagram represents a series circuit. Because each electron passes through each of these resistors, the overall resistance of the circuit increases from its value as if only one resistor were in the circuit. The total resistance in a series circuit is the sum of the individual resistances. Thus

$$R_{total} = R_1 + R_2 + \cdots + R_n + \cdots,$$

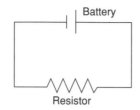

Figure 1 Simple electrical circuit.

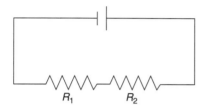

Figure 2 Series circuit with two resistors.

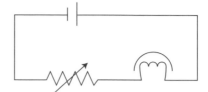

Figure 3 Dimmer mechanism.

where R_n represents the resistance of the nth resistor. When combined with Ohm's law, the equation for resistances in series states that, as we add resistors in series, the current in the circuit must decrease because the total resistance in the circuit is increasing.

A practical application for this concept is a simple way to dim a light bulb. The circuit for this dimmer mechanism is shown in Fig. 3, where the resistor with the arrow through it indicates that the resistance of that device may be changed. When the resistance of this variable resistor is increased, the current in the entire circuit decreases. Because the current decreases, the power and energy delivered to the light bulb decreases, and the bulb becomes dimmer. When the resistance of the variable resistor decreases, the current increases, and the bulb becomes brighter. (A modern dimmer switch, however, uses a solid-state device known as a silicon controlled rectifier (SCR) to minimize heat generation and power loss inherent in a resistor-type dimmer.)

Notice that the current in a series resistor is the same everywhere in its circuit. Every electron that passes through any component must pass through all other components. If the electrons were to build up at any point they would, by electrostatic repulsion, stop the electrons following behind them. Thus, the rate of movement of the electrons does not change anywhere in the circuit, and the current remains the same everywhere.

A useful and sometimes frustrating property of a series circuit is that if any device in the circuit stops working all devices stop working because the current stops everywhere in the circuit. This property can be frustrating when a light is removed from a string of Christmas tree lights, and all of them cease to work. On the other hand, the turning off and on of one light will control all others that are connected in series with it. This property is used in flashing Christmas tree lights. Other practical uses of series circuits include switches, fuses, and circuit breakers, all of which are connected in series with other components and which stop the motion of the electrons (current flow) through all components in the circuit.

For alternating current circuits other components, such as inductors and capacitors, can be connected in series with each other or with resistors. In this case one must work with impedances of the components instead of just the resistances.

See also: BATTERY; CAPACITOR; CIRCUIT, AC; CIRCUIT, DC; CIRCUIT, PARALLEL; CURRENT, ALTERNATING; CURRENT, DIRECT; ELECTRICAL RESISTANCE; ELECTRICAL RESISTIVITY; INDUCTOR; OHM'S LAW; RESISTOR

Bibliography

NILSSON, J. W. *Electric Circuits* (Addison-Wesley, Reading, MA, 1990).

RIDSDALE, R. E. *Electric Circuits for Engineering Technology* (McGraw-Hill, New York, 1984).

DEAN ZOLLMAN

CIRCULAR DICHROISM

See DICHROISM, CIRCULAR

CIRCULARLY POLARIZED LIGHT

See POLARIZED LIGHT, CIRCULARLY

CLAUSIUS, RUDOLF JULIUS EMMANUEL

b. Köslin, Prussia, January 2, 1822; *d.* Bonn, Germany, August 24, 1888; *thermodynamics, kinetic theory, electrodynamics.*

Clausius was the sixth son in a family of eighteen children. His father, a Lutheran minister, was a Prussian Schools Inspector in Pomerania. Clausius attended the gymnasium (academic high school) in Stettin and in 1840 entered the University of Berlin. He was a gymnasium teacher while pursuing his Ph.D., completing it in 1848 at Halle with a dissertation on the reflection of light in the atmosphere. In 1850 he habilitated (passed examination, by producing a second dissertation, to qualify to teach at the university level) at the University of Berlin and began a series of papers on heat theory. In 1855 he moved to Zurich Polytechnic as a full professor and to Würzburg twelve years later. He was called to Bonn in 1869 and taught there until his death, holding an oral examination on his death bed. Clausius led an ambulance brigade in the Franco-Prussian War and was honored by the Prussian state. He was also honored by scientific societies across Europe for his physics.

Clausius's research spanned several domains of physics. All of them were attempts to understand how the microscopic structure of matter determined its macroscopic properties. He explained the colors of the sky by reflections within hollow particles of the atmosphere. With his work establishing the two laws of thermodynamics he tried to investigate the molecular motion "we call heat." To accomplish this, Clausius had to displace the idea that heat was a substance ("caloric") that was conserved and replace it with the principle that heat and work were equivalent. The latter principle, the conservation of energy, or the first law of thermodynamics, was not well-established even by 1850. Using this equivalence as a principle, Clausius could explain many known properties of gases, and he reinterpreted the workings of the heat engine. In the next decade Clausius applied the first law to many phenomena of gases and to thermoelectricity. He then expressed the second law of thermodynamics as the equivalence between transforming heat Q into work at a constant absolute temperature T and transforming heat at one temperature into heat at a lower temperature. He expressed this "equivalence value" as $dS = dQ/T$. (He named S entropy only in 1865.) These changes always occurred in the cycle of a heat engine and entropy was either conserved, or, for real engines, entropy was increased.

For two decades Clausius tried to fathom the meaning of the two laws of thermodynamics in molecular terms. In 1857 he examined the motions of gas molecules that randomly interacted with each other and their container. James Clerk Maxwell developed these ideas into the kinetic theory of gases, introducing probability into physics. Clausius could not accept this approach. For him the motions of molecules were determined by mechanics, not probability. Clausius's mechanical approach led to the virial theorem and connected gas theory to Hamilton's principle of least action. However, it only worked for reversible, thermal operations and not for irreversible ones.

Clausius also worked in electrodynamics. He explained the interaction of current-carrying wires by the absolute, not the relative, velocities of charged particles moving through the wires. He also brought his theoretical ideas to bear on technical problems, developing a theory of the steam engine, the critical power source of the nineteenth century, and a theory of the dynamo.

In Clausius's work we see both the power and limitations of a strictly mechanical view of nature. While he worked in critical areas of physics after his work on thermodynamics, it was never as compelling or successful as his first papers on heat.

See also: ELECTROMAGNETISM, DISCOVERY OF; KINETIC THEORY; LEAST-ACTION PRINCIPLE; THERMODYNAMICS, HISTORY OF

Bibliography

BRUSH, S. G. *Kinetic Theory,* Vol. 1: *The Nature of Gases and Heat* (Pergamon, New York, 1965).

CLAUSIUS, R. J. E. "On the Moving Force of Heat and the Laws Which Can Be Deduced Therefrom." *Philos. Mag.* **2,** 1–21, 102–119 (1851).

Clausius, R. J. E. "On the Kind of Motion We Call Heat." *Philos. Mag.* **14**, 108–127 (1857).

Gibbs, J. W. *The Scientific Papers of Josiah Willard Gibbs* (Dover, New York, [1906] 1961).

Jungnickel, C., and McCormmach, R. *The Intellectual Mastery of Nature: Theoretical Physics from Ohm to Einstein* (University of Chicago Press, Chicago, 1986).

Elizabeth Garber

CLIFFORD ALGEBRA

Clifford algebra extends the real number system to include vectors $\mathbf{u}$, $\mathbf{v}$, . . . , and their products $\mathbf{uv}$, $\mathbf{uvw}$, ... It is well suited to modeling geometry in physics, and its vector products, representing surfaces and higher-dimensional objects, allow simple but rigorous descriptions of rotations, reflections, and other geometric transformations. The algebra is named for nineteenth-century English mathematician William Kingdon Clifford, who called it geometric algebra. Complex numbers and objects known as quaternions (hypercomplex numbers) form two examples of a Clifford algebra.

A Clifford algebra, like the field of real numbers, is closed under addition and multiplication; this means that all sums and products of elements are themselves elements of the algebra. Products of three or more vectors are associative, that is, independent of the order of multiplication: $\mathbf{u(vw)} = \mathbf{(uv)\,w} \equiv \mathbf{uvw}$. The basic axiom of Clifford algebra is that the product of any vector $\mathbf{v}$ with itself is its square length $\mathbf{v} \cdot \mathbf{v}$:

$$\mathbf{v}^2 = \mathbf{vv} = \mathbf{v} \cdot \mathbf{v}. \tag{1}$$

Suppose $\mathbf{v}$ is the sum of two other vectors: $\mathbf{v} = \mathbf{a} + \mathbf{b}$. Relation (1) becomes $(\mathbf{a} + \mathbf{b})^2 = (\mathbf{a} + \mathbf{b}) \cdot (\mathbf{a} + \mathbf{b})$, whose expansion, followed by the elimination of $\mathbf{a}^2 = \mathbf{a} \cdot \mathbf{a}$ and $\mathbf{b}^2 = \mathbf{b} \cdot \mathbf{b}$, leaves the basic result

$$\mathbf{ab} + \mathbf{ba} = 2\mathbf{a} \cdot \mathbf{b}. \tag{2}$$

However, if $\mathbf{a}$ and $\mathbf{b}$ are not collinear, $\mathbf{ab}$ cannot be a pure scalar (number) since $\mathbf{aab} = \mathbf{a(ab)} = \mathbf{(aa)b}$

would then equate vectors pointing in different directions. Consequently

$$\mathbf{ab} - \mathbf{ba} = 2(\mathbf{ab} - \mathbf{a} \cdot \mathbf{b}) \tag{3}$$

vanishes only if $\mathbf{a}$ and $\mathbf{b}$ are aligned. Thus, unlike scalars, vector products generally do not commute: $\mathbf{ab} \neq \mathbf{ba}$. Otherwise, however, vectors multiply like scalars. In particular, products are linear and distributive over addition. The multiplication rules for vectors are the same as for square matrices.

A Clifford algebra common in physics is the Pauli algebra, based on vectors in three-dimensional Euclidean space $\mathbb{R}^3$. In the Pauli algebra, a conventional representation uses the 2×2 Pauli spin matrices

$$\sigma_1 = \begin{pmatrix} 0 & 1 \\ 1 & 0 \end{pmatrix}, \sigma_2 = \begin{pmatrix} 0 & - \\ i & i \end{pmatrix}, \sigma_3 = \begin{pmatrix} 1 & 0 \\ 0 & -1 \end{pmatrix}, \tag{4}$$

to represent unit vectors along the Cartesian axes. A vector $\mathbf{v}$ with Cartesian components v_x, v_y, v_z, is thus represented by the matrix $v_x\sigma_1 + v_y\sigma_2 + v_z\sigma_3$. (In physics literature, this matrix is often written $\mathbf{v} \cdot \boldsymbol{\sigma}$, but one should understand that it represents a vector, not a scalar.) The unit scalar is represented by the unit matrix. An infinite number of different matrix representations may be used, but only the algebra of the products is physically significant.

From Eq. (3), products of any two real n-dimensional vectors are sums

$$\mathbf{ab} = \mathbf{a} \cdot \mathbf{b} + \mathbf{a} \wedge \mathbf{b} \tag{5}$$

of a dot product and a wedge (or exterior) product defined by

$$\mathbf{a} \wedge \mathbf{b} := \tfrac{1}{2}(\mathbf{ab} - \mathbf{ba}). \tag{6}$$

The dot product $\mathbf{a} \cdot \mathbf{b}$ is a scalar, but $\mathbf{a} \wedge \mathbf{b}$ is neither a scalar nor a vector; it is a new element called a bivector. In three dimensions, $\mathbf{a} \wedge \mathbf{b}$ is related to the vector cross product $\mathbf{a} \times \mathbf{b}$ (see below), but unlike cross products, wedge products are associative and well defined in spaces of higher dimension.

Just as n-dimensional vectors $\mathbf{a}$ and $\mathbf{b}$ can be expanded in a Cartesian basis $\{\mathbf{e}_1, \mathbf{e}_2, \dots, \mathbf{e}_n\}$, so substitution of such expansions into Eq. (6) shows that

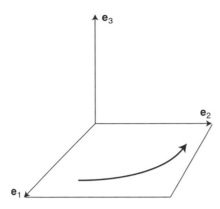

Figure 1 The bivector $\mathbf{e}_1\mathbf{e}_2 = -\mathbf{e}_2\mathbf{e}_1$ represents the plane patch of unit area that contains the perpendicular unit vectors $\mathbf{e}_1$ and $\mathbf{e}_2$. Its orientation is the twist which rotates $\mathbf{e}_1$ into $\mathbf{e}_2$. In $\mathbb{R}^3$, one can set $\mathbf{e}_1\mathbf{e}_2\mathbf{e}_3 = i$ and relate bivectors to their vector dual, for example $\mathbf{e}_1\mathbf{e}_2 = i\mathbf{e}_3$.

any bivector $\mathbf{a} \wedge \mathbf{b}$ can be expanded in a basis of $n(n-1)/2$ unit bivectors. In three dimensions,

$$\mathbf{a} \wedge \mathbf{b} = (a_xb_y - a_yb_x)\mathbf{e}_1\mathbf{e}_2 + (a_yb_z - b_ya_z)\mathbf{e}_2\mathbf{e}_3$$

$$+ (a_zb_x - a_xb_z)\mathbf{e}_3\mathbf{e}_1. \quad (7)$$

Whereas the vector $\mathbf{e}_1$ corresponds to a directed line segment of unit length, the bivector $\mathbf{e}_1\mathbf{e}_2 = \mathbf{e}_1 \wedge \mathbf{e}_2 = -\mathbf{e}_2\mathbf{e}_1$ represents an oriented plane patch of unit area that contains both $\mathbf{e}_1$ and $\mathbf{e}_2$ (see Fig. 1). Its orientation is the sense of the rotation from $\mathbf{e}_1$ directly to $\mathbf{e}_2$. In three dimensions, the bivectors $\mathbf{e}_1\mathbf{e}_2$, $\mathbf{e}_2\mathbf{e}_3$, $\mathbf{e}_3\mathbf{e}_1$ span a linear space distinct from the original vector space spanned by $\mathbf{e}_1$, $\mathbf{e}_2$, $\mathbf{e}_3$. The unit vectors square to 1 [see Eq. (2)], but the basis bivectors square to -1, for example, $(\mathbf{e}_1\mathbf{e}_2)^2 = \mathbf{e}_1\mathbf{e}_2\mathbf{e}_1\mathbf{e}_2 = -\mathbf{e}_1\mathbf{e}_1\mathbf{e}_2\mathbf{e}_2 = -1$. (One can identify Hamilton's hypercomplex units $\mathbf{i}, \mathbf{j}, \mathbf{k}$ as unit bivectors: $\mathbf{i} = \mathbf{e}_3\mathbf{e}_2$, $\mathbf{j} = \mathbf{e}_1\mathbf{e}_3$, $\mathbf{k} = \mathbf{e}_2\mathbf{e}_1$, but Hamilton used them as basis vectors in a notation still common today. The confusion of vectors and bivectors persists in the vector cross product.)

To uncover further geometrical significance of $\mathbf{e}_1\mathbf{e}_2$, note from Eq. (2) that it is an operator that rotates both $\mathbf{e}_1$ and $\mathbf{e}_2$ by the same right angle,

$$\mathbf{e}_1(\mathbf{e}_1\mathbf{e}_2) = \mathbf{e}_2, \ \mathbf{e}_2(\mathbf{e}_1\mathbf{e}_2) = -\mathbf{e}_1. \quad (8)$$

Any vector $\mathbf{v} = v_x\mathbf{e}_1 + v_y\mathbf{e}_2$ in the xy plane is similarly rotated. Furthermore, a linear combination of the identity and a right-angle rotation can produce a rotation by an arbitrary angle θ in the xy plane:

$$\mathbf{v}' = \mathbf{v}(\cos \theta + \mathbf{e}_1\mathbf{e}_2 \sin \theta) = \mathbf{v} \exp (\mathbf{e}_1\mathbf{e}_2 \theta). \quad (9)$$

Terms can be expanded in powers of θ to prove the last equality. To rotate a vector $\mathbf{r} = x\mathbf{e}_1 + y\mathbf{e}_2 + z\mathbf{e}_3$ with a nonvanishing component z perpendicular to the rotation plane, one can use

$$\mathbf{r}' = \exp (\mathbf{e}_2\mathbf{e}_1\theta/2) \ \mathbf{r} \exp (\mathbf{e}_1\mathbf{e}_2\theta/2). \quad (10)$$

[Elements of the standard matrix representation of $\exp(\mathbf{e}_1\mathbf{e}_2\theta/2)$ are known in mechanics as Caley–Klein parameters.] If $\theta = \omega t$, Eqs. (9) and (10) describe rotations at constant angular velocity ω.

Bivectors can also be used for reflections. For example,

$$(\mathbf{e}_1\mathbf{e}_2) \ \mathbf{r} \ (\mathbf{e}_1\mathbf{e}_2) = x\mathbf{e}_1 + y\mathbf{e}_2 - z\mathbf{e}_3 \quad (11)$$

is the reflection of $\mathbf{r}$ in the xy plane. Two successive reflections in intersecting planes can be seen to be equivalent to a rotation by twice the angular opening between the planes.

The product $\mathbf{e}_1\mathbf{e}_2\mathbf{e}_3$ is a trivector representing an oriented unit volume. There is only one linearly independent trivector in $\mathbb{R}^3$ since from Eq. (2), $\mathbf{e}_1\mathbf{e}_2\mathbf{e}_3 = -\mathbf{e}_2\mathbf{e}_1\mathbf{e}_3 = \mathbf{e}_2\mathbf{e}_3\mathbf{e}_1 = \ldots$. Its square is -1 and it commutes with all the basis vectors, and hence with all elements of the algebra. It can be identified with the unit imaginary,

$$\mathbf{e}_1\mathbf{e}_2\mathbf{e}_3 = i. \quad (12)$$

The identification (12) associates bivectors with imaginary vectors directed normal to the plane, for example, $\mathbf{e}_1\mathbf{e}_2 = \mathbf{e}_1\mathbf{e}_2\mathbf{e}_3\mathbf{e}_3 = i\mathbf{e}_3$ (see Fig. 1). The vector $\mathbf{e}_3$ is called the vector dual to the bivector $\mathbf{e}_1\mathbf{e}_2$; it is the axis of the rotation plane. More generally [see Eq. (7)], the vector cross product $\mathbf{a} \times \mathbf{b}$ is the vector dual to the bivector $\mathbf{a} \wedge \mathbf{b}$:

$$\mathbf{a} \wedge \mathbf{b} = i\mathbf{a} \times \mathbf{b}. \quad (13)$$

As an application, note how easily any vector $\mathbf{v}$ is split into parts parallel and perpendicular to an arbitrary unit vector $\mathbf{e}$:

$$\mathbf{v} = \mathbf{vee} = (\mathbf{v} \cdot \mathbf{e})\mathbf{e} + (\mathbf{v} \wedge \mathbf{e})\mathbf{e}$$
$$= (\mathbf{v} \cdot \mathbf{e})\mathbf{e} - (\mathbf{v} \times \mathbf{e}) \times \mathbf{e}. \quad (14)$$

Like the cross product itself, the dual relationship between bivectors and vectors is meaningful only in three dimensions. Bivectors, however, are used for rotations and reflections in Cartesian spaces of any dimension $n \geq 2$.

Because of Eq. (12), every element p in the Pauli algebra is the sum of a complex scalar and a complex vector in a four-dimensional space. It can be expressed as a complex linear combination of what is called the paravector basis $\{\mathbf{e}_0, \mathbf{e}_1, \mathbf{e}_2, \mathbf{e}_3\}$, where for convenience one defines $\mathbf{e}_0 = 1$:

$$p = p_0\mathbf{e}_0 + p_x\mathbf{e}_1 + p_y\mathbf{e}_2 + p_z\mathbf{e}_3. \quad (15)$$

If the coefficients p_0, p_x, p_y, p_z are real, p is a real paravector, and such elements can represent spacetime vectors in relativity. Their boosts (velocity transformations) are simply rotations [Eq. (10)] in spacetime planes containing the time axis $\mathbf{e}_0$. The spacetime metric is built into the scalar norm

$$p\bar{p} = p_0^2 - p_x^2 - p_y^2 - p_z^2, \quad (16)$$

where $\bar{p} = p_0\mathbf{e}_0 - (p_x\mathbf{e}_1 + p_y\mathbf{e}_2 + p_z\mathbf{e}_3)$. The momenta of light signals are spacetime vectors p with null norms ($p\bar{p} = 0$); such p can be written as a real scalar times $(1 + \hat{\mathbf{p}})$, where $\hat{\mathbf{p}}$ is a unit vector: $\hat{\mathbf{p}}\hat{\mathbf{p}} = 1$. They are nonzero elements whose inverse $p^{-1} = \bar{p}/(p\bar{p})$ does not exist. They have no counterpart in the field of real numbers.

The center (commuting part) of the Pauli algebra is the complex field, and the even subalgebra, containing only products real scalars and bivectors, is the quaternion algebra. Clifford algebras are also readily found for n-dimensional vector spaces with basis vectors $\mathbf{e}_j$ satisfying $\mathbf{e}_j \cdot \mathbf{e}_k = \frac{1}{2}(\mathbf{e}_j\mathbf{e}_k + \mathbf{e}_k\mathbf{e}_j) = 0$, $j \neq k$ and

$$\mathbf{e}_j^2 = \begin{cases} +1, & j = 1, \ldots, p \\ -1, & j = p+1, \ldots, n \end{cases}. \quad (17)$$

Clifford himself worked only on cases with $p = n$ and $p = 0$. Elements of the algebra, sometimes called cliffors or Clifford numbers, reside in a 2^n-dimensional real vector space with parts comprising a scalar, an n-dimensional vector, an $n(n-1)/2$-dimensional bivector, and higher-order multivectors.

The field of complex numbers is the Clifford algebra with $n = 1$ and $p = 0$, and the quaternion algebra is that with either $n = 2$ or $n = 3$ (the two algebras are equivalent) and $p = 0$. The Pauli algebra discussed above has $n = p = 3$. Physicists commonly use representations of the algebra in relativity, electromagnetism, and quantum mechanics, especially in treatments of light polarization and electron spin. In Dirac's theory of the electron, the algebra of the Dirac matrices is the Clifford algebra with $n = 4$ and $p = 1$. Only Clifford algebras with $2p - n = 3 + 4N$ for any integer N share the Pauli-algebra property that they possess a natural complex structure in a space of 2^{n-1} dimensions.

See also: ELECTROMAGNETISM; PAULI; WOLFGANG; POLARIZED LIGHT; SCALAR; SPACETIME; SPIN, ELECTRON; VECTOR

Bibliography

BAYLIS, W. E. "Why i?" *Am. J. Phys.* **60**, 778–797 (1992).

HESTENES, D. *Space-Time Algebra* (Gordon and Breach, New York, 1966).

HESTENES, D., and SOBCZYK, G. *Clifford Algebra to Geometric Calculus* (Reidel, Dordrecht, 1984).

PORTEOUS, I. *Clifford Algebras and Classical Groups* (Cambridge University Press, Cambridge, Eng., 1995).

RIESZ, M. *Clifford Numbers and Spinors: With Riesz's Private Lectures to E. Folke Bolinder and a Historical Review by Pertti Lounesto* (Kluwer, Dordrecht, 1993).

WILLIAM E. BAYLIS

CLOCK

See ATOMIC CLOCK

CLOUD, GRAY

Often, on a summer day, towering cumulus clouds can be seen from some distance away. These thunderheads typically appear as billowing white folds extending upward from a flat base that appears gray. An observer directly under the cloud would look up to see a dark gray cloud. The same cloud can appear white or gray, depending on the relative positions of the cloud, the sun, and the observer. In the example above, the base of the cloud is dark because the top of the cloud is reflecting almost all the sunlight illuminating it, so very little passes through.

In general, a cloud appears white if it is illuminated from the side viewed by the observer, and gray if illuminated from the opposite side. The shade of gray would depend on the thickness and density of the cloud. At the edges of a dark cumulus cloud, where the cloud is thinner, the sunlight may shine through, giving a "silver lining." However, there are many complicating factors that can alter the situation. For example, a dark cloud, or the dark bottom of a thunderhead, may appear bluish when viewed from a great distance. This coloration is due to the "air light," or atmospheric scattering between the cloud and the observer; blue light is preferentially scattered into the line of sight. Such an effect is complementary to the yellowish appearance of white clouds at a great distance. Gray storm clouds can be further darkened by dust that has been carried aloft by high winds, and also tinged with a yellowish or brownish color by that suspended dust.

Even more subtle coloration effects occur when clouds are illuminated by light reflected from the surface of the earth beneath them. Most people have seen the nighttime glow of cloud layers over urban areas, which, of course, is simply the reflection of artificial lighting back toward the surface. Similar effects can occur even without artificial lighting. Cloud layers can reflect the sharp demarcation between snow-covered land areas and adjacent darker bodies of water. Apparently, in arctic oceans, this lighter area in the clouds can warn ships of the approach of pack ice even before the ice itself is visible. The underside of clouds also can be tinged with distinct colors from various surfaces below them, in the case of desert sands or greenish ocean colors or even the purple of heather in bloom over extensive areas.

See also: CLOUD, WHITE; REFLECTION; SKY, COLOR OF

Bibliography

FALK, D.; BRILL, D.; and STORK, D. *Seeing the Light* (Harper & Row, New York, 1986).
MINNAERT, M. *The Nature of Light and Colour in the Open Air* (Dover, New York, 1954).
WALDMAN, G. *Introduction to Light* (Prentice Hall, Englewood Cliffs, NJ, 1983).

GARY WALDMAN

CLOUD, WHITE

Billowing white clouds in the sky are a common sight. Although many people have appreciated their beauty, few have really thought about why they appear as they do. We know they are composed of small droplets of water suspended in the air, but we also know that a single droplet of water, or indeed a large drop of water, is nearly transparent. Higher, wispy looking clouds, like cirrus clouds, are composed of ice crystals. But single ice crystals are translucent, also passing most of the light hitting them. Why then do the clouds appear to be an opaque white?

At the boundary between air and water, only about 2 percent of the incident unpolarized light reflects if the light strikes perpendicular to the surface; the remainder of the light passes through the surface. This small reflectance is why water appears relatively transparent. At other angles of incidence, the reflected portion of the light is larger, but unless the light impinges nearly parallel to the surface, more will pass through than reflect. However, when sunlight striking suspended water droplets is considered, it must be remembered that the light transmitted at the first surface will eventually hit another side of the droplet where another fraction will be reflected. When there are myriads of water droplets in every small volume of space, the compounding of small fractions of the light reflected at each surface quickly amounts to nearly total reflection. It is the many reflecting surfaces rather than the reflectance of each that produce the effect. Since the scattering particles are much larger than a wavelength of light, there is no wavelength preference in the reflection, and the volume of space filled with suspended droplets appears an opaque white. It may be noted that the same principles apply to other finely divided transparent or translucent particles, such as snow, salt, and sugar; piled together they are opaque white. Even white paper on a microscopic scale is composed of translucent cellulose fibers piled on top of each other.

The white color of clouds can be modified by the scattering action of the air itself. Everyone has seen the beautiful pink, red, and orange colors of clouds at sunset or sunrise. At those times, when the sun is low in the sky, the rays of the sun must travel a long path through the atmosphere before they strike clouds near the observer. The atmosphere selectively scatters short wavelengths from the sunlight (which is why the sky is blue), so that the light that reaches the clouds directly is relatively deficient in shorter wavelengths (violet and blue) and relatively rich in longer wavelengths (red and orange). This reddish light is then reflected by the clouds, imparting its color to them. Even when the sun is high in the sky, atmospheric scattering can change the white color of clouds if those clouds are very far from the observer on the horizon. In this case the sunlight reflecting from the clouds is white, but in the long path from the cloud to the observer some of the light at shorter wavelengths is removed by scattering so that what remains is yellowish in color. That color is seen as the color of the clouds.

See also: CLOUD, GRAY; REFLECTION; SKY, COLOR OF

Bibliography

FALK, D., BRILL, D., and STORK, D. *Seeing the Light* (Harper & Row, New York, 1986).

MINNAERT, M. *The Nature of Light and Colour in the Open Air* (Dover, New York, 1954).

WALDMAN, G. *Introduction to Light* (Prentice Hall, Englewood Cliffs, NJ, 1983).

GARY WALDMAN

CLOUD CHAMBER

The cloud chamber, invented by Charles T. R. Wilson in 1912, played a vital role in the early development of particle physics. This detector produces visible trajectories (tracks) when charged particles traverse the apparatus. It was used to discover the positron in 1933 and was used extensively in the early studies of mesons seen in cosmic rays. This type of particle detector works by suddenly decom-

pressing a container filled with a vapor, cooling the vapor adiabatically so that it becomes supersaturated. Once the gas is supersaturated, any areas of nonuniformity can form regions of condensation that can be observed visually or photographed.

The two important requirements for drop formation in the cloud chamber are that the gas is supersaturated and that nucleation centers are present. It was Lord Kelvin who realized that because of surface tension, the vapor pressure is greater at the surface of a liquid drop than at a plane surface of the same liquid. When a drop is very small, evaporation will rapidly make the drop become smaller, thereby reducing the surface area. This means that under normal conditions, any small drops formed will quickly disappear even when the vapor pressure of the gas under consideration is high. If the drop carries a surface charge, the surface tension is modified so that small drops grow rather than evaporate. The polar nature of the vapors used will influence this process. For example, water vapor will condense preferentially on negative ions and ethyl alcohol will condense on positive ions. Such effects cause the degree of supersaturation needed to form visible tracks to be significantly higher for water than for ethyl alcohol.

When a charged particle of sufficient energy traverses a gas, it will ionize gas molecules along its path. These ions will be the nucleation centers on which droplets will form, giving rise to the visible track seen in the cloud chamber. Since the ionization is velocity dependent, a measurement of the amount of ionization can determine the velocity of the particle, allowing one to calculate its mass if the momentum of the charged particle is also known. The momenta of the charged particles can be measured by placing the chamber in a magnetic field and measuring the curvature of the photographed tracks. It is important that the vapor in the cloud chamber be dust free, or the vapor will condense around the dust particles rather than the ionized gas regions.

The expansion can be accomplished by the sudden movement of a piston or membrane. There are four important time periods for a cloud chamber:

1. Expansion time, which should be as short as possible, typically 10 to 20 ms.
2. Drop growth time, after which a flash photograph would be taken, typically around 100 ms.
3. Recovery time needed to bring the chamber back to its original state and be ready for the next expansion. This time ranges from 1 to 10 min. and is one of the main drawbacks of the cloud cham-

ber. This time can be shortened to about one-half minute by using a compression cycle to heat up the gas and speed the re-evaporation of the drops that have formed.

4. Sensitive time, the time after expansion during which the chamber is capable of producing droplets from charged particles passing through the vapor. This time varies from a fraction of a second for small chambers to seconds in high-pressure chambers. Since the sensitive time is long, the expansion of the chamber can be triggered by the particle being photographed by a fast electronic detector (e.g., a Geiger–Mueller tube). Tracks obtained in this manner are usually not high quality due to nonuniformity of the gas expansion and gas motion in the time interval between the triggering of the expansion and the photographing of the tracks. Ions from previous events can be removed by an electrostatic clearing field produced by wires strung through the chamber with high voltage applied to them.

A variation on the cloud chamber is known as the diffusion cloud chamber, which was originally suggested by Alexander Langsdorf Jr. in 1939. This type of detector remains sensitive continuously by setting up a temperature gradient so that vapor will diffuse from a heated top portion of the chamber to a cooled bottom portion. The sensitive region of these devices will be the region just above the cooled bottom of the chamber, which is usually a few inches in extent. Here the problem is to maintain a stable and turbulence-free system.

A simple recipe for building a diffusion cloud chamber can be found in *Scientific American* (January 1951, p. 29):

1. Fill a flat, round container with dry ice.
2. Cover with a metal disk 5–17 in. in diameter, with a black velvet cloth on top.
3. Place a glass cylinder (open on both ends) on this disk.
4. Cover the glass cylinder with a second metal disk, with a piece of alcohol-soaked felt attached to the bottom of it.
5. Place a tray of water on top of the second metal disk (4).

The chamber will be continuously sensitive to cosmic rays passing through it. The tracks should be visible against the black velvet background when the chamber is illuminated with a flashlight.

Clearing electrodes can be placed in the diffusion cloud chamber that can be turned off when one wants the chamber to allow tracks to be photographed. These type chambers can operate at roughly 5-s cycle times, significantly higher than expansion-type chambers. This cycling time was important for the devices to be used in particle beam accelerator experiments. Since the sensitive region was horizontally extended in size, it was well suited to use in such particle beams.

The cloud chamber was replaced by the bubble chamber in particle physics due to the bubble chamber's higher density liquid media and its higher repetition rates. While the diffusion chamber has reasonable cycle times, they have only a relatively thin sensitive area, and the interaction rate is significantly lower in a gas as compared to a liquid. The one advantage that remained of the cloud chamber over the bubble chamber was the fact that the cloud chamber could be triggered by the particle that traverses the detector. This is not possible for the bubble chamber due to the very short time the superheated liquid is sensitive to charged particles. Thus, while the cloud chamber is rarely used today, it was an important device in the early development of the field of particle physics and can still be used to visually verify the presence of cosmic rays with a fairly simple apparatus.

See also: COSMIC RAY; IONIZATION; IONIZATION CHAMBER; PARTICLE PHYSICS

Bibliography

HENDERSON, C. *Cloud and Bubble Chambers* (Methuen, London, 1970).

MARTON, L. *Methods of Experimental Physics*, Vol. 5: *Nuclear Physics* (Academic Press, New York, 1961).

RITSON, D. M. *Techniques of High-Energy Physics* (Interscience, New York, 1961).

WILSON, C. T. R. *The Principles of Cloud-Chamber Technique* (Cambridge University Press, Cambridge, Eng., 1951).

L. DONALD ISENHOWER

CMBR

See COSMIC MICROWAVE BACKGROUND RADIATION

COAXIAL CABLE

A coaxial cable is a type of transmission line that consists of two cylindrical conductors whose centers are collinear, separated by air or some other dielectric. (Cross-sectional and cutaway views are shown in Fig. 1.) Its most important characteristics are a lack of susceptibility to the effects of nearby external electromagnetic fields and negligible external fields. Coaxial cables have been used extensively up to the present time for radio-frequency transmission lines and also for multichannel telephone carriers and television lines. The development of fiber optic cables is making the use of coaxial cables for many applications obsolete.

The exact construction of a coaxial cable depends on the intended application. The inner conductor could be made of solid annealed copper wire or of stranded wire. The signal is attenuated less by the solid wire, but the stranded wire permits greater flexibility of the finished product.

The dielectric (insulating) core material is generally made of polyethylene-based or Teflon derivatives. For the various types of polymers used,

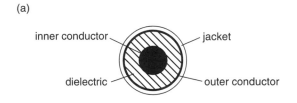

(a)

inner conductor — jacket

dielectric — outer conductor

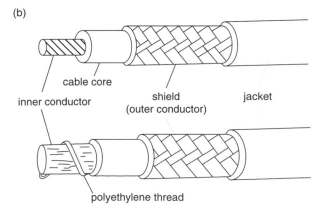

(b)

cable core

inner conductor

shield (outer conductor)   jacket

polyethylene thread

Figure 1 Coaxial cable in both (a) cross-sectional and (b) cutaway views.

dielectric constants vary from about 1.5 to 2.5. The dielectric constant is important because it determines the mass and size of the finished cable. Lower dielectric constants usually result in smaller and lighter cables. The dielectric constant can be made lower by using air-spaced (semisolid) cores. These are constructed by wrapping a filament of polyethylene in a spiral fashion around the inner conductor and then adding a thin tube of the same material leaving air spaces around the conductor. This is particularly advantageous if the cable is to be suspended in air.

The outer conductor usually consists of groups of very small diameter copper wire that are plated together and wrapped around the dielectric core. This is covered by a polyethylene or vinyl jacket that is generally moisture resistant and able to withstand a fairly wide range of temperatures and pressures so that, if necessary, the cable could be laid underground.

Electrical Characteristics

All coaxial cables have a characteristic impedance, which is the total resistance that the cable offers to the flow of current. This is due to the resistance of the conductors themselves and to the fact that the cable acts as both a capacitor and an inductor. For the cable to transmit the signal effectively, the transmitter, the cable, and the receiver of the signal should have impedances that match. This requirement also applies to the devices (plugs, jacks, etc.) that are used to terminate the cables. If this is not the case there may be significant losses due to reflection. This is similar to the partial reflection that occurs when a traveling wave on a slinky meets a boundary to another slinky of different density and stiffness. Part of the energy of the wave passes into the new medium, but part is reflected back where it may interfere with the incoming waves.

Other important electrical characteristics include the speed of propagation of the signal, which is inversely proportional to the dielectric constant of the core insulator and the cable attenuation. Loss of energy is due mainly to the resistance of the conductors, which can be decreased by using larger diameter wire at the cost of extra weight.

See also: CAPACITOR; CONDUCTOR; DIELECTRIC CONSTANT; DIELECTRIC PROPERTIES; FIBER OPTICS; INDUCTOR; OPTICAL FIBER

Bibliography

LORRAIN, P., and CORSON, D. *Electromagnetic Fields and Waves,* 2nd ed. (W. H. Freeman, San Francisco, 1970).

MICHELS, W. C. *Electrical Measurements and Their Applications* (Van Nostrand, New York, 1957).

GRACE A. BANKS

COBE

See COSMIC BACKGROUND EXPLORER SATELLITE

COHERENCE

Coherence refers to the spatial or temporal correlation between electromagnetic waves, or between the quantum states of particles. The most familiar examples of coherence are found in optical experiments on interference, diffraction, and the statistical properties of laser light, but the study of coherent states of atoms and subatomic particles is also an active field of modern physics.

The basic concepts of coherence are exemplified in the two-slit interference experiment. A monochromatic light source, wavelength λ, illuminating two slits separated by a distance d, will produce a pattern of regularly spaced minima and maxima on a screen a distance L from the slits; the maxima are separated by a distance $\Delta = \lambda L/d$. The sharpness of these interference fringes can be characterized by the fringe visibility, defined by

$$V = \frac{I_{max} - I_{min}}{I_{max} + I_{min}},$$

where I_{max} and I_{min} are, respectively, the maximum and minimum intensities in the interference pattern. Light for which $V = 0$ is said to be incoherent; a light source with $V = 1$ is completely coherent;

and visibility values between 0 and 1 indicate partial coherence (see Fig. 1.)

A beam of light is spatially coherent when the phase difference between any two points on the wave front remains constant in time, even if the phase fluctuates randomly at any one point. An extended source consisting of many randomly fluctuating point sources can also produce spatially coherent light if the interference fringes from nearby point sources accidentally overlap. The starlight of wavelength λ is coherent over a circular area of diameter d_{coh} given by

$$d_{coh} = 0.16 \frac{\lambda R}{\rho} = 0.16 \frac{\lambda}{\nu},$$

where the star diameter is ρ, the distance to the observer is R, and ν is the angle subtended by the object at the observer. The coherence diameter of the Sun is approximately $d_{coh} = 0.02$ mm, while light from the distant star Betelguese has $d_{coh} = 80$ cm.

When the phase relationships within a light beam are constant over time, the light is said to be temporally coherent; this implies that it is also spectrally pure. For a source emitting light of frequencies ranging from $\nu_0 - \delta\nu/2$ to $\nu_0 + \delta\nu/2$, and a bandwidth of $\delta\nu$, the crests and troughs of the extreme

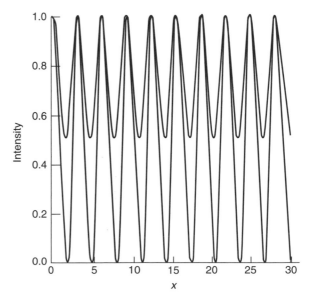

Figure 1 Graph of a two-slit interference pattern for the case of complete coherence ($V = 1$, thin line) and partial coherence ($V = 0.5$, dark line).

221

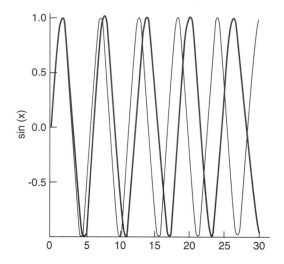

Figure 2 Amplitude of sinusoidal waves, representing optical waves from two atomic emitters that differ in frequency from one another by 10 percent. Note that after only five periods the waves are completely out of phase with each other.

frequencies will eventually get out of synchronization or out of phase with respect to each other, as illustrated in Fig. 2. The coherence time for a light source is usually defined as

$$\tau_{\text{coh}} = \frac{1}{2\pi\delta\nu}.$$

For a typical thermal light source, the coherence time is of the order of 10^{-10} s, and the coherence length, defined as $c\tau_{\text{tcoh}}$, may be a few centimeters. For carefully constructed lasers, on the other hand, coherence lengths may be many kilometers!

The physics of light emission largely determines the coherence properties of a light source. In a thermal source (incandescent bulb or a gas-discharge lamp), light is produced by spontaneous emission at random times. In such a source, the wave packets from different emission events are almost uncorrelated, and the degree of coherence is low. In lasers, however, light is produced by stimulated emission, guaranteeing a high degree of spatial as well as temporal coherence.

All light sources, including lasers, emit light with a characteristic frequency bandwidth $\delta\nu$. When the bandwidth satisfies $\delta\nu/\nu \ll 1$, it is said to be quasi-monochromatic. For a laser, the bandwidth $\delta\nu/\nu \approx 10^{-7}$, while for an arc-lamp, at intensities sufficient to carry out a reasonable experiment, $\delta\nu/\nu \approx 0.1$. This spread in frequencies tends to "wash out" the fringes in interference experiments and reduce the visibility V. In a laser cavity, temporally and spatially independent transverse and longitudinal spatial modes can also coexist simultaneously. This multimode laser emission is less coherent than single-mode laser light.

Applications

Holography is one of the most important applications of coherence, both in science and technology. It is now possible to do holography with electron beams as well as light. Interferometry is widely employed in industry to measure surface quality, film thicknesses, and small variations in thickness; it is an indispensable technique in radio and visible astronomy.

When an expanded laser beam strikes a surface with small-scale irregularities, the reflected light (see Fig. 3) exhibits sharply defined, bright and dark patterns originating in constructive and destructive interference between neighboring reflective microfacets of the surface. The speckle patterns can be analyzed to determine surface roughness, follow surface deformation, crack growth and vibrations in real time, and even search for surface irregularities in integrated circuits. Speckle interferometry is now a well-established industrial diagnostic technique.

History

The foundation for the idea of coherence was laid by Thomas Young's studies of interference in the early nineteenth century. Albert Michelson's pathbreaking interferometry studies led him to connect the visibility of interference fringes to both spatial and spectral coherence in the light from both terrestrial and astronomical sources. Coherence phenomena in particles were observed in experiments on electron-surface and ion-surface scattering. Major contributions to our understanding of higher-order optical coherence effects were made by Robert Hanbury Brown, Richard Twiss, Leonard Mandel, and Emil Wolf. The idea of coherent quantum states introduced by Roy Glauber put the quantum statistics of light waves and particles on essentially equal footing.

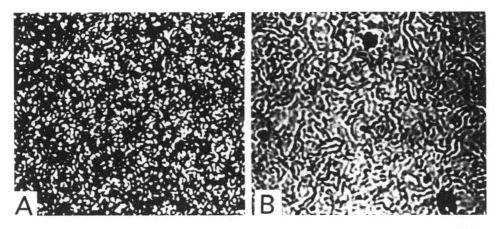

Figure 3 Speckle patterns formed by a helium-neon laser scattering from the rough surfaces of two ground-glass samples: (a) roughness 1.3 μm; (b) roughness 0.2 μm.

Experiments

Interference or diffraction experiments can reveal the degree of coherence of either incoherent or laser sources, by measuring the fringe visibility V. Coherence lengths of both lasers and incandescent sources are measurable by splitting a light beam into two parts, introducing a relative delay between the split beams, and determining what path delay is required to extinguish interference fringes. Studies of optical coherence now frequently use correlated pairs of photons generated by converting one photon of frequency 2ω to two photons of frequency ω in nonlinear crystals in a process called down-conversion.

See also: DIFFRACTION; DOUBLE-SLIT EXPERIMENT; ELECTROMAGNETIC WAVE; EMISSION; HOLOGRAPHY; INTERFEROMETRY; MICHELSON, ALBERT ABRAHAM; PHOTON; QUANTUM STATISTICS; YOUNG, THOMAS

Bibliography

BERTOLOTTI, M. *Masers and Lasers: An Historical Approach* (Briston, England, 1983).

HORGAN, J. "Quantum Philosophy," *Sci. Am.* **270** (7), 94–101 (1992).

MEYER-ARENDT, J. R. *Introduction to Classical and Modern Optics*, 4th ed. (Prentice Hall, Englewood Cliffs, NJ, 1994).

MILONNI, P. W., and EBERLY, J. H. *Lasers* (Wiley, New York, 1988).

RICHARD F. HAGLUND JR.

COHESION

Cohesion is the property of a substance to hold itself together as a result of the intermolecular, primarily electrostatic, forces between its constituent molecules. Solids exhibit the greatest cohesion; gases under normal conditions have negligible cohesion. Liquids, as may be expected from their amorphous nature, are characterized by weak bonding forces between their molecules. The cohesion of liquids is, therefore, intermediate between that of gases and that of solids. Molecules of a liquid in a container may be more attracted to the molecules of the container than they are to the like molecules of the liquid. This attraction to a dissimilar substance is called adhesion. Cohesion and adhesion are thought to be essentially the same phenomenon differing only in whether the attraction is between molecules of the same or different kinds.

If the atoms or molecules of a liquid experience a greater Coulomb (electrical) force toward the different atoms or molecules of a material dipped into the liquid than they do toward their own kind, then the adhesion is greater than the cohesion and the liquid is said to "wet" the other substance. Molecules in contact with the dissimilar substance break away from their liquid to attach themselves to the other material.

Left to itself in space, a liquid will draw together into a spherical shape as a result of cohesion pulling its molecules into the most compact configuration.

A sphere is also the shape that minimizes the surface area enclosing a volume. It is as if the surface had its own tension forces that resisted stretching, and pulled toward a minimum area.

The concept of surface tension in relation to the forces of cohesion and adhesion is nicely illustrated by a liquid in a glass tube. If the liquid wets the glass, as does water, its surface will be drawn up where it contacts the glass and the surface will be concave as viewed from above. In the less common case where the cohesion of a liquid is much greater than its adhesion to the glass, the surface will be convex, as is the case for mercury.

To break a body apart into its constituent atoms or molecules requires the expenditure of work against the forces of cohesion. The total work required to break a substance completely into dissociated atoms in their ground states is defined as the cohesive energy.

See also: ADHESION; ELECTROSTATIC ATTRACTION AND REPULSION; LIQUID; SURFACE TENSION

Bibliography

MACPHERSON, A. K. *Atomic Mechanics of Solids* (North-Holland, Amsterdam, 1990).

RICHARD B. HERR

COLLECTIVE EXCITATION

See EXCITATION, COLLECTIVE

COLLISION

A collision is a relatively brief but intense interaction between particles or extended objects while they are in close proximity. In physics, this term does not imply that the colliding objects necessarily come into physical contact; rather, it applies to situations where interactions between colliding objects are weak except during a relatively short time interval when the objects interact much more strongly with each other than with anything external. This general definition thus applies not only to two billiard balls colliding on a pool table but also to subatomic particles interacting in a particle physics experiment, a space probe passing a planet, two galaxies passing through each other, and so on.

Conservation of Energy and Momentum

If the system of colliding objects is isolated from its surroundings (so that external forces on the objects are negligible), momentum and energy are conserved, meaning that both the total momentum of the system and its total energy have the same value after the collision as before. These conservation laws allow one to determine much about the motion of the objects in the system without having to know anything about the interactions involved in the collision. For example, if two isolated objects have (nonrelativistic) velocities $\mathbf{u}_1$ and $\mathbf{u}_2$ before colliding and velocities $\mathbf{v}_1$ and $\mathbf{v}_2$ afterwards, then conservation of momentum and energy imply that:

$$m_1 u_{1x} + m_2 u_{2x} = m_1 v_{1x} + m_2 v_{2x} \tag{1a}$$

$$m_1 u_{1y} + m_2 u_{2y} = m_1 v_{1y} + m_2 v_{2y} \tag{1b}$$

$$m_1 u_{1z} + m_2 u_{2z} = m_1 v_{1z} + m_2 v_{2z} \tag{1c}$$

$$\tfrac{1}{2} m_1 u_1^2 + \tfrac{1}{2} m_2 u_2^2 + U_i = \tfrac{1}{2} m_1 v_1^2 + \tfrac{1}{2} m_2 v_2^2 + U_f, \tag{2}$$

where m_1 and m_2 are the masses of the interacting objects; U_i represents the internal energy of the system (which includes potential energy, thermal energy, rotational energy, and so on) before the collision and U_f the same afterwards; u_{1x}, u_{1y}, and u_{1z} are the x, y, and z components of the velocity $\mathbf{u}_1$; and so on. If one is given $\mathbf{u}_1$, $\mathbf{u}_2$, and $\mathbf{v}_1$, these four equations allow one to determine the three unknown components of $\mathbf{v}_2$ and the change $U_f - U_i$ in the internal energy of the system without knowing anything about the details of the collision process.

These conservation laws apply to any isolated system, but realistic systems are often not well-isolated from their surroundings. However, the special nature of a collision means that we can often apply these conservation laws even when the colliding objects are not well isolated. Because colliding objects interact very intensely and for only a short period of time, the effect of external interactions during the collision process is negligible compared to the effect of the colliding objects on each other. Therefore, even if a system is not very well isolated from its surroundings, its energy and momentum just before a brief collision process will be approximately the same as just after the collision.

Inelastic Collisions

A collision is said to be elastic if the internal energy U of the system is unchanged by the collision process: Eq. (2) implies that the kinetic energy of the system is conserved in an elastic collision. A collision process that converts kinetic energy into internal energy (or vice versa) is called an inelastic collision.

An inelastic collision typically converts some of the kinetic energy of the system to thermal energy. For example, imagine a system of two identical balls of putty that approach each other with equal speeds but opposite directions, collide, and stick together, forming a single mass of putty. Because the balls initially have equal masses but opposite velocities, they have zero total momentum before the collision. Conservation of momentum requires that the resulting single mass must have zero momentum as well, which is possible only if it is at rest. Therefore, this system has nonzero kinetic energy before the collision, but zero kinetic energy afterward, implying that kinetic energy must have been been converted to internal energy in this collision. In this case, the kinetic energy gets converted almost entirely to thermal energy: One would find the temperature of the final mass to be larger than the average temperature of the original balls.

Other examples of inelastic collisions include a diver leaping into a pool of water, an automobile running into a brick wall, a dropped book slamming into the floor, comet Shoemaker–Levy colliding with Jupiter, and the collision of galaxies. Collisions where the colliding objects stick together after the collision are sometimes called perfectly inelastic collisions.

While inelastic collisions typically convert kinetic energy primarily into thermal energy, inelastic collisions can convert some kinetic energy to sound, light, or other forms of energy. For example, at least some of the kinetic energy of a book dropped on the floor is carried away by sound waves moving through both the air and the floor (as evidenced by the loud "slam" that we hear and vibrations in the floor that we may feel through our feet).

Sometimes an inelastic collision converts internal energy to kinetic energy. For example, the collision of two molecules might initiate a chemical reaction that converts the internal energy of the reacting molecules partly into kinetic energy of the product molecules. An inelastic collision that releases internal energy is sometimes called a superelastic collision.

Elastic Collisions

A collision may convert kinetic energy into potential energy during the collision process, but if it is elastic, this potential energy will be converted back to kinetic energy before the process ends. For example, a rubber ball thrown against a wall deforms during impact, converting the kinetic energy of the ball into the potential energy of deformation. But as the ball rebounds, this potential energy is converted (almost entirely) back into kinetic energy, and the ball springs back from the wall with almost the same kinetic energy as when it hit.

Few collisions in the real world are perfectly elastic: most entail some net conversion of the kinetic energy of the system into other forms of energy. Even so, some realistic collision processes approach this ideal. Examples include two rubber balls colliding in midair, two billiard balls colliding on a pool table, or a space probe passing near a planet.

Quantum mechanics constrains some kinds of microscopic collisions to be perfectly elastic. For example, the kinetic energy of two colliding helium atoms can be converted to internal energy only if there is enough energy to bump at least one of the atoms up to its next-higher quantum energy level. At normal temperatures, helium atoms have nowhere near enough kinetic energy to do this, so the collision of any two helium atoms in a normal sample of helium gas will be perfectly elastic. Similar quantum effects make it impossible for electrons flowing in a superconductor to lose any kinetic energy in collisions with atoms in the superconductor, making it possible for the electrons to flow without resistance.

See also: ELASTICITY; ELASTIC MODULI AND CONSTANTS; ENERGY, CONSERVATION OF; MOMENTUM, CONSERVATION OF; QUANTUM MECHANICS

Bibliography

FRENCH, A. P. *Newtonian Mechanics* (W. W. Norton, New York, 1971).

MARION, J. B., and THORNTON, S. T. *Classical Dynamics* (Harcourt Brace Jovanovich, Orlando, FL, 1988).

THOMAS A. MOORE

COLOR

Light is the visible part of the electromagnetic spectrum. If white-appearing light from an incandescent light source such as the sun is passed through a prism, it will appear spread out as a spectrum—much like a rainbow. Color describes the different responses of the eye to different parts of this spectrum. For instance, the response of the eye and the brain to one end of the spectrum is called the color red. Then in order will appear the colors yellow, green, blue, and finally violet. Red is at the long-wavelength end and violet at the short-wavelength end of the visible part of the spectrum.

Objects illuminated by white light may appear colored either because they selectively reflect or transmit certain colors of light. For instance, if a transparent red dye is coated on glass, it will transmit red light and appear red when placed between an observer and a white light source. If, on the other hand, the dye is coated on white paper, it will appear red by reflection when illuminated by white light. In both instances the dye has eliminated all the light except red. The backing in this case determines whether the light is transmitted or reflected.

An object can exhibit color in one of at least four ways. The most common is absorption, in which an object exhibits color because the pigments or dyes (colorants) on the object's surface absorb all of the white light incident on it except that which is its color. For instance, if white light illuminates a green object such as a leaf, all of the light except the green light is absorbed by the pigment in the leaf and the green light is reflected from the leaf, making it appear green. Metals such as copper and gold and metallic paints also absorb light but in a way different from the pigments and dyes, so that metals generally appear more shiny (specular).

Objects can appear colored because of selective (Rayleigh) scattering of light. The sky appears blue and the sunrise or sunset red because the molecules of the atmosphere scatter light of short wavelength (blue) more than these molecules scatter light of long wavelength (red).

Most colorants when illuminated by ultraviolet light (shorter wavelengths than the visible part of the spectrum) are not visible. However, some colorants are fluorescent. When they are illuminated by ultraviolet light they emit light in the visible part of the spectrum.

Oil spread on water sometimes appears colored because of the interference of light reflected from the top surface of the oil and light reflected from the interface between the oil and the water. The visible color depends on the thickness of the film of oil. Because the film usually is not of uniform thickness or seen at different angles, a rainbow of colors is often seen.

The color of an incandescent light source such as a light bulb depends on the temperature of its filament. As the temperature increases, its color changes from red to yellow to white as relatively more short wavelength radiation is emitted by the filament. As the color of the light source changes, the color of objects illuminated by it also change, generally becoming more blue as the light becomes whiter.

The color of a light source consisting of vapor confined in a transparent envelope depends on the kind of gas and the pressure in the envelope. If the light from such a source is passed through a prism, the resulting spectrum primarily will be at discrete wavelengths separated by dark regions in the spectrum. Sodium vapor under modest pressure emits a bright yellow light; under higher pressure it becomes orange. Mercury vapor under high pressure emits a blue colored light. Fluorescent lamps consist of mercury vapor confined in a transparent tube at modest pressure. The walls of the tube are coated with a phosphor. When the mercury vapor is excited it emits radiation in the ultraviolet part of the spectrum, which excites the phosphor, causing it to emit radiation (fluoresce) in the visible part of the spectrum. The color of this light depends on the phosphor and typically ranges from blue-white to yellow-orange. Colored objects illuminated by any of these vapor sources can be markedly different from the way they appear in incandescent white light. For instance, an orange would appear orange illuminated in white light, orange to yellow in sodium light, black in mercury vapor light, and different in fluorescent light depending on the phosphors in the lamp.

Color professionals use at least three variables to characterize a color. In one system these are called hue, saturation, and lightness. Hue corresponds to the color of the dominant electromagnetic radiation in the reflected or transmitted radiation of the colorant. Saturation denotes the degree to which the color is pure and not diluted by gray. Lightness denotes the degree to which the color is not diluted by black. For instance, pink is red in hue, low in saturation, and high in lightness. The term "brightness" is often substituted for "lightness" when discussing colored lights.

The cones in the retina of the eye are responsible for the ability of a human to see color. The cones can be divided into three groups according to the spectral response of the pigment in the cones. One group of cones has a peak response toward the blue end of the spectrum, one peaks toward the red end of the spectrum, and the third in the green part of the spectrum. The relative stimulation of each group of cones serves to identify the color being observed. Because of the way the eye sees color, it is possible to take three colored lights, called additive primaries (e.g., red, green, and blue), and mix them in proper proportions to form most of the other colored lights that the eye can detect. The phosphors in a television picture tube are red, green, and blue as can be seen by looking carefully at the screen with a magnifying glass. A system of negative primaries can be used for mixing colorants to construct most of the colorants that the eye can detect. In photography and four-color printing these are the subtractive primaries yellow, magenta, and cyan.

See also: ELECTROMAGNETIC SPECTRUM; LIGHT; LIGHT, ELECTROMAGNETIC THEORY OF; NEWTON, ISAAC; RAINBOW; SCATTERING, RAYLEIGH

Bibliography

FALK, D.; BRILL, D.; AND STORK, D. *Seeing the Light* (Harper & Row, New York, 1986).

MITCHELL, E. *Photographic Science* (Wiley, New York, 1984).

OVERHEIM, R. D., and WAGNER, D. L. *Light and Color* (Wiley, New York, 1982).

WILLIAMSON, S. J., and CUMMINS, H. Z. *Light and Color in Nature and Art* (Wiley, New York, 1983).

EARL N. MITCHELL

COLOR CENTER

Color centers are the defects in some otherwise transparent solids that give them their characteristic coloration.

In general, light can do three things when it encounters a solid medium. It can be scattered, transmitted, or absorbed. All of these effects depend on the wavelength of the light and the characteristics of the medium.

In metals such as aluminum or iron, the almost free electrons that are responsible for the electrical conductivity also act as powerful scatterers of visible light. This is why most metals are shiny and reflective. In many insulating materials such as quartz, however, there are no microscopic mechanisms for either scattering or absorbing visible light, and the light can freely pass through the material. For this reason, perfect single crystals of materials such as quartz or table salt (NaCl) are transparent. The light may however change its direction slightly due to a different index of refraction inside the material; this effect is responsible for the properties of lenses. Furthermore, the index of refraction may depend on the wavelength of the light. It is for this reason that white light passing through a prism forms a rainbow on exiting: the different wavelengths, corresponding to different colors, come out at slightly different angles.

When point defects are introduced into an insulating crystal, they may absorb light passing through the crystal. Such defects commonly take several forms: lattice sites from which an atom is missing (vacancies); atoms occupying sites between lattice points (interstitial atoms); or lattice sites where an atom of one chemical species has been replaced by another element (impurities). Such defects absorb light of definite wavelengths in much the same manner that isolated atoms absorb and emit light at specific frequencies. If some of the wavelengths are removed from white light as it passes through the crystal, the exiting light will appear colored.

Strictly speaking, the term "color centers" refers to light-absorbing defects in alkali-halide crystals (crystals composed of a 1:1 mixture of elements from columns I and VII of the periodic table, such as NaCl or LiF). The defects may be induced by the introduction of chemical impurities, by x-ray, γ-ray, neutron, or electron bombardment, or by electrolysis. The simplest such defect, known as an "F Center" (from the German word for color, *Farbe*), is produced by heating an alkali halide in excess alkali vapor or by x-ray irradiation, and consists of a negative ion vacancy to which a free electron is weakly bound. The color of the crystal depends on the chemical makeup of the crystal and the type of defect. The color of the crystal depends on the chemical makeup of the crystal and the type of defect. The color centers produced by x-ray damage in LiF result in a deep yellow color. NaCl turns red after prolonged heating in the presence of Na vapor, while KCl turns blue after heating.

Similar effects are responsible for the colors of natural gems. Sapphire and ruby are both forms of the mineral corundum, which has chemical composition Al_2O_3 and is naturally colorless. The replacement of a very small number of aluminum atoms by chromium turns corundum into a red ruby, while addition of iron or titanium impurities yields the many colors of sapphire. Likewise, color in gems may result from radiation. A heavy dose of γ radiation will turn a diamond green.

See also: CRYSTAL; CRYSTAL DEFECT; CRYSTALLOGRAPHY; CRYSTAL STRUCTURE

Bibliography

MARKHAM, J. J. *F-Centers in Alkali Halides* (Academic Press, New York, 1966).
O'NEIL, P. *Gemstones* (Time-Life Books, Alexandria, VA, 1983).

PAUL A. HEINEY

COLOR CHARGE

Color charge is the property that distinguishes particles that participate in strong interactions. Just as particles with nonzero electric charges interact with electromagnetic fields so particles with color charge produce and respond to strong force fields. Leptons have no color charge and hence do not participate in any strong interaction process.

All particles that carry color charge are confined, which means that they are only found inside composite objects that are color neutral. It took some time for physicists to understand this property, and it was one of the reasons that the quark model was at first regarded with considerable skepticism. Quarks have electric charges $2/3$ or $-1/3$ of a proton charge. Nobody had ever seen particles with such charges. However all the combinations of quarks that are color neutral have integer electric charge in proton charge units; their fractionally charged constituents cannot be isolated.

The theory of the fundamental strong interactions is called quantum chromodynamics (QCD), so named because of its similarity to the theory of electromagnetism—quantum electrodynamics (QED). The term "chromo" in quantum chromodynamics refers to color charge. Mathematically QCD is described by an algebra quite different from that of ordinary numbers. The technical name for this algebra is $SU(3)$. The properties of color charge described below are the consequences of the mathematics of $SU(3)$, and of the quantum field theory based on that mathematical structure. [The $SU(3)$ of color has nothing to do with the $SU(3)$ of the three flavors in the original quark model, though the mathematics is of course the same.]

In QCD the $SU(3)$ means that there are three possible color charges for quarks; let us name them after the three primary colors, red, blue, and green. Every flavor of quark occurs in each of the three colors. Flavor $SU(3)$ is an approximate symmetry only, because the different flavors of quarks have different masses. Color $SU(3)$ is an exact symmetry; quarks of the same flavor but different colors have identical masses. Particle physicists tend to say there are six quark types, meaning six different masses. If we count quarks with different color labels as distinct then there are eighteen (3×6) quark types.

There are two quite different ways to combine color-charged quarks to make a color-neutral object. The first way is to take one quark of each color. This property is the reason for the name "color charge," because it provides an analogy to the fact that mixing the three primary colors in equal amounts gives a white or colorless object. Aside from this limited analogy, color charge has nothing to do with ordinary color, the three color names are just labels to distinguish strong charge properties. Color-neutral objects formed from three quarks are called baryons. Each of the three quarks can, in principle, have any flavor. The "in principle" is because the top quark is so heavy that it decays very rapidly, so it never has time to form any quasi-stable bound states with other quarks.

Antiquarks also come in three colors, but the three anticolors are not the same as the three colors. Let us call them antired, antigreen, and antiblue. An antibaryon contains one antiquark of each anticolor and is also a color-neutral object.

The second way to make a color-neutral particle from quarks is to have one quark and one antiquark in a quantum state that has equal probability of being red and antired, or blue and antiblue, or green and antigreen. Such a particle is called a meson. In principle, mesons also come in all possible flavor combinations.

To understand how such a state can exist we first need to know a little about the color charge properties of gluons. Gluons also carry color charges, unlike the photons for the electric case which have no electric charge. We will return shortly to discuss some important consequences of this, but let us first describe the possible gluon charges and what happens when gluons are emitted and absorbed.

There are eight types of color charge for gluons. Six of these are easily described, they each carry a color and an unmatched anticolor. When a quark emits or absorbs such a gluon the quark changes color, but the rules of color conservation ensure that only color-matched processes can occur. Thus, for example as shown in Fig. 1, if a red quark emits a red-antigreen gluon then the quark becomes a green quark, and similarly when an antired antiquark (the antiparticle of a red quark) absorbs such a gluon its color is changed to antigreen. Thus a quark-antiquark state which is red-antired can be converted to green-antigreen by exchange of a gluon. Similarly exchanges of red-antiblue and blue-antigreen gluons can convert these states to blue-antiblue. The color neutral state has equal probability of being found at any instant in any of the three matched color-anticolor configurations, and such a state persists no matter how many gluons are exchanged.

We still have to understand the last two gluon types. These are gluons with matching color and anticolor. Again the quantum state idea comes into play. A gluon that is red-antired initially does not stay that way. For example there is a process in which it can split into two gluons, one red-antiblue, the other blue-antired. Now these can recombine in such a way that the resulting state is blue-antiblue. It turns out that there are three possible states that are combinations of matched color-anticolor which are unchanged by such processes. However one of these is the color neutral combination discussed above which has no associated gluon, since gluons are not color neutral. The other two correspond to the two remaining gluons.

This follows directly from the mathematics of $SU(3)$. The reader who is familiar with matrix multiplication can follow the mathematical logic: Consider the three quark colors as the three coordinate directions of an abstract three-dimensional space. The eight gluons then correspond to the eight possible unitary orthogonal traceless (3×3) matrices

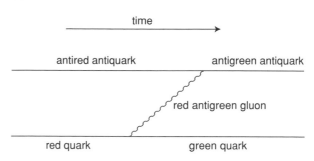

(a) Particle Process

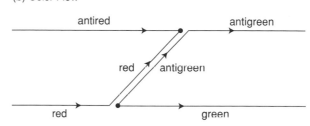

(b) Color Flow

Figure 1 An example showing the "color-flow" in a process where a gluon is emitted by a quark and absorbed by an antiquark, thereby changing the red-antired combination to green-antigreen. Such changes do not alter the net color of the system.

that generate rotations in this space. The color singlet state is the ninth unitary orthogonal matrix, namely the unit matrix (which is not traceless), which of course does not cause any rotation. There is no gluon corresponding to this matrix; it does nothing to any quark state.

Now let us return to the property of confinement. Particles with color charge are never observed in isolation. Suppose we produce a quark and an antiquark that are flying apart with very large momentum and energy. This happens for example when a high energy electron and positron collide and annihilate to form a (virtual) photon or a Z particle which then decays to produce the quark-antiquark pair. Why do we not see them separate after some time, and observe two separated fractional electric charges? The reason lies in the laws of thermodynamics—namely, that disorder is more likely than order.

As the color charged particles separate a region of nonzero color force field is produced between them. You can think of this as acting like an elastic string stretching between the quark and antiquark;

the string can only stretch by drawing energy from the separating quarks and slowing them down a bit. The energy density in this string (the color force field region) is sufficiently high that it can produce a light mass quark and antiquark (according to the rule $E = mc^2$). If the initial quark and antiquark are sufficiently energetic many such additional quark-antiquark pairs may be produced. Since any region of color-force field stores energy, the newly produced quarks and antiquarks will tend to arrange their color charges to screen or neutralize the color force field between them. Eventually a number of color-neutral particles are formed. These are the only things that get far enough to be observed flying out into any detector placed around the original collision point.

What does all this have to do with order versus disorder? The state with a widely separated quark and antiquark pair and a lot of energy in the color-force field between them is a very ordered state compared to all the possible states that can form if the color field produces many additional quark-antiquark pairs. It is a perfectly possible physical state of the system, but one that you will never observe. An analogy may help. Think of the state of a room in which all the air molecules are within 1 cm of the ceiling. This is a possible, though ordered, state for the air in the room. However you never expect to observe such a state because there are so many more arrangements of the air molecules allowed in disordered states in which the air fills the room. The probability of the ordered arrangement is so small that for all practical purposes we can say it is zero. The same is true of quark states with widely separated fractional charges.

Why is the same story not true when electric charges are separated? The difference lies in the fact that the photon is not charged. When a region of intense electromagnetic field is produced it can disorder itself by emitting photons, which go off in any direction. The energy density of electromagnetic fields is typically so low that this process dominates, though in sufficiently intense fields there can be enough energy density to produce additional charged particle-antiparticle pairs. The strong force field can likewise radiate gluons, but because these do have color charge they leave an intense region of color force field behind them much as do the initial quark and antiquark. So gluons too are confined.

Production of quark-antiquark pairs very quickly wipes out much of the information about the initial rapidly separating quarks and/or gluons. However not all information is lost. For example, the produced particles tend to come out grouped in direction, reflecting the momentum of the initiating particles. Physicists call these grouping of particles "jets." The underlying QCD theory can be used to predict the patterns of jets of particles found in high energy physics experiments. The success of these predictions in many different experiments is the major evidence in favor of the QCD theory of color charge.

Finally we come to the question that began the whole subject of particle physics. What are the forces between protons and neutrons in the nucleus—originally called the strong nuclear force? Nowadays particle physicists use the term strong force for the forces between particles with color charge, but protons and neutrons, and any other observable particles are color neutral, though composed of mutually interacting quarks. The forces between protons and neutrons are thus more like the van der Waals forces in the electrical case—they are residual forces between neutral objects due to their charged substructure. Just as electrically neutral atoms interact to form molecules due to electrical effects, so color-neutral hadrons affect one another if they are close enough together because each feels the color substructure of the other. The binding forces in nuclei are just this kind of residual interaction. The term "hadron" is used to mean any particle with a color substructure, whether a baryon or a meson. All hadrons interact through the residual strong interaction due to their color-charged constituents.

See also: ANTIMATTER; FLAVOR; HADRON; INTERACTION, STRONG; LEPTON; PARTICLE PHYSICS; QUANTUM CHROMODYNAMICS; QUANTUM ELECTRODYNAMICS; QUARK; SYMMETRY

Bibliography

Calder, N. *Key to the Universe* (Viking, New York, 1977).

HELEN R. QUINN

COMPLEMENTARITY PRINCIPLE

The idea of complementarity appears in philosophy and natural sciences when one chooses between two conflicting but equally valid standpoints. There are two approaches to the same object that are equally essential to understand its nature but are mutually exclusive. These two complementary aspects are also so closely associated that one refers to both of them by the same word simply because of the ambiguity of language. This kind of ambiguity always presents a remarkable epistemological problem. Such a duality was recognized for light by the early 1920s. Two different aspects of the phenomena—light waves and light quanta—were in that relationship of mutual exclusiveness. In different experiments the light manifests either the wave or the particle character: the wave character of light is revealed in phenomena of diffraction and interference while the photon corpuscular behavior is shown in such photon-electron interaction phenomena as the Compton and photoelectric effects. The original contribution of the Danish physicist Niels Bohr was to take the dualism between the wave and particle pictures as a suitable starting point for an interpretation of quantum mechanics.

This dualism is not confined to the properties of radiation only. Louis de Broglie showed that each moving particle (electron, neutron, proton, atom, ion, molecule) is associated with wave propagation; the complementary wave–particle nature is observed for these objects. Photons, electrons, and the other objects listed above never act at the same time as particles and waves. Both the wave and particle concepts are valid, but neither by itself is sufficient to describe the physical system under all conditions. The complementary role played by the wave and particle character of radiation and matter is included in Bohr's principle of complementarity (1928), which states that both concepts complement each other in describing the behavior of a physical system, but they cannot be revealed simultaneously. Among the complementary aspects of radiation, the wave properties of light are dominant at low energies (low light frequencies) while the corpuscular aspect manifests itself at high energies (high light frequencies). At low energies, in visible light, for instance, the wave theory of light gives the accurate description of the interference and diffraction phenomena. When the energy of light increases, the wave characteristics become less pronounced and the radiation takes a more corpuscular character. At high radiation energies, as in gamma rays, the photons act as individual particles. Werner Heisenberg's uncertainty principle describes the consequence of the wave–particle character of light and matter, and

provides an example of the complementarity relation. Heisenberg showed that it is not possible to measure simultaneously and with unlimited precision a position and momentum of a particle.

The complementarity principle evolved in the 1920s at the time of the formulation of quantum theory of the atom and during numerous discussions between Niels Bohr, Albert Einstein, Werner Heisenberg, Erwin Schrödinger, Wolfgang Pauli, George Gamow, and others. An important role in this discussion was played by the idealized thought experiments (gedankenexperiment). Einstein conducted experiments which he thought clearly revealed the inner contradictions of Bohr's interpretations.

The occasion for these duels between Einstein and Bohr were the Solvay Conferences in 1927 and 1930. One of the classical examples was an experimental arrangement that used three rectangular parallel screens mounted vertically at some distance from each other. There was a single horizontal slit in the first screen, two such slits in the second screen, and the third screen functioned as the photographic plate. A stream of electrons (or photons) fell from the left onto the first screen. The collimated beam passed through the single slit and proceeded toward the second screen with two slits. After leaving the two slits of the second screen the waves interfered with each other, resulting in an interference pattern that consisted of alternate light and dark horizontal stripes on the photographic plate. This picture was created by the joint effect of a large number of individual processes (electrons), each of which produced a tiny spot on the photographic plate where the electron had struck it. The results of this experiment showed that on the one hand the individual electrons struck the plate at a specific point in accordance with the particle conception, while on the other hand the distribution of the spots on the plate followed a law that can be understood on the basis of the electron wave interference conception.

Einstein's idea was that a control of the momentum transferred to the screen by incoming electrons would determine through which of the two apertures the electron had passed before reaching the plate. He suggested it would be possible to find the location and motion of a particle with unlimited precision. Bohr replied that if one were really able to determine the velocity and location of an electron with great precision, before it passes through the slits, then according to the law of causality one must also know through which slit in the screen it would go, and at what place it would strike the plate. According to Bohr, the electrons that passed through the lower slit would remain unaffected whether the upper slit was closed or not. But if the upper slit were covered, the interference pattern would disappear because the electrons that are known to pass through the lower slit are unable to "observe" whether the other slit is covered or not. Any measurement that could determine which slit the electron had passed through would involve a change in the experimental setup that makes the observation of the interference pattern impossible. The recoil movement imparted to the screen when the electron had its direction of motion changed is also subject to the uncertainty relations and it is no longer possible to investigate whether the electron was moving upward or downward.

In 1930, during the Sixth International Solvay Congress on Physics, Einstein proposed another experiment in which the energy of light quantum could be determined by weighing the radiation source before and after the quantum's emission. Make an ideal box, lined with perfect mirrors that could hold radiant energy for an indefinitely long time. Weigh the box. Then at a chosen point in time, controlled by a clock, open an ideal shutter to release some light. Weigh the box again. The change of mass determines the energy of the emitted light. According to Einstein, this experiment could measure the energy emitted and the time it was released with any desired precision, in contradiction to the uncertainty principle. Bohr responded with the same ideal box but on the spring scale. Bohr stated that since the box moves vertically with a change of its weight, there will be uncertainties in its vertical velocity and in its height above the table. Additionally, the uncertainty of its elevation above the earth's surface will result in an uncertainty in the rate of the clock, since according to the Einstein's theory of gravity the rate depends on the clock's position in the gravitational field. Bohr showed that the uncertainties of time and of the change in the mass of the box obey the uncertainty relation that was put in question.

The complementarity principle had important consequences in physics. On the one hand, using x-ray diffraction, the structure of materials is determined by taking advantage of the wave nature of photons. On the other hand, in x-ray photoelectron spectroscopy, the photoelectrons emitted from the sample surface as a result of the photoelectric process have a kinetic energy characteristic for the elements in the sample and provide us with information on the elemental surface composition and chemical state of atoms. Using wave properties of electrons, analytical tools such as reflection high-energy electron diffraction or low-energy electron diffraction are applied to study the structure of surfaces. The electron microscope is using the electron waves of much shorter wavelength than in the visible light and at 100 keV kinetic energy gives the insight into the sample details with magnification of the order of 250,000.

See also: BOHR, NIELS HENRIK DAVID; COMPTON, ARTHUR HOLLY; COMPTON EFFECT; DIFFRACTION, ELECTRON; EINSTEIN, ALBERT; HEISENBERG, WERNER KARL; PAULI, WOLFGANG; PHOTOELECTRIC EFFECT; QUANTUM MECHANICS; SCHRÖDINGER, ERWIN; UNCERTAINTY PRINCIPLE

Bibliography

BLAEDEL, N. *Harmony and Unity: The Life of Niels Bohr* (Science Tech. Publishers, Madison, WI, 1988).

GAMOW, G. *The Great Physicists From Galileo to Einstein* (Dover, New York, 1988).

ROZENTAL, S., ed. *Niels Bohr. His Life and Work as Seen by His Friends and Colleagues* (Interscience, New York, 1967).

PAWEL MROZEK

COMPOUND LENS

See LENS, COMPOUND

COMPTON, ARTHUR HOLLY

b. Wooster, Ohio, September 10, 1892; *d.* Berkeley, California, March 15, 1962; *x rays, cosmic rays.*

Compton's father, Elias, was a Presbyterian minister and professor of philosophy at Wooster College; his mother, Otelia, was a graduate of the Western Female Seminary in Oxford, Ohio. He and his two older brothers and older sister were raised in a home that emphasized education, discipline, and religious training. All four children became distinguished educators and together with their parents became known as America's first family of learning.

Compton received his elementary and secondary education in Wooster and his B.S. degree from Wooster College in 1913. His interest in science was awakened as a teenager in 1905 when he observed the constellation Orion and the Dog Star, Sirius, in the winter sky. In 1910 he built and flew for a short distance a triplane glider with a 27-foot wingspan. In college he could not decide whether to devote his life to science or to the ministry, but with his father's encouragement he chose science and followed his brother Karl to Princeton University for graduate study in physics. He completed his Ph.D. degree in 1916 and immediately thereafter married Betty Charity McCloskey, a former classmate at Wooster. They had two sons, Arthur Alan and John Joseph.

For his doctoral thesis, Compton began a program of x-ray researches that he continued as an instructor in physics at the University of Minnesota in Minneapolis (1916–1917), as a research engineer in the Westinghouse Lamp Company in Pittsburgh (1917–1919), as a National Research Council Fellow at the Cavendish Laboratory in Cambridge, England (1919–1920), and as Wayman Crow Professor and Chairman of the Department of Physics at Washington University in St. Louis (1920–1923). By the fall of 1922 these researches had led him to his most famous discovery, the Compton effect, for which he received the Nobel Prize in physics in 1927, sharing it with the inventor of the cloud chamber, C. T. R. Wilson.

Compton's discovery resulted from more than six years of concentrated experimental and theoretical work. The central problem confronting him was to understand how energetic radiation such as x rays was scattered by electrons in matter. As his researches progressed, he found more and more evidence that disagreed with the accepted classical theory of scattering, and ultimately he proposed a radically different one. He envisioned the incident x ray as a particle or quantum of radiation that strikes an electron, a particle of matter, in a billiard-ball-collision process, projecting it forward at a high relativistic velocity. The incident x-ray quantum thus loses some of its energy to the electron, so that the scattered x-ray quantum has a lower energy; in other words, a lower frequency or higher wavelength than

the incident one. Compton calculated the difference in wavelength between the scattered and incident x-ray quantum by setting up the equations expressing conservation of energy and conservation of momentum and solving them simultaneously, and he found that it agreed well with the value he had measured experimentally using a Bragg x-ray spectrometer. Compton's discovery constituted the first conclusive evidence for the validity of Einstein's light-quantum hypothesis, which most physicists had viewed with great skepticism after Einstein had proposed it almost two decades earlier in 1905.

In 1923 Compton transferred as professor of physics to the University of Chicago, where he was (as he was throughout his life) an inspiring teacher, and where he continued his x-ray researches. Then, in 1931, he turned to a new field of research, cosmic rays, whose nature was in dispute at the time. Robert A. Millikan at the California Institute of Technology argued that these highly penetrating rays coming from outer space consisted of light quanta (or photons), while others, including Compton, believed them to be charged particles. To settle this dispute, Compton organized a world survey of cosmic rays between 1931 and 1934. He sent out nine different expeditions—one consisting of himself, his wife, and his eldest son—to all parts of the globe to determine if cosmic rays were deflected in the magnetic field of the earth. Compton and his coworkers found that they were, which proved that they were charged particles, not photons, thus ending a dispute that had made headlines by pitting one Nobel Prize winner against another.

During the 1930s Compton also rekindled his lifelong interest in philosophical and religious concerns. He sought to elucidate questions such as the nature of free will and the relationship of man to God in light of modern science, especially relativity and quantum theory. His goal was to seek a rational and satisfying Christian philosophy, and he expounded his beliefs in many lectures and books.

After the outbreak of World War II, Compton shared the common fear that German scientists would place a nuclear weapon in the hands of Hitler, and he played a key role in initiating the Manhattan Project in the United States. He was appointed director of the Metallurgical Laboratory at the University of Chicago and brought Enrico Fermi there to build the first atomic pile, which achieved criticality on December 2, 1942. In 1945, despite his pacifist convictions, he voted with the majority of his colleagues on a committee appointed by President

Harry S. Truman to recommend that the awesome new weapon be used against Japan.

After the war Compton returned to academic life, becoming Chancellor of Washington University from 1945 to 1953 and Distinguished Service Professor of Natural Philosophy from 1954 to 1961. He died the following year while serving as a visiting professor at the University of California in Berkeley. Through his life and work he played a central role in raising physics in America to a position of preeminence in the world.

See also: COMPTON EFFECT; COSMIC RAY; MILLIKAN, ROBERT ANDREWS; NUCLEAR BOMB, BUILDING OF; X RAY

Bibliography

BLACKWOOD, J. R. *The House on College Avenue: The Comptons at Wooster* (MIT Press, Cambridge, MA, 1968).

COMPTON, A. H. *The Freedom of Man* (Yale University Press, New Haven, CT, 1935).

COMPTON, A. H. *The Human Meaning of Science* (University of North Carolina Press, Chapel Hill, 1940).

COMPTON, A. H. *Atomic Quest: A Personal Narrative* (Oxford University Press, New York, 1956).

JOHNSTON, M. ed. *The Cosmos of Arthur Holly Compton* (Knopf, New York, 1967).

SHANKLAND, R. S., ed. *Scientific Papers of Arthur Holly Compton* (University of Chicago Press, Chicago, 1973).

STUEWER, R. H. *The Compton Effect: Turning Point in Physics* (Science History Publications, New York, 1975).

ROGER H. STUEWER

COMPTON EFFECT

The Compton effect is the scattering of electromagnetic radiation by charged particles that changes the wavelength of the radiation. Physicists prior to 1923 tried in vain to explain the change in wavelength in terms of the classical picture of light as a wave. In that year Arthur H. Compton explained the scattering by treating radiation as a quantum, a particle-like packet of energy and momentum. His success was one of the most dramatic pieces of evidence that radiation can behave like a particle.

Albert Einstein, in 1905, was the first to propose that light energy came in packets, or quanta. The energy of each quantum is $h\nu$, where ν is the fre-

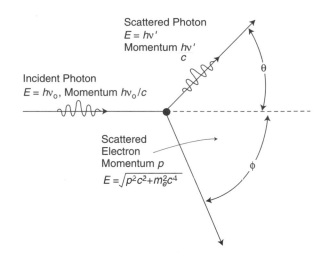

Figure 1 Compton's idea of scattering as a collision between a quantum of electromagnetic radiation and a free electron.

quency of the light and h is Planck's constant. Many physicists remained skeptical about the validity of Einstein's quantum hypothesis until the discovery of the Compton effect.

Compton pictured the scattering as a collision between a quantum of electromagnetic radiation (later called a "photon") and a free electron, as shown in Fig. 1. The incident photon has frequency ν_0 and wavelength λ_0; the corresponding quantities for the photon scattered at angle θ are ν' and λ'. The electron is at rest before the collision and moves with momentum p after the collision. The frequencies and wavelengths of the photons are related: $\nu_0 = c/\lambda_0$, and $\nu' = c/\lambda'$, where c is the velocity of light.

Compton used the laws of conservation of energy and momentum to analyze this collision. From special relativity, the energy of an electron at rest is $m_e c^2$, where m_e is the mass of a stationary electron. The energy of the moving electron is $\sqrt{p^2 c^2 + m_e^2 c^4}$. Therefore, for this collision the law of conservation of energy is written $h\nu_0 + m_e c^2 = h\nu' + \sqrt{p^2 c^2 + m_e^2 c^4}$.

Conservation of momentum holds for both the x and y directions in the diagram. The momentum of the electron before the collision is zero; after the collision it is p. The photon momentum can be found from the theory of special relativity, which states that the total energy of a particle with momentum p and rest mass m_0 is $\sqrt{p^2 c^2 + m_0^2 c^4}$. This formula leaves open the possibility of a "massless" particle, something traveling at c with zero rest mass but with energy and momentum related by $E = pc$. Since the photon energy is $h\nu$, the photon momentum is $h\nu/c$. Therefore, the conservation of momentum equations become

$$\frac{h\nu_0}{c} = \left(\frac{h\nu'}{c}\right)\cos\theta + p\cos\phi,$$

in the x direction, and

$$0 = \left(\frac{h\nu'}{c}\right)\sin\theta - p\sin\phi,$$

in the y direction.

These three equations relate ν_0, ν', p, θ, and ϕ. Elimination of p and ϕ leaves the following relationship between ν_0, ν', and θ:

$$\frac{1}{\nu'} - \frac{1}{\nu_0} = \frac{h}{m_e c^2}(1 - \cos\theta).$$

Since $\nu = c/\lambda$, this equation becomes

$$\lambda' - \lambda_0 = \frac{h}{m_e c}(1 - \cos\theta),$$

which is Compton's relationship between the wavelengths of the incident and scattered photons and the scattering angle θ.

The quantity $h/(m_e c)$, which is called the Compton wavelength of the electron, is often labeled λ_C. Its value is 2.43×10^{-12} m = 2.43 pm.

The formulas for the Compton wavelength and the scattered photon wavelength reveal much. The wavelength difference is greatest when $\theta = 180°$; for this angle the electron gains the maximum kinetic energy. The wavelength difference is independent of λ_0 for any angle θ. Therefore, the fractional change in wavelength is greatest at short wavelengths; x rays or gamma rays (tens of picometers or shorter) are needed to produce an easily detectable change in the wavelength. For scattering from a particle of rest mass m_0 the Compton wavelength is $h/(m_0 c)$, so the maximum wavelength change for scattering from protons is 0.00264 pm and is significant only for very high energy gamma rays.

The experimental setup that Compton used to confirm his wavelength equation is shown in Fig. 2. In the x-ray tube, electrons at high speed slammed into a piece of molybdenum, creating x rays. Some of these x rays penetrated the graphite block R and scattered from the electrons of carbon atoms within the block. The assembly of slits picked out only those scattered x rays traveling in a certain direction. The crystal, ionization chamber, and intermediate slits were used to detect those x rays and measure their wavelengths. In this experiment Compton used the K_α ray of molybdenum with wavelength 70.8 pm.

Figure 3 shows Compton's results. As the angle of scattering increased, the change in wavelength increased precisely as Compton predicted. Figure 3 also shows that at each angle there are scattered photons of unchanged wavelength $\lambda' = \lambda_0$. Such scattering is called Rayleigh scattering; in the classi-

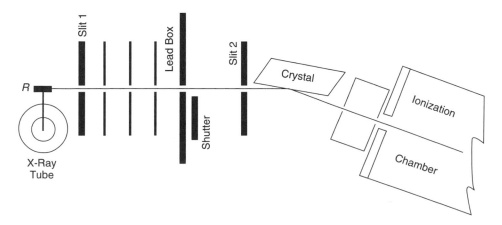

Figure 2 Experimental apparatus that Compton used to confirm his wavelength equation.

produces a background continuum of gamma rays. Theorists need to take the Compton effect into account in modeling astrophysical plasmas, laboratory fusion plasmas, and nuclear weapons effects. In nuclear medicine, measurements of bone density and bone mineral content use Compton scattering of gamma rays from radioactive isotopes. Engineers can deduce the electron density distribution within some materials from the Compton scattering spectrum.

Compton effect detectors are used in gamma ray telescopes to observe nuclear reactors in satellites and astrophysical gamma sources. On April 5, 1991, Space Shuttle Atlantis launched the Gamma Ray Observatory (GRO), which cost $600 million. Most of the experiments on this satellite use the Compton effect, and the data from the GRO are changing our ideas about some of the most energetic objects and violent events in the universe.

See also: COMPTON, ARTHUR HOLLY; GAMMA RAY OBSERVATORY; PHOTOELECTRIC EFFECT; PHOTON; RELATIVITY, SPECIAL THEORY OF; SCATTERING, RAMAN; SCATTERING, RAYLEIGH

Bibliography

APRILE, E., ed. *Gamma Ray Detectors: Proceedings of the Meeting, San Diego, CA, July 21–22 (SPIE Proceedings Vol. 1734)* (Society of Photo-Optical Instrumentation Engineers, Bellingham, WA, 1992).

BEISER, A. *Concepts of Modern Physics,* 5th ed. (McGraw-Hill, New York, 1987).

COMPTON, A. "A Quantum Theory of the Scattering of X-Rays by Light Elements." *Phys. Rev.* **21,** 483–502 (1923).

COMPTON, A. "The Spectrum of Scattered X-Rays." *Phys. Rev.* **22,** 409–413 (1923).

EISBERG, R., and RESNICK, R. *Quantum Physics of Atoms, Molecules, Solids, Nuclei and Particles,* 2nd ed. (Wiley, New York, 1985).

GEHRELS, N.; FICHTEL, C.; FISHMAN, G.; KURFESS, J.; and SCHONFELDER, V. "The Compton Gamma Ray Observatory." *Sci. Am.* **269** (Dec.), 68–77 (1993).

HURLEY, K. "Probing the Gamma-Ray Sky." *Sky and Telescope* **84,** 631–636 (1992).

O'NEILL, J.; KERRICK, A.; AIT-OUAMER, F.; TUMER, O.; ZYCH, A.; and WHITE, R. "Observations of Nuclear Reactors on Satellites with a Balloon-Borne Gamma-Ray Telescope." *Science* **244,** 451–454 (1989).

STUEWER, R. H. *The Compton Effect: The Turning Point in Physics* (Science History Publications, New York, 1975).

THOMAS GREENLEE

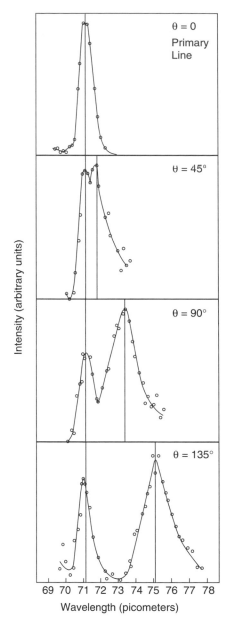

Figure 3 The results of Compton's experiments, showing the relationship between an increase in scattering and an increased change in wavelength.

cal theory of light as a wave it is the only scattering that should occur.

Almost any application of gamma rays or x rays involves the Compton effect as a useful detection process, as an unwanted complication, or as a consideration in the theoretical modeling of a system. In gamma-ray detection in nuclear physics, Compton scattering within the scintillation counter

COMPUTATIONAL PHYSICS

Traditionally, the study of physics is pursued in two different ways: theoretical and experimental. The experimental approach consists of probing, perturbing, and then carefully observing physical systems directly to uncover their properties. The experimental physicist makes quantitative measurements of the properties of these systems to describe their behavior. The theoretical approach seeks to describe all physical behavior in terms of an underlying set of laws, which are usually expressed in mathematical terms. The theoretical physicist solves the mathematical equations arising from these laws to confirm the observed measurements and to predict new physical phenomena. By verifying the new predictions the experimentalist can test the validity of the laws. The theoretical and experimental approaches to doing physics thus work hand in hand and both are necessary for a full understanding of the physical world.

The advent of high-speed computer technology has created a new approach to exploring physical phenomena and a new field of physics called computational physics. This new approach has features in common both with traditional theoretical and with experimental physics. The computational approach begins by identifying the appropriate laws that describe the physical system under study and then representing the system by the mathematical equations corresponding to those laws. This process constitutes mathematical modeling of a physical system. The next step involves solving the equations with the aid of a computer. Now the equations describing very simple systems are themselves often simple and can be solved in terms of algebraic expressions for the unknown variables without the use of a computer. Such mathematical solutions are called analytic solutions. But complex systems usually require sets of complicated, coupled equations for a realistic description. Such equations can usually be solved only in terms of explicit numerical values for the variables, as no analytic solutions exist. In this case the solutions are called numerical solutions and the computer is usually essential for performing the detailed calculations required to obtain them. A numerical solution is often represented by a list or table of numbers that describe the values of the key dependent physical parameters of the system. Once these values are obtained they can be compared with direct measurements of the actual physical system. By this comparison the validity of the mathematical model describing the system can be evaluated. Given a valid model of a physical system, the numerical values assigned to the independent parameters in the equations can be varied and the equations re-solved on the computer to obtain a new solution. Such a variation of the independent parameters describing a system is analogous to placing the actual system in a new physical state. The new solution reveals the behavior of the system in that new state. The whole process of mathematical modeling and computer evaluation of the model equations represents the exploration of a physical system by computer simulation.

In its adoption of mathematical models and evaluation of mathematical equations to describe physical phenomena, computational physics is essentially theoretical physics. In its perturbation of the independent parameters in these equations and observation of the subsequent behavior of the solutions, computational physics is very much like experimental physics. It is akin to experimental physics in another respect: Discovering reliable and efficient techniques for solving complicated equations on the computer often requires, in addition to mathematical insight, a huge amount of trial and error.

The great appeal of computational physics is that solving a particular problem in one field often yields insight into solving another problem in a totally unrelated field. The reason is that when expressed in computational terms, different physical problems often share many features in common.

The physics problems that are often the most challenging are those involving many dimensions, particles, and/or parameters. The corresponding solutions naturally involve huge arrays of numerical data. To better assess these solutions it is sometimes better to plot the data in the form of multidimensional images on a computer screen rather than to study the numerical output in tabular form. Thus, modern computational physics often takes full advantage of the tools and techniques of computer visualization (e.g., color graphics, computer-generated video, etc.) to sift through and evaluate the computational data.

Applications

Virtually every area of modern physics has spawned a computational sector. This is not surprising since every area deals with complicated physical systems that are described by complex equations requiring computers for solution. Thus there are the fields of computational atomic and molecular physics, computational fluid dynamics, computational nuclear and particle physics, computational solid-state physics, and computational astrophysics to name but a few of the areas in which there has been an explosion of activity. There are even scientific journals devoted to computational physics, such as the *Journal of Computational Physics,* and there are physicists whose primary focus is solving physics problems by computational means.

Consider some of the forefront fields of computational physics. Computational molecular dynamics seeks to explain the structure, bonding, motion, and reactions of molecules in a many-body medium. Applications range from understanding the propagation of microcracks and fracture in solids to the computer-aided molecular design of new materials. Computational fluid dynamics seeks to understand the laminar and turbulent motions of gases and liquids. These motions are described by the Navier–Stokes equations, a complicated set of nonlinear, coupled, multidimensional partial differential equations in space and time. Solving these equations on a computer is a key aspect of oceanographic and atmospheric studies as well as weather forecasting. Computational plasma physics involves the study of plasmas (ionized gases) in electric and magnetic fields and requires the solution of the equations of magnetohydrodynamics (MHD). The MHD equations comprise another set of partial differential equations, and they describe the motion of charged particles in electromagnetic fields as functions of space and time. Computer simulations of high-temperature plasmas are essential to our understanding of such phenomena as solar magnetic flaring, cosmic jets in radio galaxies, and controlled nuclear fusion.

Computational particle physics is the numerical implementation of lattice gauge theory, the discretized version of quantum chromodynamics (QCD). QCD describes all the interactions between the elementary particles in terms of quarks and gluons. By replacing the continuum of space and time by discrete points in a four-dimensional, spacetime lattice, lattice gauge theory yields equations that can determine, at least in principle, the masses of composite particles like protons and neutrons and the structure of atomic nuclei.

Numerical relativity solves Albert Einstein's equations of general relativity on a spacetime lattice by computer simulation. Einstein's equations describe the behavior of matter and gravitational fields when particle motions approach the speed of light and the gravitational field is strong enough to cause light to bend significantly. These equations are nonlinear, coupled partial differential equations and the computer is essential for obtaining general solutions. Applications include the collapse of stars to black holes, the collision and coalescence of binary black holes, and the generation and propagation of gravitational radiation.

Computer simulations play a particularly vital role in certain fields of physics where experiments are difficult or even impossible to perform. Consider astrophysics, for example. Here the computer serves as a new kind of telescope, permitting the "observer" to see cosmic phenomena that are either too distant, too short-lived, or too obscured to be seen by conventional telescopes. The computer also serves as a new form of physics laboratory that allows one to build, probe, and perturb exotic physical systems like black holes or colliding galaxies, which are otherwise inaccessible in terrestial labs.

Future Prospects

From a historical perspective, most of the activity in computational physics is in its infancy. The discipline did not exist prior to the advent of high-speed computers. Some of the first significant work in computational physics took place at Los Alamos during World War II in connection with the Manhattan Project. Much of this early work involved gas dynamics and dealt with the physics of a nuclear blast wave. Computational physics did not really become a widespread research endeavor, however, until many years later, when computers became smaller, faster, and more accessible to the general physics community. The importance of the discipline was officially acknowledged in 1982 when the Nobel Prize in physics was awarded to Kenneth Wilson for his invention of the renormalization group, the basis for many computer algorithms for studying critical phenomena. Soon after this, the National Science Foundation (NSF) established national computing centers at several sites across the United States to make state-of-the-art computing in the physical sciences broadly available to researchers.

One focus of computational physics is learning how to exploit parallel computing. Parallel computing involves the simultaneous use of many computers or processors to tackle a complicated problem. The problem has to be carefully split up into separate pieces so that each piece can be submitted to a different processor and solved simultaneously. To stimulate progress in this new and potentially powerful form of computing, the NSF has launched a Grand Challenge program to support collaborative efforts to solve some of the most difficult problems in computational physics by using parallel machines and parallel algorithms.

See also: CRITICAL PHENOMENA; EXPERIMENTAL PHYSICS; MAGNETOHYDRODYNAMICS; MODELS AND THEORIES; NAVIER–STOKES EQUATION; QUANTUM CHROMODYNAMICS; RELATIVITY, GENERAL THEORY OF; THEORETICAL PHYSICS

Bibliography

HOCKNEY, R. W., and EASTWOOD, J. W. *Computer Simulation Using Particles* (McGraw-Hill, New York, 1981).

KALOS, M. H., and WHITLOCK, P. A. *Monte Carlo Methods* (Wiley, New York, 1986).

KOONIN, S. E. *Computational Physics* (Benjamin/Cummings, Menlo Park, 1986).

POTTER, D. *Computational Physics* (Wiley, Chichester, Eng., 1973).

RICHTMEYER, R. D., and MORTON, K. W. *Difference Methods for Initial-Value Problems*, 2nd ed. (Wiley, New York, 1967).

ROACHE, P. J. *Computational Fluid Dynamics* (Hermosa, Albuquerque, NM, 1976).

SHAPIRO, S. L., and TEUKOLSLY, S. A. "Building Black Holes: Supercomputer Cinema." *Science* **241,** 421 (1988).

STUART L. SHAPIRO

CONDENSATION

Condensation is the process whereby a substance changes from its vapor phase to its liquid or solid phase at the same temperature and pressure. Evaporation (phase change from liquid to vapor) and sublimation (phase change from solid to vapor) are the reverse processes.

Figure 1 shows a schematic phase diagram for a simple substance. The lines separate the pressure–temperature plane into regions where a single phase is the most stable thermodynamic state. The lines

are called coexistence curves because they define temperatures and pressures at which two phases exist in equilibrium at the same time. The triple point (*TP*) is where the coexistence curves intersect, and all three phases exist at the same time. Condensation takes place when the system crosses the coexistence curve from the vapor phase to either the liquid or solid phases. The plot shows that condensation can take place over a range of temperatures; the lower the vapor pressure, the lower the temperature at which it condenses.

Examples of condensation are evident all around. On a cool morning, the grass may be wet with dew even if it has not rained. Dew forms when moisture-laden warm air cools near the cold ground. The condensation temperature (dew point) depends on the partial pressure of water vapor (related to the humidity). In a steamy bathroom, where the humidity is high, droplets of water form on a mirror at a much higher temperature than in a dry room. Clouds form when warm, moist air rises and cools as it expands in the lower pressures of the upper atmosphere. At the lower temperatures, water vapor condenses to form the tiny water droplets that make up the clouds.

Condensation is an example of a first-order phase transition. This means that heat is evolved during the transition, even though the temperature does not change. This heat is called the latent heat, and the energy comes from the ordering of the vapor molecules as they condense to the liquid or solid phase. There is also a discontinuous change in certain properties, like density. If steam condenses to

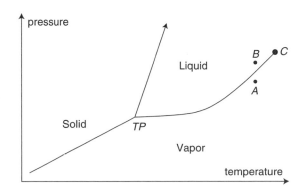

Figure 1 Schematic pressure-temperature phase diagram of a simple substance showing the coexistence curves that separate solid, liquid, and vapor phases. *TP* marks the triple point and *C* marks the critical point.

water at 100°C and 1 atm, for example, the density changes abruptly from approximately 0.001 g·cm^{-3} to about 1 g·cm^{-3}. This corresponds to a path *directly* from B to A in the diagram. It is also possible for the vapor to become liquid without a phase change taking place. In this case, the density changes continuously and we do not use the term "condensation." If the system's pressure and temperature change in such a way that it goes from B to A along a path that goes around C, the coexistence curve is avoided, and the density changes smoothly from about 0.001 g·cm^{-3} to about 1 g·cm^{-3}. Point C, the critical point, marks the end of the coexistence curve between liquid and vapor. There is no known critical point associated with either the vapor-to-solid or the liquid-to-solid transition.

Condensation need not always take place at the coexistence line. For example, very clean water vapor at 1 atm may cool below 100°C and *not* condense to liquid water. The vapor is "supercooled" because its temperature is lower than it should be for that particular pressure. (In the same way, very clean liquid water may exist above 100°C at 1 atm. It is then "superheated.") If some impurity or even a drop of liquid is introduced into a supercooled vapor, condensation occurs immediately and dramatically. The reason is that water needs a nucleation site on which to condense. If there are no nucleation sites, the vapor persists even though the liquid may be the more stable state. In a very clean system, the only possible nucleation site is a droplet of liquid, formed when random fluctuations bring a number of vapor molecules into close proximity so that they coalesce. If the number of molecules in the droplet is very small, it quickly evaporates again, but if the number is large enough, the droplet is stable and more molecules condense on it, so it grows. In the supercooled state, the system reduces its energy if a liquid droplet forms because the liquid phase is a lower energy state than the vapor phase, but it costs energy to form the surface between the liquid droplet and the vapor. The energy reduction from droplet formation is proportional to the volume of the droplet, and the energy increase from the surface formation is proportional to the surface area. These forces balance at a critical size below which nucleation does not occur and above which it does. For water at room temperature, the critical radius of a droplet is about 10^{-9} m, or about 200 molecules. The energy to form such a droplet is relatively large (about 100 times $k_B T$), so that water vapor often exists in a supercooled state.

A cloud chamber is an example of a device that exploits supercooling to detect high-energy particles. Supercooled alcohol vapor in a large container awaits passage of a cosmic ray. When the high-energy particle travels through the vapor, it ionizes some vapor molecules, and the ions form nucleation sites for the condensation of liquid alcohol. Thus the particle leaves a visible trail of liquid in its track. Cloud chambers have been replaced by more sophisticated detectors but are often on display in science museums.

See also: CLOUD CHAMBER; DENSITY; EVAPORATION; PHASE; PHASE, CHANGE OF; PHASE TRANSITION; TRIPLE POINT

Bibliography

KITTEL, C., and KROEMER, H. *Thermal Physics,* 2nd ed. (W. H. Freeman, New York, 1980).

WALDRAM, J. R. *The Theory of Thermodynamics* (Cambridge University Press, Cambridge, Eng., 1985).

JANET TATE

CONDENSATION, BOSE–EINSTEIN

At very low temperatures a gas of essentially noninteracting particles known as bosons comes together in a single, macroscopic quantum state, a Bose–Einstein condensate. It is no longer a gas in the sense of individual atoms moving randomly, but the atoms lose their separate identities in the new, coherent state. This is a unique form of matter, predicted in 1925 and first observed directly in 1995.

Bosons are particles of integer spin (e.g., photons or helium nuclei). The quantum mechanical wave functions for a set of identical bosons are exactly the same when one exchanges one boson for another; that is, they are symmetric under particle interchange. Because of this wave function symmetry, bosons behave quite differently at low temperature from the other major class of particles, the fermions. Whereas fermions (e.g., electrons) tend to avoid each other and cannot occupy the same quantum state, bosons prefer to stay together and, under the right conditions, actually condense into a single quantum state.

Bose–Einstein condensation occurs when the temperature is low enough and the density of the gas is high enough so that the individual particles can no longer exist separately without quantum mechanical overlap. This occurs when the average separation of bosons in the gas becomes smaller than a critical value, $\Lambda/1.377$, where $\Lambda = h/\sqrt{2\pi mkT}$ is the thermal de Broglie wavelength, h is Planck's constant, m is the boson mass, k is Boltzmann's constant, and T is the absolute temperature. Λ is a measure of the spread of the quantum mechanical wave function of the bosons, so when the average separation becomes less than this critical length, the wave functions of the particles overlap to such an extent that the gas enters a new phase with many bosons in the ground state, forming a single, coordinated or coherent quantum state. The quantum mechanical wave function of this new phase involves all the bosons in the condensate, so they should no longer be thought of as individual particles.

The coherent state of matter in a Bose–Einstein condensate is analogous to the coherent state of light in a laser beam. A gas of ordinary atoms consists of a collection of unrelated quantum waves, like the light from an ordinary lightbulb, the atoms bouncing randomly around in the gas. But in a Bose–Einstein condensate the individual atoms lose their separate identities in a single, all-encompassing wave function and therefore create a form of coherent matter, like the coherent light of a laser. The condensate and the laser are both macroscopic systems which exhibit overall quantum behavior. Such quantum mechanical behavior is usually confined to microscopic systems like a single atom.

Satyendranath Bose proposed the quantum theory of bosons in 1924 for a photon gas, and Albert Einstein, using the approach of Bose, first predicted what we now call Bose–Einstein condensation for a particle gas in 1925. This theory deals with an "ideal" gas of bosons, that is, noninteracting point particles whose wave functions satisfy boson symmetry rules. Because there are always forces between particles in real gases that tend to interfere with or obscure the condensation at low temperature, the experimental realization of the Bose–Einstein condensation required the production of a gas dilute enough that the interparticle forces are negligible, yet dense enough that the thermal de Broglie wavelengths overlap.

Bose–Einstein condensation plays a role in many physical phenomena, but it cannot usually be observed directly. For example, a form of Bose–Einstein condensation, complicated and obscured by other effects, occurs in superfluidity of liquid helium, the Higgs boson condensation in the early universe which gave rise to the masses of particles we now observe, and weakly interacting exciton gases in materials. These manifestations of Bose–Einstein condensation are obscured primarily by interactions that cause phase transitions, molecule formation, or other effects before the purely quantum effects due to identical bosons can be observed. In addition, stable bosons are composite particles, unlike the electron, which is a stable fermion. Particles of integer spin are typically made up of an even number of stable fermions (e.g., helium nuclei, hydrogen atoms, etc.).

Bose–Einstein condensates were created and directly observed experimentally for the first time on June 5, 1995, by a group of physicists from the National Institute for Standards and Technology at Boulder, Colorado, and the University of Colorado. The group was led by Eric A. Cornell and Carl F. Wieman. They and their coworkers used a laser cooling technique, followed by evaporative cooling of a gas of spin-polarized rubidium atoms to lower the temperature of the gas to about 20 nK. Below about 170 nK they observed the Bose–Einstein condensate by a laser time-of-flight measurement of the velocity distribution of the cold atoms. At their coldest temperatures the interparticle spacing was much smaller than the thermal de Broglie wavelength. They had about 2,000 atoms in the condensate, contained in an ingeniously designed spinning magnetic trap. The condensate could be preserved for more than 15 s. In July 1995, Randall G. Hulet and coworkers at Rice University observed a Bose–Einstein condensate in a gas of lithium, using a different set of techniques. In October 1995, Wolfgang Ketterle and coworkers at the Massachusetts Institute of Technology succeeded in creating a Bose–Einstein condensate consisting of about 500,000 sodium atoms, using yet a third approach.

The creation of this new state of matter promises to open up a new field of physics, perhaps as the invention of the laser did. Using these Bose–Einstein condensates, physicists hope to study new phenomena in macroscopic and mesoscopic quantum systems, as well as quantum statistical phenomena that have not been accessible by using prior techniques. These studies might include spontaneous symmetry breaking and decay processes of unstable macroscopic states. In addition, we might hope to make extremely "bright" sources of atoms (i.e., very large numbers in the single quantum state) and actually produce something like an "atom laser" (a source providing a coherent quantum state at high particle population).

See also: ABSOLUTE ZERO; BOSON, HIGGS; BROGLIE WAVELENGTH, DE; FERMIONS AND BOSONS; GROUND STATE; LIQUID HELIUM; SPIN AND STATISTICS; SYMMETRY BREAKING, SPONTANEOUS; WAVE FUNCTION

Bibliography

ANDERSON, M. H.; ENSHER, J. R.; MATTHEWS, M. R.; WIEMAN, C. E.; and CORNELL, E. A. "Observation of Bose–Einstein Condensation in a Dilute Atomic Vapor." *Science* **269**, 198 (1995).

BRADLEY, C. C.; SACKETT, C. A.; TOLLETT, J. J.; and HULET, R. G. "Evidence of Bose–Einstein Condensation in an Atomic Gas with Attractive Interactions." *Phys. Rev. Lett.* **75**, 1687 (1995).

BURNETT, K. "An Intimate Gathering of Bosons." *Science* **269**, 182 (1995).

DAVIS, K. B.; MEWES, M.-O.; ANDREWS, M. R.; VAN DRUTEN, N. J.; DURFEE, D. S.; KURN, D. M.; and KETTERLE, W. "Bose–Einstein Condensation in a Gas of Sodium Atoms." *Phys. Rev. Lett.* **75**, 3969 (1995).

GOODSTEIN, D. L. *States of Matter* (Prentice Hall, Englewood Cliffs, NJ, 1975).

HUANG, K. *Statistical Mechanics*, 2nd ed. (Wiley, New York, 1987).

TAUBES, G. "Physicists Create New State of Matter." *Science* **269**, 152 (1995).

WILLIAM E. EVENSON

CONDENSED MATTER PHYSICS

The subfield of physics known as condensed matter physics is the largest and most active field of physics even though it is not clearly defined nor well delineated. The subfield tends to encompass an increasing number of topics as the body of knowledge grows. Condensed matter physics grew out of solid-state physics, which has been classically defined as the study of properties of solid bodies with shapes that are elastically restored after weak deformation forces are removed. With time, this constrained definition of solid-state physics became inadequate as the practitioners in this subfield branched into new

research areas, some of which transcended traditional solid phases. Eventually this digression lead to the name change. Thus condensed matter physics today includes a host of topics not included in traditional solid-state physics, including the study of the following: quantum liquids, such as liquid helium, which exhibits cooperative phenomena common to superconductors; liquid crystals, which exhibit ordering phenomena somewhat related to crystalline phases; bodies undergoing plastic deformation that do not return to their initial shapes; disordered and porous materials, which may not have well-defined bulk shapes; and gels, that can greatly expand in volume upon undergoing phase transitions.

Prior to the twentieth century, the physics of materials dealt primarily with the study of the macroscopic properties of crystalline phases. This limited scope persisted as x-ray diffraction techniques were introduced, bringing new quantitative understanding to the periodic structure of crystalline phases on an atomic scale. The discovery of quantum mechanics in the 1920s greatly influenced the physics of materials, leading to new understanding of the electronic properties of periodic structures, the free electron theory of metals, and the properties of defects in alkali halides.

In these early years, the field of physics was very broad and was not divided into subfields. Many giants in the physics of the early twentieth century (e.g., Enrico Fermi, Wolfgang Pauli, Lev Landau, Hans Bethe, Eugene Wigner) made significant contributions to the physics of materials as well as to nuclear physics, high energy physics, astrophysics, and other subfields of physics. The first physicist to delineate the subfield of solid-state physics explicitly was Frederick Seitz, whose classic book *Modern Theory of Solids* (1940) defined the field of solid-state physics for several decades. Starting in 1955, Seitz and David Turnbull edited a series of books for Academic Press entitled *Solid-State Physics,* which was influential in shaping the subfield during its dramatic growth during the 1960s and 1970s.

Many advances in condensed matter physics were stimulated by practical societal needs, leading to the expansion of the scope of the subfield. Early studies of semiconductors were stimulated by the need for sensitive detector materials for wireless communication signals, and somewhat later by the needs for radar detector materials during World War II. The discovery of the transistor in 1948 ushered in the age of semiconductor electronics, communications, and digital computers. Interest in condensed matter physics rose sharply because of the rapid development of the science base in this subfield, the promise seen by industry for commercial exploitation of this science, and the demands of other fields of science and technology for materials with special properties and controlled performance.

In the early post–World War II period, the main focus of solid-state physics was on semiconductors and periodic structures. When it became clear that many of the techniques developed for periodic solids were equally applicable to liquids and to amorphous materials, the boundaries of the subfield expanded and the term "condensed matter physics" came into general use. The relative popularity enjoyed by this subfield of physics can be judged by the abundance of journals and publications and the large number of members of the American Physical Society identified with the subfield of condensed matter physics.

Condensed matter physics today is the largest subfield of physics and relates strongly to other subfields of physics as well as to other fields of science. Most strongly linked to the subfield of condensed matter physics is the subfield of atomic, molecular, and optical (AMO) physics. There are blurred boundaries between these subfields, due in part to the high incidence of interdisciplinary research, especially in the use of materials in AMO physics of particular interest to the condensed matter physics community, such as the use of quantum wells to study quantum electronics phenomena. Likewise, optical techniques, spectroscopy, and time-resolved studies are all traditional techniques of condensed matter physics. Fuzzy boundaries also exist between condensed matter physics and the field of materials science, which focuses on the study of structure-property relations of materials. To the extent that such studies focus on the physical phenomena, they would tend to be classified as condensed matter physics, while emphasis on the microstructure responsible for the manifestation of a specific physical phenomenon would tend to be classified as materials science. As materials research has become increasingly interdisciplinary, the boundary between condensed matter physics and materials science has become increasingly blurred, as is evidenced by the rise of the new subfield of physics called materials physics, which sits at the interface between these fields.

Condensed matter physics also has strong ties to chemistry (physical chemistry and polymer chemistry in particular), to electrical engineering (through semiconductor, magnetic, and optical devices, as well as theoretical issues such as noise and solitons), to Earth and planetary science in terms of materials found on the planet Earth and in the solar system, especially materials under conditions of extreme temperature, pressure, radiation, and magnetic fields. More recently, strong ties have been made to biology and biophysics in terms of techniques, instrumentation, and models originating in condensed matter physics now being applied to studies of proteins, nucleic acids, membranes, the human genome, and brain functions.

Characteristics of the Field

Condensed matter physics, like other physics subfields, has both experimental and theoretical aspects. In the early decades of the twentieth century, the ratio of theoretical to experimental papers in condensed matter physics was relatively greater than it is today. The greater prevalence of experimental work accompanied the growth of the field, as did increasing industrial and societal interest in this science for potential practical spin-offs. Condensed matter physics has a basic science component that focuses on the detailed understanding of physical phenomena, as well as an applied science component that focuses on the discovery of phenomena with potential for practical applications, and in the use of basic physics to develop materials with desired properties. Condensed matter physics, particularly in the industrial environment, tends to be part of an interdisciplinary materials research effort, with participation from condensed matter physicists, chemists, materials scientists, computer scientists, and others to solve significant large problems.

Condensed matter physics is replete with examples of the synergistic relation between science and technology. The impetus to purify semiconductors, grow better crystals, and study semiconductor properties initially came from the needs of the wireless communication industry and radar requirements of the military in World War II. The availability of superior materials and practical knowledge about semiconductors after World War II led to systematic scientific studies and to increased basic understanding of semiconductor physics. These scientific advances in turn led to further technological advances, including the discovery of the transistor and the birth of semiconductor electronics. Increased technological interest further stimulated major advances in semiconductor physics, such as the discovery of cyclotron resonance phenomena, and elegant studies of impurity levels in semiconductors. Many examples may be found of major advances in both science and technology arising from the push and pull between condensed matter physics and the technology supporting the semiconductor, magnetics, optoelectrons, and communications industries.

As condensed matter physics research became more sophisticated, the requirements on materials synthesis, materials characterization, and property measurements became increasingly demanding, resulting in the establishment of materials research laboratories to provide facilities and technical expertise that were unavailable to individual investigators but necessary to carry out state-of-the-art materials research. Such materials research laboratories are now prevalent at universities, industrial laboratories, and national laboratories. In addition, national facilities have been established in the areas of synchrotron radiation, neutron scattering, and high magnetic field research, and they are widely used by condensed matter physicists. In other areas where national facilities are inappropriate, sophisticated centers developed by individual investigators or a small group of investigators are necessary to carry out research at very low temperatures, at high pressure, or on mesoscopic structures, to mention only a few examples.

Current Topics

The topics of interest to condensed matter physics are continually changing. Topics that comprise the underpinnings of the field, such as spectroscopy, electronic structure, magnetism, and optical properties, have long time horizons. Other topics have shorter time horizons. One of the most important recent developments in semiconductor physics, both from the point of view of physics and for the purpose of device applications, has been the achievement of structures in which the electronic behavior is essentially two-dimensional (2D). This means that, at least for some phases of operation of the device, the carriers are confined in a potential such that their motion in one direction is restricted and thus is quantized, leaving only a two-dimensional momentum, or k-vector, that characterizes motion in a plane normal to the confining potential. The major sys-

tems where such 2D behavior has been studied are metal-oxide-semiconductor (MOS) structures, quantum wells, and superlattices. More recently, quantization has been achieved in one dimension (the quantum wires) and "zero" dimensions (the quantum dots). Many examples of topics of current interest can be identified, with significant breakthroughs in the subfield occurring on an annual basis.

Some insight into the scope of condensed matter physics can be found in the sorting categories of the annual meeting of the Division of Condensed Matter Physics of the American Physical Society. These sorting categories consist of a matrix of topics relating to the phenomena and properties of materials and topics relating to the structures giving rise to these phenomena and properties. Examples of phenomena and properties of current interest include electronic structure, transport properties, magnetism, phonons, phase transitions, optical properties, nonlinear phenomena (e.g., chaos and noise), superconductivity, many-body theory, simulation and numerical methods, surface science and interfaces, and mesoscopic physics. The topics are also organized from the standpoint of materials classes and structures, such as clusters, complex fluids (e.g., liquid crystals, microemulsions and micelles, colloids), quantum fluids, disordered materials (e.g., glasses, porous media, fractals), metals and alloys, highly correlated metals, semiconductors, ferroelectrics, magnetic materials, layered systems (e.g., superlattices, sandwiches, intercalated systems), heterostructures, polymers, and conducting polymers. Another method of classification of condensed matter physics topics is in accordance with experimental or theoretical techniques, such as x-ray diffraction, neutron scattering, high-pressure physics, photoemission studies, Raman scattering, molecular beam epitaxy, scanning tunneling microscopy. These cross-cutting themes, together with the rapid growth of condensed matter physics, have led to fragmentation of the subfield and a high emphasis on topical symposia.

Because of the continued introduction of new materials in new geometric arrangements and the importance of materials to society, condensed matter physics can be expected to remain an active and challenging subfield of physics for the indefinite future.

See also: ATOMIC PHYSICS; BIOPHYSICS; LIQUID CRYSTAL; LIQUID HELIUM; MAGNET; MATERIALS SCIENCE; MOLECULAR PHYSICS; QUANTUM MECHANICS; SEMICONDUCTOR; SUPERCONDUCTIVITY; TRANSISTOR, DISCOVERY OF

Bibliography

ASHCROFT, N. W., and MERMIN, N. D. *Solid-State Physics* (Holt, Rinehart, and Winston, New York, 1976).

KITTEL, C. *Introduction to Solid-State Physics,* 7th ed. (Wiley, New York, 1996).

MILDRED S. DRESSELHAUS

CONDUCTION

Conduction refers to a large variety of processes in which a physical quantity is conducted or transported across a sample in response to some external condition. Such processes can all be described in a similar fashion, in spite of the disparity of physical quantities under consideration. We will focus here on three major cases: diffusion, electrical conduction, and thermal conduction.

Perhaps the simplest example of conduction is that in which atoms or molecules, or other particles, migrate or diffuse through a material as a result of a nonuniform concentration of particles. In their random thermal motion, particles collide with each other exchanging kinetic energy and a net flow is generated from the high to the low concentration regions, as easily visualized when two liquids of different colors mix. As illustrated in Fig. 1, the number of particles N that diffuse through a cross-sectional area A per unit time is proportional to the variation in concentration n per unit length in a direction x perpendicular to the cross section:

$$\frac{\Delta N}{\Delta t} = -DA\frac{\Delta n}{\Delta x}.$$

The diffusion coefficient, or diffusivity, D characterizes each particular process, and depends on the materials involved and other parameters such as the temperature. Its units in the International System (SI) are m^2/s. The equation above is usually reformulated in a more general and rigorous form, expressing the particle current density (number of particles crossing a unit area in the unit of time) in terms of the concentration gradient:

$$J = -D\,\nabla n.$$

The diffusivity, giving the current density generated per unit of concentration gradient, is thus a measure of the "intensity" of the diffusion process. For example, the diffusion coefficient for gases is approximately 10^5 times larger than for liquids, since the close proximity of neighboring particles in liquids makes diffusion much slower than in gases. Diffusion occurs not only in gases and liquids, but also in solids, where it results in a rearrangement of atoms in the lattice sites. In the case of solids, different coefficients are defined for interstitial, vacancy, chemical, and surface diffusion, as well as self-diffusion. It is also possible to have interphase diffusion, the most common case being that of a gas diffusing in a solid.

Electrical conduction refers to the transport of electrical charges when an electric field, or an electric potential difference, is applied to a sample. For instance, if a piece of wire is connected to a battery, the potential difference (or voltage) provided by the battery generates an electric field in the wire. The electrons in the metal, under the effect of the resulting force, move toward the positive terminal of the battery establishing an electric current. The mathematical representation of this process is analogous to that of diffusion; the amount of electrical charge q traversing a cross-sectional area A per unit time is proportional to the change in electric potential Φ per unit length in a direction perpendicular to the surface:

$$\frac{\Delta q}{\Delta t} = -\sigma A\frac{\Delta \Phi}{\Delta x}.$$

In the more general form of this equation, the charge current density is expressed in terms of the applied electric field E:

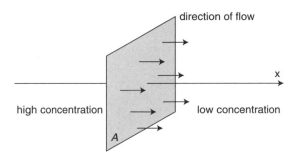

Figure 1 Particles flow through a cross-sectional area A due to a difference in concentration.

$$J = -\sigma\nabla\Phi = \sigma E.$$

$$J = -\kappa\nabla T.$$

Therefore, electrical conductivity σ, being the charge current density produced per unit of electric field, reflects the ease with which a material conducts electrical charges. Its SI units are $\Omega^{-1}m^{-1} = S\ m^{-1}$. The electrical conductivity spans the widest range of values of all conduction coefficients: At room temperature, the conductivity of pure metals is approximately 10^{25} times larger than the conductivity of the best insulators.

In metals, many properties are determined by the presence of electrons which can flow quite freely in the material; these conduction electrons are responsible for the transport of electrical charge and the high value of electrical conductivity. Semiconductors, in contrast, have few intrinsic free carriers and a much lower conductivity when in pure form; holes (positively charged) or electrons introduced by the addition of impurity atoms provide the main carriers of charge, resulting in a p- or n-type semiconductor, respectively. In an electrolytic solution as well as in a gas ionized by a high electric field, both positive and negative ions contribute to electrical conduction.

Not all conduction phenomena involve the transport of matter. In thermal conduction, it is energy which is conducted from one end of the sample to another, when those regions are at different temperatures. Since temperature is a quantitative indication of the kinetic energy of the atoms or molecules in a material, particles in a high temperature region possess larger kinetic energy. Through random collisions, they transfer part of their energy to neighbors possessing less energy. The outcome of this process, repeated throughout the substance, is a net transfer of thermal energy (conduction of "heat") from the high to the low temperature regions. The same form of the conduction equation as before is valid for thermal conduction; now the time rate of heat transfer, or, alternatively, the heat current density, is related to the temperature gradient:

$$\frac{\Delta Q}{\Delta t} = -\kappa A \frac{\Delta T}{\Delta x},$$

where Q is the quantity of heat and T is temperature, or, in the most generally valid form,

In this equation, as well as in previous ones, the negative sign indicates the direction of flow; for thermal conduction, it is from hot to cold, or from high to low temperature. The thermal conductivity κ has SI units of watt/(meter °C); however, values are frequently cited in tables in cal/s m °C. Good thermal conductors are characterized by a high value of κ (e.g., $\kappa_{silver} = 99$ cal/s m °C), whereas poor thermal conductors are characterized by a low value of κ (e.g., $\kappa_{down} = 0.0046$ cal/s m °C). The lowest values of κ correspond to gases, and the highest to metals, spanning a range of four decades.

The spacing between molecules in liquids is considerably smaller than in gases, facilitating collisions and making the transfer of energy faster. In all crystalline solids, lattice vibrations provide the mechanism for the transport of heat energy; in metals, the additional contribution by free electrons enhances the thermal conductivity, just as occurs in the case of electrical conductivity. The free electrons can travel past many atoms (about one hundred) before suffering any collisions, so they are more efficient at transporting energy than the ions that vibrate in localized lattice sites. The relationship between the electrical and thermal conductivities in metals is explicitly recognized in the Wiedemann–Franz law:

$$\frac{\kappa}{T\sigma} = \text{constant.}$$

This law had special significance in lending early support to the free-electron gas model for metals.

See also: DIFFUSION; ELECTRICAL CONDUCTIVITY; ELECTRON, CONDUCTION; JOULE HEATING; SUPERCONDUCTIVITY; THERMAL CONDUCTIVITY

Bibliography

GRAY, D. E., ed. *American Institute of Physics Handbook,* 3rd ed. (McGraw-Hill, New York, 1972).

KITTEL, C., and KROEMER, H. *Thermal Physics,* 2nd ed. (W. H. Freeman, New York, 1980).

OHANIAN, H. C. *Physics,* 2nd ed. (W. W. Norton, New York, 1989).

MARIA CRISTINA DI STEFANO

CONDUCTION ELECTRON

See ELECTRON, CONDUCTION

CONDUCTOR

A conductor is an object that contains an enormous supply of electric charge that is free to move about within the material, producing electric currents. The most familiar conductors are metals, in which one or more electrons are detached from each atom and flow in response to electric fields. (In an insulator, by contrast, the electrons are attached to specific atoms.) Some liquids (such as salt water) also conduct electricity—in this case the free charges are ions.

When an isolated conductor is placed in an electric field, the free charges move: positive charges in the direction of the applied field, or (more commonly) negative charges (electrons) in the opposite direction. The effect is to produce accumulations of charge at the surface. This "induced" charge generates a field of its own, which is opposite in direction to the original field. Charge continues to flow until this secondary field exactly cancels the original one, at which point there is no net force, and the motion ceases. In practice, this equilibrium is established virtually instantaneously. In the static case, then, (1) the net charge is zero everywhere except on the surface, (2) the total electric field within the conductor is zero, and (3) the electric potential (voltage) is constant throughout the conductor.

To sustain a current in an ordinary conductor (such as a piece of wire) it must be hooked up to a source of electromotive force (a battery, say, or a generator). In that case a continuous current will flow, in accordance with Ohm's law: $I = V/R$, where V is the voltage difference between the two ends, and R is the resistance of the wire. The electric field inside a current-carrying conductor is not zero, nor is the electric potential constant. Even though current is flowing, there is still no *net* charge anywhere (except possibly at the surface); in any small volume there is just as much (stationary) positive charge as there is (moving) negative charge.

A perfect conductor offers no resistance to the flow of current ($R = 0$). In such a material, current, once established, keeps flowing indefinitely without any electromotive force. Inside a perfect conductor the electric field is zero and the electric potential is constant. A superconductor is a perfect conductor with the additional property that the magnetic field inside also vanishes (the Meissner effect).

In the quantum theory of solids a crystal lattice gives rise to "allowed" energy bands for the electrons, separated by "forbidden" gaps. In an insulator at equilibrium, the highest-energy occupied band is entirely filled, and it takes a substantial jump in energy to "liberate" an electron for purposes of conduction. In a conductor, by contrast, the highest occupied band is only partly filled, and very little energy is required to promote an electron to a conducting state. A semiconductor is an insulator that has been doped with a relatively small number of atoms of a different species, which carry either an extra electron (which is forced into the conduction band), or one fewer electron (leaving a "hole" in the otherwise filled band, into which electrons can easily move). The result is conduction properties intermediate between those of an insulator and a true conductor.

See also: DOPING; ELECTRIC POTENTIAL; FIELD, ELECTRIC; INSULATOR; OHM'S LAW; SEMICONDUCTOR; SUPERCONDUCTIVITY

Bibliography

ASHCROFT, N. W, and MERMIN, N. D. *Solid-State Physics* (Saunders, Philadelphia, 1976).

PURCELL, E. M. *Electricity and Magnetism*, 2nd ed. (McGraw-Hill, New York, 1985).

DAVID GRIFFITHS

CONSERVATION LAWS

A conservation law is a statement of constancy in nature. One quantity (the conserved quantity) remains constant while other quantities may change. For example, in a collision of elementary particles, the total momentum is conserved (i.e., remains constant) while other quantities, such as the speeds and directions of the particles, and even the number of particles and their masses, may change.

Conservation laws are regarded by scientists as particularly significant, in part because they embody the simple idea of constancy, and in part because of their relation to symmetry principles.

Conservation laws are useful in analyzing processes in which the details are complex and perhaps not measurable or calculable. In an elementary particle collision, for example, conserved quantities such as energy, momentum, angular momentum, and electric charge are the same after the collision as before, even though the details of what happened during the collision are beyond measurement.

Conservation laws apply in isolated systems, those for which external influences are absent or too small to be significant. A particle collision is isolated in this sense because gravity, electric fields, magnetic fields, and neighboring atoms have no appreciable effect during the brief moment of the collision. The solar system as a whole is nearly isolated because the nearest stars are very far away compared with the size of the solar system. The system of the Sun and the planet Mercury is not quite isolated because other planets have a small influence on the motion of Mercury. (However, the Sun together with any single planet is so nearly isolated that Johannes Kepler was able to deduce a law that is equivalent to the conservation law of angular momentum from data on planetary orbits.)

The quantities believed to be absolutely conserved in all processes are energy (including mass), momentum, angular momentum (including spin), and charge. The conservation laws of energy, momentum, and angular momentum are related to the uniformity of spacetime—the fact that laws of nature do not change with time nor depend on location or on direction in space. The conservation of charge is related to a more subtle symmetry of quantum waves.

Momentum is a vector (directed) quantity. Following a collision, be it of automobiles or particles, the vector sum of the momenta has the same direction and the same magnitude as before the collision. Nonrelativistically, the momentum of an object is its mass times its velocity ($\mathbf{p} = m\mathbf{v}$). Relativistically, the definition is different but the conservation law remains valid.

Energy is a scalar (numerical) quantity that takes many forms. Kinetic energy and potential energy are forms of mechanical energy. Heat is a mode of energy transfer, often from the ordered mechanical energy of a system to the disordered internal energy of a body. Tires skidding on pavement, for example, dissipate energy in the form of heat from the kinetic energy of the vehicle to the internal energy of the pavement and the tire (which become warmer).

Mass locks up energy. When mass decreases, energy is released—as in a nuclear reactor, where the masses of the atoms after fission are less than the original masses of the atoms, and in the Sun, where the helium formed by fusion has less mass than the fusing hydrogen. In some particle collisions, the masses of the products are greater than the masses of the original particles, and energy conversion goes in the other direction. At the deepest level, mass is the total measure of energy. The mass of an isolated system is a complete measure of the energy content of the system (in a frame of reference where the system's center of mass is at rest).

Angular momentum, like momentum, is a directed quantity. A spinning gyroscope free of all outside influences will spin forever with both its axis of spin and its magnitude of angular momentum preserved unchanged. Earth's angular momentum is approximately constant, but the Sun and the Moon, tugging on its nonspherical mass, cause its axis of spin to wobble (precess) along an arc that repeats itself every 26,000 years. For particles, as well as for Earth, it is useful to distinguish *spin* angular momentum and *orbital* angular momentum. Spin results from rotation of the body about its own axis. Orbital angular momentum results from the motion of the body's trajectory relative to some other point.

Benjamin Franklin first stated a law of charge conservation. It has withstood the test of time. Its conservation in the large-scale world is believed due to the conservation of charge in every particle interaction. The conservation of charge stabilizes the electron, the lightest of all charged particles. If the electron were to decay spontaneously, it would have to do so into uncharged particles, which would violate the law of charge conservation.

Other conservation laws that may be absolute include laws of particle number, such as baryon number. The number of neutrons plus protons plus other "heavy" particles (called baryons), minus the number of antibaryons, is, according to all present evidence, a constant. This law accounts for the stability of the proton, which is the lightest baryon. Modern particle theory suggests the possibility that the proton may be unstable, but with the probability of its decay being almost infinitesimal. Using sensitive experiments, scientists have searched for proton decay but have not found it. The total numbers of each of three families of leptons also appear to be constant.

Approximate, or partial, conservation laws govern quantities that are conserved in some processes, but not in all. In the hierarchy of interaction strengths (strong, electromagnetic, weak, and gravitational), the stronger interactions are restricted by more conservation laws than are the weaker interactions. Particle properties called strangeness, charm, and isospin, for example, are conserved in strong interactions, but not in weak interactions. Whether gravity is restricted by even fewer conservation laws than weak processes is not known.

One of the first conservation laws to be discovered (late in the eighteenth century) was the conservation of mass in chemical reactions. We now know that this law is not exact. Rather, it is the number of baryons in the nuclei of the reacting atoms that is constant. But changes of mass in chemical reactions are so small that the law of mass conservation remains true to excellent approximation in chemistry.

According to the reductionist view that dominates modern science, conservation laws in the large-scale world result from the action of such laws in the submicroscopic world. There is extensive evidence that this is the case; often, elementary particle interactions provide the clearest and most accurate tests of the conservation laws.

See also: BARYON NUMBER; CONSERVATION LAWS AND SYMMETRY; ENERGY, CONSERVATION OF; MOMENTUM, CONSERVATION OF

Bibliography

BERNSTEIN, J. *The Tenth Dimension* (McGraw-Hill, New York, 1989).

FORD, K. "Conservation Laws" in *The World of Physics*, edited by J. H. Weaver (Simon & Schuster, New York, 1987).

PAGELS, H. *The Cosmic Code: Quantum Physics as the Language of Nature* (Simon & Schuster, New York, 1982).

KENNETH W. FORD

CONSERVATION LAWS AND SYMMETRY

Many physical systems have symmetry. What this means is that the system looks the same if it is transformed in some way. The symmetry is the set of transformations that leave the physics of the system unchanged. Continuous symmetries are those in which the transformations that leave the physics unchanged depend on one or more continuous variables. These are closely related to conservation laws—principles that dictate that some physical quantity not change with time. To understand this connection, we must begin with a discussion of how physical parameters do change with time.

Physicists describe the systems they study in terms of a set of parameters called "coordinates" that specify the configuration of the system. For example, for a particle moving along a track (and, therefore, constrained to move in only one dimension) there is one coordinate: the position along the track, which we often call x. The dynamics of the system is then the set of rules that determine how such coordinates change in time. In a wide class of physical systems, it is useful to discuss this time dependence by asking how the energy of the system changes when one of the coordinates changes. If the energy of the system can be decreased by a small change in one of the coordinates, then there is a force on the system that tends to change the coordinate and decrease the energy. Such a force, in turn, produces time dependence by causing acceleration. In the example of the particle on the track, if the track is tilted in the gravitational field of the earth, there is a downward force of gravity, and the contribution of gravity to the energy of the particle can be described by a potential energy $V(x)$ that decreases as the particle moves down the track.

For each coordinate, there is a quantity called a momentum, whose rate of change with time is equal to the force. (The function that describes the total energy, kinetic plus potential, in terms of the coordinates and the momenta is called the Hamiltonian. This function plays a crucial role in the mathematical description of dynamics and in quantum mechanics.) For a particle of mass m moving in one dimension, the momentum (called p; there is only one because there is only one dimension) is the mass times the velocity:

$$p = mv = m\frac{dx}{dt}.$$

Then the dynamics is described by the equation

$$\frac{dp}{dt} = F,$$

where F is the force. This is equivalent to Newton's second law, $F = ma$. (Actually, it is slightly more general; true even if the mass depends on x.)

It is now simple to state the connection between continuous symmetries and conservation laws. If there is a continuous symmetry of the system, then there is a family of transformations that can be made on the system, labeled by a set of continuous parameters, that leave the physics unchanged. It is convenient to choose these labels so that if all are zero, the transformation is the trivial one in which nothing is changed. Then small values of the parameters correspond to small transformations for which the coordinates of the transformed system are only slightly different from those of the original system. Then, for each parameter in the continuous symmetry, there is a transformation for which the parameter is small and all the other parameters vanish. This corresponds to a small change of some combination of the coordinates of the system. But because it is a symmetry, the physics is unchanged and, therefore, the energy is unchanged. Thus the force associated with a change of this combination of the parameters is zero. Then the corresponding combination of momenta does not change with time! That is, this combination of momenta is conserved.

For example, for a particle moving in one dimension, suppose that the system has a symmetry under translations. That is, the physics is the same if x is transformed to $x + a$ for some fixed translation a. Here, only the single coordinate x changes (because it is the only coordinate there is) and, thus, according to the argument above, the momentum corresponding to x must be conserved. And this is right. If the physics is the same for any a, then the potential energy $V(x)$ must equal $V(x + a)$ for any a, which is only possible if $V(x)$ is constant. But then the potential energy does not change when the coordinate changes; there is no force on the particle, and the momentum does not change with time, in which case we say it is conserved.

This seems rather trivial, but consider the situation for more than one particle. For simplicity, we will discuss two, with masses m_1 and m_2, described by coordinates x_1 and x_2, and some potential energy $V(x_1, x_2)$. Now translation symmetry implies that the physics is unchanged under the transformation

$$x_1 \rightarrow x_1 + a,$$

$$x_2 \rightarrow x_2 + a.$$

All the coordinates change in the same way, so the momentum corresponding to a translation is the sum of the momenta for each particle. Thus the total momentum $P = m_1 v_1 + m_2 v_2$ is conserved.

Again, we could work this out step by step. Translation symmetry implies that the potential energy actually only depends on the difference $x_1 - x_2$, which does not change under a translation. This in turn implies that the force on particle 1 is equal in magnitude but opposite in sign to the force on particle 2. Thus the corresponding momenta change in opposite directions with time, and the total momentum is constant. The connection between the symmetry and the conservation law is a shortcut that allows us to get directly to the conserved quantity. Note that translation symmetry is related to Newton's third law: for every action there is an equal and opposite reaction.

This shortcut is often a huge help in simplifying the analysis of the dynamics. A nice example is the motion of a planet around the Sun. If we ignore the small nonuniform motion of the Sun (because the Sun is so much heavier than any of the planets), and the slight oblateness of the Sun, the physics of the planet is unchanged if we make any rotation about the center of the Sun. The corresponding conserved quantity is the angular momentum $\mathbf{r} \times \mathbf{p}$ [where $\mathbf{r}$ is the vector from the Sun to the planet and $\mathbf{p}$ is the momentum. This immediately implies that the motion of the planet is in a single plane (containing the center of the Sun)]. It also implies Kepler's second law: The line joining the planet and the Sun sweeps out equal areas in equal times.

Our discussion so far has been Newtonian—ignoring Einstein's special relativity. For this reason, one familiar conservation law does not fit the general description above. The conservation of total energy is related to time translation symmetry just as the conservation of total momentum is related to space translation symmetry. Here we cannot use quite the same argument, because time is not a coordinate. Rather, this conservation law underlies the whole system of Hamiltonian dynamics. However, in special relativity, time and the space coordinates are related. The transformations of Lorentz symmetry, going from one inertial frame (i.e., a coordinate system that is not accelerating) to another, mix up the space coordinates with time. It is possible to reformulate physics in a way that treats space and time symmetrically. However, because information cannot be transmitted faster than the speed of light, this reformulation is difficult (or impossible) in a formalism that includes concepts like the potential energy that describe forces transmitted over finite distances at a fixed time. The only way we know to build Lorentz invariant interactions is to use a field theory, such as Maxwell's description of electricity and magnetism, in which the fundamental quantities, the coordinates, are fields such as the electric and magnetic fields, which depend on space and time.

In a field theory, where the physical quantities depend on space and time, a conservation law is a bit more complicated because it describes physics at each point in space and time. Consider a quantity such as electric charge. In a field theory, charge conservation means that the rate of change of charge in any volume of space must equal the rate at which charge flows into the volume. The connection was worked out in 1918 by the great mathematician Emmy Noether. She showed that a continuous symmetry implies the existence of a conserved current such that the rate of change of charge equals the flow into the region. This is the form in which we use the connection today—it is referred to as Noether's theorem.

Because of Noether's theorem, Lorentz symmetry and translation symmetry in space and time are related to the existence of a conserved energy momentum tensor, which plays a critical role in Einstein's general theory of relativity.

The symmetry that is related to the conservation of electric charge is particularly important in field theory, although it is somewhat bizarre. The theory must be unchanged when each field is multiplied by a complex number of the form $e^{iQ\theta}$, where Q is the charge of the particle described by the field and θ is an arbitrary real angle. In fact, the theory has a much larger symmetry. The angle θ may depend on space and time if, at the same time, we make a compensating change in the electric and magnetic potential fields. This huge symmetry, called gauge invariance, actually determines the structure of electrodynamics, the interactions of electric and magnetic fields with matter. Because θ can be different at each point in space and time, there must be a quantity that is separately conserved at each point of space and time. At a fixed time, this quantity is the combination of the electric charge and the flux of the electric field. The easiest way to think about flux is in terms of field lines. The point is that the contribution of any positive electric charge to the electric field can be described by a set of field lines that radiate out from the charge and can end only on a nega-

tive electric charge. The electric field at any point in space is parallel to the field line through that point, and its magnitude is proportional to the density of field lines. The flux of the electric field through any surface is then proportional to the number of field lines that cut through the surface (at least if all the lines pass through in the same direction, otherwise you must subtract the number going the wrong way). Flux conservation is the statement that the lines can start only at positive charges and end at negative charges. For example, a single positive charge produces straight field lines that radiate out uniformly in all directions from the charge. The electric field drops off like $1/r^2$ (where r is the distance from the charge) because the field lines at radius r are spread out over a sphere of area $4\pi r^2$, and thus their density falls like $1/r^2$. The conservation of flux allows us to determine the electric charge in any region just by measuring the electric field on the surface of the region—counting the field lines that come out. Gauge invariance is a crucial concept in modern physics. All of the forces that we know of are associated with gauge symmetries.

In addition to energy, momentum, angular momentum, and electric charge, there are a number of quantities that at least so far appear to be conserved, and thus may be associated with symmetries of the dynamical theory.

Quantities that have not yet been shown to be mutable include baryon number (the total number of proton, neutrons, and related particles, minus the number of their antiparticles), electron number (the total number of electrons and electron neutrinos, minus the number of their antiparticles), muon number (the total number of muons and muon neutrinos, minus the number of their antiparticles), and tau number (the total number of taus and tau neutrinos, minus the number of their antiparticles). (The muon and the tau are heavier, unstable versions of the electron.) These would-be conservation laws are on a rather different footing than the conservation of electric charge because they are not related to gauge symmetries. Many physicists would be surprised if any of these conservation laws were exact. However, it is clear that if there are interactions that change these quantities, they are very weak. Otherwise we would have seen them.

Often, symmetries are useful even if they are only approximate. A beautiful example in particle physics is the approximate $SU(3)$ symmetry that rotates the fields describing the three light quarks up, down, and strange into one another (in a complex space). If we could somehow turn off the electromagnetic and weak interactions, and if the masses of the quarks were exactly equal, this symmetry would be an exact symmetry of the remaining strong quantum chromodynamic (QCD) interactions. Since we cannot do this, the symmetry is broken. However, because the weak and electromagnetic interactions are much weaker than the QCD interactions, and because the differences between the different quark masses are small compared to the proton mass and other such particle masses, $SU(3)$ is a good approximate symmetry. It shows up dramatically in the appearance of families of strongly interacting particles with related properties, and it was crucial historically to our understanding of QCD.

Ironically, the approximate $SU(3)$ symmetry was also crucial in the development of our understanding of the weak interactions. Noether's theorem focused attention on the approximately conserved currents related to $SU(3)$ symmetry. It is from these currents, and similar objects involving other fields, that the first successful phenomenological descriptions of the weak interactions were constructed. The connection with symmetry was an important clue, in turn, to the symmetry structure of the electroweak interactions that unify the weak interactions with electromagnetism in a gauge theory. Still today, Emmy Noether's connection between symmetries and conservation laws is a critical part of the arsenal of the physicist.

See also: CONSERVATION LAWS; ELECTRIC FLUX; GAUGE INVARIANCE; GAUGE THEORIES; GRAND UNIFIED THEORY; INTERACTION, ELECTROWEAK; INTERACTION, STRONG; QUANTUM CHROMODYNAMICS; QUANTUM FIELD THEORY; SYMMETRY

HOWARD GEORGI

CONSTANT

See BOLTZMANN CONSTANT; COSMOLOGICAL CONSTANT; DIELECTRIC CONSTANT; ELASTIC MODULI AND CONSTANTS; FINE-STRUCTURE CONSTANT; GAS CONSTANT; GRAVITATIONAL CONSTANT; HUBBLE CONSTANT; MATHEMATICAL CONSTANTS; PLANCK CONSTANT; RYDBERG CONSTANT

CONVECTION

When a fluid (gas or liquid) is warmed, it typically expands, becoming less dense. As it is lifted by surrounding cooler, denser fluid, the warmer fluid carries thermal energy (heat) along with it. This movement of thermal energy, carried by matter, is called convection (literally, to "carry with"). When the motion is caused by, or aided by, a pump or fan, it is called forced convection.

Convection is the primary method of heating homes. Warm air rises, is cooled, and sinks, providing circulation. Similarly, cooled air falls, is warmed, and rises. Convection mixes the water in lakes and oceans, distributing oxygen, nutrients, and thermal energy. Because water reaches its maximum density at 4°C, convection stops at this temperature and water then freezes from the top (surface) down, rather than from the bottom up. This prevents lakes from freezing solid, so fish can survive during the winter.

Convection in the atmosphere carries energy upward, at the same time dissipating polluting dust and gases from cities, farms, and forests. When a warm layer of air collects above a surface (especially likely above geologic bowls or valleys), the convection process is arrested and pollutants remain trapped. This is called a thermal inversion.

Although atmospheric physics involves many complexities, some basic properties are easily understood as manifestations of convection. When the surface of the earth is heated by the Sun, air is warmed and rises, where it is then cooled (by expansion and by radiation) to form clouds and thence rain. Because hot air is less dense, cooler air tends to move in under the heated air, causing hot and cold cycles over periods that are often a few days in length. If the heating is sufficiently intense, the incoming cooler air may pick up sufficient speed to produce hurricanes or tornadoes, especially if driven by hot, wet air from the ocean surface. Condensation of moisture, as the air rises, releases energy that keeps the air warmer than its surroundings, so it continues to rise, aiding the storm-building process.

When a layer of fluid is heated from below, as at the surface of the Sun or of a cup of coffee (with its surface cooled by evaporation), a thermal instability may occur that leads to convection cells, known as Bénard cells. These may take the form of symmetrical, close-packed, hexagonal cells, with warm fluid rising in their centers and cooled fluid descending at their edges.

Combustion of a substance relies on convection to supply oxygen, brought in by cool air, and to remove hot waste products that would otherwise smother the flame. Although convection helps achieve approximately uniform temperatures, and aids in the growth of crystals by transporting dissolved materials to the site of the crystal growth, it also limits the degree of perfection of the growing crystals because of the mixing in of the dissolved materials. A major success of space flight has been to grow better crystals in the "weightless" environment where convection is avoided.

By analogy, an electric current carried by uncompensated moving charges, such as an electron beam, is sometimes called a convection current.

See also: ATMOSPHERIC PHYSICS; CONDENSATION; HURRICANE; TORNADO

Bibliography

ALLEN, O. E., *Planet Earth: Atmosphere* (Time-Life Books, Alexandria, VA, 1983).

TARBUCK, E. J., and LUTGENS, F. K. *Earth Science*, 6th ed. (Macmillan, New York, 1991).

ROBERT P. BAUMAN

COOLING

See LASER COOLING; MAGNETIC COOLING

COOPER PAIR

Many metals exhibit the phenomenon of superconductivity when they are cooled below a certain temperature called the superconducting critical temperature T_c. In the superconducting state, such a metal loses all of its resistance to the flow of an electrical current (i.e., it becomes a perfect conductor of electricity). The value of T_c is characteristic of the particular superconducting material. Prior to 1986, the highest value of T_c exhibited by any superconductor was 23 K for the intermetallic compound Nb_3Ge. Since 1986, the maximum value of T_c has risen dramatically in metallic copper oxides. For example, a value of about 133 K has been obtained for the compound $HgBa_2Ca_2Cu_3O_{8+\delta}$.

To understand why the electrical resistance of a metal vanishes in the superconducting state, it is first necessary to see how it originates in the normal (nonsuperconducting) state. A metal is a good conductor of electricity because it contains negatively charged conduction electrons that flow freely throughout its volume and carry an electrical current. The conduction electrons are the loosely bound outermost electrons of the atoms that comprise the metal. These electrons have been removed from the atoms, thus transforming the atoms into positively charged ions. The positively charged ions form the crystal lattice, a regular spatial arrangement of the ions. A normal metal has a finite electrical resistance because the mobile conduction electrons are scattered by ions that are displaced from their equilibrium positions in the crystal lattice by thermal excitations, and by defects in the crystal lattice [e.g., impurity atoms, vacancies (missing atoms), and dislocations (long arrays of atoms located in improper positions in the lattice)].

In contrast, the electrical resistance of a metal vanishes in the superconducting state because the conduction electrons form pairs, called Cooper pairs, that move in concert throughout the crystal lattice *without scattering*. The attractive interaction that binds the two electrons of a Cooper pair together, overcoming the strong Coulomb repulsion between them, utilizes the lattice vibrations, or phonons, which ordinarily scatter the electrons in the normal state. The way the attractive electron-phonon interaction works can be visualized in the following way: A negatively charged conduction electron passing in the vicinity of a positively charged ion attracts the ion and displaces it from its equilibrium position in the lattice, producing a net positive charge that attracts a second negatively charged electron traveling through that region. This results in a weak attraction between the two electrons that can overcome the Coulomb repulsion because the electrons move much faster than the ions (i.e., the lattice distortion persists for a relatively long time compared to the transit time of the electron that produced it). The attractive electron-phonon interaction is schematically illustrated in Fig. 1.

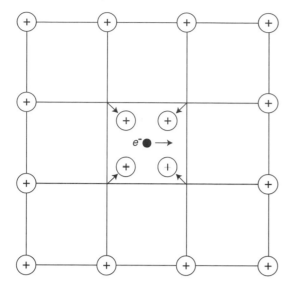

Figure 1 Schematic illustration of the electron-phonon interaction responsible for binding the Cooper pairs in conventional superconductors. A negatively charged conduction electron passing in the vicinity of a positively charged ion attracts the ion and displaces it from its equilibrium position in the lattice. This produces a net positive charge that attracts a second negatively charged electron traveling through that region, resulting in an attraction between the two electrons.

The Cooper pairs are named after Leon N. Cooper who showed in 1956 that electrons in a metal tend to form bound pairs under the influence of a weak attractive interaction. Electron pairing is an essential element of the microscopic theory of superconductivity proposed by John Bardeen, Cooper, and J. Robert Schrieffer (BCS) in 1957. The BCS theory successfully accounts for the properties of conventional superconductors and provides an expression for T_c in terms of properties of the normal state. This expression is given by

$$k_B T_C \approx 1.13 h f_D \exp[1/N(E_F)V], \qquad (1)$$

where k_B is Boltzmann's constant, h is Planck's constant, f_D is the Debye frequency—a characteristic frequency of the quantized vibrations (phonons) of the crystal lattice, $N(E_F)$ is the density of electronic states at the Fermi level E_F (the maximum energy of the conduction electrons), and V is the strength of the attractive electron-phonon interaction. The theory

predicts the existence of an energy gap $\Delta(T)$, the minimum energy required to break a Cooper pair, which has a maximum value at zero temperature of $\Delta(0) = 3.5\ k_B T_c$ and vanishes at T_c.

Two phenomena that provide evidence for the existence of Cooper pairs in superconductors are the quantization of magnetic flux and the Josephson effect. The magnetic flux passing through a hole (or normal region) in a superconductor subjected to a magnetic field is quantized in integral multiples of a fluxoid quantum $\phi_o = hc/e^* = 2.07 \times 10^{-7}$ G cm². The charge $e^* = 2e$ ($e = 1.60 \times 10^{-19}$ C is the charge on a single electron) indicates that a superconducting charge carrier contains two electrons. In 1965 Brian D. Josephson predicted the existence of a zero-voltage supercurrent through a tunnel junction due to the tunneling of Cooper pairs, a phenomenon known as the dc Josephson effect. He also predicted that the application of a dc voltage difference V across a tunnel junction would produce an alternating current of frequency $f = e^*V/h$ where $e^* = 2e$, again implying the existence of electron pairs. A Josephson tunnel junction is formed by two superconductors separated by a thin insulating barrier or a weak link that consists of a short constriction in the cross section of one of the superconductors, a point contact, or a thin layer of normal metal.

An important aspect of Cooper pairs is illustrated by the following analogy: Rather than performing a "tango" in the superconducting state, the electron pairs participate in a "square dance," exchanging partners on a time scale of order $\tau_c \approx h/k_B T_c$. The average separation of electrons in a pair is the coherence length $\xi \approx v_F \tau_c \approx h v_F / k_B T_c$, where v_F is the Fermi velocity, the velocity of the conduction electrons at the Fermi level. For conventional (BCS) superconductors such as Pb, alloys of Nb and Ti, and the intermetallic compound Nb₃Sn, $\xi \approx 100$-1000 Å, while for the new high T_c oxide superconductors, $\xi \approx 10$ Å.

Superconductors can be divided into two classes, type I and type II, depending upon their behavior in an external magnetic field. A type-I superconductor exhibits complete expulsion of a magnetic field from its interior (Meissner effect) until the magnetic field reaches the critical field H_c at which superconductivity is quenched. In contrast, a type-II superconductor allows the magnetic field to penetrate into its interior through the normal cores of magnetic fluxoids that form in the vortex phase or Schubnikov phase between the lower critical field

H_{c1}, below which the magnetic field is expelled, and the upper critical field H_{c2}, above which bulk superconductivity is destroyed. Certain superconducting properties of a type-II superconductor depend upon the superconducting coherence length ξ. The radius of the normal core of a fluxoid is about equal to ξ, and the maximum value of the upper critical field is given by $H_{c2}(0) \approx \phi_o/\pi\xi o^2$, where ξ_o is the coherence length at $T = 0$ K. In the vortex phase, the magnetic fluxoids arrange themselves in a triangular structure known as the Abrikosov lattice, which was predicted on theoretical grounds and experimentally observed by decorating the surface of the superconductor with fine ferromagnetic particles (e.g., iron particles) that accumulate at the locations of the fluxoids and viewing the particles with scanning electron microscopy.

In addition to electrical charge, an electron carries a spin which behaves in a magnetic field like a microscopic bar magnet. The two electrons that form a Cooper pair in a conventional (BCS) superconductor have oppositely oriented spins so that net spin S of the Cooper pair is zero, while the orbital angular momentum L of the electron pair about its center of mass is also zero. The $S = 0$ spin state is called a spin-singlet (Fig. 2a), and the $L = 0$ orbital angular momentum state is termed an s-wave state (the values of S and L are in units of $h/2\pi$). The superconductivity exhibited by conventional superconductors is often referred to as s-wave superconductivity.

Another example of BCS pairing is provided by superfluid ^{3}He. The viscosity and thermal resistivity of a superfluid are zero, analogous to the vanishing electrical resistivity of a superconductor. In the superfluid state of ^{3}He that occurs below ~0.002 K, the ^{3}He atoms form pairs as in Fig. 2b, where the spins of their nuclei (each with $S = 1/2$) are aligned parallel to one another (the two electrons of a He atom fill up the $1s^2$ shell, yielding a net electronic spin of zero); the "glue" which binds the resultant spin-triplet ($S = 1$) p-wave ($L = 1$) pairs together are magnetic excitations of the surrounding liquid. In certain unconventional superconductors, such as the so-called heavy fermion f-electron cerium and uranium intermetallic compounds and the high T_c copper oxides, there is considerable evidence that electrons are paired in states with opposite spin (spin-singlet) in some materials and parallel spin (spin-triplet) in other materials, with nonzero angular momentum about the center of mass. For a spin-singlet, the lowest nonzero orbital angular momentum corresponds to a d-wave ($L =$

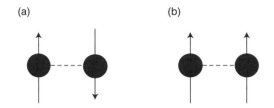

Figure 2 (a) Spin-singlet and (b) spin-triplet Cooper pairs. The solid circles represent the electrons in a superconductor or the nuclei in a superfluid, and the arrows are the electronic or nuclear spins.

2) state, while for a spin-triplet, it is a p-wave state. Some researchers believe that the pairing in heavy fermion f-electron and high T_c cuprate superconductors is mediated by magnetic excitations, rather than phonons.

See also: BARDEEN, JOHN; CRYSTAL DEFECT; ELECTRICAL RESISTANCE; ELECTRON, CONDUCTION; JOSEPHSON EFFECT; PHONON; SPIN; SPIN, ELECTRON; SUPERCONDUCTIVITY; SUPERCONDUCTIVITY, HIGH-TEMPERATURE

Bibliography

COX, D. L., and MAPLE, M. B. "Electronic Pairing in Exotic Superconductors." *Phys. Today* **48** (Feb.), 32 (1995).
GINZBURG, V. L., and ANDRYUSHIN, E. A. *Superconductivity* (World Scientific, Singapore, 1994).
LYNTON, E. A. *Superconductivity* (Wiley, New York, 1962).
TINKHAM, M. *Introduction to Superconductivity* (McGraw-Hill, New York, 1975).

M. BRIAN MAPLE

COORDINATE SYSTEM, CARTESIAN

If you are trying to locate someone in a city, you need an address. Assume that the city streets are straight and all run either east–west or north–south. You need two numbers, one to tell you how far east or west to go and one to tell you how far north or south to go. These numbers are called coordinates and the system of streets is a coordinate system. There must also be two special streets from which you start measuring north–south and east–west.

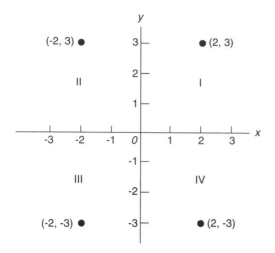

Figure 1 Illustration of the Cartesian coordinate system.

These two streets are called axes and their intersection is called the origin. Now imagine a map on which we have drawn the two axes. Label the origin O and label the point where we want to go as P. It is customary to call the horizontal (or east–west) axis the x axis, and to call the vertical (or north–south) axis the y axis. We want to go from O to P. The number that tells us how far to go east or west is then called the x coordinate, or the abscissa, of the point P, and the number that tells us how far to go north or south is called the y coordinate, or ordinate, of P. We can go a step further with this illustration and suppose that the person we want to locate is in a tall building; then we need to know the floor number. We imagine a third axis drawn perpendicular to the other two and passing through their intersection (the origin). This is called the z axis, and the z coordinate tells us how high to go in the building.

This illustration shows one use of a coordinate system, namely, to locate a point in space. There are other applications, but first we need to discuss in a little more detail the mathematics and terminology of rectangular Cartesian coordinate systems. First consider two dimensions, that is, a plane or a sheet of paper or your computer screen. It is convenient to use (or at least imagine) rectangular coordinate graph paper, that is, paper that is covered by two sets of parallel lines equally spaced with one set running horizontally across the paper and one set running vertically. Select one line of each set; these will be the x and y axes and their intersection is the origin. Mark the horizontal axis x and the vertical axis y. On

each axis we mark a scale of evenly spaced numbers using positive numbers to the right on the x axis and up on the y axis, and using negative numbers to the left on the x axis and down on the y axis. (Fig. 1). We can indicate a point P by the notation (x, y). Then the origin is the point $(0, 0)$. The point $(-2, 3)$ means $x = -2$, $y = 3$; we find this point by going from the origin two units to the left and three units up and plot it (that is, place a dot at the point).

The two axes divide the plane into four parts called quadrants, these quadrants are usually labeled by roman numerals I, II, III, and IV. The first quadrant I is the area where x and y are both positive. The second quadrant II corresponds to negative x and positive y, the third quadrant III to negative x and negative y, and the fourth quadrant IV to positive x and negative y (Fig. 1).

It is useful to see the geometric meaning of equations. For example, the equation $x = -2$ means all the points where $x = -2$ and y has all values; this means all the points on the vertical straight line two units to the left of the y axis. Similarly, the equation $y = 0$ means the x axis and the equation $y = x$ means the line through the origin bisecting quadrants I and III. More complicated equations may represent curves. The lines $x = $ constant and $y = $ constant are called coordinate lines (these are the lines on a sheet of graph paper).

So far we have considered using coordinate systems to locate points in space. Another use for a rectangular coordinate system is to plot a graph of data. Suppose we know the distance D a car has gone at time t for each t from $t = 0$ to $t = 20$ min. We would then label the horizontal axis t instead of x and label the vertical axis D. It is customary to call the variable plotted on the horizontal scale (x or t) the independent variable and to call the other variable the dependent variable. Now in this problem we no longer keep the restriction (mentioned above) of using the same scale along both axes. The units are not even the same (say time in minutes and distance in kilometers) so we mark each axis not only with a scale but also with a unit. Be very careful about equations when the scales are different. If we had the equation $D = t$ in this problem, this would really mean $D = vt$, where v is a constant equal to 1 km/min.

Coordinate systems in three dimensions can be used for the same purposes as in two dimensions. Using a three-dimensional rectangular coordinate system we can locate any point by specifying its coordinates (x, y, z). Equations of the form $x = $ constant (and similarly for y and z) now represent planes called coordinate surfaces, and more complicated equations may correspond to slanted planes or curved surfaces. Suppose we know the temperature $z = T(x, y)$ at each point (x, y) of a sheet of material. We can draw a graph of the temperature by plotting the points (x, y, z) satisfying this equation (or given in a table of measured values); the result will be some kind of surface.

There are many uses of rectangular coordinates besides plotting points and graphs. Consider a particle moving under gravity near the surface of the earth or a charged particle in a constant electric field (such as in a parallel-plate capacitor). Here the forces are easily expressed in terms of x, y, z so rectangular coordinates would be appropriate to use in discussing the motion. Any problem involving rectangles or rectangular boxes or objects in these shapes would use rectangular coordinates.

We have described a three-dimensional rectangular coordinate system in which the x axis points east, the y axis points north, and the z axis points up from the surface of the earth. This is called a right-handed coordinate system. To see whether a system is right-handed, imagine grasping the z axis with your right hand so that your thumb points in the direction of positive z. Then if your fingers curl in the direction of rotation of the positive x axis toward the positive y axis, the system is right-handed. Interchanging any two axes of a right-handed system makes it a left-handed system. The reason for concern about which system you have is that many of the formulas in vector analysis are different (by a minus sign) in the two systems. Since it is a nuisance to deal with two sets of formulas, we practically always use right-handed systems. It is interesting to note that the reflection in a mirror of a right-handed system is a left-handed system.

Physical space is three-dimensional; however, many physical problems involve more than three variables. For example, if we want to describe the motion of an object, there are four variables, its position (x, y, z) and the time t. Or, if it is an object that is rotating as well as translating, there are some more variables to describe its state of rotation. It has proved very useful for problems involving two or three variables to think of each set of values of the variables as coordinates of a point in space. We can extend this to think of a set of values of four variables as a point in a space of four dimensions. Although we cannot actually plot the point in physical space, it is useful to generalize many of the concepts

and methods developed in two and three dimensions. For example, we learn to find the scalar product of two vectors $\mathbf{A}$ and $\mathbf{B}$ as $A_x B_x + A_y B_y + A_z B_z$ in three dimensions. Then in four dimensions, we simply add another term for the fourth components. This generalization continues indefinitely. We say that an ordered set of n numbers corresponds to a point in a space of n dimensions. The term "ordered" is essential here. For example, the set of three numbers 1, 2, and 3 can be arranged in six ways: $(1, 2, 3)$, $(2, 1, 3)$, and so on. Each of these six arrangements corresponds to a different point in three-dimensional space and each is a set of the three numbers in a particular order.

In the most general usage, the term "Cartesian" simply means that the coordinate surfaces are planes (rather than curved surfaces as in, say, spherical coordinates). The coordinate planes (and axes) may intersect at right angles (rectangular coordinates), or at some other angles (oblique coordinates). Oblique coordinates are useful in applications (for example, in crystallography and in special relativity). But since rectangular coordinate systems are the ones used most of the time, you will find that when people say Cartesian, they usually mean rectangular Cartesian.

See also: COORDINATE SYSTEM, CYLINDRICAL; COORDINATE SYSTEM, POLAR; COORDINATE SYSTEM, SPHERICAL; RIGHT-HAND RULE

MARY L. BOAS

COORDINATE SYSTEM, CYLINDRICAL

Many physical systems and geometric figures have an axis of symmetry. Consider, for example, a cylinder, a cone, an ellipsoid, a torus (doughnut), or any other geometric figure that can be obtained by rotating a plane figure around an axis. Also consider physics problems such as the rotation of an object about an axis, the magnetic field around a long, straight, current-carrying wire, or any problem involving surfaces or solids with axial symmetry (for example, finding their volumes or moments of inertia). For such problems, it is convenient to use cylindrical coordinates r, θ, z. As shown in Fig. 1, the

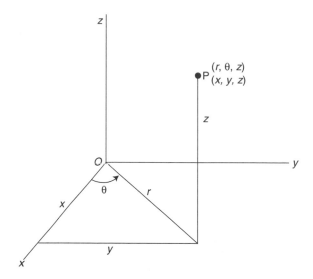

Figure 1 Illustration of the cylindrical coordinate system.

cylindrical coordinates of point P, relative to the origin O, are simply a combination of rectangular and polar coordinates; in the x, y plane we use polar coordinates r, θ, but we keep z as the third coordinate. Then the equations relating rectangular and cylindrical coordinates are just the ones we had in polar coordinates for x, y, and r, θ; the coordinate z is the same in both systems. Thus we have

$$x = r\cos\theta \qquad\qquad r^2 = x^2 + y^2$$

$$y = r\sin\theta \qquad\qquad \tan\theta = \frac{y}{x}$$

$$z = z \qquad\qquad\qquad z = z.$$

For calculus problems, the formulas for the volume element dV and the arc length element ds in cylindrical coordinates are easily written down from the work we have already done in polar coordinates:

$$dV = r\,dr\,d\theta\,dz$$

$$ds^2 = dr^2 + r^2\,d\theta^2 + dz^2.$$

The coordinate surfaces in rectangular coordinates were the three sets of planes $x =$ constant, $y =$ constant, and $z =$ constant. The three planes through any point were mutually perpendicular. In cylindrical coordinates the coordinate surfaces $r =$ constant are circular cylinders with axis along the z axis. The surfaces $\theta =$ constant are half-planes extending out from the z axis, and the surfaces $z =$ constant are planes parallel to the x, y plane. The three coordinate surfaces passing through a point again intersect at right angles. We call such a system an orthogonal curvilinear coordinate system; orthogonal because the surfaces are perpendicular, and curvilinear because they are not all planes.

We have discussed some problems for which it is convenient to use cylindrical coordinates because there is axial (often called cylindrical) symmetry. Another way to decide on an appropriate coordinate system is to see whether surfaces of interest in the problem are coordinate surfaces. Suppose we are given the temperatures at all points on the surface of some solid and we are asked to find the temperatures at all interior points. If the given solid is a rectangular box, we would use rectangular coordinates, but if the given solid is a cylinder cut off at the top and bottom by planes parallel to the x, y plane, then we would use cylindrical coordinates. (These are problems in partial differential equations known as boundary value problems.)

See also: COORDINATE SYSTEM, CARTESIAN; COORDINATE SYSTEM, POLAR; COORDINATE SYSTEM, SPHERICAL

MARY L. BOAS

COORDINATE SYSTEM, POLAR

Consider the motion of a planet around the sun, neglecting all forces except the force of gravity between the sun and the planet. We know from Kepler's first law (or our study of mechanics) that the path of the planet is an ellipse with the sun at one focus. The gravitational force between the sun and a planet depends on the distance between the two bodies. In discussing this problem we would like to have a coordinate system with the origin at the sun and with the distance between the sun and the planet as one of our coordinates. We can do this by introducing a system known as polar coordinates.

Figure 1 shows the relation between polar coordinates r, θ, and rectangular coordinates x, y, of a point P. The coordinate r is the distance from the

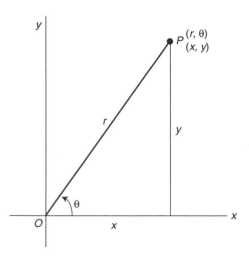

Figure 1 Illustration of the relation between polar coordinates and rectangular coordinates.

origin O to the point P. The angle θ is the angle from the positive x axis to the line OP. If the rotation from the positive x axis to OP is counterclockwise, then θ is positive; if this rotation is clockwise, then θ is negative. For example, if $\theta = -\pi/4$, then the point P is in the fourth quadrant. If $\theta = \pi$ or if $\theta = -\pi$, then P is on the negative x axis. We see here an important difference between rectangular coordinates and polar coordinates. In rectangular coordinates there is just one set of coordinates for a given point, but in polar coordinates there are many sets. To avoid this we often restrict the values of θ to a 2π interval, say $0 \le \theta < 2\pi$, or $-\pi < \theta \le \pi$. Negative values of r are sometimes allowed with the understanding that $(-r, \theta)$ means the same point as $(r, \theta + \pi)$. However, in applications it is better to consider r as positive since it usually represents a distance.

Just as in the case of rectangular coordinates, equations in polar coordinates represent curves or straight lines. Some curves have simpler equations in polar coordinates (and some are simpler in rectangular coordinates). For example, in polar coordinates, the equation $r = $ constant is a circle with center at the origin, and the equation $\theta = \pi/2$ represents the positive y axis.

From Fig. 1 we can write the relations between polar and rectangular coordinates of a point P:

$$x = r \cos \theta \qquad r^2 = x^2 + y^2$$

$$y = r \sin \theta \qquad \tan \theta = \frac{y}{x}.$$

Note that the formula for $\tan \theta$ does not determine θ until we consider a sketch to determine the quadrant of P. For example, if $\tan \theta = 1$, θ could be either $\pi/4$ if P is in the first quadrant, or $5\pi/4$ if P is in the third quadrant.

For problems using calculus, we need the area element and the arc length element in polar coordinates. From Fig. 1 we can see that if we increase r by dr, then the distance P moves is dr. If we increase θ by $d\theta$, then the line segment OP of length r swings through the angle $d\theta$ so P moves the distance $r \, d\theta$. Thus the formulas for the area element dA and the arc length element ds in polar coordinates are

$$dA = r \, dr \, d\theta$$

$$ds^2 = dr^2 + r^2 \, d\theta^2.$$

As we indicated at the beginning, polar coordinates are useful for applied problems in which a natural variable in the problem is the distance in a plane measured from a fixed point in the plane. We can think of many such examples in addition to the motion of planets: spacecraft moving around the earth, a charged particle in an accelerator, a pendulum oscillating in a plane, or the vibration of a circular membrane such as a drumhead. In evaluating double integrals, it is sometimes useful to change from rectangular to polar coordinates. Here is another kind of problem for which we need polar coordinates. Suppose we know the temperatures at all points of the circular rim of a disk with insulated faces and we want to find the temperature at all points of the disk. Such problems are called boundary value problems and they can be solved most easily if the boundary corresponds to a constant value of one of the variables. If we center the disk at the origin, then the equation of the rim is $r = $ constant, so we would use polar coordinates. As a final example we might mention the importance of polar coordinates when we are using complex numbers and the complex plane. In rectangular coordinates in the complex plane we identify the point (x, y) with the complex number $z = x + iy$, where $i = \sqrt{-1}$. Changing this to polar coordinates, we have $z = r(\cos \theta + i \sin \theta)$. Using Euler's formula, we obtain the simple and extremely useful equation $z = r \exp(i\theta)$.

See also: COORDINATE SYSTEM, CARTESIAN; COORDINATE SYSTEM, CYLINDRICAL; COORDINATE SYSTEM, SPHERICAL

MARY L. BOAS

COORDINATE SYSTEM, SPHERICAL

On the surface of the earth, we can use rectangular coordinates to locate a point only if the distances we are considering are very small compared with the radius of the earth (say for a map of a city). For locating more widely separated points, the system that is actually used on the surface of the earth (latitude and longitude) is closely related to spherical coordinates. Consider semicircles on the surface of the earth connecting the North and South Poles; these semicircles are called meridians. The meridian through Greenwich, England is called the prime meridian. Also draw the meridian through a point P on the surface of the earth. Then the longitude of the point P is the angle from the plane of the prime meridian to the plane of the meridian through P, measured east or west up to 180°. The latitude of the point P is the arc measured from the equator along the meridian to P (or the corresponding angle at the center of the earth). The latitude is measured north or south from the equator to the North Pole or the South Pole.

From Fig. 1 we can find the relations between spherical coordinates r, θ, ϕ, and rectangular coordinates x, y, z. The coordinate r is the distance OP from the origin to the point P. The angle θ is measured from the positive z axis to the line OP.

The angle ϕ is measured from the positive x axis to the line in the x, y plane directly under OP. Be careful to distinguish the spherical coordinates r and θ from the r and θ used in polar and cylindrical coordinates. If necessary, change one r to R or ρ. The labeling of angles is a more serious problem; check carefully the meaning of θ and ϕ in any reference materials you use since the notation used here is common in physics and applied mathematics, but θ and ϕ are interchanged in calculus books.

In Fig. 1 it may be helpful to think of the plane containing the z axis and the point P as a door hinged along the z axis (or the cover of a book standing upright on a desk). Then the height of the door is $r \cos \theta$; this is the z coordinate of P. The width of the door is $r \sin \theta$; the x and y projections of this width give $x = r \sin \theta \cos \phi$ and $y = r \sin \theta \sin \phi$. From Fig. 1 we can write equations giving the spherical coordinates of a point in terms of its rectangular coordinates. Thus we have

$$x = r \sin \theta \cos \phi \qquad r = (x^2 + y^2 + z^2)^{1/2}$$

$$y = r \sin \theta \sin \phi \qquad \tan \theta = \frac{(x^2 + y^2)^{1/2}}{z}$$

$$z = r \cos \theta \qquad \tan \phi = \frac{y}{x}.$$

As in polar and cylindrical coordinates, many possible values of the spherical coordinate angles correspond to the same point P so we need some restrictions on the allowed values of these angles. The usual requirements are $0 \leq \theta \leq \pi$ and $0 \leq \phi < 2\pi$. Note that θ is related to latitude but is measured down from the positive z axis instead of up and down from the x, y plane; ϕ is related to longitude but is measured in the positive angle direction instead of both east and west.

For calculus, we need the arc length, volume, and surface area elements. To find these we consider from Fig. 1 the elements of distance corresponding to dr, $d\theta$, and $d\phi$. When r changes by dr with θ and ϕ constant, P moves the distance dr. When θ changes by $d\theta$ with r and ϕ constant, P moves the distance $r\, d\theta$ (in Fig. 1, OP, of length r, rotates about O through the angle $d\theta$). When ϕ changes by $d\phi$ with r and θ constant, the door in Fig. 1 rotates through an angle $d\phi$ so the top of the door rotates through an angle $d\phi$ about the z axis. The width of the door is

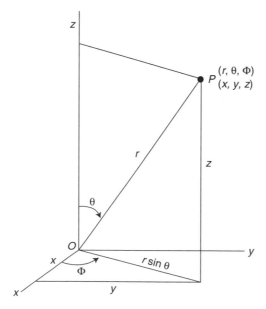

Figure 1 Illustration of the relationship between spherical coordinates and rectangular coordinates.

$r \sin \theta$, so the distance that P moves is $r \sin \theta \, d\phi$. Using these three elements of distance, we can write the formulas for the arc length element ds and the volume element dV in spherical coordinates, and for the surface area element dA on the sphere $r = a$:

$$ds^2 = dr^2 + r^2 \, d\theta^2 + r^2 \sin^2 \theta \, d\phi^2$$

$$dV = r^2 \sin \theta \, dr \, d\theta \, d\phi$$

$$dA = a^2 \sin \theta \, d\theta \, d\phi.$$

The coordinate surfaces are of interest because they tell us the kind of problem most easily handled in spherical coordinates. The surface $r = $ constant is a sphere with center at the origin. The surface $\theta = $ constant is (one nappe of) a right circular cone with vertex at the origin and axis along the z axis. For example, the equation $\theta = \pi/4$ gives the upper half (or nappe) of a cone, and the equation $\theta = 3\pi/4$ gives the lower nappe of the same cone; the two equations together give the whole cone $x^2 + y^2 = z^2$. The surfaces $\phi = $ constant are half-planes containing the z-axis. The three coordinate surfaces through a point P intersect at right angles, so this is an orthogonal curvilinear coordinate system.

We have discussed one use of spherical coordinates, namely, to locate a point on a sphere such as the earth. There are many other applications. Since the spherical coordinate r is the distance from the origin to the point P, we would use spherical coordinates for any problem involving this distance. Two very important forces depend on the distance between the interacting objects: the force of gravity between two masses and the electrostatic force between two charges. Thus spherical coordinates are convenient to use in discussing the motion of one mass around another or the motion of one charged particle around another. For example, in quantum mechanics we use spherical coordinates to discuss the hydrogen atom since the interaction between the proton and the electron depends on the distance between them. Other examples using spherical coordinates are problems involving geometric shapes formed by coordinate surfaces such as a sphere, a hemisphere, the shape of the section of an orange, and so on. For such surfaces or the volumes inside them we might solve boundary value problems, or we might find surface areas, volumes, or moments of inertia.

See also: COORDINATE SYSTEM, CARTESIAN; COORDINATE SYSTEM, CYLINDRICAL; COORDINATE SYSTEM, POLAR

MARY L. BOAS

COPERNICAN REVOLUTION

During the Copernican Revolution (1500-1700), Western conceptions of the earth and its role in the cosmos changed dramatically—from the ancient and medieval notion of an unmoving Earth at the center of a finite spherical world (the Aristotelian–Ptolemaic system) to the vision of a planetary Earth that rotated on its axis and orbited the Sun in an infinitely extended universe. This intellectual revolution encouraged deep-seated changes in technical astronomy and the physics of motion. It accompanied the emergence of the experimental method and the establishment of new scientific organizations as a crucial component of the broader "Scientific Revolution." Its wider cultural significance is evident in early modern European literature, philosophy, and socioeconomic thought. Indeed, the Copernican Revolution shares with the Protestant Reformation, the rise of the national state, and the transformation of the economy a critical role in this turbulent transitional period of European history.

Though the revolution in cosmology bears his name, Nicolaus Copernicus was not a thoroughgoing revolutionary. Copernicus was not the first person to posit a moving Earth (certain ancient Pythagoreans had done so), a Sun-centered planetary system (Aristarchus had envisioned this in 250 B.C.E.), or a rotating Earth (Jean Buridan and Nicole Oresme had considered, though ultimately rejected, earthly rotation in the 1300s). In his technical astronomy, he continued to combine circular motions in his models of planetary movements against the stars for the purpose of accurate prediction—the approach pioneered by ancient Greek astronomers and dominant for the previous 1,500 years. Copernicus was, however, the first European since ancient times to base astronomical calculations on a Sun-centered system and to posit a moving, rotating Earth as the physical reality. His treatise *De Revolutionibus Orbium Coelestium* (*On the Revolution of the Heavenly Orbs*), published in 1543, introduced the

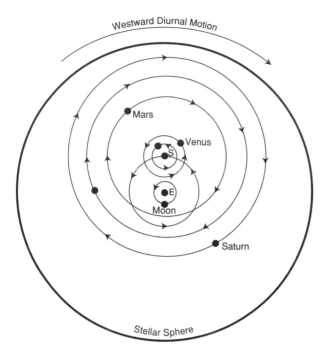

Figure 1 The Tychonic system.

new cosmology to a wider public and presented his techniques for calculating planetary positions to Europe's astronomers.

Between 400 and 500 copies of the first edition of *De Revolutionibus* were printed, and the work was widely read in the latter sixteenth century. Many people absorbed little more than the general cosmological discussion in Book I, but technically proficient astronomers studied the subsequent books carefully. The typical response was to ignore or reject the Copernican cosmological theory, but astronomers respected and used some of Copernicus's technical models to calculate planetary and lunar motions.

The leading astronomer of the late sixteenth century, Tycho Brahe, was impressed enough by the advantages of the Sun-centered system to propose yet a third alternative system. Probably influenced by the mathematician Paul Wittich, Tycho asserted that the Sun moved about an immobile Earth as in the traditional system, but the planets orbited the Sun as in the Copernican system (Figure 1). This "Tychonic" system was destined to be the chief competitor of the Copernican system by the early seventeenth century because it embodied the advantages of the Copernican system but not the Copernican disadvantages that derived from hypothesizing a moving Earth.

Tycho also contributed crucially to European astronomy by accumulating a large amount of very ac-

curate observational data in the final decades of the sixteenth century. Working with this data in the first decade of the seventeenth century, Johannes Kepler instituted a technical revolution in astronomy. Kepler had accepted Copernican cosmology from his student days at the University of Tübingen when he studied with one of the very few astronomers who believed in the Copernican system, Michael Mästlin. During the early years of the seventeenth century, Kepler tried intensively to accommodate a Sun-centered astronomy embodying circular orbits to Tycho's planetary data. Enduring periods of great frustration while working with the data of Mars, he finally discovered that the planetary orbit was a simple ellipse, and the planet's changing speed on the orbit was accurately described by assuming that an imaginary line between it and the Sun swept out equal areas in equal times. By 1627, with the publication of his *Rudolphine Tables,* Kepler demonstrated that predictive accuracy using his "ellipse law" and "area law" was far greater than any set of previous planetary tables had attained. As he realized, his achievement did not demonstrate the truth of Copernican cosmology, since his laws were as applicable to the Tychonic system as to the Copernican system. But Kepler's own advocacy of Copernican cosmology, combined with the polemical defense of it by Galileo in the 1630s, persuaded more and more people to accept it in the mid-seventeenth century.

Along with Kepler, Galileo Galilei was a pivotal figure in the transformation of astronomy and physics in the Scientific Revolution. Not a technical astronomer, Galileo nevertheless significantly affected astronomy by carrying out and publicizing observations of celestial phenomena with the newly invented telescope in the years following 1609. One of his most important series of observations revealed that Venus shows reflected light from the Sun and appears at various times in the full range of phases from crescent to full. Because Venus never appears to our sight farther than 46° east or west of the Sun, Venus should not show the full range of phases according to the geometry of the Aristotelian–Ptolemaic system, in which both Venus and the Sun orbit Earth. If Venus's orbit is closer to Earth than is the Sun's orbit in the Aristotelian–Ptolemaic system Venus should never appear in a full or near-full phase. Earthly observation of the fully illuminated hemisphere of Venus would require a position for Venus either 180° opposite to the Sun (which would contradict the fact that it never departs more than 46° from the Sun) or at a position farther from Earth than the Sun (which is at odds with the assumption of its closer proximity to Earth). Galileo's observations plainly demonstrated that Venus orbits the Sun (Figure 2), thus eliminating the traditional Aristotelian–Ptolemaic system from serious consideration and narrowing the competition to the Copernican versus the Tychonic systems.

Galileo contributed monumentally to the science of motion. In the first decade of the seventeenth century, he asserted that all bodies fall with the same uniformly accelerated motion in the absence of resistance, regardless of their weights, shapes, densities, or sizes. He also conceived and analyzed what he claimed to be a novel type of motion—a movement at uniform speed that would continue interminably in the absence of resistance. In his writings, Galileo offered examples of a new type of motion that included a ball rolling at uniform speed along a frictionless path around the earth and a cannonball projected linearly and imagined free of gravitational influence. His writings do not present a consistent and sharply defined concept of this motion at constant speed. In some of his works the motion is circular, while elsewhere he treats it as rectilinear; at some places in his writings the motion is the effect of a cause, while elsewhere Galileo assumes it to proceed in the absence of causal maintenance,.

This new motion served Galileo well in his effort to convince contemporaries that the earth moves. He used it to argue that a body falling from a tower should appear to land at the base of the tower whether or not the earth rotated or moved through space. Traditional common sense and Aristotelian physics had assumed that the body should fall to the west of the tower's base if the earth rotated from west-to-east during the time of the ball's descent.

But neither Galileo's telescopic observations nor his analyses of the physics of motion furnished him definitive proof that the earth moves. No significant advantage appeared to favor either the Copernican or the Tychonic system in the early-to-mid 1600s. Indeed, it is noteworthy that in Galileo's most famous work, *Dialogue on the Two Chief World Systems* (1632), the Copernican system vied against the traditional Aristotelian–Ptolemaic system. He ignored the Tychonic system. The *Dialogue* was not written as a "text" to expose objectively the current state of the cosmological issue; it was a frankly polemic work designed to persuade its readers to embrace the Copernican system.

Building on Galileo's brilliant but inconsistent analyses of his new type of motion, the Frenchmen

René Descartes and Pierre Gassendi essentially redefined the very concept of motion in the mid-1600s. "Motion" for them no longer was an effect requiring a cause as in ancient and medieval physics (largely due to the influence of Aristotle's analysis of motion in the *Physics*). Now they regarded "motion" as a state, analogous to rest, requiring no causal sustenance—and they conceived "motion" as rectilinear and uniform (constant speed), assuming the absence of resistance. To the post-Newtonian mindset, this definition of motion implies that any type of movement other than uniform rectilinear proceeds under the imposition of an external agency. But to exploit the new definition of motion incisively to construct a widely applicable and internally consistent science of mechanics awaited the activity of Isaac Newton in the 1670s and 1680s. Neither Descartes nor Gassendi developed the full implications of their novel approach to motion.

The Copernican Revolution culminated with acceptance of a Sun-centered planetary system largely due to Isaac Newton's theoretical achievements in physics. Empirical evidence for the motions of the

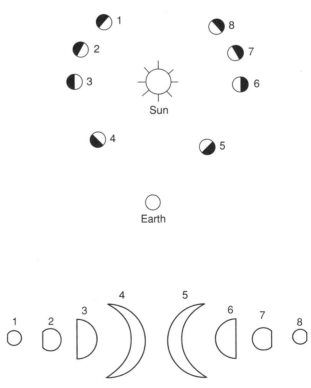

Figure 2 The phases of Venus. Shapes are given as viewed from Earth at the numbered positions, respectively.

earth came significantly later with James Bradley's detection of the aberration of light in the late 1720s showing that the earth moves, and Jean Foucault's pendulum in 1851 demonstrating that the earth rotates on its axis. Isaac Newton's new physics of the 1680s, including his idea of universal gravitation, eventually provided physical explanation for Kepler's laws and won virtually universal approval for Copernican cosmology.

The key to Newton's achievement was his explicit concept of "force" as embodied in what is now known as his second law of motion. By the early 1680s, he conceived of force as active in changes of motion. Newton regarded motion itself as a state of uniform and rectilinear movement per Descartes and Gassendi, and he later expressed this in his first law of motion. Applying his concepts of motion and force, Newton was able to formulate an internally consistent science of mechanics that dealt successfully with its three most critical issues: collision, fall, and the motion of a body in a closed path (notably the orbital motion of a planet). Indeed, conceiving that fall and a planet's orbital motion were effects of the same force (later in the 1680s a universal gravitational force), he proceeded to show the intimate relation between Galileo's law of falling bodies and Kepler's three laws of planetary motion. In this, he demonstrated the identity of celestial and terrestrial physics, thus finally discrediting the ancient Aristotelian belief that the physics of the terrestrial and celestial regions essentially differed.

The power of Newton's theoretical accomplishment published in his *Principia Mathematica* (1687) decidedly swayed Europeans to accept the Copernican system. Implications of this for science and for the wider culture were profound. In science, Newton's stunning success with the new concept of force in mechanics and celestial physics encouraged him and his successors to posit forces as explanations for chemical, optical, electric, magnetic, and thermal phenomena. Ultimately, however, the concept of force would not prove as fruitful for the theoretical basis of these other physical sciences as would the mid-nineteenth-century concept of energy.

The transition Alexandre Koyré termed "from the closed world to the infinite universe" also left deep imprints in the wider literary culture of Europe and America. The uncertainty of living in an endless centerless universe could frighten the early seventeenth-century English poet John Donne, who wrote that the

...new Philosophy calls all in doubt,
The Element of fire is quite put out;
The Sun is lost, and th'earth, and no mans wit
Can well direct him where to looke for it.
And freely men confesse that this world's spent,
When in the Planets, and the Firmament
They seeke so many new; then see that this
Is crumbled out againe to his Atomies.
'Tis all in peeces, all cohaerence gone;
All just supply, and all Relation.

Yet the vastness of the cosmos and the revelations of the telescope could encourage as well a sense of adventure with prospects of interminable novelty and discovery. In the latter seventeenth century, Bernard de Fontenelle speculated joyfully about the existence of other worlds with living inhabitants possessing a rationality superior to our own. His *Entretiens sur la Pluralite des Mondes* (*Conversations on the Plurality of Worlds*) influenced literary and popular thought throughout the eighteenth century.

The Copernican Revolution became an inspiration and model for people who sought analogous fundamental changes in other departments of life in the eighteenth century. Sociopolitical critiques of Voltaire in pre-revolutionary France and laissez-faire economic theories of Adam Smith drew strength from the triumph of the Newtonian world view. The German philosopher Immanuel Kant asserted in his *Critique of Pure Reason* (1781) that the human mind is endowed with "a priori categories" (time, space, Euclidean geometry) that shape our apprehension of experience. He speculated that these categories limit our understanding and that "things-in-themselves" (things as they really are) can never be known with certainty by human thought. This assertion was to exert enormous influence on subsequent philosophy—Kant termed it his "Copernican Revolution" in philosophy.

See also: COPERNICUS, NICOLAUS; GALILEI, GALILEO; KEPLER, JOHANNES; KEPLER'S LAWS; NEWTON, ISAAC; NEWTONIAN MECHANICS

Bibliography

COHEN, I. B. *The Birth of a New Physics* (W. W. Norton, New York, 1985).

KOYRÉ, A. *From the Closed World to the Infinite Universe* (Johns Hopkins University Press, Baltimore, MD, 1957).

KUHN, T. *The Copernican Revolution: Planetary Astronomy in the Development of Western Thought* (Harvard University Press, Cambridge, MA, 1957).

NICHOLSON, M. *Science and Imagination* (Archon, Hamden, CT, 1956).

TATON, R., and WILSON, C., eds. *Planetary Astronomy from the Renaissance to the Rise of Astrophysics, Part A: Tycho Brahe to Newton* (Cambridge University Press, Cambridge, Eng., 1989).

VAN HELDEN, A. *Measuring the Universe: Cosmic Dimensions from Aristarchus to Halley* (University of Chicago Press, Chicago, 1985).

ROBERT K. DeKOSKY

COPERNICUS, NICOLAUS

b. Torun, Poland, February, 19, 1473; *d.* Frauenburg, Poland, May 24, 1543; *astronomy, cosmology.*

Copernicus was a "Renaissance man"—canon lawyer, physician, scholar and translator of Greek literature, monetary theorist, and astronomer. He was born the youngest of four children into a German merchant family living in the Polish city of Torun. At the age of ten, his father died, and the children were adopted by maternal uncle Lucas Watzenrode, who was building a career begun in Italian academia and now proceeding within the Church—eventually he would attain a bishopric. Under his uncle's watchful eye, Copernicus attended secondary school and then the University of Cracow before traveling to Italy for his professional education. In the late 1490s and early 1500s, he obtained a doctorate in canon law after study at the University of Bologna, studied medicine at the prestigious University of Padua, learned Greek, and familiarized himself with the technical intricacies of astronomy. When he returned to his native land by 1506 at the latest, his family connections had eased him into what would be a lifetime occupation in Church administration.

Astronomy became Copernicus's passionate avocation. Astronomers strove to formulate models that accurately predicted planetary motions against the distant fixed stars. Calendar reckoning required models to describe with accuracy the relation between the Sun and the Moon. At the turn of the sixteenth century, European astronomy was dominated by the approach of Claudius Ptolemy, whose *Almagest* had been the most influential text on technical astronomy for 1,500 years. The writings of Aristotle heavily influenced discussions about cosmology and physics.

Some time before 1514, speculatively between 1508 and 1510, Copernicus composed a short tract titled *Commentariolus,* in which he envisioned a new system of astronomy containing the essential ideas that would underly his later treatise *De Revolutionibus Orbium Colestium* (*On the Revolution of the Heavenly Orbs*). *Commentariolus* circulated only in manuscript among a select few individuals and remained unpublished. Copernicus did not even acknowledge his authorship in the work. He did not consent to publish his ideas until much later: *De Revolutionibus* was published shortly before his death in 1543.

De Revolutionibus (as had the earlier *Commentariolus*) offered essentially independent cosmological and technical innovations to the European astronomer. The cosmological changes have been most responsible for the enduring fame of Copernicus. He asserted that the Sun was the body around which the planets orbited (Fig. 1), rather than Earth, which in traditional astronomy and cosmology had been assumed an absolutely unmoving body at the center of the universe. The Moon orbited Earth in the Copernican system as it had in the Aristotelian–Ptolemaic planetary system. To Earth Copernicus assigned two major motions—a yearly motion around the Sun and a daily rotation to account for the apparent east-to-west motions of the Sun, Moon, planets, and stars. The stars did not move in the Copernican system; in the Earth-centered system, the stars rotated about Earth once per day. Copernicus overviewed the new cosmology in Book I of *De Revolutionibus,* starkly different in its non-technical format from subsequent books of the treatise that spoke to the technical and mathematical concerns of the working astronomer.

For people living in the sixteenth century, three categories of thought discouraged the theory of a moving Earth. Theologically, a mobile Earth seemed inconsistent with passages in Holy Scripture that implied the Sun moved about the Earth daily. A second obstacle to the new system was the absence of any "stellar parallax"—expected differences in the measured angular separations of stars over the time of the year if Earth moved about the Sun. This effect due to the changing position of the Earth-bound observer would not be detected until the late 1830s by the German astronomer Friedrich Bessel with the aid of increasingly powerful telescopes. Until then, a Copernican had to posit without means of verification that the parallax effect was too small to be ob-

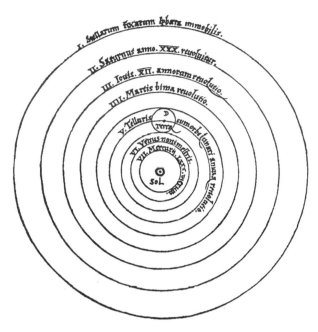

Figure 1 The Copernican system, from *De Revolutionibus Orbium Colestium* (1543).

served—that is, the distance between Earth and the stars was immensely greater than the diameter of Earth's orbit.

Physics, the third category of thought, posed the most significant challenges to the new system. Common sense and the more systematic discussions in Aristotle's *Physics* led to anticipation that clouds and birds would lag behind a moving Earth. Moreover, according to Aristotelian philosophy, each of the traditional four elements between the universal center and the Moon (earth, water, air, fire) moved spontaneously toward its "natural place." The most important "natural motion" was the fall of a heavy body explained as the spontaneous effort of the element earth to achieve rest in its natural place at the center of the universe. To put Earth in motion was to destroy the concept of natural place and thus to demand the foundations of a new physics. Copernicus spoke briefly and inadequately to these issues in Book I. Not until the innovations of Galileo Galilei, René Descartes, and Isaac Newton in the seventeenth century would a new physics consistent with Copernican cosmology emerge.

Despite the difficulties, Copernicus embraced the new system because of three overriding considerations. First, calculation of the relative distances of the planets from the center of the system—based exclusively on observational data—was not possible

assuming that Earth was at the center, but the calculation was possible in the Sun-centered system. Second, Ptolemy had had to make special provision in each of his planetary models to account for retrograde motion (change in direction of the planet's motion against the background of the stars). In the Copernican system, retrograde motion was expected as a visual effect of the different speeds of the planet and Earth on their respective orbits (the planet does not actually change direction on its orbit).

Finally, the Sun-centered system by its very geometry accounted for long-known, peculiar affiliations between planetary phenomena and the Sun. Venus and Mercury were visible only in the early evening or early morning just before sunrise—never around midnight. Their orbits lay inside Earth's orbit in the Copernican system. The planets outside Earth's orbit in the Copernican system were Mars, Jupiter, and Saturn. Whenever any of them exhibited its occasional periods of retrograde motion, it was in "opposition" to the Sun (approximately 180° from the Sun). To account for these phenomena, Ptolemy had been forced to impose an awkward condition on his planetary model that was unnecessary in the Copernican arrangement.

Were these "advantages" of the Copernican system enough to overcome the disadvantages of a moving Earth? Obviously for Copernicus they were, but for the great majority of his latter sixteenth-century readers—astronomers and non-astronomers alike—they were not. This was a choice made in the sixteenth century on the basis of aesthetic and philosophical criteria. Criteria of accuracy or overall simplicity favored neither system. Copernicus realized that his astronomy was equivalent to Ptolemaic astronomy in its predictive ability; it promised neither more nor less accurate predictions of planetary motions against the stars than its Ptolemaic competitor. Moreover, the technicalities of Copernican astronomy were just as complex as those of Ptolemaic astronomy.

Absence of proof for his cosmology notwithstanding, Copernicus introduced technical features to European astronomy that were appropriate for an Earth-centered or Sun-centered perspective of the planets. As much or more than the cosmological question, these features attracted the attention of astronomers in the latter sixteenth century.

In the *Almagest*, Ptolemy had successfully applied the ancient Greek belief that the geometry of astronomical phenomena was fundamentally circular. His aim was to devise a model for a planet that combined simple circular motions to generate a complex, irregular resultant motion that replicated the planet's actual movement against the background of the stars. In Ptolemy's astronomy, the planet moved about a smaller circle termed the "epicycle" (Fig. 2). As the planet moved on the epicycle, the center of the epicycle moved about a larger circle called the "deferent." The (unmoving) Earth was near the center of the deferent. For each planet, Ptolemy had to assign specific ratios of sizes and ratios of angular speeds to the two circular motions in his effort to match the model's resultant motion with the actual motion of the planet. By refining this basic model, he did indeed accurately reproduce some planets' motions (though a small difference between model and reality would expand to disturbing size over the centuries, aggravated by inevitable inclusions of erroneous observational data). But one of Ptolemy's crucial refinements drew the ire of numerous medieval critics and motivated Copernicus to seek astronomical reform. To attain the necessary predictive accuracy, Ptolemy had made the motion of the epicyclic center (point F in Fig. 2) non-uniform; he had placed the "equant" point (from which the motion of the epicyclic center would appear to proceed at constant angular velocity) at a position off of the geometric center of the deferent (point C in Fig. 2). His critics, including Copernicus, despised this violation of the classical Greek precept that combining simple, uniform circular motions was proper for astronomical model making.

This philosophical objection induced Copernicus to "purify" astronomy by purging it of the Ptolemaic equant. In the *Commentariolus* and later in Book V of *De Revolutionibus,* he presented models that substituted for the Ptolemaic equant equivalent constructions that reproduced a planet's movement by combining additional uniform circular motions. Historians of astronomy suspect that Copernicus did not independently formulate them. In the thirteenth and fourteenth centuries, the astronomers of Marāgha in northwest Iran had discovered and used them. Possibly these innovations of medieval Islamic astronomers were exposed to Copernicus during his period of study in Italy; they most likely came to Italy by way of Byzantine sources. It merits emphasis that the Marāgha astronomers constructed their models assuming an Earth-centered planetary system. The Ptolemaic equant or an equivalent was required in both Earth-centered and Sun-centered planetary systems to account for an irregularity in each planet's motion.

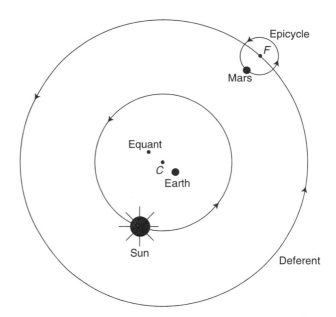

Figure 2 Ptolemaic planetary astronomy. In the Ptolemaic model, the planet moves around a smaller circle (the epicycle), which concurrently moves about a larger circle (the deferent). The angular velocity of the planet's motion along the epicycle is greater than the angular velocity of the epicycle around the deferent. Therefore, the planet appears to move retrograde from the perspective of Earth when on the inner segment of the epicycle. To insure that periods of retrograde motion for Mars, Jupiter, or Saturn always occur when the planet is in opposition to the Sun in the planetary model, the Sun is in opposition to the planet (180° apart as viewed from Earth) when the planet is on the inner segment of its epicycle, and the Sun and planet move such that an imaginary line between Earth and the Sun remains parallel to an imaginary line between the planet and the center of its epicycle (F).

Actually, the planetary models of Book V dealt only with the prominent, astrologically significant, east-west movements of planets against the stars (motion in longitude). But the planets also moved within a range of 8° north and south of the "ecliptic" plane. (In the traditional system, the ecliptic plane was defined by the path of the Sun against the stars; in the Copernican system, it was the plane of Earth's orbit.) Satisfactory predictions of these changes in planetary "latitudes" had eluded all previous astronomers. Changes in latitude correspond in Figs. 2 and 3 to movement above and below the plane of the page. In

Book VI, Copernicus's attempt to account for latitudes achieved no advantage over his predecessors. In fact, it depended so heavily on Ptolemy's treatment of latitudes that Johannes Kepler later commented how here Copernicus represented Ptolemy rather than nature.

The publication of Copernicus's *De Revolutionibus* was a crucial event in the opening phase of the early modern "Scientific Revolution." The treatise both reflected the past and portended a new agenda for physical science. Preoccupied with uniform circular motions in technical astronomy, it introduced a sophisticated, medieval-Islamic rendition of that approach to European astronomers and thus retained an ancient Greek philosophical ideal. For this, Copernicus won praise from astronomers such as the German Erasmus Reinhold. Reinhold ignored the theory of a moving Earth, but he used Copernicus's alternative for the equant in Book V as the basis for production of his *Prutenic* planetary tables, which astronomers referenced consistently in the latter sixteenth century. Not Copernicus, but Johannes Kepler became the major technical revolutionary in early modern astronomy; by the middle of the first decade of the 1600s, Kepler had rejected the circle as the chief geometrical tool of the astronomer and shown that accurate predictions of Martian longitudes and latitudes result from assuming the orbit is an ellipse.

Technical conservatism notwithstanding, by endorsing a Sun-centered cosmology featuring a moving Earth, *De Revolutionibus* threatened prevailing concepts of physics that the work of successors Galileo, Descartes, and Newton would destroy. Around the time Copernicus wrote *De Revolutionibus*, Andreas Vesalius was reforming the study of human anatomy. Paracelsus was advocating a chemical, laboratory-based foundation for medicine. Learned academies and courtly patronage were challenging the universities' monopoly of higher learning. In concert, *De Revolutionibus* and these other agents of change initiated profound intellectual, methodological, and institutional alterations that would shape European scientific culture into its modern form by the close of the seventeenth century.

See also: COPERNICAN REVOLUTION; GALILEI, GALILEO; KEPLER, JOHANNES; NEWTON, ISAAC

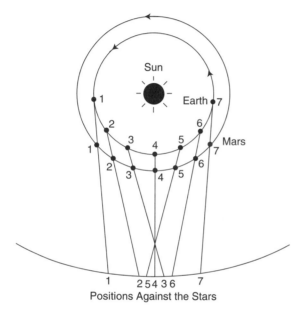

Positions Against the Stars

Figure 3 Retrograde motion and the Copernican arrangement. In the Copernican model, the perception from Earth that the planet moves retrograde results from the motions of the observer and the observed body. By drawing lines of sight from Earth through the planet to the stars for seven sets of Earth–planet positions, we recognize that the effect of retrograde motion is demanded by the Copernican arrangement and does not require special explanation (the planet does not actually change direction). From positions 1–3, the planet appears to move against the stars in its overall direction. From positions 3–5, the planet appears to move retrograde. At position 5, the planet appears to resume its overall motion. Note that at the heart of retrograde motion (position 4), the Sun and the superior planet (Mars, Jupiter, or Saturn) are in opposition (180° apart as viewed from Earth)—a condition demanded by the Copernican arrangement and requiring no special explanation.

Bibliography

ARMITAGE, A. *Copernicus* (A. S. Barnes, New York, 1962).

GINGERICH, O., ed. *The Nature of Scientific Discovery: A Symposium Commemorating the 500th Anniversary of the Birth of Nicolaus Copernicus* (Smithsonian Institution Press, Washington, DC, 1975).

SWERDLOW, N., and NEUGEBAUER, O. *Mathematical Astronomy in Copernicus's "De Revolutionibus"* (Springer-Verlag, New York, 1984).

WESTMAN, R., ed. *The Copernican Achievement* (University of California Press, Berkeley, 1975).

ROBERT K. DeKOSKY

CORIOLIS FORCE

Newton's laws of motion only hold in an inertial reference frame. It is convenient and intuitive to use equations of motion having the same form in a rotating reference frame, such as the surface of a planet. Newton's first law states that an object in motion in the absence of outside forces follows a straight path. However, in a rotating reference frame, that straight path appears curved as the reference frame itself changes direction. The curvature of the path can be accounted for by imaginary forces of the correct strength and direction. The centrifugal force accounts for the apparent motion of a particle actually at rest as the reference frame rotates out from under it, while the Coriolis force accounts for the change in apparent velocity of a moving particle due to the simultaneous motion of the reference frame in a different direction.

The Coriolis force was first described in print in 1835 by Gaspard G. Coriolis, a French civil engineer and physicist. He invented the concept as part of his theory of the composition of accelerations, an outgrowth of his work on water wheels. The most frequent use of Coriolis forces occurs in considering motion on the rotating surface of the earth. The imaginary force appears to act as a force proportional to the speed of motion and perpendicular to the direction of motion. In the Northern Hemisphere, the deflection is to the right of the direction of travel, while in the Southern Hemisphere the deflection is to the left.

Derivation

This derivation of the strength of the Coriolis force uses simple concepts from vector calculus, but it may be read fruitfully by ignoring the vector nature of the equations and treating them as algebraic equations.

Consider two coordinate systems with common origin, one rotating with angular velocity $\boldsymbol{\omega}$, and one fixed. (For our purposes we may ignore linear motion of the rotating coordinate system and assume that $\boldsymbol{\omega}$ remains constant in time.) The velocity of an object moving in the fixed coordinate system $\mathbf{v}_f$ is the sum of its velocity in the rotating coordinate system $\mathbf{v}_r$ and the motion of the coordinate system,

$$\mathbf{v}_f = \mathbf{v}_r + \boldsymbol{\omega} \times \mathbf{r}_r. \qquad (1)$$

The second term uses the vector cross product that yields a vector at right angles to the two factor vectors.

The acceleration of the object in the fixed coordinate system $\mathbf{a}_f$ likewise is the sum of the acceleration in the rotating coordinate system $\mathbf{a}_r$ and the acceleration due to the motion of the rotating coordinate system,

$$\mathbf{a}_f = \mathbf{a}_r + \boldsymbol{\omega} \times \mathbf{v}_f. \qquad (2)$$

The equation of motion in the fixed coordinate system from Newton's second law,

$$\mathbf{F} = m\mathbf{a}_f \qquad (3)$$

gives the acceleration in terms of the mass of the object m and the real force acting on it $\mathbf{F}$.

We can now describe this same motion in the rotating coordinate system, attempting to find an equation of motion having the same straightforward form as Eq. (3). We proceed by algebraically substituting Eq. (2) into Eq. (3), and then further substituting using Eq. (1). Rearranging, we can write an equation of motion in the rotating coordinate system that again depends on the product of mass and apparent acceleration,

$$\mathbf{F} - m\boldsymbol{\omega} \times (\boldsymbol{\omega} \times \mathbf{r}_r) - 2m(\boldsymbol{\omega} \times \mathbf{v}_r) = m\mathbf{a}_r. \qquad (4)$$

The two new terms on the left-hand side can be treated as forces, although they actually represent the rotation of the coordinate system.

The first of these new force-like terms is the centrifugal force, depending only on the position and mass of the object and the rotational velocity of the coordinate system, and acting perpendicular to the axis of rotation. The second term is the Coriolis force, depending on the velocity and mass of the object as well as the rotational velocity of the coordinate system.

The magnitude of the Coriolis force, which is $-2m\omega v_r \sin \phi$, depends on the angle ϕ between the rotation axis and the velocity of the object. The force acts in a direction perpendicular to the plane containing the rotation axis and the velocity vector. For horizontal motion with velocity V at a latitude θ on the surface of a planet rotating with angular velocity Ω, the Coriolis force has a magnitude of $2m\Omega V \sin \theta$, directed perpendicular and to the right of the direction of motion in the Northern Hemisphere ($\theta > 0$) and to the left of the direction of motion in the Southern Hemisphere ($\theta < 0$).

Applications

Coriolis forces are most commonly used to model systems on the earth's surface. They remain quite weak there, due to the relatively slow rotation of the earth, only becoming important for large distances or high velocities. For example, although they act on water draining from a bathtub, they are overwhelmed by local forces so that the vortex formed by the water going down the drain may spin in either direction.

Meteorology. Coriolis forces do become important for atmospheric motions traversing hundreds or thousands of kilometers over periods of many hours or days. In their absence, winds would blow directly from regions of high pressure to regions of low pressure. The Coriolis forces deflect the winds perpendicularly until the Coriolis and pressure forces balance. The wind then blows clockwise around a region of low pressure in the Northern Hemisphere, producing a cyclonic flow. In the Southern Hemisphere, the flow is counterclockwise, producing anticyclonic flow. The flow around hurricanes is an extreme example of such winds.

Foucault pendulum. The Foucault pendulum is a famous demonstration of Coriolis forces. A swinging pendulum will continue swinging in the same plane regardless of the rotation of the earth beneath it. As a result, during a full rotation of the earth, a properly isolated pendulum will appear to have its plane of motion spin completely around. This may be understood using the concept of a Coriolis force. The magnitude of the Coriolis force depends on the velocity of the pendulum, and it appears to act to deflect the pendulum by exactly the right amount to cause its plane of motion to rotate once per day.

See also: CENTRIFUGAL FORCE; FRAME OF REFERENCE, INERTIAL; FRAME OF REFERENCE, ROTATING; HURRICANE; NEWTON'S LAWS; PENDULUM, FOUCAULT; VORTEX

Bibliography

BARGER, V. D., and OLSSON, M. G. *Classical Mechanics: A Modern Perspective* (McGraw-Hill, New York, 1973).

MARION, J. B. *Classical Dynamics of Particles and Systems*, 2nd ed. (Academic Press, New York, 1970).

WALLACE, J. M., and HOBBS, P. V. *Atmospheric Science: An Introductory Survey* (Academic Press, New York, 1977).

MORDECAI-MARK MAC LOW

CORRESPONDENCE PRINCIPLE

The correspondence principle was introduced by Niels Bohr in the development of quantum mechanics and presented in his publications between 1918 and 1923. In general terms, the principle states that quantum mechanics must give the results of classical mechanics in the limit of large quantum numbers. The principle could then be used to verify the validity of results from the early version of quantum mechanics. More specifically, Bohr postulated that in the limit of large quantum numbers, results from the quantal theory of atomic spectra must coincide with the classical theory of radiation by atoms. This allowed him to provide a prescription for the calculation of the intensities of atomic spectral lines.

In accordance with Bohr, the allowed states of electrons in atoms are stationary states with energies E_n, labeled by integer quantum numbers $n = 1, 2, 3, \ldots$. Transitions between two stationary states with quantum numbers n and m are related to the frequency ν of light emitted or absorbed during the transition by the Bohr relation

$$E_n - E_m = h\nu,$$

where h is the Planck constant. In the limit of large values for n and m, the energies are almost continuous functions of the quantum numbers and, writing that $n = m + \Delta m$, where Δm is a small increment, one can approximate the above energy difference by the relation between increments,

$$\Delta E = (E_{m+\Delta m} - E_m) = \Delta m \left(\frac{dE}{dm} \right).$$

A second relation of the early quantum theory may now be used to relate classical radiation frequencies ν_{cl} to the frequencies ν of the Bohr relation. The mechanical action I, obtained by integrating over areas of phase space, is quantized by the relation $I = hm$. In addition, classical mechanics shows that the change of E with I equals the frequency ν_{cl} of a particle in orbital motion. Dividing both sides of the equation for ΔE by h one then finds that

$$\nu = (\Delta m)\ \nu_{cl},$$

which establishes the relation between quantal and classical radiation frequencies for large quantum numbers.

The modern theory of quantum mechanics is self-contained and does not need to resort to the correspondence principle to derive results. However, the connection established by the principle between classical and quantal descriptions has emerged as a fruitful way of solving complicated many-particle systems in quantum mechanics, by constructing quantal wave functions from bundles of classical trajectories. This connection can be rigorously established starting from the Schrödinger wave equation of quantum mechanics and taking the limit of short de Broglie wavelengths, also called the eikonal limit of quantum mechanics.

In atomic systems containing many particles (electrons and nuclei), the description of the classical orbits is complicated by particle interactions that may lead to chaotic or irregular motions, covering whole regions of phase space. The connection between quantum and classical mechanics, and their description of spectra in the presence of chaotic motion, is an active area of research.

See also: BOHR, NIELS HENRIK DAVID; BOHR'S ATOMIC THEORY; QUANTUM MECHANICS; QUANTUM MECHANICS, CREATION OF; QUANTUM NUMBER; SPECTRAL SERIES; WAVE FUNCTION

Bibliography

CHILD, M. S. ed. *Semiclassical Methods for Molecular Scattering and Spectroscopy* (Reidel, Dordrecht, 1980).
HEISENBERG, W. *The Physical Principles of the Quantum Theory* (Dover, New York, 1949).
JAMMER, M. *The Conceptual Development of Quantum Mechanics* (McGraw-Hill, New York, 1966).
TABOR, M. *Chaos and Integrability in Nonlinear Dynamics* (Wiley, New York, 1989).

DAVID A. MICHA

COSMIC BACKGROUND EXPLORER SATELLITE

In the first decade after its discovery the cosmic microwave background radiation (CMBR) had been studied from the ground, from balloons, and even with interstellar CN molecular thermometers. By 1975 those measurements were limited by interfering emission by oxygen and water vapor in the Earth's atmosphere. In response to a NASA announcement of opportunity several groups proposed satellite concepts, the successful ones were amalgamated to form the Cosmic Background Explorer (COBE) science working group that defined the science requirements and designed the science instruments. The team was made up of scientists from Berkeley, the Goddard Space Flight Center, the Jet Propulsion Laboratory, Massachusetts Institute of Technology, and Princeton University. Goddard Space Flight Center was assigned the responsibility for building the instruments and designing and building the spacecraft.

COBE (pronounced co-bee) was designed to study three questions of critical importance to cosmology. (1) Does the cosmic microwave background radiation have a blackbody spectrum over the crucial wavelength range of its peak brightness? The big bang theory of the evolution of the universe predicts a microwave heat remnant with a blackbody spectrum. (2) Are there tiny fluctuations in the CMBR temperature distribution across the sky? Fluctuations of about 1 part in 100,000 are predicted by the big bang theory. (3) Is there a cosmic background of infrared radiation? Some models for how galaxies first form predict a large amount of starlight, which by now has been redshifted to infrared wavelengths.

The COBE instruments and the mission design had to solve some difficult problems. The spectrum measuring experiment, called FIRAS (for far-infrared absolute spectrometer) needed to make accurate absolute measurements. Therefore, COBE had to carry a precision calibrator—a radiation source whose temperature is known to a precision of 0.01 K (K = Kelvin degrees above absolute zero). The instruments for measuring CMBR temperature fluctuations across the sky (anisotropy) were called differential microwave radiometers (DMRs). They were designed to measure the difference in the temperatures from two sky positions with an accuracy of 30 μK (0.00003 K). For such sensitive measurements, the DMR antenna beams could never

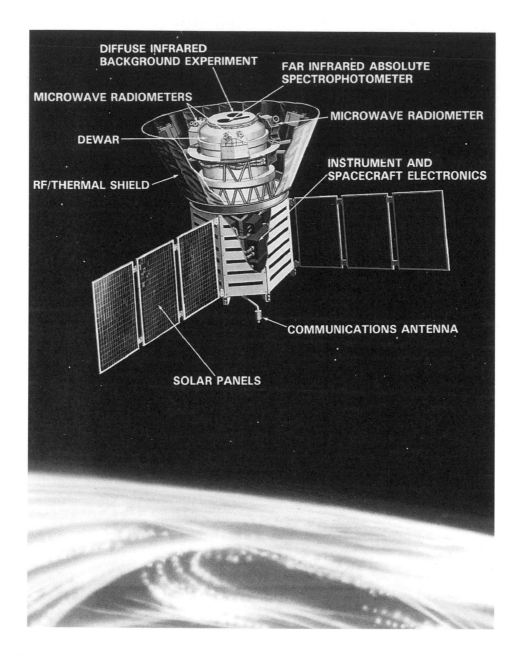

Figure 1 An artist's conception of COBE in orbit. The cone-shaped shield used to exclude Earth and Sun radiation is cut away to show the instrument package. FIRAS and DIRBE are inside the large cylindrical Dewar, launched with 500 liters of liquid-helium coolant. Their antennas look out through the openings in the top. The DMRs are fastened to the upper Dewar ring; their horn antennas are aimed 30° to either side of the spacecraft axis. The entire satellite rotates about once a minute, always pointing away from the Earth to keep it below the radiation shield. The spacecraft axis is also kept about 94° away from the Sun by an ingenious orbit and pointing control system provided by NASA.

point close to the Sun or Earth. The third COBE instrument, DIRBE, was a specially designed infrared telescope with detectors in 10 wavelength bands between 1 and 240 μm. DIRBE stands for diffuse infrared background experiment. It too could never be pointed close to the Sun or Earth or its liquid-helium coolant would be rapidly boiled away, increasing detector noise. The need to avoid pointing at these sources placed a severe requirement on the COBE orbit and the spacecraft attitude control system. NASA engineers had a clever solution.

COBE was put into a "polar, sun-synchronous" orbit. This means that on each orbit the satellite passes over the Earth's poles, and the plane of the orbit precesses once a year to follow the Sun as it appears to move around the sky. Furthermore, the orbit plane was made perpendicular to the Earth–Sun line by launching at sunrise (sunset would also work). In this orbit, a satellite oriented to always point away from the Earth will never point at the Sun. The orbit is made to precess (as a top does when not standing straight up) by the torque exerted by the Earth's equatorial bulge. The launch of COBE at sunrise on November 18, 1989, was perfect. The nominal orbit was achieved, and all instruments and spacecraft systems subsequently worked.

COBE's instruments had to be very stable and well shielded to achieve the science goals. They succeeded largely because earlier versions of the instruments had been used for a decade on the ground, in balloons, and in aircraft. The instrumental subtleties and pitfalls were understood. The FIRAS and DIRBE instruments were inside a large Dewar, cooled to 1.5 K by liquid helium. (The 500 liter Dewar is the large metal tank shown in Fig. 1.) Cold instruments were needed to avoid contamination of the feeble cosmic signals by radiation from the instruments themselves. The use of liquid-helium coolant greatly complicated the COBE mission, but the cold temperature was also needed to cool the special FIRAS calibrator, whose temperature had to be adjusted close to the expected CMBR temperature, 2.73 K. The calibrator was actually inserted into the FIRAS antenna, replacing the sky signal with that of a blackbody (the calibrator) at a precisely known temperature. The careful design of this special calibrator was largely responsible for FIRAS's very successful measurement of the CMBR spectrum.

COBE carried six DMRs, two redundant radiometers at each of three different wavelengths, 9.5, 5.7, and 3.3 mm. The DMR boxes are shown in Fig. 1,

fastened to the outside of the COBE Dewar. Notice that each radiometer has two horn antennas pointed 60° apart in the sky. As the spacecraft rotates (about once a minute) the beams scan around and the DMR records the difference in sky temperature shown by the two horns. COBEs ingenious pointing system keeps the rotation axis always pointing away from the Earth. Orbital motion then moves the rotation axis around the sky in 108 minutes, and on every orbit the antenna beams scan out a 60° wide helix. The Sun-synchronous orbit then causes COBE to scan the whole sky about every six months.

COBE made two important measurements of the CMBR. The FIRAS found that to an accuracy of ±0.03 percent the CMBR spectrum matches that of a blackbody, a result uniquely predicted by the big bang theory. FIRAS also measured the temperature of the CMBR to be 2.73 ± 0.01 K. The DMR made maps of the sky in three wavelength bands. The three bands were used to independently assess the contaminating radiation from our galaxy. It was found that galactic emission is not strong enough to contaminate COBE's measurement of CMBR anisotropy if the region within 20° of the galactic plane is ignored in the data analysis. COBE's maps at wavelengths of 5.7 and 3.3 mm were then shown to contain a small, but significant sky signal with the spectrum expected for bumpiness due to CMBR anisotropy. The root-mean-square magnitude of the signal was 30 μK, at the low end of the range acceptable to standard versions of the big bang theory.

The DIRBE instrument made maps of the entire sky at 10 wavelengths between 1 and 240 μm. The data give valuable information about the zodiacal light from interplanetary dust in our solar system, infrared emission from stars and interstellar dust in our galaxy, and the intensity of cosmic infrared background radiation.

NASA's COBE satellite was an unqualified success. The science instruments met or exceeded their sensitivity goals, and the spacecraft performed its complex pointing and instrument support functions throughout the mission lifetime. Liquid helium ran out nine months after launch, and the DMRs were finally shut off after four years of observations. COBE's results strongly support the Big Bang theory of cosmic evolution and point the way for even more detailed studies of the early universe.

See also: BIG BANG THEORY; COSMIC MICROWAVE BACKGROUND RADIATION; COSMIC MICROWAVE BACKGROUND

RADIATION, DISCOVERY OF; COSMOLOGICAL CONSTANT; COSMOLOGICAL PRINCIPLE; COSMOLOGY; COSMOLOGY, INFLATIONARY; RADIATION, BLACKBODY; RADIATION, THERMAL; UNIVERSE

Bibliography

CHOWN, M. *Afterglow of Creation: From the Fireball to the Discovery of Cosmic Ripples* (Arrow, London, 1993).

GULKIS, S.; LUBIN, P. M.; MEYER, S. S.; and SILVERBERG, R. F. "The Cosmic Background Explorer." *Sci. Am.* **262** (1), 131–139 (1990).

PEEBLES, P. J. E., and WILKINSON, D. T. "The Primeval Fireball." *Sci. Am.* **216** (6), 28–37 (1968).

PEEBLES, P. J. E.; SCHRAMM, D. N.; TURNER, E. L.; and KRON, R. G. "The Evolution of the Universe." *Sci. Am.* **271** (4), 53–57 (1994).

DAVID WILKINSON

COSMIC MICROWAVE BACKGROUND RADIATION

The big bang theory of the beginning and evolution of the universe unequivocally predicts that the universe should be filled with heat radiation left over from its earliest moments. The discovery (in 1964) and subsequent measurements of this radiation—the cosmic microwave background radiation (CMBR)—lends strong support to the big bang theory. In the early universe the radiation was extremely hot (10 billion degrees, a minute after the bang), but the expansion of the universe cooled the radiation to its current temperature of only 2.73 K (K = Kelvin degrees above absolute zero). By calculating the detailed interactions that such a universal radiation field should have over the entire history of the universe, theoretical physicists reached the remarkable conclusion that the initial blackbody spectrum of the radiation would be preserved to a high degree of accuracy (about 1 part in 100,000). The characteristic shape of that spectrum is shown in Fig. 1. Radiation with a temperature of 2.73 K has its peak intensity at a wavelength of 2 mm. The CMBR spans the microwave spectral band, so, instruments to measure the CMBR resemble radar sets and microwave communications links. Indeed, the CMBR was discovered at AT&T Bell Labora-

tories, using apparatus designed to study space communications.

The CMBR can be measured with high accuracy, providing our clearest direct view of the big bang. According to the standard big bang model, the CMBR had its last strong interaction with matter (scattering from free electrons) at a time only 300,000 years after the bang. At the time the temperature of the universe was about 4,000 K, too hot for stars or galaxies to start to form, and the universe was quite smooth, in contrast to the lumpiness of matter that we see today. The CMBR photons come straight to us from that early time, carrying valuable information about an epoch never before observed, an epoch thousands of times earlier than we can reach with optical telescopes. (Since electromagnetic radiation travels at finite speed, looking deeper into space also means looking farther back in time and seeing events closer to the bang.)

NASA's Cosmic Background Explorer (COBE) satellite was designed specifically to study the CMBR. Two questions were of primary interest. (1) Does the

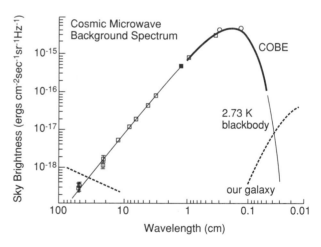

Figure 1 The blackbody spectrum for a temperature of 2.73 K (wavelength increases to the left). The lines show the theoretical shape, but the thicker line is also the measurement made by the COBE satellite. The measurement errors are smaller than the line's thickness. Clearly the CMBR spectrum traces that of a 2.73 K blackbody exceedingly well. The results of some terrestrial experiments (made before COBE) are shown as squares and circles. Radiation from our galaxy is shown by the dashed lines. Note that our galaxy has a fortuitous window through which we can view the cosmic microwave background radiation.

CMBR spectrum have the same shape as that of a blackbody as predicted by the big bang theory? (2) Is there any bumpiness (anisotropy) in the distribution of CMBR temperature across the sky? Both of these questions had been studied from the ground and from high-flying scientific balloons. However, the accuracy of the measurements were limited by atmospheric effects, partial sky coverage, and short observation times—problems solved by a satellite mission. COBE was an extraordinary success. Figure 1 shows the COBE measurement of the CMBR spectrum as a heavy line going over the peak of the predicted curve. The agreement with a blackbody curve is excellent, and the error bars ($\pm 0.03\%$) are too small to show on the graph. Not only was the COBE result very accurate, the measured wavelength range spanned the important peak region, much of which is obscured from the ground by atmospheric radiation from oxygen and water vapor. Most scientists now regard the COBE measurement of the CMBR spectrum as strong evidence that our universe expanded from a hot early phase, like that envisioned by the big bang model. No other cosmological model has yet explained the CMBR and its blackbody spectrum.

Before COBE flew, ground and balloon based experiments had shown that the distribution of CMBR temperature (intensity) is uniform across the sky to 1 part in 10,000 except for one predicted effect. Earth's motion through the sea of CMBR photons causes the forward hemisphere (in the direction of motion) to appear warmer than the backward hemisphere. The effect is similar to the Doppler effect that shifts the frequency of a train whistle up for an approaching train and down for a receding train. The temperature distribution due to motion has a cosine shape on the sky, $T(\theta) = 2.73(v/c)\cos(\theta)$ K, where θ is the angle measured from the direction of motion, v is Earth's speed, and c is the speed of light. The magnitude of the effect is measured to be about 0.003 K, so Earth is moving through the CMBR at about 0.1 percent of the speed of light; the direction is toward the constellation Leo. The magnitude and direction of this motion is due to a combination of Earth's orbit around the Sun, the Sun's orbit around the center of our galaxy, and the "peculiar" motion of the Galaxy through the CMBR. The latter turns out to be surprisingly large, 600 km/s.

However, the most interesting bumps in the CMBR temperature are expected to be about 100 times weaker than the motional effect. Theorists cal-culate that there must have been very small fluctuations (1 part in 100,000) in the CMBR temperature even when the universe was still hot (4,000 K). The fluctuations in temperature and density "seed" the formation of galaxies and other large cosmic structure that we see today. In 1992 the COBE team announced the discovery of CMBR fluctuations with a root-mean-square magnitude of 30 μK. The fluctuations that COBE found have angular sizes ranging from 7° to 90°; they are too large to have been formed at the time that the CMBR photons were last scattered. It is currently believed that these fluctuations are remnants of the very early universe, possibly formed by quantum fluctuations later processed by an inflationary expansion.

Accurate measurements and a well-developed theory of the CMBR have convinced most scientists that the big bang theory is correct in its essential features. The measured spectrum and fluctuations match the theory surprisingly well, and no other theory comes close to explaining the data. COBE and subsequent experiments probing smaller angular sizes are giving us our first glimpse of structural detail in the 300,000-year-old universe, an age equivalent to 14 hours of human life.

See also: ATMOSPHERIC PHYSICS; BIG BANG THEORY; COSMIC BACKGROUND EXPLORER SATELLITE; COSMIC MICROWAVE BACKGROUND RADIATION, DISCOVERY OF; COSMOLOGICAL PRINCIPLE; COSMOLOGY; DOPPLER EFFECT; ELECTROMAGNETIC RADIATION; LIGHT, ELECTROMAGNETIC THEORY OF; PHOTON; RADIATION, BLACKBODY; RADIATION PHYSICS; TEMPERATURE SCALE, KELVIN; UNIVERSE; UNIVERSE, EXPANSION OF; UNIVERSE, EXPANSION OF, DISCOVERY OF

Bibliography

CHOWN, M. *Afterglow of Creation: From the Fireball to the Discovery of Cosmic Ripples* (Arrow, London, 1993).

GAMOW, G. "Modern Cosmology." *Sci. Am.* **190** (3), 55–63 (1954).

GUTH, A. H., and STEINHARDT, P. J. "The Inflationary Universe." *Sci. Am.* **250** (5), 116–128 (1984).

PEEBLES, P. J. E., and WILKINSON, D. T. "The Primeval Fireball." *Sci. Am.* **216** (6), 28–37 (1968).

PEEBLES, P. J. E.; SCHRAMM, D. N.; TURNER, E. L.; and KRON, R. G. "The Evolution of the Universe." *Sci. Am.* **271** (4), 53–57 (1994).

WILKINSON, D. T. "Anisotropy of the Cosmic Blackbody Radiation." *Science* **232**, 1517–1522 (1986).

DAVID WILKINSON

COSMIC MICROWAVE BACKGROUND RADIATION, DISCOVERY OF

The discovery of the cosmic microwave background radiation in 1965 was one of the great scientific discoveries of the twentieth century and earned the 1978 Nobel Prize in physics for Arno A. Penzias and Robert W. Wilson. We live in a bath of microwave and infrared photons whose characteristic temperature is 2.7 K. This discovery provided crucial data for cosmology and convinced physicists and astronomers that we live in an expanding, evolving universe that originated in a hot big bang about 10 or 15 billion years ago. It also provided a means for gathering further observational evidence about the early universe and stimulated great interest that continues to this day in investigating the properties of the very early universe.

Cosmology of the Time

Einstein's theory of general relativity was applied to the spacetime structure of the universe as early as the 1920s by Wilhelm de Sitter, Alexander Friedmann, and Georges Lemaître, but until the discovery by Penzias and Wilson, cosmology was considered much more of a mathematical exercise than an observational science. One knew that all galaxies appear to recede from us with a velocity proportional to their distance (Hubble's Law) and that the sky was dark at night, but the list of unassailable "facts" stopped there. This paucity of data meant that cosmology in the early 1960s was widely viewed as a speculative field. Nevertheless, various theoretical cosmologies were developed and a few astronomers were interested in testing these theories as best they could. The leading theories at this time were the big bang universe, whose development extended back to Lemaître and was continued primarily by George Gamow and his collaborators, and the steady-state universe, invented in England in 1948 by Hermann Bondi, Thomas Gold, and Fred Hoyle. The big bang was a universe that began in a highly condensed state at a definite time in the past and was now evolving. The nature of the condensed state ranged from Lemaître's "primeval atom" to Gamow's hot stew of neutrons. The steady-state universe removed the problem of a special time when spacetime began by postulating that the universe has always existed and throughout all time remains the same in its overall properties. To accomplish this in the face of the observed continual separation of galaxies from each other, the theory proposed a slow, ongoing creation of matter at all places. As important as their findings were, however, Penzias and Wilson were *not* conducting an experiment to test any model of the universe—their discovery was completely serendipitous.

The Discovery Story

Bell Telephone Laboratories in Holmdel, New Jersey, was one of the premier industrial research laboratories in the world, noted for its excellence in both developing practical devices (e.g., the transistor in the 1940s) and fostering fundamental research (e.g., the first detection of extraterrestrial radio waves in 1932, which began the field of radio astronomy). During the heady Cold War "space race" days launched along with the Soviet Sputnik satellite in 1957, there was much effort devoted to the development of long-distance communication systems based on satellites. Bell Labs played a key role in the American program with the Echo satellite, a large balloon launched in 1960 and designed to reflect radio signals passively, and Telstar, the first true communications satellite, launched in 1963. Successful transmission of microwave signals by these satellites required extremely high sensitivity in both the satellites and the ground stations, and for this purpose an antenna of unprecedented size (for its type) was built at Holmdel and then supplied with the most sensitive receiving electronics in the world. The antenna was a horn-reflector with an aperture 6 m in diameter and of a type that had been earlier designed for microwave relay of telephone calls. The large size made it able to detect extremely faint signals, but even more important was its ability, unlike a standard "dish" antenna, to reject any interfering signal from all directions other than the small patch of sky that it directly faced. The microwave receiver, based on recent designs at Bell Labs, had at its heart a maser amplifier.

This then was the environment in which Penzias and Wilson found themselves in the early 1960s. After Arno Penzias's family had fled Nazi Germany in 1939, he had grown up in New York City, obtained his Ph.D. in physics at Columbia University under Charles Townes (inventor of the maser), and begun working at Bell Labs in 1961. For his thesis he had built a maser receiver mounted on a radio telescope to search for hydrogen gas in clusters of galaxies.

Robert Wilson had been raised in Houston, Texas, and obtained his Ph.D. in physics at the California Institute of Technology under John Bolton, a pioneering Australian radio astronomer who had established a radio observatory at Caltech. Wilson's thesis had involved a sensitive mapping of the microwave emission from our Milky Way galaxy, after which he joined Bell Labs in 1963. Once the horn-reflector was released from its communications duties, Penzias and Wilson began to exploit it for sensitive radio astronomy projects. The first observations, at a wavelength of 7 cm (or frequency of 4,080 MHz), were designed to search for a possible "halo" of emission from the outermost parts of the Milky Way. Their maser amplifier was augmented with two essential components, a reference "cold load" chilled by liquid helium to 4 K and a switch allowing one to compare this load with any signal from the sky. Virtually all previous observations in radio astronomy had measured sky intensities *relative* to a comparison (such as a nearby region of sky) whose value was constant but only approximately known in magnitude. In contrast Penzias and Wilson's setup allowed them to deduce accurate *absolute* measurements of sky intensity.

They began observations in the spring of 1964 and immediately measured an excess signal of 3 to 4 K no matter where they pointed the antenna. (Radio astronomers often express intensities in terms of a temperature; the convention is that the measured intensity is specified by the temperature to which a perfectly emitting body, a so-called blackbody, would have to be raised to emit the observed intensity.) This intensity was much more than expected from the Milky Way, which in any case would not have been constant from all sky directions, so they suspected that some part of their antenna or receiver was misbehaving and consistently supplying a spurious signal. Many months were spent in attempting to track down this problem, but in the end suspects ranging from pigeon droppings to New York City radars were all eliminated from the list. As Penzias (1972) later recalled, "Having explored and discarded a host of terrestrial explanations, and 'knowing' at the time that no astronomical explanation was possible, we frankly did not know what to do with our result."

The solution only came in a roundabout fashion in early 1965 when Penzias happened to learn of a preprint written by James Peebles, a theoretical astrophysicist at nearby Princeton University. Peebles had developed the ideas of his Princeton colleague Robert Dicke, who argued for an oscillating universe that renewed itself each time in an expanding hot big bang. Peebles found that such a universe should now be evidenced as background thermal radiation at a temperature of about 10 to 30 K. Two other physicists at Princeton, Peter Roll and David Wilkinson, were even then building a sensitive radio receiver and small horn antenna at a wavelength of 3 cm to search for this predicted radiation. When the Princeton group learned of the excess signal detected by Penzias and Wilson, they paid a visit to Bell Labs and quickly became convinced of its reality. Although Penzias and Wilson were leery of the cosmological ideas (Wilson in fact had learned the steady-state theory from Hoyle himself at Caltech), the two groups agreed to publish adjacent articles.

The history-making articles duly appeared in July 1965 in the *Astrophysical Journal* (Penzias and Wilson, 1965; Dicke, Peebles, Roll, and Wilkinson, 1965). In only 800 words Penzias and Wilson reported an "excess antenna temperature" of 3.5 ± 1 K at 7-cm wavelength that was "isotropic [the same in all directions], unpolarized, and free from seasonal variations." At somewhat greater length Dicke and his colleagues discussed this phenomenon as "cosmic blackbody radiation" from a "primordial fireball" and pointed out that it placed important constraints on such questions as whether the universe was open or closed. But they could not be sure that it was truly blackbody radiation until its intensity had been measured at many other wavelengths. The Princeton group redoubled its efforts and soon had determined a temperature at 3-cm wavelength of 3.0 ± 0.5 K. Many other measurements followed from around the world, and by the early 1970s the presence of the microwave background radiation, interpreted as the remnant of the early hot universe, was an accepted part of cosmology. Moreover, cosmologists no longer considered the steady-state universe as a viable model.

Prior Intimations

There are many fascinating aspects to this story of the discovery of the cosmic microwave background radiation. Described above are its accidental nature, the setting in an industrial research lab, and the circumstance that a nearby group was simultaneously and independently in the process of *purposely* looking for the radiation. Furthermore, in retrospect one can find several earlier measurements by radio astronomers that were on the edge of making the discovery but were not pursued because of technological limitations or preferences for other interpretations. Even at Bell Labs, satellite communication engineers had written several reports (using similar equipment to that of Penzias and Wilson) listing a possible excess intensity of unknown origin in their equipment (but not discussed in any detail). Finally, it was soon realized that the existence of this bath of 3-K radiation nicely explained a 25-year-old puzzle concerning the observed excitation of the CN molecule in interstellar regions.

Perhaps most interesting and surprising is that more than fifteen years earlier theorists had calculated, based on Gamow's idea of a hot big bang universe, that "the temperature in the universe at the present time is found to be about 5 K" (Alpher and Herman, 1948). Ralph Alpher and Robert Herman were physicist colleagues at the Applied Physics Lab of Johns Hopkins University in Baltimore, and Alpher was also working on his Ph.D. thesis under the esteemed (and unconventional) nuclear physicist George Gamow at George Washington University in nearby Washington, D.C. In a series of papers over a five-year period Alpher and Herman worked out the details of the physical conditions in the evolution of the early universe. They focused particularly on forming the elements, with their correct relative abundances, in the first few minutes of the universe (when temperatures were of order 10^9 K) by the processes of neutron capture and beta decay. In several of these papers Alpher and Herman mentioned that the universe today should have a temperature calculated on various assumptions to be in the range of 5 to 28 K, while Gamow cited in his own publications a range of 3 to 50 K. This point was never particularly stressed in these papers, although they did try unsuccessfully to persuade a few radio astronomers to attempt a measurement; in retrospect perhaps it would have been possible with the technology of the 1950s, but only if someone had been sufficiently motivated to carry out a major experiment. Further hurting any such motivation was that the calculation was a small part of a complex theory of element production that lost favor through the 1950s. In any case, neither Dicke nor Peebles (nor certainly Penzias and Wilson) had read the papers—instead they derived all over again the existence of a background radiation from a different, although related, cosmological perspective.

See also: BIG BANG THEORY; COSMIC BACKGROUND EXPLORER SATELLITE; COSMIC MICROWAVE BACKGROUND RADIATION; COSMOLOGY; NUCLEOSYNTHESIS; OLBERS'S PARADOX; RADIATION, BLACKBODY; RADIATION, THERMAL; UNIVERSE, EXPANSION OF; UNIVERSE, EXPANSION OF, DISCOVERY OF

Bibliography

ALPHER, R. A., and HERMAN, R. "Evolution of the Universe." *Nature* 162, 774–775 (1948).

ALPHER, R. A., and HERMAN, R. "Reflections on Early Work on 'Big Bang' Cosmology." *Phys. Today* 41 (8), 24–34 (1988).

BERNSTEIN, J. *Three Degrees Above Zero* (Scribners, New York, 1984).

BRUSH, S. J. "Prediction and Theory Evaluation: Cosmic Microwaves and the Revival of the Big Bang." *Perspectives on Science* 1, 565–602 (1993).

DICKE, R. H.; PEEBLES, P. J. E.; ROLL, P. G.; and WILKINSON, D. T. "Cosmic Black-Body Radiation." *Astrophys. J.* 142, 414–419 (1965).

KRAGH, H. *Cosmology and Controversy* (Princeton University Press, Princeton, NJ, 1996).

NORTH, J. D. *The Measure of the Universe: A History of Modern Cosmology* (Oxford University Press, Oxford, Eng., 1965).

PENZIAS, A. A. "Cosmology and Microwave Astronomy" in *Cosmology, Fusion, and Other Matters—George Gamow Memorial Volume,* edited by F. Reines (Colorado Association University Press, New York, 1972).

PENZIAS, A. A. "The Origin of the Elements [Nobel Address]." *Science* 205, 549–554 (1979).

PENZIAS, A. A., and WILSON, R. W. "A Measurement of Excess Antenna Temperature at 4080 Mc/s." *Astrophys. J.* 142, 419–421 (1965).

WEINBERG, S. *The First Three Minutes* (Basic Books, New York, 1977).

WILSON, R. W. "The Cosmic Microwave Background Radiation [Nobel Address]." *Science* 205, 866–874 (1979).

WOODRUFF T. SULLIVAN III

COSMIC RAY

Cosmic rays are high-energy charged particles, originating in outer space, that travel at nearly the speed of light and strike the earth from all directions. Most cosmic rays are the nuclei of atoms, ranging from the lightest to the heaviest elements in the periodic table. Cosmic rays also include high-energy electrons, positrons, and other subatomic particles. The term "cosmic rays" usually refers to galactic cosmic rays, which originate in sources outside the solar system and are distributed throughout our Milky Way galaxy. However, this term has also come to include other classes of energetic particles in space, such as nuclei and electrons accelerated in association with energetic events on the Sun (called solar energetic particles) and particles accelerated in interplanetary space.

Discovery and Early Research

Cosmic rays were discovered in 1912 by Victor Hess, when he found that an electroscope discharged more rapidly as he ascended in a balloon. He attributed this to a source of radiation entering the atmosphere from above, and in 1936 he was awarded the Nobel Prize for his discovery. For some time it was believed that the radiation was electromagnetic in nature (hence the name cosmic "rays"), and some textbooks still incorrectly include cosmic rays as part of the electromagnetic spectrum. However, during the 1930s it was found that cosmic rays must be electrically charged because they are affected by Earth's magnetic field.

From the 1930s to the 1950s, before human-made particle accelerators reached very high energies, cosmic rays served as a source of particles for high-energy physics investigations and led to the discovery of subatomic particles that included the positron and muon. Although these applications continue, the main focus of cosmic ray research is currently directed toward astrophysical investigations of where cosmic rays originate, how they get accelerated to such high velocities, what role they play in the dynamics of our galaxy, and what their composition tells us about matter from outside the solar system. To measure cosmic rays directly, before they have been slowed down and broken up by the atmosphere, research is carried on by means of particle detectors carried on spacecraft and high-altitude balloons.

Cosmic Ray Energies and Acceleration

The energy of cosmic rays is usually measured in units of mega-electron volts (MeV) or giga-electron volts (GeV). (One electron volt is the energy gained when an electron is accelerated through a potential difference of 1 V). Most galactic cosmic rays have energies between 100 MeV (corresponding to a velocity for protons of 43 percent of the speed of light) and 10 GeV (corresponding to 99.6 percent of the speed of light). The number of cosmic rays with energies beyond 1 GeV decreases by about a factor of 50 for every factor of 10 increase in energy. Over a wide energy range the number of particles per square meter per steradian per second with energy greater than E (measured in giga-electron-volts) is given approximately by $N(>E) = k(E + 1)^{-a}$, where $k \approx 5,000$ per square meter per steradian per second and $a = 1.6$. The highest energy cosmic rays measured to date have had more than 10^{20} eV, equivalent to the kinetic energy of a baseball traveling at approximately 100 mph!

It is believed that most galactic cosmic rays derive their energy from supernova explosions, which occur approximately once every fifty years in our galaxy. To maintain the observed intensity of cosmic rays over millions of years requires that a few percent of the more than 10^{51} ergs released in a typical supernova explosion be converted to cosmic rays. There is considerable evidence that cosmic rays are accelerated as the shock waves from these explosions that travel through the surrounding interstellar gas. The energy contributed to the Galaxy by cosmic rays (about 1 eV per cm^3) is about equal to that contained in galactic magnetic fields and in the thermal energy of the gas that pervades the space between the stars.

Cosmic Ray Composition

Cosmic rays include essentially all of the elements in the periodic table; about 89 percent of the nuclei are hydrogen (protons), 10 percent helium, and about 1 percent heavier elements. The common heavier elements (such as carbon, oxygen, magnesium, silicon, and iron) are present in about the same relative abundances as in the solar system, but there are important differences in elemental and isotopic composition that provide information on the origin and history of galactic cosmic rays. For example, there is a significant overabundance of the rare elements Li, Be, and B produced when heavier cosmic rays such as carbon, nitrogen, and oxygen fragment into lighter nuclei during collisions with the interstellar gas. The isotope ^{22}Ne is also overabundant, showing that the nucleosynthesis of cosmic rays and solar system material have differed. Electrons constitute about 1 percent of galactic cosmic rays. It is not known why electrons are apparently less efficiently accelerated than nuclei.

Cosmic Rays in the Galaxy

Because cosmic rays are electrically charged they are deflected by magnetic fields, and their directions have been randomized, making it impossible to tell where they originated. However, cosmic rays in other regions of the Galaxy can be traced by the electromagnetic radiation they produce. Supernova remnants such as the Crab Nebula are known to be a source of cosmic rays from the radio synchrotron radiation emitted by cosmic ray electrons spiraling in the magnetic fields of the remnant. In addition, observations of high-energy (10 MeV to 1,000 MeV) gamma rays resulting from cosmic ray collisions with interstellar gas show that most cosmic rays are confined to the disk of the Galaxy, presumably by its magnetic field. Similar collisions of cosmic ray nuclei produce lighter nuclear fragments, including radioactive isotopes such as ^{10}Be, which has a half-life of 1.6 million years. The measured amount of ^{10}Be in cosmic rays implies that, on average, cosmic rays spend about 10 million years in the Galaxy before escaping into intergalactic space.

Very-High-Energy Cosmic Rays

When high-energy cosmic rays undergo collisions with atoms of the upper atmosphere, they produce a cascade of "secondary" particles that shower down through the atmosphere to Earth's surface. Secondary cosmic rays include pions (which quickly decay to produce muons, neutrinos, and gamma rays), as well as electrons and positrons produced by muon decay and gamma ray interactions with atmospheric atoms. The number of particles reaching Earth's surface is related to the energy of the cosmic ray that struck the upper atmosphere. Cosmic rays with energies beyond 10^{14} eV are studied with large "air shower" arrays of detectors distributed over many square kilometers, which sample the particles produced. The frequency of air showers ranges from about 100 per square meter per year for energies greater than 10^{15} eV to only about 1 per square kilometer per century for energies beyond 10^{20} eV.

Most secondary cosmic rays reaching Earth's surface are muons, with an average intensity of about 100 per square meter per second. Although thousands of cosmic rays pass through our bodies every minute, the resulting radiation levels are relatively low, corresponding, at sea level, to only a few percent of the natural background radiation. However, the greater intensity of cosmic rays in outer space is a potential radiation hazard for astronauts, especially when the Sun is active and interplanetary space may suddenly be filled with solar energetic particles. Cosmic rays are also a hazard to electronic instrumentation in space; impacts of heavily ionizing cosmic ray nuclei can cause computer memory bits to "flip" or small microcircuits to fail. Cosmic ray interaction products such as neutrinos are also studied by large detectors placed in deep underground mines or under water.

Cosmic Rays in the Solar System

Just as cosmic rays are deflected by the magnetic fields in interstellar space, they are also affected by the interplanetary magnetic field embedded in the solar wind (the plasma of ions and electrons blowing from the solar corona at about 400 km/sec), and therefore have difficulty reaching the inner solar system. Spacecraft venturing out toward the boundary of the solar system have found that the intensity of galactic comic rays increases with distance from the Sun. Just as solar activity varies over the 11-year solar cycle, the intensity of cosmic rays at Earth also varies, in anti-correlation with the sunspot number.

The Sun is also a sporadic source of cosmic ray nuclei and electrons that are accelerated by shock waves traveling through the corona and by magnetic energy released in solar flares. During such occurrences the intensity of energetic particles in space can increase by a factor of 10^2 to 10^6 for hours to days. Such solar particle events are much more frequent during the active phase of the solar cycle. The maximum energy reached in solar particle events is typically 10 to 100 MeV, occasionally reaching 1 GeV (roughly once a year) to 10 GeV (roughly once a decade). Solar energetic particles can be used to measure the elemental and isotopic composition of the Sun, thereby complementing spectroscopic studies of solar material.

A third component of cosmic rays, comprised of only those elements that are difficult to ionize, including He, N, O, Ne, and Ar, was given the name "anomalous cosmic rays" because of its unusual composition. Anomalous cosmic rays originate from electrically neutral interstellar particles that have entered the solar system unaffected by the magnetic field of the solar wind, been ionized, and then accelerated at the shock wave formed when the solar wind slows as a result of plowing into the interstellar gas, which is thought to occur somewhere between 75 and 100 AU from the Sun (1 AU is the distance from the Sun to the Earth). Thus, it is possible that the Voyager-1 spacecraft, which should reach 100 AU by the year 2007, will have the opportunity to observe an example of cosmic ray acceleration directly.

See also: ELECTROMAGNETIC RADIATION; INTERSTELLAR and INTERGALACTIC MEDIUM; MUON; POSITRON; SOLAR WIND; SUPERNOVA

Bibliography

GAISSER, T. K. *Cosmic Rays and Particle Physics* (Cambridge University Press, Cambridge, Eng., 1990).

JOKIPII, J. R., and McDONALD, F. B. "Quest for the Limits of the Heliosphere." *Sci. Am.* **272,** 58–63 (1994).

LONGAIR, M. S. *High-Energy Astrophysics,* 2nd ed., Vol. 1: *Particles, Photons, and Their Detection* (Cambridge University Press, Cambridge, Eng., 1992).

SIMPSON, J. A. "Elemental and Isotopic Composition of the Galactic Cosmic Rays." *Annu. Rev. Nucl. Part. Sci.* **33,** 323–381 (1983).

R. A. MEWALDT

COSMIC STRING

Cosmic strings is the name given to a class of objects that may have been produced in the early universe. Cosmic strings are just one of many kinds of hypothetical cosmological topological defects, which also include monopoles and domain walls. Cosmic strings differ from particulate matter in that they are essentially one-dimensional objects. They have a microscopic width in two dimensions, but they can reach astronomical lengths in the third dimension. In contrast, monopoles are more particulate in nature, while domain walls are more like two-dimensional surfaces. One can also form hybrid topological defects, such as strings bounded by monopoles, walls bounded by strings, and even strings attached to other types of strings. The simplest types of cosmic strings exist apart from other defects and have no ends. They either close on themselves, forming

loops, or extend to infinity. The latter possibility can only occur if the universe is itself infinite in extent. If the universe is infinite, then most of the string is in the infinite length strings and not the loops.

Formation

It is now thought that matter is made of fields that have reached their present configuration by a sequence of phase transitions involving spontaneous symmetry breaking. As in condensed matter systems, symmetry breaking phase transitions can proceed imperfectly, leaving behind topological defects. For example, string-like defects can be formed in nematic liquid crystals when the pressure is suddenly increased. These liquid crystal strings are observed to behave in much the same way as cosmic strings are believed to behave. Unlike in condensed matters systems, cosmological phase transitions occur in an essentially infinite medium, which guarantees that topological defects will form if the field topologies allow them to exist. At present, we cannot be sure whether the fields that make up the matter in our universe allow for cosmic strings or any other type of topological defect. If the universe undergoes an inflationary phase after formation of cosmic strings, then the strings may be diluted to such an extent that there would be no strings in the observable part of our universe.

Dynamics

Cosmic strings are in tension, just as is a stretched elastic band; however, the tension is enormous. The large tension causes cosmic strings to move with velocities approaching that of the speed of light. With this enormous velocity, pieces of string will often collide with one another. During these collisions different pieces of string may reconnect. This reconnection allows a small loop of string to become part of an infinite string, or an infinite string to chop off a small loop of string. In cosmology, the latter process will dominate so that most of the string will eventually end up in small loops. The small loops will themselves shrink due to energy loss to gravitational radiation. While most objects that move produce some small amount of gravitational radiation, cosmic strings are copious producers of gravitational radiation because they move at velocities close to the speed of light. A cosmic string loses most of its energy to gravitational radiation in only 10,000 oscillations. Most of the energy that was in cosmic strings in the past would now be in the form of a gravitational radiation background. Cosmic string models produce more gravitational radiation than most other cosmogenic theories.

Grand Unified Strings

Cosmic strings may have formed at the grand unified scale, where the weak, strong, and electromagnetic forces are unified, when the universe, soon after the big bang, had a temperature of 10^{29} K. These grand unified strings have a mass per unit length of 10^{21} kg/m, which is massive enough so that their gravitational attraction could be responsible for the formation of galaxies and the other structures we see in the universe. They would also be responsible for the anisotropy of the temperature of the microwave background radiation that was observed by the cosmic background explorer (COBE) satellite. Cosmic strings lighter than grand unified strings would not produce these observable effects and would be very difficult to detect.

See also: BIG BANG THEORY; COSMIC BACKGROUND EXPLORER SATELLITE; COSMIC MICROWAVE BACKGROUND RADIATION; COSMOLOGY; COSMOLOGY, INFLATIONARY; GALAXIES and GALACTIC STRUCTURE; GRAND UNIFIED THEORY; SYMMETRY BREAKING, SPONTANEOUS; UNIVERSE

Bibliography

GIBBONS, G.; HAWKING, S.; and VACHASPATI, T., eds. *The Formation and Evolution of Cosmic Strings* (Cambridge University Press, Cambridge, Eng., 1990).

KOLB, E. W., and TURNER, M. S. *The Early Universe* (Addison-Wesley, Reading, MA, 1990).

PEEBLES, P. J. E. *Principles of Physical Cosmology* (Princeton University Press, Princeton, NJ, 1993).

VILENKIN, A. "Cosmic Strings." *Sci. Am.* **257** (6), 94–102 (1987).

ALBERT STEBBINS

COSMOLOGICAL CONSTANT

"Cosmological constant" is the term Albert Einstein introduced into his equations of general relativity in 1917 when attempting to apply his new theory to the evolution of the whole universe. He was guided by the then accepted notion that the universe was static. As no solution to his original equations could be found that allowed a static universe, he modified them by adding what amounts to a universal repulsion of matter. Specifically, Einstein's equations relate the curvature of spacetime, represented by the tensor $G_{\mu\beta}$, to the energy-momentum tensor of matter $T_{\mu\beta}$ by the equation

$$G_{\mu\beta} = 8 \pi G T_{\mu\beta}, \tag{1}$$

where G is Newton's constant. Then the addition proposed by Einstein that allowed a static universe was of the form

$$G_{\mu\beta} - \lambda \, g_{\mu\beta} = 8 \pi G T_{\mu\beta}, \tag{2}$$

where λ is the cosmological constant and $g_{\mu\beta}$ is the metric tensor desribing spacetime.

Since Einstein's original suggestion, we now know the universe is not static but, rather, is expanding. The equations for an expanding universe do not require the addition of a cosmological constant. Nevertheless, having been introduced, it was not clear why such a term should have to be equal to zero in the observed universe. In particular, one can rewrite Eq. (2) in the following fashion:

$$G_{\mu\beta} = 8 \pi G T_{\mu\beta} + \lambda \, g_{\mu\beta} = 8 \pi G T'_{\mu\beta}. \tag{3}$$

When expressed this way, it is clear that the cosmological constant is equivalent to adding a new term to the energy momentum tensor of matter. This new term has the same form as a term that would correspond to some nonzero energy density for empty space. While it may sound ludicrous to expect empty space to possess nonzero energy density, in the context of modern quantum field theory it can be shown that this can occur, and indeed is quite generic.

One can probe for the existence of a cosmological constant by attempting to measure the geometry of the universe on large scales, in particular by probing the relation between physical distance and the observed redshift of distant galaxies. Such measurements suggest that the actual energy density of empty space, which would correspond to a cosmological constant, cannot be much greater than the energy density that would be required to stop the observed expansion of the universe, about 10^{-29} g/cm^3. This value is so much smaller than any a priori expectation based on quantum field theoretic arguments that it is generally expected that some mechanism must exist in the universe that ultimately forces the cosmological constant to be identically zero. No known mechanism has yet been explicitly derived, although arguments based on our poorly defined notions of quantum gravitational effects have suggested that one might be able to demonstrate that a zero value for a cosmological constant is far more probable than any other value.

To make matters more confusing, observations now suggest that a nonzero value of the cosmological constant might help resolve various outstanding problems in modern cosmology, in particular the apparent discrepancy between the age of the universe and the age of our galaxy. By slowing the deceleration of the expansion of the universe, a cosmological constant would make a universe with a fixed expansion rate as observed today older, as would seem to be required by estimates of the age of our galaxy based on the inferred age of the oldest globular clusters in the halo of our galaxy.

The resolution of what has become known as the cosmological constant problem is considered by many physicists to be one of the most important outstanding problems in particle physics and cosmology. New observations, such as might be possible by observations of gravitational lensing of distant quasars by intervening galaxies, measurements of the angular size of distant objects, the relationship between apparent luminosity and redshift of distant supernovas, or detailed measurements of the spectrum of fluctuations in the observed cosmic microwave background radiation will help determine if indeed the cosmological constant is nonzero in the universe. At the same time, new efforts to understand the nature of quantum gravity may someday explain why the cosmological constant is so close to zero, or indeed, exactly equal to zero today.

See also: COSMIC MICROWAVE BACKGROUND RADIATION; COSMOLOGICAL PRINCIPLE; COSMOLOGY; COSMOLOGY, INFLATIONARY; GRAVITATIONAL LENSING; QUANTUM FIELD THEORY; QUASAR; REDSHIFT; RELATIVITY, GENERAL THEORY OF; SPACETIME; UNIVERSE, EXPANSION OF; UNIVERSE, EXPANSION OF, DISCOVERY OF

Bibliography

KRAUSS, L. M., and TURNER, M. S. "The Cosmological Constant Is Back." *J. Gen. Rel. Grav.* **27,** 1135 (1995).
WEINBERG, S. "The Cosmological Constant Problem." *Rev. Mod. Phys.* **61,** 1 (1989).

LAWRENCE M. KRAUSS

COSMOLOGICAL PRINCIPLE

The cosmological principle holds that we do not occupy a privileged position in the universe. It is the logical extension of Copernicus's displacement of the earth from the center of the cosmos, and it provides an important starting point for modern cosmological models. However reasonable, it is of course, just a hypothesis.

By assuming that we occupy an "ordinary" position in the universe, one can infer much about the structure of the universe on the very largest scales. While the Universe abounds with structure—stars, galaxies, clusters of galaxies, superclusters, voids, and great walls—it does have a simplicity first recognized by the American astronomer Edwin Powell Hubble. From our vantage point, the universe appears the same in any direction and is said to be isotropic. From this and the cosmological principle, one can infer that on very large scales the universe is also homogeneous; that is, averaging on a sufficiently large scale (more than 100 Mpc) smooths out local inhomogeneities, and the properties of the universe (e.g., density of matter, types of galaxies) do not vary from place to place. Isotropy and homogeneity are the starting points for the highly successful big bang model of the universe. As it so happens, Albert Einstein and others had assumed both properties to make the equations of general relativity tractable before Hubble's discovery.

Around 1948 British astrophysicists Fred Hoyle and Hermann Bondi and American astrophysicist Thomas Gold went a step further and proposed the perfect cosmological principle. This extension of the cosmological principle holds that we also do not live at a special time and implies that the universe takes on the same appearance at all times. The perfect cosmological principle formed the basis for the steady-state cosmological model. Because the universe is expanding, Hoyle, Bondi, and Gold had to postulate the continuous creation of matter, albeit at a very tiny rate, in order to fill the expanding space between galaxies with new galaxies. Although many find the steady-state theory philosophically attractive, there is now overwhelming evidence that we live in an evolving universe. Two examples are the existence of cosmic microwave background radiation and the abundances of the light elements D, ^{3}He, ^{4}He, and ^{7}Li, produced during the first seconds after the big bang.

While the cosmological principle allows the inference of homogeneity, the homogeneity of the observable universe has also been confirmed by observations. First, there is the uniformity of the cosmic microwave background radiation: The temperature measured across the sky in different directions is the same to about one part in 10^4. This implies that the distribution of matter in the universe at early times was similarly homogeneous. (There are variations in the temperature of the cosmic microwave background radiation at the level of a few parts in 10^5, which indicate that small inhomogeneities did exist and seeded the structure seen today.) Second, surveys of the distribution of matter in the form of bright galaxies indicate that on the large scale the universe today is homogeneous.

While the cosmological principle is a very sensible scientific response to the anthropocentric theories that preceded Copernicus, it is a hypothesis. Because we have no direct knowledge of the universe outside the boundaries of the observable universe (about 15 billion light-years across), we cannot test the cosmological principle on the very largest scales. In fact, cosmic inflation, proposed by American physicist Alan H. Guth, suggests that the universe on extremely large scales may be very different from the part that we are able to observe.

Inflation holds that tiny regions of the universe can undergo a rapid period of expansion—called inflation—and grow to a size that can encompass all that we can see today and much, much more. If true, our observable universe traces its history (and its big bang) back to a particular patch (or bubble). Other bubbles could well have had very different histories. The Russian cosmologist Andre Linde has speculated that the spawning of bubble universes has and will go on forever and further, that the properties of different bubble universes could be very different— different realizations of the laws of physics and even different numbers of spatial dimensions! However, one might still expect that the cosmological princi-

ple is satisfied in a grander sense; namely, that we occupy the most probable kind of bubble universe.

See also: BIG BANG THEORY; COPERNICAN REVOLUTION; COSMIC MICROWAVE BACKGROUND RADIATION; COSMOLOGICAL CONSTANT; COSMOLOGY; COSMOLOGY, INFLATIONARY; HUBBLE, EDWIN POWELL; HUBBLE CONSTANT; UNIVERSE; UNIVERSE, EXPANSION OF

Bibliography

BONDI, H., and GOLD, T. "The Steady-State Theory of the Expanding Universe." *Mon. Not. R. Astron. Soc.* **108,** 252–270 (1948).

HOYLE, F. "A New Model for the Expanding Universe." *Mon. Not. R. Astron. Soc.* **108,** 372–382 (1948).

LINDE, A. "The Self-Reproducing Inflationary Universe." *Sci. Am.* **271** (5), 48–55 (1994).

WEINBERG, S. *First Three Minutes* (Basic Books, New York, 1977).

MICHAEL S. TURNER

COSMOLOGY

Cosmology is the scientific study of the origin and evolution of the universe. It has attracted the attention of some of the greatest scientists, including Albert Einstein, Isaac Newton, Copernicus, and Ptolemy. Since gravity plays the fundamental role in the evolution of the universe, and an unbounded system like the universe cannot be treated self-consistently in the framework of Newtonian gravity, modern cosmology can be dated to the advent of Einstein's general theory of relativity in 1916.

Einstein attempted to find static solutions to his field equations but could not; his models either expanded or contracted. Because there was no evidence for such behavior and there was a strong belief that the universe was static, Einstein introduced an additional term (the cosmological constant) into his field equations to obtain nonevolving models. In the early 1920s a number of other physicists studied models of the universe, including the Russian Alexander Friedmann, the Belgian Georges Lemaitre, and the Dutch Willem de Sitter. Their work, together with that of Einstein's, ultimately led to the big bang model of the expanding universe.

At about the same time, the American astronomer Edwin Powell Hubble used the newly commissioned and then largest telescope in the world, the 100-in. Hooker telescope at Mt. Wilson, to make three startling discoveries that would change our view of the cosmos. In 1924 he demonstrated that most of the nebulae, wispy patches of light on the sky, contained individual stars and established that they were "island universes" (now known as galaxies) and not gas clouds within our own Milky Way galaxy. Hubble's discovery settled a debate that had raged for decades and enlarged the known universe more than a billion fold. In 1929, building upon the work of Vesto Milo Slipher, Hubble announced evidence for a linear relationship between the recessional speeds of galaxies and their distances,

$$v = H_0 d,$$

where v is the speed of the galaxy and d is its distance, and the proportionality constant H_0 (called "K" by Hubble) is the Hubble constant. Finally, he showed that galaxies were roughly distributed uniformly across the sky in all directions and uniformly throughout space.

Hubble's three discoveries provide the basis for the modern big bang cosmological model of the expanding universe. It took a few years for theorists to catch up to Hubble's discoveries and correctly interpret Hubble's law as being due to the expansion of the universe. Eventually, Einstein called his introduction of the cosmological constant his greatest blunder—though more would be heard about it, even today. Hubble and his assistant Milton Humason continued using the 100-in. Hooker telescope to probe the far reaches of the universe. However, they did not come close to seeing to the edge of the observable universe, some 15 billion light-years away. In part, this quest led to the building of the 200-in. Hale telescope at Mt. Palomar Observatory, which began operating in the early 1950s. The 200-in. Hale telescope did indeed take us to the edge of the observable universe.

In 1964 two American radio astronomers, Arno A. Penzias and Robert Wilson, made an extraordinary discovery which, together with Hubble's contributions, led to the present "standard model" of cosmology, the hot big bang model. Quite by accident, though through careful and meticulous work to be sure, they discovered that the universe is filled with the kind of microwave radiation associated with

a blackbody at a temperature of about 3 K. They had measured the temperature of the universe. This microwave radiation is now known as the cosmic microwave background radiation.

Because any system cools as it expands, the existence of the cosmic background radiation implies that the universe began from a hot big bang. This has profound consequences for the study of the earliest moments. It means that the universe began as a hot soup of the fundamental particles—quarks and leptons—and as it expanded and cooled layer upon layer of structure developed. First, around 10^{-5} s the quarks condensed to form neutrons and protons; next, when the universe was seconds old, neutrons and protons formed the nuclei of the simplest chemical elements, D, ^{3}He, ^{4}He, and ^{7}Li. When the universe was about 300,000 years old, the nuclei and electrons present combined to form neutral atoms. When this occurred, the matter in the universe became transparent to the thermal radiation present (ionized matter is very opaque), and matter and radiation "decoupled." For this reason, the surface of last scattering for the cosmic microwave background radiation is the universe at 300,000 years, when its size was about one-thousandth of its present size and the temperature was around 3,000 K. Because of the subsequent expansion, the radiation we detect today has a temperature of only about 3 K (diminished by the thousand-fold expansion since decoupling). Next, very small primeval variations in the density of matter (about a part in 10^5) were amplified by the attractive influence of gravity, eventually leading to the "macro-structure" that is so conspicuous today—stars, galaxies, clusters of galaxies, superclusters, voids, and so on. The existence of primeval lumpiness was confirmed in 1992 when the Cosmic Background Explorer (COBE) satellite detected similarly small variations in the temperature of the cosmic microwave background radiation coming from different directions.

The hot big bang model embodies our present understanding of the evolution of the universe, which extends back to within a fraction of a second of the bang. Four pillars provide the observational basis for the standard hot big bang cosmology. First, there is the expansion of the universe itself. By now the velocities of more than 50,000 galaxies have been measured, extending out to the edge of the observable universe. While a precise determination of the Hubble constant still eludes cosmologists, the velocities and distances of these galaxies are consistent with Hubble's linear relationship, which itself is in accord with that predicted in an expanding universe. The Hubble constant sets the time back to the bang, by current measures between about 10×10^9 years and 20×10^9 years, an age that is consistent with other determinations of the age of the universe (e.g., the ages of the oldest stars and the radioactive elements).

Second, there is the existence of the cosmic microwave background radiation. The COBE satellite measured its temperature with remarkable precision, $2.728 \text{ K} \pm 0.002 \text{ K}$. Further, COBE measurements show it to be the most perfect blackbody ever studied; any deviations from a blackbody spectrum are less than 0.03 percent. While many have sought explanations for the cosmic microwave background radiation other than a hot big bang, there exists no other viable alternative.

Third, there are the abundances of the light elements D, ^{3}He, ^{4}He, and ^{7}Li. The predictions of big bang nucleosynthesis are in agreement with the measured abundances of these elements in the most primitive samples of the cosmos. Deuterium and ^{4}He are striking confirmations of the big bang: There is no plausible explanation for deuterium other than the big bang as it is destroyed in virtually all astrophysical processes, and while stars do indeed produce ^{4}He by nuclear reactions that fuse four protons, stars cannot come close to accounting for the large amount of ^{4}He (by mass, about 25 percent) in the universe.

Big bang nucleosynthesis is the earliest test of the hot big bang model. Moreover, American astrophysicist David Schramm and his collaborators have taken it one important step further. They have shown that agreement between all four light elements is only obtained if the mass density of ordinary matter is between 1 and 15 percent of the critical density. This is the most accurate determination of the density of ordinary matter and has profound implications for the dark matter problem.

Fourth, there are the tiny variations in the temperature of the cosmic microwave background radiation seen in different directions. These temperature fluctuations, which were first detected by COBE and since have been confirmed by a number of other experiments, provide the evidence for the inhomogeneities in the matter that seeded the macrostructure in the universe.

The success of the hot big bang model has allowed cosmologists to ask even more fundamental questions about the universe. Is there enough matter to halt the expansion? What is the nature of the ubiquitous dark matter? What is the origin of the primeval lumpiness needed to seed the formation of large-scale structures? When did the first galaxies form and what are the largest structures in the universe? Why is the universe composed of matter alone and not equal amounts of matter and antimatter? Why is the observable universe so "smooth" (homogeneous and isotropic)? What is the origin of the expansion and of the big bang itself?

Cosmologists believe that the answers to these questions involve events that took place during the earliest moments and involve in a fundamental way the unification of the forces and particles of nature. The connection between the very small (elementary particle physics) and the very large (cosmology) has become a most important and exciting one. It owes its existence to the fact that the energies corresponding to temperatures reached during the earliest moments match and exceed those energies achieved in the most powerful particle accelerators. Knowledge of how the fundamental particles behave under the most extreme conditions—high energies and high densities—is crucial to understanding the earliest moments. Conversely, the study of the evolution of the universe may reveal clues concerning the unification of the forces and particles of nature.

The most attractive and most expansive idea about the earliest moments is inflation. Inflation holds that very early on, the universe underwent a rapid period of expansion driven by an unusual form of energy predicted to exist by unification theories (false-vacuum energy). During inflation the universe grows in size by a factor much greater than it has since. Because of this extraordinary growth, a very small portion of the universe, which is smooth and flat, grows to a size that encompasses our observable universe and much, much more. Further, quantum mechanical fluctuations that only occur on the tiniest of length scales are stretched to cosmological length scales and ultimately lead to the primeval lumpiness. While inflation is highly speculative, its consequences are far reaching and testable.

Inflation makes three fundamental predictions: (1) that the universe has the critical density; (2) that the spectrum of density inhomogeneity is scale-invariant; and (3) that a scale-invariant spectrum of gravitational waves exists. Since ordinary matter can only contribute between 1 and 15 percent of critical density, inflation demands the existence of an "extraordinary" form of matter. Remarkably, there are independent reasons to believe that such matter exists. Their attempts to unify the forces and particles of nature have led particle physicists to predict the existence of new, as yet undiscovered, particles—and further—calculations indicate that sufficient numbers of several of these hypothetical particles should be left over from the earliest moments to contribute critical density. The two most promising candidates are the axion, a particle that is almost a million times lighter than the electron, and the neutralino, a particle that is around a hundred times heavier than the proton. In addition, while cosmologists have yet to determine precisely the mass density of the universe, they have shown that most of the matter is dark (not in the form of stars) and indications are that the mass density is significantly greater than that which ordinary matter can account for. The inflationary prediction of exotic dark matter and scale-invariant density inhomogeneities has led to a very detailed picture for the formation of structure in the universe, known as cold dark matter. Cold dark matter provides the most detailed and most successful description of the formation of structure, as well as many predictions that can and are being tested.

Inflation could extend our understanding of the universe back to within 10^{-43} s of the bang. What about events that happened even earlier—like the bang itself? A fundamental problem arises when trying to discuss events earlier than 10^{-43} s: Einstein's theory of gravity must be modified to include the effects of quantum mechanics and thus the exploration of the earliest moments—the quantum gravity epoch—awaits a quantum description of gravity. Many approaches to this very difficult problem have been explored; the most successful is superstring theory. Superstring theory unifies all the forces of nature, including gravity, and holds that the fundamental particles are actually small loops of string (about 10^{-33} cm in diameter) and the number of spacetime dimensions is ten (time + nine spatial dimensions). Superstring theory opens the door for the first sensible discussion of the very earliest moments and at the same time raises new questions—among them, where did the other six spatial dimensions go?

Inflation and other ideas about the early universe will continue to be tested by a flood of diverse observations and experiments. They include the search for the axions or neutralinos that may comprise the dark matter in our own galaxy, mapping of the cosmic background radiation fluctuations with more than 30 times the angular resolution of COBE, large surveys of the distribution of matter in the universe

(e.g., the Sloan Digital Sky Survey, which will consist of 10^6 galaxy positions), deep images of the universe from the Hubble Space Telescope and large ground based telescopes such as the Keck Observatory on Mauna Kea, x-ray and infrared observations of the universe, and on and on. At the very least, the most attractive and compelling ideas about the earliest history of the universe will be tested. It is possible that some of our ideas about the universe will be confirmed and our "standard model" of cosmology will be extended back to even earlier times, answering some of the most fundamental questions about the universe and shedding light on the unification of the forces and particles of nature.

See also: BIG BANG THEORY; COSMIC BACKGROUND EXPLORER SATELLITE; COSMIC MICROWAVE BACKGROUND RADIATION; COSMIC MICROWAVE BACKGROUND RADIATION, DISCOVERY OF; COSMIC STRING; COSMOLOGICAL CONSTANT; COSMOLOGICAL PRINCIPLE; COSMOLOGY, INFLATIONARY; DARK MATTER; DARK MATTER, BARYONIC; DARK MATTER, COLD; DARK MATTER, HOT; HUBBLE, EDWIN POWELL; HUBBLE CONSTANT; NUCLEOSYNTHESIS; SUPERSTRING; UNIVERSE, EXPANSION OF

Bibliography

CHRISTIANSON, G. *Edwin Hubble: Mariner of the Nebulae* (Farrar, Straus, and Giroux, New York, 1995).

TURNER, M. S. "Why is the Temperature of the Universe 2.726 Kelvin?" *Science* **262**, 861–867 (1993).

WEINBERG, S. *First Three Minutes* (Basic Books, New York, 1977).

MICHAEL S. TURNER

COSMOLOGY, INFLATIONARY

Inflationary cosmology describes the very early stages of the evolution of the universe, and its structure at extremely large distances from us. For many years, cosmologists believed that the universe from the very beginning looked like an expanding ball of fire. This explosive beginning of the universe was called the big bang. At the end of the 1970s a different scenario of the universe's evolution was proposed. According to this scenario, the universe at the very early stages of its evolution came through the stage of inflation, exponentially rapid expansion in a kind of unstable vacuum-like state (a state with large energy density, but without elementary particles). This stage could be very short, but the universe within this time became exponentially large. At the end of inflation the vacuum-like state decayed, the universe became hot, and its subsequent evolution could be described by the standard big bang theory. The existence of the inflationary stage is necessary to resolve many difficult problems that arise when one tries to unify modern theory of elementary particles and the standard big bang theory.

There were several different attempts to develop inflationary theory. The first realistic version was proposed in 1979 by Alexei Starobinsky of the L. D. Landau Institute of Theoretical Physics in Moscow. His model was rather complicated, based on the theory of anomalies in quantum gravity. A somewhat simpler model based on the theory of cosmological phase transitions with supercooling was suggested in 1981 by Alan Guth of the Massachusetts Institute of Technology. This model was very popular, but, as Guth soon realized, it did not work. In 1982 Andrei Linde of the P. N. Lebedev Physical Institute in Moscow introduced the so-called new inflationary universe scenario, which Andreas Albrecht and Paul J. Steinhardt of the University of Pennsylvania also later discovered. This scenario was free of the main problems of Guth's model, but it was still not very realistic.

Most of the inflationary models that exist at present are based on the chaotic inflation scenario suggested by Linde in 1983. This scenario does not require either complicated quantum gravity effects or phase transitions with supercooling, and it can be realized in a wide class of models of particle physics. The simplest version of this scenario describes the universe filled by a massive scalar field. An example of such a field is the Higgs field, which is used in the standard model of electroweak interactions. It can be shown that if the scalar field originally was sufficiently large and homogeneous, it could change its value only very slowly. Therefore, for a long time its energy density remained almost constant. According to Einstein's theory of gravity, the universe filled by matter with a constant energy density should expand exponentially, which corresponds to the stage of inflation. However, gradually the scalar field became smaller and began oscillating. The oscillating scalar field created elementary particles that came into a state of a thermal equilibrium, and after that the universe expanded in accordance with the standard big bang theory.

According to the simplest model of chaotic inflation, the universe during inflation could expand approximately $10^{10^{12}}$ times. This number is model-dependent, but it is huge in all realistic inflationary models. Such an expansion makes the universe extremely large and stretches all its inhomogeneities. Geometric properties of our space become similar to the properties of an almost flat surface of a huge inflating balloon. This explains why our universe is so big, why it is so homogeneous and isotropic, and why its geometric properties are so close to the properties of a flat space.

Simultaneously with removing old inhomogeneities, inflation provides a mechanism of generation of small density perturbations, which are responsible for galaxy formation. This happens because of exponential stretching and amplifying of quantum fluctuations of the scalar field during inflation. Perturbations of density predicted by inflationary theory should lead to a specific anisotropy of the primordial background radiation. Anisotropy of this type was indeed discovered in 1992 by the Cosmic Background Explorer (COBE) satellite.

In certain cases quantum fluctuations amplified during inflation may become extremely large. They may considerably increase the initial value of the scalar field, which drives inflation. The probability of such events is very small, but those rare parts of the universe where it happens begin expanding with a much greater speed. This creates a lot of new space where inflation may occur, and where large quantum fluctuations become possible. As a result, the universe enters a stationary regime of self-reproduction: Inflationary universe permanently produces new inflationary domains, which in their turn produce new inflationary domains. Such a universe looks not like a single expanding fireball created in the big bang, but like a huge self-reproducing fractal consisting of inflationary domains of all possible types.

According to many inflationary models, quantum fluctuations at the stage of inflation may be so large that they may change properties of the vacuum state in different parts of the universe. As a result, the universe becomes divided into many exponentially large parts where the laws of low-energy physics and even dimensionality of spacetime may be different.

Inflationary cosmology is still in the process of active development. It changes in parallel with the rapid development of the theory of elementary particles. There exist some theories of elementary particles where an inflationary regime cannot be realized. It may happen also that some popular inflationary models will be in conflict with future astronomical observational data. At present, however, inflationary theory provides the best explanation of the properties of our universe, and many cosmologists believe that inflation, or something very similar to it, should be a necessary ingredient of any internally consistent cosmological theory.

See also: BIG BANG THEORY; COSMIC BACKGROUND EXPLORER SATELLITE; COSMOLOGY; UNIVERSE; UNIVERSE, EXPANSION OF; UNIVERSE, EXPANSION OF, DISCOVERY OF

Bibliography

KOLB, E. W., and TURNER, M. S. *The Early Universe* (Addison-Wesley, New York, 1990).

LINDE, A. D. *Particle Physics and Inflationary Cosmology* (Harwood Academic, Chur, Switzerland, 1990).

LINDE, A. D. "The Self-Reproducing Inflationary Universe." *Sci. Am.* **271** (5), 48–55 (1994).

LINDE, A. D.; LINDE, D. A.; and MEZHLUMIAN, A. "From the Big Bang Theory to the Theory of a Stationary Universe." *Phys. Rev. D* **49** (4), 1783–1826 (1994).

ANDREI LINDE

COULOMB, CHARLES AUGUSTIN

b. Angoulême, France, June 14, 1736; *d.* Paris, France, August 23, 1806; *mechanics, electricity, magnetism.*

Coulomb was a physicist and engineer of the first rank. He is remembered especially for his determination of the force laws for magnetism and electricity, known as Coulomb's laws, and for his innovations in instrument design, the Coulomb torsion balance. He contributed importantly also in studies of friction and applied mechanics. Coulomb's family came from Montpellier, in the south of France, where the family had been involved in law, administration, and the military. Coulomb's father, Henry, and mother, Catherine Bajet, had at least two daughters, but little is known of the details of Charles's sisters or of his family life. The family moved to Paris, where Charles probably attended school. Charles's father left Paris to return to Montpellier and Charles later followed

his father there. His mother wanted Charles to study medicine, but he wanted to study mathematics. He joined the *Montpellier Société des Sciences* (Science Society) at the age of twenty-one and was active there until he decided to go back to Paris so that he could be tutored for entrance to the *Ecole du génie* (School of Military Engineering) at Mézières, France, which at that time was the best technical school in Europe. Coulomb was not a member of the nobility but was probably admitted to the school on the basis of his ability plus his family's tradition in the military. At Mézières, Coulomb and the other students studied geometry and algebra; architectural drafting and design; mapping; surveying; mechanics, including hydraulics; and other practical engineering elements including stonecutting, fortification design, bookkeeping, and cost estimating. The students spent much time outside in on-the-job training constructing small bridges, retaining walls, and arches while directing local laborers. Coulomb graduated from the engineering school in November 1761 and then spent twenty years rising through the ranks of the military engineers, serving in Martinique, West Indies, and then at military forts all around France.

Coulomb always seemed stimulated to work on physics and engineering ideas wherever he was stationed. In Martinique he turned his observations and work experience into an important paper on statics. This paper, presented in 1773, covered major problems in applied mechanics of concern to engineers: strength of materials, flexure and rupture of beams, rupture of masonry piers, earth pressure theory, and design of arches. Coulomb also presented papers on the efficiency of methods of work (ergonomics). During his years serving near Cherbourg, Coulomb designed a magnetic compass, using a magnet suspended by a thread. This project was the beginning both of his later work in magnetism and electricity, and his work in torsion. Near Bordeaux, Coulomb wrote on the theory and practice of friction devices, and at Besançon, he wrote on floating drydock design and the principles of dredging and excavations underwater. Several of these projects he submitted to contests at the Académie Royale des Sciences (Royal Academy of Sciences) in Paris, then the most important scientific academy in Europe. He was quite successful, winning two contests there and having several of his papers published by the Academy. This served eventually to gain him election to the Academy in December 1781 and allowed him to take up residence permanently in Paris, where he could turn all his studies to the physics of torsion and of electricity and magnetism. He remained as a sort of engineering consultant to the Corps of Military Engineers until his retirement in 1790; in fact, technically he was engineer in charge of the Bastille when it fell on July 14, 1789. Coulomb was very active in the Academy of Sciences until his death in 1806, and he held various commissions both under the monarchy before the French Revolution and later under Napoleon. He married late in life, and his young wife, Louise Desormeaux, bore him two sons, Charles Augustin II, who later emigrated to the United States and married, establishing the Coulomb family there, and Henry Louis, who died in midlife leaving no children.

Coulomb's studies in mechanics preceded his researches in physics. His application of calculus and other mathematics to studies in friction, the flexure of beams, and the rupture and shear of brittle materials contributed to these areas of engineering and mechanics and influenced strongly his conception of the molecular state of electricity and of magnetism. His rigorous analysis of the physics of torsion in thin cylinders allowed him to construct a torsion balance that could quite precisely measure extremely small forces. A wire or thread is suspended vertically; at its lower end is attached a crosspiece. The crosspiece of this magnetic or electric balance, in simple form, was a long, thin magnet suspended from a thread, or a straw suspended with a pith ball glued to each end, with electric charge then applied to each pith ball. Coulomb had showed both theoretically and experimentally that the angle of twist for a given length and thickness of thread could be used to measure the torque upon the thread. He discovered that the magnitude of the twisting force on the thread or thin wire is directly proportional to the angle through which the thread is twisted. Now torque is force applied, say, at right angles to a lever. Coulomb's torsion balance could measure forces of less than 9×10^{-4} dynes.

In a general sense, Coulomb followed Isaac Newton in articulating and extending Newton's theory of forces to electricity and magnetism. Coulomb wrote, "One must necessarily resort to attractive and repulsive forces of the nature of those which one is obliged to use in order to explain the weight of bodies and celestial physics." And one must obtain exact quantification of these laws. He worked on studies of electricity and of magnetism throughout the 1780s and 1790s. He made many discoveries, although he was unable to find a connection between electricity and magnetism (electromag-

netism). His most famous work was his proof and experimental demonstrations of the inverse square laws for electricity and magnetism. These state (1) that the force between two electrically charged bodies varies as the product of the electrical strength (charge) of the bodies and varies inversely as the square of the distance between the bodies; (2) for magnetism the force varies as the product of the magnetic pole strengths of the two magnetic bodies and varies inversely as the square of the distance between the two magnetic poles. These two famous laws (Coulomb's laws) are identical in form (homologous) to Newton's law of gravitational attraction $[F \propto (M_1 M_2)/R^2]$, where the gravitational force is proportional to the product of the masses of two bodies and inversely proportional to the square of the distance between the bodies.

Coulomb imagined that each particle of a magnet ("magnetic molecule" he wrote) was polarized, that is, was a tiny molecular magnet with a positive and negative end. Coulomb's fundamental researches in electricity and in magnetism represented the extension of Newtonian mechanics to new areas of physics, and his work influenced Jean Baptiste Biot, Siméon Denis Poisson, André Marie Ampère, Pierre Simon Laplace, and other nineteenth-century physicists. Coulomb's work in mechanics, interestingly enough, influenced the great English physicist Thomas Young in his work on mechanics and on light. Some have written recently that Coulomb may not have performed his experiments quite as he wrote that he did. They base this on the facts that they have been unable today to recreate Coulomb's experiments to give exactly the results he claimed. It is interesting to note here that current experiments on the characteristics of eighteenth-century iron and brass wires have shown values closer to those determined by Coulomb than to values assumed by those one century after Coulomb. Coulomb was considered a master in experimental physics by those who knew him.

See also: COULOMB'S LAW; GRAVITATIONAL ATTRACTION; INVERSE SQUARE LAW; LAPLACE, PIERRE-SIMON; YOUNG, THOMAS

Bibliography

DICKMAN, S. "Could Coulomb's Experiment Result in Coulomb's Law?" *Science* **262,** 500–501 (1993).
GILLMOR, C. S. *Coulomb and the Evolution of Physics and Engineering in Eighteenth-Century France* (Princeton University Press, Princeton, NJ, 1971).

GOODWAY, M., and SAVAGE, W. R. "Coulomb's Data on Harpsichord Wire." *MRS Bulletin* **17,** 24–27 (1992).

ROLLER, D., and ROLLER, D. H. D. "The Development of the Concept of Electric Charge: Electricity From the Greeks to Coulomb" in *Harvard Case Histories in Experimental Science,* edited by J. B. Conant (Harvard University Press, Cambridge, MA, 1966).

C. STEWART GILLMOR

COULOMB'S LAW

Coulomb's law is also known as the fundamental law of electrostatics. This law is named after Charles Augustin Coulomb, who, in 1785, published the claim of having verified this law by experiment. The law gives the mathematical description of the electrostatic force between two electrical charges Q_1 and Q_2.

According to Coulomb, the force F is directly proportional to the product of the two charges Q_1 and Q_2; and inversely proportional to the square of their respective distance d.

$$F = \frac{1}{4\pi\varepsilon_o\varepsilon_r} \frac{Q_1 Q_2}{d^2},$$

where

$$\frac{1}{4\pi\varepsilon_o\varepsilon_r}$$

is a proportional factor defined as $10^{-7}c^2$, c being the velocity of light. From this definition ε_o is a universal constant; its value is (in SI units) 8.854×10^{-12} AsV^{-1}m^{-1}. The ε_r is a factor that indicates the electrical behavior of the material between the two charges. (In the case of vacuum between the two charges, $\varepsilon_r = 1$.)

The charges Q_1 and Q_2 can be either positive or negative. If both have the same sign, F will be positive and the force is repulsive. If the charges have the opposite signs, F will be negative and the force is attractive. The force F is acting along the direct line between the two charges.

Consequences of the Inverse Square Law

Because of Coulomb's law, it is possible to give a mathematical description of an electrical field that is produced by each electrical charge. Thus, the electrical field E can be written as:

$$E = \frac{F_{\text{Coul}}}{q_1} = \frac{1}{4\pi\varepsilon_o} \frac{q_2}{d^2}.$$

This means that the electrical field is a vector field. If there is more than one charge, the resulting electrostatic field is given by the vector sum of the electrostatic fields resulting from each of these charges. This is also called the "principle of superposition."

The similarities between Coulomb's law and the gravitational force law are obvious, although in the latter there occurs neither a negative mass nor a repulsive force. As the product of the charges served as a definition of the charge, it is obvious that the analogy between Coulomb's law and the gravitational force law was designed. The electrical field also can be seen as analogous to a gravitational field. The work done to move a charge in an electrical field depends only on its starting point and its end, but not on the path.

If a charged sphere is considered, a radially symmetric field is noted outside the ball (which is also called the Coulomb field). Mathematically we can treat the charge in this case as being a point charge in the center of the ball. As a result of the inverse square law the charge of this ball is distributed on its surface, but there is neither a charge nor an electrical field to be detected inside the ball. Another consequence of the inverse square nature of Coulomb's law is Gauss's law. The flux through the surface of any body in any electrical field is proportional to the charge inside the body.

Historical Background

In 1785 Coulomb published his first of seven papers on electricity and magnetism. In this first article he described experiments with the instrument for which he is famous: the torsion balance (Fig. 1). In this setup there are two pith balls that are charged. One of the balls is fixed, the other one is fastened to the end of a needle that is suspended by a thin silver wire. The twist of the wire produces a force that is proportional to the angle of torsion. The silver wire was clamped in the torsion microme-

law in the case of electrostatic attraction. Before Coulomb's publication, several other scientists such as Henry Cavendish, Joseph Priestley, and Charles Stanhope had already assumed that the force between two charges should be the inverse square. Coulomb, however, was the first to claim the precise experimental verification of the relation. His result was not generally adopted; in Germany it was highly contested. It has to be remarked that the focus of these investigations was on the force–distance relation. Coulomb formulated the proportionality between the force and the product of the charges in his second paper, too, but he did not give an experimental verification. From the point of physics, the important aspect in Coulomb's law is its inverse square character.

Henry Cavendish had deduced the inverse square law from theoretical considerations. Cavendish also succeeded in giving an experimental proof of the law. In his experiment he used a charged metal sphere and closed two metal semispheres around it. Both spheres were briefly brought in contact and the two semispheres were removed. If the inverse square law was to be true, there should be no charge left on the inner sphere, which Cavendish found to be the case. However, Cavendish did not publish either his experiments or his results. (In 1879 James Clerk Maxwell edited some of his manuscripts and thereby learned of the experiment.) Cavendish's experiment is remarkable because even in the second half of the twentieth century his setup served as the most accurate experiment for the determination of the force-distance relation. This is due to the fact that the accuracy of the measurement depends only on the sensitivity of the galvanometer that is used to detect any remaining charges.

Although Coulomb's law is fundamental in the field of electrostatics, it serves as a mathematical description of just a few of the electrodynamic phenomena. This is understandable in the context of the formulation of this law: It was published about fifteen years before the construction of the Voltaic pile, which can be taken as the first step in the development of electrodynamics. The electrodynamic phenomena are described by Maxwell's equations in which Coulomb's law is integrated.

See also: COULOMB, CHARLES AUGUSTIN; ELECTROSTATIC ATTRACTION AND REPULSION; GAUSS'S LAW; GRAVITATIONAL ATTRACTION; INVERSE SQUARE LAW; MAXWELL'S EQUATIONS

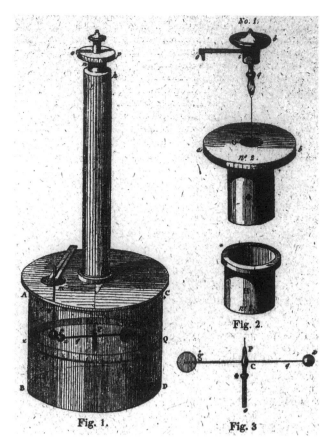

Figure 1 Original illustration of Coulomb's electrical torsion balance.

ter, which made it possible to produce a definite additional torsion (which means an additional force). The counterbalance of the electrostatic and the torsion force is the general idea of Coulomb's measurement. With these investigations he claimed to have proven to the utmost precision that the force between two repulsive charges is the inverse square. Although Coulomb's torsion balance was the most sensitive existing instrument for the determination of forces, he did not make very accurate measurements. Furthermore, during a replication of the torsion balance experiment it was not possible to determine the inverse square law with the setup Coulomb described. This was due to the electric charge of the observer. Only when its influence was eliminated by the use of a Faraday cage, data corresponding to the inverse square law could be obtained. One year later Coulomb described another experiment that served as a verification of this

Bibliography

HEILBRON, J. L. *Elements of Early Modern Physics* (University of California Press, Berkeley, CA, 1982).

MAGIE, W. F. *A Source Book in Physics* (McGraw-Hill, New York and London, 1935).

MAXWELL, J. C., ed. *The Electrical Researches of the Honorable Henry Cavendish, FRS Written Between 1771 and 1781* (Cambridge University Press, Cambridge, Eng., 1879).

PETER HEERING

COVALENT BOND

Early in the twentieth century a fairly accurate picture of the structure of the atom had been pieced together. Joseph J. Thomson of the Cavendish Laboratory had discovered that electrons were contained in all atoms, and that these electrons were negative particles of electricity that could be deflected by magnets. Ernest Rutherford, who was then at McGill University in Montreal, discovered the tiny yet massive positively charged center of atoms and named it the atomic nucleus. Niels Bohr and others developed the picture of an atom as a tiny central core surrounded by a swarm of electrons, then spent the next fifteen years trying to understand the physical laws governing the interactions between the nucleus and the swarming electrons. As the structure of the atom began to take shape, chemists of the day were working to understand how these atoms comprised of electrons distributed around nuclei are held together to form molecules. This was the beginning of the science of chemical bonding.

For certain kinds of molecules the forces responsible for holding the atoms together were thought to be electromagnetic in nature. It was thought that a neutral atom might transfer an electron to another neutral atom; leading to one positively and one negatively charged ion. These two ions would then be attracted to each other through coulombic attraction and be held together. The bonds between such atoms were termed ionic bonds, and molecules which were thought to be held together by these bonds were gases such as HCl and solids such as NaCl. However, the forces responsible for holding together molecules such as H_2 and O_2 could not be explained, as both atoms of these molecules are identical, thus there cannot be one positively charged atom and another negatively charged atom.

It was to this problem that Gilbert Newton Lewis of the University of California directed his attention in his 1916 paper "The Atom and the Molecule." Lewis conceived of a nonionic bond as being the result of atoms *sharing* electrons. The atoms sharing these electrons were said to be linked by a covalent bond. Lewis had earlier developed the idea of a cubical atom, with a nucleus at the center of a cube and electrons on its corners. For atoms where there were an insufficient number of electrons to fill all corners, a molecule could be formed with another atom by sharing corners, edges, or faces of the cubes to fill all corners with electrons. A shorthand notation for describing this sharing, known as Lewis dot structures, was developed. This notation is still widely used, forming the basis for describing the bonding of organic molecules.

Though the cubical atom has no theoretical or experimental basis, the concept of electron sharing has proved tremendously useful. In the early 1930s John Slater used the new theories of quantum mechanics to recast the covalent bond in terms of a theory known as directed valence. In this theory, electrons of a molecule spend some time associated with one atom, some time with another atom, and some time being shared. Later, Richard Feynman, who was then an undergraduate at MIT, showed that electrons that are shared between nuclei will produce a force that acts to hold the nuclei together. Thus, shared electrons can be thought of as the bonds between atoms and are known as covalent bonds.

See also: ATOM; BOHR, NIELS HENRICK DAVID; ELECTRON; FEYNMAN, RICHARD PHILLIPS; IONIC BOND; MOLECULE; THOMSON, JOSEPH JOHN

Bibliography

FEYNMAN, R. P. "Forces in Molecules." *Phys. Rev.* **56,** 340–345 (1939).

LEWIS, G. N. "The Atom and the Molecule." *J. Am. Chem. Soc.* **38,** 762–782 (1916).

SLATER, J. C. *Quantum Theory of Molecules and Solids,* Vol. 1 (McGraw-Hill, New York, 1963).

M. E. EBERHART

CPT THEOREM

Symmetry plays an important role in our current understanding of the fundamental laws of nature. The most familiar symmetries are continuous symmetries, such as symmetry under rotations (associated with the conservation of angular momentum), under translations in space (conservation of momentum), and translations in time (conservation of energy). These symmetries are called continuous because they are described by a continuous parameter. For example, one can translate a system by any vector distance **a.** In contrast, there are several symmetries that have a fundamental significance that are discrete. A simple model for such a phenomenon is provided by a perfect crystal lattice. If one translates the lattice by an arbitrary amount, the system looks different, but if one translates by an integer multiple of the lattice spacing, the system looks the same.

In elementary particle physics, the most prominent discrete symmetries are charge conjugation *C*, parity *P*, and time reversal *T*. A system is invariant under parity if the equations describing it are unchanged when one replaces all vectors, such as the the electric field **E,** with their minus vectors, while leaving pseudovectors, such as the magnetic field **B,** alone. A theory is invariant under charge conjugation if, upon replacing all of the particles by their antiparticles and changing the sign of the electric and magnetic fields, the equations are unchanged. (In more complicated cases, such as Yang–Mills theories, slightly more involved transformations of the various fields are required.) Finally, time reversal is the operation of changing the sign of time in the equations of physics.

All three symmetries are good, to a very high degree of accuracy, in atomic physics; they are also good symmetries of strong interactions. But *C*, *P*, and *T* are all known to be violated by the weak interactions. For example, the neutrino is a left-handed particle, which means that its spin is always found to be parallel to its momentum. This violates parity, since under parity the momentum changes sign but the spin does not. The antineutrino always has its spin antiparallel to its momentum. This violates charge conjugation, since charge conjugation of the neutrino should yield an antineutrino of the same spin. If one charge conjugates a neutrino and performs a parity transformation, however, one gets an antineutrino of the correct spin. So potentially, this

operation, which is referred to as the product of the symmetries $C \times P$, or simply *CP*, is potentially a good symmetry. However, it is known from the study of certain elementary particles (*K* mesons) that *CP* is also violated by the weak interactions. In addition, time reversal *T* has been shown to be violated in this system. These violations are quite small, and in fact this is the only system where, to date, such symmetry violations have been observed.

It is tempting to guess, from this, that there are no exact discrete symmetries in nature. There is reason to believe, however, that the product $C \times P \times T$ (*CPT*) is an exact symmetry of the laws of nature.

This belief follows, in part, from experimental tests, but it is also motivated by the success of quantum field theory in describing the interactions of elementary particles. Quantum field theory is the framework that combines the laws of quantum mechanics with those of Einstein's special relativity. It embodies the principle of locality of interactions promulgated by Einstein. There is no action at a distance in these theories; interactions are transmitted from point to point by fields, such as the electric and magnetic fields. At the present time, a quantum field theory referred to as the standard model of elementary particles successfully describes the interactions of all of the known elementary particles down to length scales of order 10^{-16} cm. This theory of the weak, electromagnetic, and strong interactions has been tested at these scales to better than 1 part in 100.

The basic ingredients of a quantum field theory are fields. In the standard model, the basic fields include the vector potential and scalar potential of electromagnetism, **A** and ϕ. When combined with the principles of quantum mechanics, these fields describe the photon. There is a field ψ that describes the electron. Other fields in the theory include those that describe the gluons (the particles that mediate the strong interactions), the quarks, the neutrinos, and so on; roughly speaking, there is one field for each type of particle in the model. All of these fields can be grouped into two types. There are Bose fields, which describe particles with integer spin such as the photon and the gluons (spin one), and the (presently) hypothetical Higgs particle (spin zero), and there are fermionic fields, which describe particles with spin $\frac{1}{2}$, such as the electron, the neutrino, and the quarks, and the intermediate vector bosons, which mediate weak interactions.

In quantum mechanics, the coordinates and momenta of particles are operators that do not commute. For example, $[x, p] = xp - px = i(h/2\pi)$. In quantum field theories, the fields themselves are operators. One of the most fundamental aspects of these theories is that they are only consistent if particles with spin $\frac{1}{2}$ obey anticommutation relations. For x and p, this would mean writing a relation of the form $xp + px = h/2\pi$. The corresponding relations for fields have an important consequence: spin $-\frac{1}{2}$ particles obey the Pauli exclusion principle, or what are known as Fermi statistics. Bose fields, on the other hand, obey commutation relations. These commutation relations build in automatically the principle of Bose statistics, which is the appropriate rule for particles such as the photon. This understanding of the connection between spin and statistics was one of the early triumphs of quantum field theory, and it can be proved with a high degree of rigor. One can prove that any quantum field theory that obeys these rules and the rules of special relativity respects *CPT*, regardless of any of its other features. The proof involves enumerating all of the terms that could possibly appear in such a theory.

The most striking consequence of this theorem is that the masses of particles are identical to those of their antiparticles, even though *C* and *CP* are known not to be conserved. This is easy to see. *CP* takes a particle to an antiparticle with the same spin and opposite momentum. If we think of describing the state of a particle in the future, acting on this with *CPT* produces an antiparticle in the past. The statement that *CPT* is a symmetry means that this particle has the same energy and mass as the original one. For particles that are unstable (undergo radiative decay) the theorem has a further consequence: the lifetimes (half-lifes) of particles are identical to those of their antiparticles (but not necessarily the decay rates to particular final states). Other consequences of the *CPT* theorem include the requirement that the magnetic moments are equal in magnitude. The theorem is experimentally well-verified. The best limits on possible deviations come from studies of the two neutral *K* mesons. Their lifetimes are known to be equal to a part in 10^3; their masses to a part in 10^{14}. The electron and positron magnetic moments are known to be equal to better than a part in 10^{10}.

While quantum field theory has encountered no obstacles in describing experiments at ever higher energies, there are good reasons to suspect that quantum field theory does not provide the ultimate form of the laws of nature. First, by its very nature quantum field theory is always an effective descrip-

tion of phenomena, valid only for some range of energies. In addition, all attempts to understand Einstein's general relativity in the framework of quantum field theory have failed. Since our understanding of the *CPT* theorem is formulated entirely in the framework of quantum field theory, it is important, both experimentally and theoretically, to ask whether the theorem breaks down at some level as well. Superstring theory is an example of a theoretical framework that might incorporate both gravity and quantum mechanics. In this theory, quantum field theory is relevant only to the description of long wavelength excitations (excitations with energies much lower than the Planck mass, about 10^{19} times larger than the proton rest energy, m_pc^2). The status of the *CPT* theorem in string theory is not completely clear. While it can be shown to hold in a certain approximation, it is not presently known if *CPT* invariance is an exact property of these theories.

See also: ANTIMATTER; CHARGE CONJUGATION; PARITY; SUPERSTRING; TIME REVERSAL INVARIANCE

Bibliography

KANE, G. *The Particle Garden, Our Universe as Understood by Particle Physicists* (Addison-Wesley, Reading, MA, 1995).
PERKINS, D. H. *Introduction to High-Energy Physics* (Addison-Wesley, Reading, MA, 1987).

MICHAEL DINE

CREEP

Creep is the time-dependent deformation of a solid subjected to a force, in contrast to elastic or plastic deformation. This phenomenon is observed in a wide range of materials, from elements, such as aluminum, to complex solids such as concrete. After the force is applied, the rate of stretch/compression is rapid and then slows. A second region of constant rate, or viscous, flow is followed by a third region in which the rate increases rapidly until failure occurs, as shown in Fig. 1. (Not all materials exhibit all three regions.)

Rapid is a relative term. For rocks at room temperature, it may take millions of years for large de-

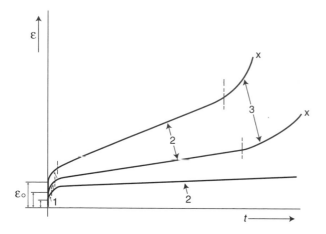

Figure 1 Idealized deformation (strain, ε) vs time curves for three increasingly larger constant loads. ε_0 is the time-independent (elastic) strain. Stage 1 is the initial transient, stage 2 is the steady stage, and stage 3 is the final or rupture period. The point of rupture is denoted with an *X*.

formations to occur. However, the sag of stone beams is detectable in historic times. Silly putty behaves as a nearly perfect solid in a short bounce, but flows under its own weight in a few hours.

The higher the temperature, the greater the creep rate. Although creep was known around the start of the twentieth century, the advent of high temperature machinery spurred its intense study. For reasons of thermodynamic efficiency, boilers and steam turbines use superheated steam. Jet engines operate at even higher temperatures. The mechanisms of creep need to be understood well enough to produce materials with creep time constants longer than the design time of these devices. Although no single model is applicable to all materials, for simple crystals, two mechanisms, atomic diffusion and dislocation motion, are primarily responsible for creep. Grain boundaries are also important for creep in the more complex materials.

Both diffusion and dislocation glide occur at exponentially increasing rates with temperature. This explains why creep is a greater problem at higher temperatures. By choosing materials with low diffusion rates and arranging grain boundaries to impede dislocation motion, the problem of creep has been subordinated to that of fatigue in designing high temperature materials.

For the design of precision parts (e.g., turbines), most effort has been placed in understanding stage 1, since by stage 2 the part would have deformed too much to operate properly.

See also: CRYSTAL; CRYSTAL DEFECT; CRYSTALLOGRAPHY; DIFFUSION; SOLID; VISCOSITY

Bibliography

BERNASCONI, G., and PIATTA, G., eds. *Creep of Engineering Materials and Structures* (Applied Science Publishers, London, 1979).

BOYLE, J. T., and SPENCE, J. *Stress Analysis for Creep* (Butterworths, London, 1983).

BRESSERS, J., ed. *Creep and Fatigue in High Temperature Alloys* (Applied Science Publishers, London, 1981).

ROY M. EMRICK

CRITICAL PHENOMENA

A physical system can have a number of phases: a gas phase, a liquid phase, and various solid phases. In daily life we are well aware of the existence of these phases and the fact that phase transitions are taking place: Water turns into ice when cooled; metals can be melted by heating; water and gasoline evaporate. The theory of critical phenomena tries to understand this behavior of matter. This is an important quest, if not for scientific curiosity, then because of the numerous technological applications in modern society.

On a phenomenological level, one makes a distinction between first-order and second-order phase transitions. Common phase transitions, like the transition of water to vapor or ice, or the solidification of molten material, are first-order transitions. This means that the material releases a nonzero quantity of heat, the so-called latent heat, when it goes through the transition temperature T_c. This is an indication of a structural change in the material. For example, in the water–ice transition a latent heat of 334 $J \cdot g^{-1}$ is released when the water molecules get ordered into a lattice, rather than moving around randomly. This is an abrupt change from the disordered fluid to an ordered solid.

In a second-order transition, also called continuous phase transition, the properties of the system do not change abruptly. For example, above the critical temperature $T_c = 1043$ K, called the Curie temperature, iron can only be magnetized by applying a magnetic field. Below T_c, iron is ferromagnetic, meaning that the material can stay magnetized even in the absence of a magnetic field. The magnitude of the magnetization continuously decreases as the Curie temperature is approached from below, vanishing entirely at T_c and all higher temperatures. In contrast to freezing, there is no abrupt change in the properties of the system. Another example of a continuous phase transition is superconductivity, discovered in 1911 by Heike Kamerlingh Onnes by cooling mercury to about 4.2 K. Since then many other materials have been found to become superconductors below transition temperatures, which can be as high as 120 K. At present this is a very active research area of great technological importance.

Continuous phase transitions are characterized by a so-called order parameter. This parameter is generally defined as a quantity that vanishes at one side of the transition (usually the high-temperature side) and differs from zero at the other side. It is a matter of physical intuition and experience to identify the proper order parameter for a given physical system. For example, the magnetization is a suitable order parameter to describe the ferromagnetic transition. Since the direction of the magnetization is arbitrary, this order parameter is a vector in space. In the case of superconductors the order parameter is a complex quantity ψ, defined such that its absolute value is a measure for the density n_s of superconducting electron:

$$n_s = |\psi|^2,$$

and its phase ϕ is the potential for the superfluid velocity:

$$v_s = \frac{h}{2\pi m} \nabla\phi,$$

where h is Planck's constant and m is the mass of the electron.

In the theory of critical phenomena the appearance of an order parameter is understood as a manifestation of a change of symmetry of the system. Without magnetization a ferromagnetic material is isotropic in all directions. However, when the magnetization appears it defines a preferred direction. One says that the rotational symmetry is broken.

The theoretical relation between continuous phase transitions and a change of symmetry was first noted by Lev Landau in 1937. Together with Evgenii Ginzburg he formulated in 1960 a general theory of continuous phase transitions that involve a broken symmetry. This important theory is referred to as the Ginzburg–Landau model.

Continuous phase transitions are characterized by the fact that certain quantities show very large fluctuations as the critical point is approached; some may even diverge. The heat capacity C, in particular, often diverges in the neighborhood of T_c according to the law

$$C \sim |T_c - T|^{-\alpha}.$$

The number α, with typical values $\simeq 0$–0.2, is called a critical exponent. Many more of these critical exponents are defined. For example, a critical exponent β with values $\simeq 0.3$–0.4 describes how the magnetization vanishes when the Curie temperature is approached from below:

$$M \sim |T_c - T|^{\beta}.$$

The surprising fact is that the numerical values found experimentally for these critical exponents for very different systems are often nearly equal to within the experimental error. This very important observation is known as universality. One may assign each system to a universality class in such a way that apparently diverse systems in the same universality class have the same critical exponents.

An important goal of the theory of critical phenomena is to explain this remarkable congruence in experimental properties. What is called the modern theory begins with the scaling hypothesis put forward by Ben Widom (1965) and the universality hypothesis of Leo Kadanoff (1967). The intuitive idea is that close to a phase transition the range of correlations between the atoms becomes very long, much longer than the range of the interaction. It seems reasonable, therefore, to suppose that the critical exponents should not depend on the fine details of the interaction but only on such general features as the dimensionality and symmetry of the interaction. This basic idea was translated into a recursive mathematical procedure, the so-called renormalization group approach, by Kenneth Wilson, who received the 1982 Nobel Prize in physics for this fundamental contribution to the theory of critical phenomena.

Many important results relating to phase transitions are being obtained by investigating theoretical models. These models are devised as an abstraction of physical reality for the purpose of being mathematically more tractable than realistic models. Universal properties should be unaffected by such modeling. The most important is the Ising model, proposed by Wilhelm Lenz in 1920 as a simple model for a ferromagnet. The model was first solved exactly by Ernst Ising (1925) for dimension one and by Lars Onsager (1944) for dimension two. No exact solution for dimension three is known, but critical exponents and other properties have been obtained numerically and by computer simulation. An important new class of models, the so-called ABF models, was proposed by George Andrews, Rodney Baxter, and Peter Forrester in 1984. Many more models, some of them exactly solvable, are known and their study is an important branch of statistical mechanics.

See also: CURIE TEMPERATURE; FERROMAGNETISM; PHASE; PHASE TRANSITION; STATISTICAL MECHANICS; SUPERCONDUCTIVITY

Bibliography

BINNEY, J. J.; DOWRICK, N. J.; FISHER, A. J.; and NEWMAN, M. E. J. *The Theory of Critical Phenomena* (Clarendon Press, Oxford, Eng., 1992).
REICHL, L. E. *A Modern Course in Statistical Physics* (Edward Arnold, London, 1980).

CHRISTIAAN G. VAN WEERT

CRITICAL POINTS

A critical point is a specific temperature, pressure, and density at which a phase transition may occur without discontinuous changes in these variables. At sufficiently high pressure, the density of a liquid can be approximately equal to that of the gas. The transformation from one state to the other can then take place without a significant latent heat transfer.

This is not common at standard temperature and pressure, where phase transitions from a liquid

to a solid or a liquid to a gas commonly involve dramatic changes in density and volume. Properties of the material, such as material strength, flexibility, malleability, viscosity, magnetization, compressibility, and specific heat, can also change suddenly as the phase changes. In contrast, at a critical point the state variables, that is, temperature, pressure, and density, change continuously as the phase changes.

In the phase diagram for water shown in Fig. 1, liquid and gas phases are indistinguishable at the critical point. In the pressure versus density diagram for xenon shown in Fig. 2, both liquid and gas coexist in the shaded region. This occurs because both temperature and pressure are constant for a range of densities in this region. Near the critical point, as derivatives in the state variables approach zero, properties such as specific heat and compressibility, which are normally constants, undergo dramatic changes, becoming infinite at the critical point.

The slowly varying state near the critical point can be described as having a high degree of order (low entropy). As well-ordered states, critical points are analogous in their degree of order to the magnetization of ferromagnets, to the spontaneous polarization of ferroelectrics, and to the coherent structure wave function of superconductors and superfluids.

Theoretical work and experiments have led to relationships among various critical exponents. For example, we say that near a critical point a thermodynamic quantity such as specific heat varies directly as $t^{-\alpha}$, where $t = (T/T_c) - 1$, and α is a measurable

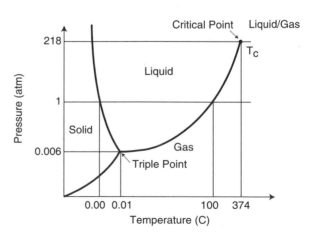

Figure 1 Phase diagram for water.

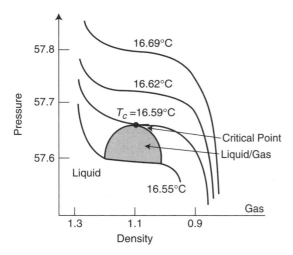

Figure 2 Pressure versus density diagram for xenon.

critical exponent for T greater than the transition temperature T_c. This formalism suggested a universal form for the equation of state expressed in reduced variables. Leo P. Kadanoff suggested that a critical point would occur when fluctuations in the order parameter (e.g., magnetization) had a correlation length greater than the range of forces responsible for the phase transition. When this occurred, the material properties became insensitive to the forcing, depending only on the dimensionality of the material property and the symmetry group of the order parameter. Kenneth G. Wilson's application of renormalization theory has revolutionized the field, describing critical phenomena in terms of scaling effects due to symmetries in two and three dimensions. This technique allows direct calculation and comparison of critical exponents, relating scaling properties in several dimensions.

See also: CRITICAL PHENOMENA; FERROELECTRICITY; FERROMAGNETISM; PHASE TRANSITION; RENORMALIZATION; STATE, EQUATION OF; SUPERCONDUCTIVITY

Bibliography

KADANOFF, L. P., et al. "Static Phenomena Near Critical Points." *Rev. Mod. Phys.* **39,** 395 (1967).

WILSON, K. G., and KOGUT, J. "The Renormalization Group and the ϵ Expansion." *Phys. Rep.* **12C,** 75–200 (1974).

ALICE L. NEWMAN

CRYOGENICS

Cryogenics is the branch of science that deals with the effects of very low temperature and the fact that the atoms move less energetically when the temperature is low. Everything we see around us is composed of atoms and these atoms are continually in motion. As the temperature goes down, the atoms move at lower speeds and some very striking phenomena occur just because the atoms move more slowly. Whether they are the atoms of aluminum in a pop can, atoms of silicon in a computer chip, or atoms of hydrogen in the human body, the atoms move around with a random energy called thermal energy. If some of this thermal energy is removed by placing a cold object in contact with it, say by placing the pop can on a chunk of ice, then the temperature goes down and the aluminum atoms in the can move more slowly. Heat flow is the name given to the flow of thermal energy, and temperature is the name given to the property that determines the direction of heat flow. Heat always flows from hot to cold or from high temperature to low temperature. For example, if you touch a hot tea kettle with your finger, heat would flow from the hot tea kettle to your finger; alternatively, if you should touch a cold pop can, heat would flow from your finger to the cold pop can.

To place all of this in perspective, it may be helpful to recall some familiar temperatures. If you start at a high temperature with water molecules in the form of steam and continually cool it, then the steam condenses to water at 100°C (212°F) and the water freezes to ice at 0°C (32°F). If you continue to cool the ice and remove all the thermal energy that can possibly be removed, the coldest that the ice can be made is −273.16°C. In fact, if you remove all the thermal energy that can possibly be removed from any substance, the coldest anything can be made is −273.16°C, the absolute zero of temperature. Because this is a special temperature, we define a new scale simply by adding 273.16 to the Celsius scale to get the Kelvin scale. On this scale, absolute zero is 0 K, the freezing point of water is 273.16 K, and the vaporization point of water is 373.16 K. The laws of science become much simpler if they are expressed in terms of the Kelvin temperature scale.

To create temperatures below room temperature, the most common method is to use a machine much like a household refrigerator. For a typical refrigerator, some working fluid, like freon, is compressed to a high density and the gas becomes hotter. This heating arises because the piston pushes against the gas doing work on it, making the molecules move faster. This hot gas flows through a long coil located on the back of the refrigerator and cools down nearly to room temperature by giving heat up to the surrounding air. High-pressure freon gas is then expanded through a special valve and it cools. In the case of a household refrigerator, this cold gas, in turn, flows through another heat exchanger and cools milk in the refrigerator; for a more sophisticated refrigerator, the cold gas is used to cool the incoming high-pressure gas to go to even lower temperatures. Details of the heat exchanger and the pressure–temperature cycle of compression and expansion may differ, but the principles of a nitrogen liquefier (77 K) or helium liquefier (4.2 K) are much like those of a household refrigerator.

As the temperature of a material is lowered, some truly remarkable properties can be observed to occur just because the atoms are more quiet and move with lower speeds. For example, if you were to take 700 ℓ of nitrogen gas (recall that nitrogen comprises about 80% of the air around us) and cool it, the gas would condense to a liquid having a volume of 1 ℓ at 77 K. This is a large reduction in volume. In the gas form, atoms are far apart compared to the size of the atoms, and bump into one another only occasionally. In the liquid state, the thermal energy is low enough that weak attractive forces between atoms can hold them adjacent to one another. If you cool the liquid nitrogen further, the liquid will freeze to a solid. In the solid, the directionality of the bonding between atoms takes over and the atoms form a crystal with very definite rigid shape with each atom on a specific site. All substances show these very characteristic phases of solid, liquid, and gas form.

The electrical and magnetic properties of a material also can change dramatically as the substance is cooled. Possibly the most dramatic change is the change in electrical resistance of some metals when they become superconducting. If you cool a metal wire, say a length of lead wire, the electrical resistance will abruptly drop to zero in a very narrow temperature range. At 7.3 K, the electrons in the wire bump into defects in the wire and scatter every which way, causing resistance to current flow. The scattering creates heat and tends to raise the temperature of the wire. At 7.2 K, the electrons suddenly lock into a new phase in which the scattering stops and the electrical resistance goes to zero. The electrons have

been transformed from normal particles that act independently (and scatter) to particles the act as a collective unit (and do not scatter). It is as though the electrons lock together in a giant molecule that flows through the crystal without scattering.

A related phenomena occurs in helium. Helium is a gas at room temperature, but if it is cooled, a liquid forms at 4.2 K. If it is cooled further, the atoms of the liquid undergo another phase transition at 2.2 K in which the viscosity, or resistance to flow through a pipe, suddenly goes to zero. The atoms form into giant coherent gangs that can flow without scattering (super fluid). In many ways it is like a giant molecule in which all the atoms condense into one coherent unit. A cup full of helium in the superfluid state held above the surface of helium will empty as the atoms creep up the side of the cup, over the top, and into the fluid outside. It is like a giant snake creeping over the edge. Water and other normal fluids will not creep out of a container.

Cryogenics and low temperatures can be useful for many medical and industrial purposes. One example is the magnetic resonance imaging (MRI) machine used in hospitals to image a tumor growing inside the body or a torn ligament in a football player's knee. The large magnet used to create the magnetic fields is made of a superconductor that has been cooled to low temperatures. A second example is the liquified natural gas ships used to transport natural gas from an oil well in Saudi Arabia to the place of use, say New York City. The natural gas that frequently is burned off at an oil well can be cooled to transform the gas to a dense liquid and ship it in giant refrigerated tanks from the oil field to the use point in refrigerated tanks in a ship. A third example is the use of liquid oxygen in a steelmaking process. An effective way to generate and deliver the oxygen at a low cost is to liquify air, separating the nitrogen from the oxygen in a distillation column. As new phenomena are discovered and more efficient refrigerators develop, the use of cryogenics will become more widespread.

See also: ABSOLUTE ZERO; LIQUID HELIUM; REFRIGERATION; TEMPERATURE; TEMPERATURE SCALE, CELSIUS; TEMPERATURE SCALE, FAHRENHEIT; TEMPERATURE SCALE, KELVIN

Bibliography

SCOTT, R. B. *Cryogenic Engineering* (Van Nostrand, Princeton, NJ, 1959).

WHITE, G. K. *Experimental Techniques in Low-Temperature Physics*, 3rd ed. (Oxford University Press, Oxford, Eng., 1979).

DOUGLAS K. FINNEMORE

CRYSTAL

In common usage, a crystal is usually considered to be a natural solid object with flat faces that meet in straight lines and sharp corners, such as a gemstone. The scientific definition of crystals is restricted to solids in which the atoms are arranged in regular repeating units (see Fig. 1). An ideal crystal consists of an infinite number of parallelepiped unit cells, each containing exactly the same arrangement of atoms, which are packed together so as to fill all space. The microscopic structure of a crystal can be determined using x-ray, neutron, or electron diffraction together with some knowledge of its chemical composition. The diffraction pattern of x rays after passing through a crystal consists of many sharp spots, with the symmetry of the diffraction pattern mirroring that of the crystal and the intensities giving information on the arrangement of atoms within each unit cell.

Most crystalline solids consist of a large number of tiny crystals called grains, which are randomly oriented with respect to each other. Such solids are said to be polycrystalline. Metals such as copper and aluminum, which we encounter in daily use, are usually polycrystalline. X-ray diffraction shows that the atoms are arranged periodically on the microscopic scale, but these features are not evident in the external form of the metal.

For some materials, large single crystals can be grown consisting of just one grain. In these cases, the regular arrangement of atoms on the microscopic level usually leads to anisotropic shapes on the macroscopic scale. If single crystals are cut or broken, they often cleave in such a way that they have flat faces, or "facets," with sharp edges between them. The form of the crystal depends on the microscopic structure. For example, table salt is composed of Na and Cl atoms arranged in a cubic lattice, and crystals of NaCl are cubic in appearance. Quartz, diamond, snow flakes (crystalline water), and NaCl are all examples of large faceted single crystals.

Almost any substance will form a crystal if cooled sufficiently slowly from a liquid state. Crystals may also be formed by precipitation of a solid from a solution. For example, if a saturated solution of sugar

(a)

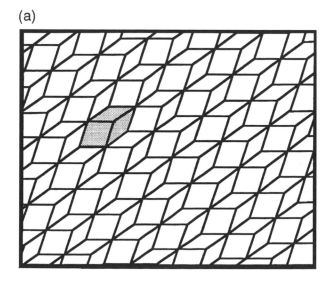

(b)

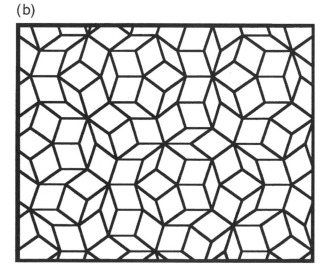

Figure 1 (a) A crystalline (periodic) array of unit cells. The unit cell in this case is composed of three smaller rhombuses. The diffraction pattern obtained from such an array consists of a set of regularly spaced spots. (b) A quasiperiodic arrangement of two units cells, known as a "Penrose tiling." This pattern is not periodic. The diffraction pattern from a Penrose tiling consists of many sharp spots in an arrangement with fivefold rotational symmetry.

in water is allowed to evaporate, sugar crystals will grow inside the container. Other methods of crystal growth include condensation from a vapor, or thermal transformation of a different solid phase.

Crystals can be classified by the nature of the forces holding them together. In an "ionic crystal" such as NaCl, the sodium atoms are positively charged and the chlorine atoms are negatively charged. The Na and Cl atoms alternate, and the crystal is held together by electrostatic attraction. Ionic crystals are most often composed of elements from opposite sides of the periodic table of elements, such as LiF or CsF. In a "covalent crystal" such as diamond or silicon, the outermost electrons on each atom are shared with adjacent atoms, leading to very strong bonding between atoms and a rigid structure. Covalent crystals are most often composed of elements from the middle of the periodic table of elements; diamond, for example, is pure carbon. In a "metallic crystal" such as copper or lead, the outermost electrons are completely detached from the atoms and move almost freely through the crystal, leading to the phenomenon of electrical conduction. Metals are also typically composed of elements from the middle of the periodic table of elements, such as iron, copper, nickel, and their alloys. In a "molecular crystal" such as ice or dry ice (solid CO_2), electrons are bound tightly to individual molecules to form electrically neutral subunits. The attractive force between molecules in this case arises from small fluctuations in the electric fields around the molecules, and they are quite weak. Thus, molecular crystals are usually much softer than ionic or covalent crystals. Molecular crystals are generally composed either of chemically unreactive elements, such as Ar or Kr, or of neutral molecules. Proteins and even viruses can form molecular crystals.

Some substances are solid without being crystalline. Examples are glass, wood, and plastics. Many noncrystalline solids have an amorphous structure. In general, amorphous (or glassy) materials, such as tar, are formed by cooling a liquid rapidly so that the atomic motion ceases without the atoms having time to arrange themselves into a crystal lattice. They can thus be thought of as extremely viscous liquids. The clear glass used in optical lenses is an example of an amorphous material. Natural glasses such as obsidian or pumice are formed when silica-rich molten rock is rapidly cooled. Plastics such as those used in children's toys or food wraps also often have an amorphous structure.

Substances with a biological origin, such as wood, wool, or cheese usually have a very complicated microstructure, with regions that are partially ordered, regions that are amorphous, and regions that are liquid crystalline. "Rocks" are natural solids composed of crystalline grains and/or glasses, and/or organic remains such as coals. In some cases, rocks may have a simple structure, or even be single crystals such as naturally occurring quartz, but more commonly they are complex aggregates of many different substances.

A crystal lattice cannot have fivefold rotational symmetry, like a pentagon or a starfish, because unit cells with fivefold symmetry cannot be assembled to fit together smoothly in all directions. Only crystals having two-, three-, four-, or sixfold rotational symmetry about any given axis are geometrically allowable. In a seeming paradox, a quasicrystal is a solid which gives rise to a sharp diffraction pattern, like a crystal, but which displays a fivefold rotational symmetry in its diffraction pattern, inconsistent with possible packings of identical unit cells. The microscopic structure of quasicrystals is believed to consist of two or more unit cells, arranged not periodically but "quasiperiodically" according to definite rules (see Fig. 1). Most quasicrystals observed have consisted of metallic alloys, such as Al–Mn or Al–Cu–Fe.

The term "crystal" has a more restricted meaning in some subfields of science and technology. In electronics, the term "crystal" is restricted to mean "piezoelectric crystal." In horology, "crystal" refers to the transparent covering of a watch.

See also: CRYSTALLOGRAPHY; CRYSTAL STRUCTURE; DIFFRACTION; DIFFRACTION, ELECTRON; ELECTROSTATIC ATTRACTION AND REPULSION; IONIC BOND; IONIZATION; LIQUID CRYSTAL; PIEZOELECTRIC EFFECT; SPECTROSCOPY, X-RAY

Bibliography

ASHCROFT, N. W., and MERMIN, N. D. *Solid-State Physics* (Holt, Rinehart and Winston, New York, 1976).

DIETRICH, R. V., and SKINNER, B. J. *Gems, Granites, and Gravels: Knowing and Using Rocks and Minerals* (Cambridge University Press, Cambridge, Eng., 1990).

DIVINCENZO, D. P., and STEINHARDT, P. J. eds. *Quasicrystals: The State of the Art* (World Scientific, Singapore, 1991).

WARREN, B. E. *X-Ray Diffraction* (Addison-Wesley, Reading, MA, 1969).

PAUL A. HEINEY

CRYSTAL DEFECT

Ideal crystals are made of atoms located in rows and columns of high symmetry. Crystal defects exist when diversions from this strict symmetry exist.

Defects in a crystal structure can be classified as point, line, and planar. Each contributes interesting properties to real crystals. Point defects are vacancies (atom missing from a crystal site), interstitials (atom between sites), and variations thereof. Such defects are present at finite temperature due to a thermodynamic equilibrium between the increase in disorder they produce and the energy required to create them. Figure 1 shows several ways they might be formed. Nonthermal-equilibrium concentrations of vacancies may be present due to radiation, defor-

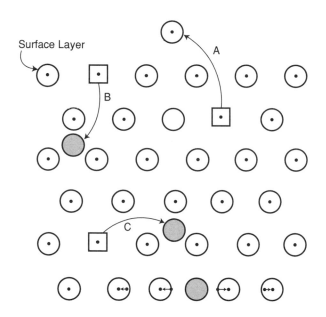

Figure 1 The dots represent the perfect lattice points, the open circles are atoms on their proper sites, shaded circles are interstitials, and squares are vacant sites (or vacancies). In A, a vacancy is formed by an atom jumping to the surface of the crystal, and in B, a surface atom jumps to an interstitial position. In C, an atom jumps from a regular site to an interstitial site, forming a Frenkel pair. A special type of interstitial, or crowdion, is illustrated in the bottom row: an extra atom squeezes its neighbors from their normal positions. Actually, vacancies are most often formed internally by a dislocation mechanism as shown in Fig. 2.

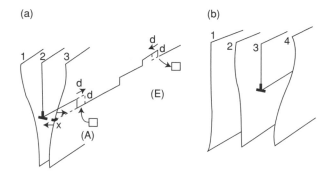

Figure 2 An edge dislocation can be thought of as the edge of the extra plane of atoms, marked by the ⊥ symbol. The top part of the crystal can move to the right relative to the bottom if plane 3 breaks at x (a), the bottom part joining plane 2 to make a whole plane while the dislocation is now in plane 3 (b). The motion of the dislocation through the crystal allows the top to move one lattice spacing relative to the bottom. A dislocation may also have jogs that move a lattice spacing d in the direction shown when a vacancy is absorbed (A) (i.e., an atom leaves the extra plane) or emitted (E) (i.e., an atom attaches to it).

mation, and other mechanisms. Vacancies are very important for diffusion. In ionic crystals, they form color centers (which, for example, cause old white glass left in the sun to turn purple). Radiation-formed vacancies can coalesce into large voids. These voids will greatly shorten the life of fusion reactor walls and other devices that operate in a high radiation environment. It has been observed that in certain irradiated materials, small voids form on a lattice like that of the crystal, but with a much greater spacing. Vacancy diffusion also results in high temperature creep.

The primary line defect is the dislocation. An edge dislocation may be thought of as an extra partial plane of atoms inserted in the lattice, as shown in Fig. 2. A screw dislocation is a spiral staircase-like arrangement of the planes about a line. A crystal would be close to its theoretical strength if it were not for dislocations, whose motion through a crystal allow it to deform. Dislocations may be present due to crystal growth and their number can be increased by bending. A copper wire that has been heated is relatively soft and bends easily because the dislocations can move freely. After more bending (or pounding), the copper is much harder (called

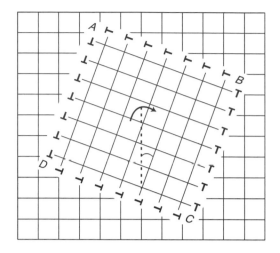

Figure 3 An idealized, small square grain, *ABCD,* in a large lattice. When the mismatch is not too big, the grain boundary behaves like a series of dislocations. Actual grains are irregular and may have varying sizes and size distributions depending on the history of the crystal.

work hardening). More deformation produces more dislocations, which become tangled and cannot move as easily. The hardness of steel can be controlled by determining the size of the crystal grains through thermal treatments. Smaller grains make it harder for dislocations to move. The motion of dislocations through a crystal leaves behind a trail of vacancies.

When a crystal forms from a molten material, little crystals form at many locations and have different orientations. The plane where one crystallite meets another is called a grain boundary, as in Fig. 3. The atoms in the grain boundary are more disordered than in the bulk. Diffusion occurs much more rapidly in grain boundaries (and along dislocations) than in the bulk. In general, the faster the melt is cooled, the smaller the grains. In complex materials like steel, different compounds form in different grains, allowing the production of materials with a wide range of properties. Nanocrystals (grain size of nanometers) have been made that have properties much different from bulk material because such a large fraction of the atoms is in the grain boundaries.

See also: COLOR CENTER; CREEP; CRYSTAL; CRYSTALLOGRAPHY; CRYSTAL STRUCTURE; LIQUID CRYSTAL; SYMMETRY; WHISKERS

Bibliography

KITTEL, C. *Introduction to Solid-State Physics,* 5th ed. (Wiley, New York, 1976).
POIRIER, J.-P. *Creep of Crystals* (Cambridge University Press, Cambridge, Eng., 1985).
READ, W. T. *Dislocations in Crystals* (McGraw-Hill, New York, 1958).

ROY M. EMRICK

CRYSTALLOGRAPHY

Crystallography is the study of crystals. It became a science that departed from mineralogy by providing a method for explaining how the shape of a material determines its physical properties—the symmetry principle. By the 1880s physicists had collected sufficient evidence to explain diverse phenomena such as birefringence, thermal expansion, pyroelectricity, and piezoelectricity as consequences of crystal form.

The study of crystals dates back to classical Greece and Rome with Teofrasto and Plinio, who studied a variety of quartz. In the seventeenth century the first attempts were made to explain the structural regularity of crystals; Robert Hooke suggested that the morphology of quartz could be explained if it were made up of a periodic arrangement of spheres and Christiaan Huygens considered that calcite was a periodic collection of ellipsoids in order to explain its birefringence. René Just Haüy hypothesized in 1784 that crystals were built by grouping "molecules" in the shape of a parallelepiped which, according to their spatial arrangement, would account for the observed macroscopic crystal forms. In 1827, Auguste Cauchy obtained the equations of elasticity theory with which he could describe, using twenty-one parameters, how a solid was strained when subjected to a known external stress. He furthered his investigations and found that crystals, by their lattice nature, needed fewer numbers to be described: fifteen for the general crystal and just two for the particular case of a cubic crystal. Five years later Franz Ernst Neumann used these results to study the interaction between light and matter in mechanical terms; he considered light to be made up of small material particles. His student, and later a professor at Göttingen, Woldemar Voigt, was the first to synthesize in a single book all the knowledge about the relationship between physical properties and crystal structure. His work, although out-dated even during his lifetime for lack of electromagnetism and the microscopic nature of matter, provided scientists with a fresh way of thinking in which the system of study was simplified because of its internal symmetry.

In 1912 modern crystallography was born. That year, Max von Laue and his group obtained a picture of diffraction of x rays by the crystal ZnS. Those experiments demonstrated the wave nature of x rays, discovered by Wilhelm Conrad Röntgen at the end of the nineteenth century, and the periodic arrangement of clusters of atoms inside the crystal. They set the framework for our understanding on the subject. William Lawrence Bragg and his father, William Henry Bragg, worked on the same line of research and obtained the famous equation

$$2d \sin \theta = n\lambda, \tag{1}$$

where d is the interplanar distance of a given family of crystal planes, n (called the order of reflection) is a natural number, λ is the wavelength of the x rays used, and θ is the angle of incidence and reflection of the beam. This equation prescribes which angles are suitable for reflection for a given wavelength and family of planes—not all reflections allowed by geometry are actually found in nature.

If one analyzes a single crystal sample using white x rays, which contain not a single wavelength but a range of them, one obtains a von Laue pattern. Under those conditions, λ in Eq. (1) can take many values, but θ, the angle between the impinging ray and the plane, is fixed for a given family of planes. In general there exists one wavelength λ, for which the Bragg equation holds and reflection can occur. If one surrounds the sample with a photographic film, or any modern detector, one gets "spots" at various locations on the film, corresponding to reflections from the different crystal planes (see Fig. 1). By mathematically processing these data, one can work back to obtain what created that diffraction pattern and thus find the microscopic structure of the crystal: what is its crystal lattice and what atoms are sitting at the lattice points. The 1912 von Laue picture falls into this category.

There are other standard methods of studying crystals by x rays. The powder method is very popular. It consists of using a sample that, instead of being a single crystal, is crushed into small crystals 1 μm or less across. A monochromatic x-ray beam is sent toward the sample, and for a given family of planes there are many crystallites that are properly oriented and will produce Bragg reflection. All the planes properly oriented form the envelope of a cone with axis on the incident beam (as with an umbrella, any segment of the sheet will form the same angle with respect to the main pole). The reflected beams will lay on a cone with 2 times the aperture of the previous one, since the reflected beam forms an angle of 2θ with respect to the original beam, whereas the plane and the original beam make an angle of θ. If one places a photographic film in front of the incident beam one obtains a circle—the intersection of the cone and the plane. Usually, the film is made in the form of a narrow strip shaped as a circle and placed on a plane that contains the incident

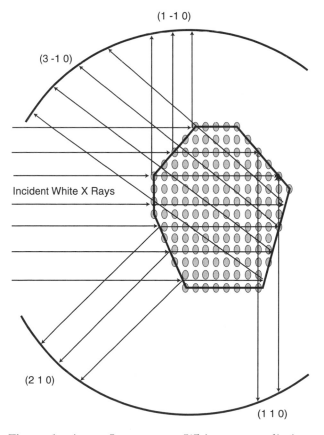

Figure 1 A von Laue setup. White x-ray radiation reaches a crystal and is diffracted to different spots on the circular film. Each spot corresponds to a Bragg reflection from the family of planes indicated.

beam. A hole is made to allow the incident beam to reach the sample. The intersection of the cones corresponding to various planes and the strip will give a line diffraction pattern.

Throughout the early 1990s there appeared new techniques that allowed the direct observation of crystalline surfaces. Low angle diffraction of electrons instead of x rays from surfaces has shed light on the understanding of morphological changes as crystals are grown for electronic applications.

Scanning tunneling microscopy has provided, for the first time, a direct observation of the reticular conformation of crystals by making it possible to observe single atoms.

It is not just the techniques for studying crystals that have evolved; the properties that scientists are studying have evolved as well. Areas of intensive research are superconductivity, ferromagnetism, and quantum Hall effects, among many.

See also: BRAGG'S LAW; CRYSTAL; CRYSTAL STRUCTURE; DIFFRACTION; DIFFRACTION, ELECTRON; FERROMAGNETISM; HALL EFFECT; SCANNING PROBE MICROSCOPIES; SCANNING TUNNELING MICROSCOPE; SUPERCONDUCTIVITY; X RAY, DISCOVERY OF

Bibliography

KITTEL, C. *Introduction to Solid-State Physics,* 6th ed. (Wiley, New York, 1986).

WOOD, E. A. *Crystals and Light,* 2nd rev. ed. (Dover, New York, 1977).

ZYPMAN, F. R. "Symbolic Programming Helps to Teach Debye–Scherrer Diffraction." *Comput. Phys.* **7** (1), 22–26 (1993).

FREDY R. ZYPMAN

CRYSTAL STRUCTURE

It has long been thought that the beauty and regularity of crystals gives evidence of an underlying regular structure. In the eighteenth and nineteenth centuries precise measurements of crystal planes were made, and after the discovery of x rays in 1895, Max von Laue, William Lawrence Bragg, and Bragg's father, William Henry, began in 1912 to use x-ray diffraction to determine the structure of crystals.

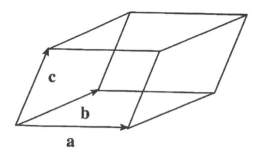

Figure 1 Unit cell for a lattice.

To describe a crystal structure we start with a lattice, a regular array of points in space. Each point of the lattice is at the corner of a parallelepiped-shaped unit cell, defined by three primitive lattice vectors **a**, **b**, **c**, as shown in Fig. 1. Space is filled with these parallelepipeds stacked together. Inside each unit cell are one or more atoms, called the basis. The lattice plus the basis gives the crystal structure. For example, Fig. 2 shows a two-dimensional honeycomb structure for which the lattice is a hexagonal layer and the basis is two atoms per unit cell, one at the corner and one inside the cell. In a crystal, various planes can be defined containing regular arrays of atoms. The directions and spacings of the planes can be determined by x-ray diffraction.

In addition to the regular spacing of atoms, called translational symmetry, crystals display symmetry under rotations and reflections. They have axes of symmetry and reflection planes. For example, in Fig. 2 each atom has a threefold axis of symmetry perpendicular to the plane, and three reflection planes that contain the symmetry axis. It is found that only twofold, threefold, fourfold, and sixfold axes of symmetry are possible for a crystal structure.

There are seven combinations of symmetry with structure called crystal classes, described by the

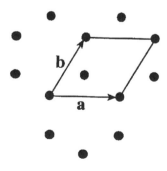

Figure 2 Lattice with two atoms per unit cell.

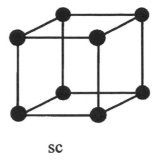 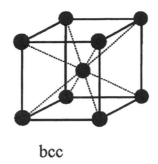 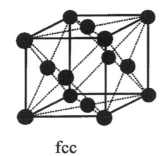

sc   bcc   fcc

Figure 3   Cubic lattices.

shape of the unit cell. In a triclinic crystal, the unit cell has no particular symmetry, that is, the sides and the angles are all different. In a monoclinic crystal, the unit cell has two right angles. In an orthorhombic crystal all three angles of the unit cell are right angles, so it is a rectangular parallelepiped. The tetrahedral crystal has in addition two equal sides ($a = b$), so the base of the unit cell is a square. In a cubic crystal, the unit cell is a cube. In the trigonal class it is a rhombus, which can be obtained from a cube by stretching along the cube diagonal. A hexagonal lattice is composed of hexagonal layers stacked one above the other. The base of the unit cell has sides $a = b$ with an angle of $60°$, while the c-axis is perpendicular to the base.

Some crystal classes have more than one lattice. For example, in the cubic case, starting with the simple cubic lattice, we can make a body-centered-cubic (bcc) lattice by putting an atom in the center

of the cube, and a face-centered-cubic (fcc) lattice by putting an atom in the center of each face of the cube. These are shown in Fig. 3. In each of these cases it is possible to define a set of primitive lattice vectors for which there is only one atom per unit cell, but the cubic unit cell is often the most convenient.

Some Simple Crystal Structures

The face-centered-cubic lattice corresponds to one way of packing spheres to occupy the smallest amount of space, that is, close packing. Many metals (Cu, Ag, Au) crystallize in the fcc structure. The planes perpendicular to the body diagonal contain hexagonal layers of atoms. Another way of arranging the layers gives a hexagonal-close-packed (hcp) structure, in which there are two atoms per unit cell in a hexagonal lattice. In Fig. 2 the atoms at the cor-

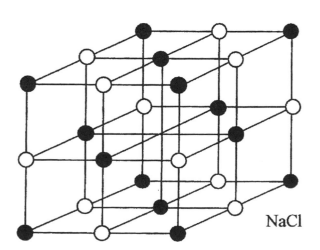 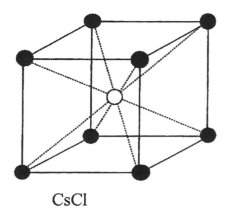

NaCl   CsCl

Figure 4   Ionic crystal structures.

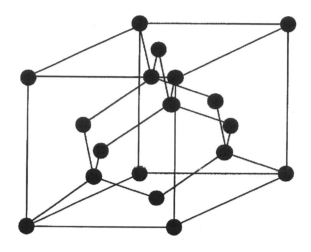

Figure 5   Diamond structure.

ners of the unit cell form a hexagonal layer; call it *A*. The atoms inside the unit cell form another hexagonal layer; call it *B*. To make the hcp structure, we put layer *A* at $z = 0$, layer *B* at $z = c/2$, layer *A* at $z = c$, and so forth, alternating layers *A* and *B*. The fcc structure can be described by stacking hexagonal layers *ABCABC*, where *C* is a hexagonal layer with atoms in the opposite position of the unit cell from *B* in Fig. 2. Zn and Cd are examples of metals that crystallize in the hcp structure. Some metals, including Fe and Nb, crystallize in the bcc structure.

Many ionic crystals have the sodium chloride (NaCl) structure, which we obtain by taking a simple cubic lattice and placing Na+ and Cl− ions alternately on the sites. The lattice is fcc; that is, the Na+ and the Cl− ions lie on two interpenetrating fcc lattices. The CsCl structure, on the other hand, has a simple cubic lattice. To describe it we place the Cs+ ions on the corners of the unit cell and the Cl− ions at the body centers. The structures are shown in Fig. 4.

The semiconductors Si and Ge crystallize in the diamond structure, which is based on the fcc lattice. This is illustrated in Fig. 5. Half the Si atoms are on the fcc lattice sites, and half are on a second fcc lattice displaced one-fourth of the way along the cube diagonal. Each Si atom is surrounded by a tetrahedron of Si neighbors, and the atoms are connected by covalent bonds. The III-V compounds such as GaAs have the zinc-blende structure, which is like the diamond structure, except that the Ga atoms are on one fcc lattice and the As on the other.

Complex Crystal Structures and Quasicrystals

Many materials crystallize in complex structures with many atoms per unit cell. An interesting example of this is $Y_1Ba_2Cu_3O_7$, which is orthorhombic and has 14 atoms per unit cell. If the oxygen content is reduced to between 6.4 and 7, the compound becomes a high temperature superconductor.

As mentioned above a regular crystal cannot have a fivefold axis of symmetry, so pentagons cannot be the basis of a regular crystal structure. However, x-ray patterns with fivefold symmetry have been seen, in structures called quasicrystals. Examples are Ga-Mg-Zn compounds. In two dimensions the plane cannot be "tiled" with pentagons; however, by combining two shapes, the plane can be tiled in a fashion that is not regular but has a consistent repetition of a fivefold pattern. Quasicrystals have similar patterns in three dimensions.

See also: CRYSTAL; CRYSTAL DEFECT; CRYSTALLOGRAPHY; DIFFRACTION; IONIC BOND; SYMMETRY

Bibliography

KITTEL, C. *Introduction to Solid-State Physics,* 6th ed. (Wiley, New York, 1986).
NELSON, D. R. "Quasicrystals." *Sci. Am.* **255** (2), 42 (1986).

LAURA M. ROTH

CURIE, MARIE SKLODOWSKA

b. Warsaw, Poland, November 7, 1867; *d.* Sancellemoz, France, July 4, 1934; *radioactivity, nuclear physics.*

Marya Sklodowska was born in Warsaw in 1867. Poland had then lost its independence and been annexed to czarist Russia. Her father, Wladislaw Sklodowski, had been lucky to study sciences in Russia and he was now a teacher of physics and mathematics in Warsaw. Her mother, Bronisłava Boguska, managed a boarding school for girls, and the Sklodowski family lived in the same building. Marya was the last of four children—three girls and one boy—and the incomes of her parents were modest. Zozia, the eldest daughter, died in 1876 from typhus, and Marya's mother died two years later from tuberculosis.

Marya began her studies in a public school in Warsaw and she succeeded brilliantly. She finished her secondary school in 1883 but, very tired, she spent the following year in the country, living with a peasant family. Back in Warsaw in 1884, she began to follow some lectures at a traveling university and was attracted by the positivist ideas of Auguste Comte in France. Marya lost her mother's religious faith and brought her ideals to science.

Marya lived in close relation with her sister Bronia, three years older, who began to earn her living by giving lessons to children. Bronia wanted to become a woman doctor and she wished to go to Paris to study. Marya decided to help her sister by sending her money that she would earn by becoming a private teacher. Marya worked first for a Warsaw family and, one year later, she left for the country to enter the service of a landowner family. Three years later she returned to Warsaw and lived there with her father. Bronia was now studying in Paris. In 1890 she announced her marriage and insisted that Marya come to Paris. Bronia could help with Marya's lodging.

Marya hesitated for a while but finally decided to go to Paris to study physics at the Sorbonne. She arrived in Paris in September 1891 and entered the Sorbonne. She lived first with her sister and her husband but she later found an independent room nearer the Sorbonne.

Despite some difficulties with language, Marya clung to her studies. Obstinate and clever, she succeeded in her enterprise by working hard, although she lived in relative poverty. In July 1893 she received a bachelor's degree in physics. In July 1894 she also received one in mathematics.

Marya (now Marie) Sklodowska met Pierre Curie for the first time in 1894 at a friend's home. Pierre, seven years older than Marie, was already well known in scientific circles. With his brother Jacques, he had discovered piezoelectricity around 1880. He had developed a piezoelectric quartz crystal, able to detect and measure very weak electric currents ($\sim 10^{-12}$ A). In 1882, before he was twenty-five, he became a lecture demonstrator at the *Ecole de Physique et de Chimie Industrielle de Paris* (Paris School of Physics and Chemistry). In that engineering school, he became professor after successful researches on the symmetries of crystals and on the magnetism of various substances. Marie and Pierre had the same passion for science and they were attracted to each other. They married in July 1895.

From that marriage, three girls were born, Irène in 1897, a baby in 1903 who lived only a few hours, and Eve in 1904. Irène with her husband Frédéric Joliot discovered artificial radioactivity in 1934. Eve worked in journalism.

Upon completion of her academic studies, Marie began to work on magnetism in the laboratory of Gabriel Lippman, her professor of physics. After their marriage, Pierre Curie succeeded in obtaining for her the possibility of working at the *Ecole de Physique et de Chimie*, where he was now professor. Marie Sklodowska Curie published her first paper at the end of 1897.

Meanwhile, at the beginning of 1896, Henri Becquerel had discovered the existence of new rays emitted by uranium salts that he named "uranic rays." This was the discovery of radioactivity. This completely new subject attracted Marie Sklodowska Curie, and she decided, on the advice of her husband, to study this new property of matter. She decided to investigate salts and minerals she obtained from various laboratories. The powder to be examined was set on a metallic plate connected to an electric potential of 100 V. Facing this plate was another metallic plate, connected to an electrometer controlled by the piezoelectric quartz crystal invented by Pierre Curie. This apparatus played the role of an ionization chamber activated by rays emitted by the substance to be studied. She verified the radioactivity of uranium salts, and she discovered the same property for other substances, namely, for thorium coumpounds. In her publication of April 1898, she noted that thorium and uranium were the two elements having the biggest atomic masses known at that time. However, she also noted that two uranium minerals under examination were more active than uranium itself. This fact lead her to the belief that the minerals might contain an element that was more active than uranium.

From that time, Pierre Curie joined Marie in trying to isolate this postulated new element. This common work led them to the discovery of two new radioactive substances, which they named polonium, in remembrance of Marie's native country (July 18, 1898, note to the Academy of Sciences) and radium (December 26, 1898, note to the Academy of Sciences).

The technique they used consisted of submitting the compound to be studied to various chemical treatments so as to obtain a stronger radioactivity for the resulting product. The new elements were not isolated, but polonium was expected to have chemical properties similar to those of bismuth, and radium similar to those of barium.

Even though small, the quantity of radium present seemed to be greater than that of polonium, but more ore was necessary to isolate radium. After various researches, the Curies received, at the end of 1898, 100 kg of a residue of pitchblend from a factory in Czechoslovakia that extracted uranium from that ore. The residue seemed promising and the Curies arranged to buy about 1 ton of the ore at a low price.

Very hard work then began for the Curies. The ore was treated in portions of 20 kg and the extracting work took about four years under difficult conditions. In 1902 about 0.1 g of radium chloride was isolated with great purity after numerous operations, notably fractional crystallizations. By dosimetry of the radium chloride, Marie Curie determined an atomic mass of radium of 225 with a precision of about 1 unit. Later, radium was extracted by industrial means and Marie Curie obtained in 1907 a sample of 0.4 g of pure radium chloride. She then published an atomic mass for radium of 226.2 ± 0.5. Radium is now known as the element $_{88}Ra^{226}$.

The discovery of radium brought international fame to the Curies. In 1900 Marie Curie was appointed as lecturer at the *Ecole Normale Supérieure des jeunes filles de Sèvres* (a college of education for young women) near Paris. In May 1903 she defended her doctoral thesis in which she described her work in isolating radium. The Nobel Prize in physics of 1903 was shared for the discovery of radioactivity; half was awarded to Henri Becquerel and half to Pierre and Marie Curie. In 1904 Pierre Curie was appointed full professor in the Faculty of Sciences of Paris and a little later Marie was made responsible for students' practical work in physics for this chair.

After the discovery of radium, Pierre Curie took great interest in the study of its physiological effects. With two medical doctors, he participated in the first attempts to treat cancer by radium. Unfortunately, in April 1906 Pierre was struck down by the horses of a carriage and thrown under the wheels of the vehicle. Marie Curie suffered a severe shock. However, obstinate and determined as she was, she managed to face this adversity and decided to continue her research work. Until then, honors seemed to have been reserved for her husband but, in May 1906, to replace her deceased husband, she was appointed as a physics lecturer in the Faculty of Sciences in Paris and she became full professor in 1908.

In August 1906 Lord Kelvin, at the age of eighty-two, published an article in which he doubted the reality of radium. Even if this idea was not taken seriously by specialists, Marie Curie was offended by it and decided to isolate "her radium" as a pure metal. With André Debierne, a devoted chemist, she decided to proceed by electrolysis of radium chloride. She overcame numerous difficulties and succeeded in her enterprise, determining in particular the melting point of radium. In 1911 she was awarded the Nobel Prize in chemistry for her work on radium. Marie Curie was the first scientist to obtain the Nobel Prize twice, an extraordinary honor.

After her husband's death, Marie Curie experienced terrible times on two occasions. Pierre Curie had been a member of the Academy of Sciences and Marie Curie was pressed to stand as a candidate by some colleagues, such as the physicists Jean Perrin and Paul Langevin, and the mathematician Emile Borel. She followed their advice but lost the election in January 1911, with a difference of only one vote in favor of Edouard Branly. Branly had devised the first radio receiver in 1895 but in 1909, the Nobel Prize was awarded to Guglielmo Marconi, who achieved the first radio connection with the United States in 1902. Even if Branly's election to the Academy could be interpreted as a compensation for this loss, some unfair remarks were made about Marie Curie, and she was very affected. She never tried again for this honor.

In 1911 some reporters discovered that there was a sentimental affection between Marie Curie and Paul Langevin, a renowned physicist and former student of Pierre Curie. This was probably true as Langevin had an unpleasant relationship with his wife. Reporters found or received some compromising letters written by Marie Curie to Paul Langevin, published them in newspapers, and the scandal burst out. It was maintained by the press: "Out foreigner!" was the cry in the street near her home. Clamors and rumors came to a stop with the divorce judgment of Paul Langevin at the end of 1911 without mention of Marie Curie. Marie Curie broke off her relationship with Paul Langevin and lived alone until her death.

Despite these events, Marie Curie enjoyed a high reputation. Before receiving her 1911 Nobel Prize she had obtained from the Sorbonne and the government funds for building a new Institut du Radium containing a physics laboratory directed by herself and a biology laboratory to study the biological effects of radioactive substances.

The construction of the Radium Institute had just been finished when the World War I began in

August 1914. Scientists such as Paul Langevin and André Debierne were called up and scientific research stopped in civilian laboratories. Marie Curie decided to contribute to the war effort. Ten days after the declaration of war, the Ministry of Defense officially asked Marie to set up operator teams for x-ray radiology services. Her idea was to take care of wounded soldiers. She succeeded in buying cars and equipping them with radiological instruments. With a driver, a surgeon, and two assistants, she undertook a first test near Paris. One of her assistants was Irène Curie, age 17, who wanted to devote herself to this task. Wounded soldiers were examined with x rays to detect the presence of bullets or shell splinters in the body. During the four years of war, Marie and Irène were occupied with that task. In 1916 Marie Curie began to train new assistants at the Radium Institute, which was then empty. In all, some 200 cars were equipped and about 1 million soldiers were examined in different hospitals.

After the war Marie Curie returned to her new laboratory at the Radium Institute, equipped it and resumed research on radioactivity. She trained new research workers, among them her daughter Irène who submitted her doctoral thesis in 1925. Marie Curie was by now very well known. In 1920, near the Radium Institute, the construction of a hospital for cancer treatment was started. It was called "Fondation Curie." In 1920 Marie Curie was visited by an American journalist, Marie (Missy) Meloney, who proposed to undertake a campaign in the United States to collect some $10,000, the sum necessary to buy 1 g of radium, to be offered to Marie. Meloney asked, in case she was successful, that Marie would promise to go to the United States to receive this radium in person. Meloney herself would be in charge of the official reception.

Thus, in 1921 Marie Curie boarded a ship for the United States with her two daughters, Irène and Eve. As soon as she landed in New York, she was received with great enthusiasm by journalists. She was rather terrified as she disliked public honors. However, she had to attend numerous official receptions, and she finally received from the hands of President Warren Harding the gift of radium. By temperament, Marie Curie was suspicious. In her discussions with Meloney, she asked to receive an official document specifying that this radium was given to herself for the needs of her laboratory.

Since the heroic epoch of radium separation, Marie had had health problems from time to time

owing to the toxic effects of radioactive substances. Nevertheless, she was an attentive director of her laboratory. She followed with great care the work of her assistants and nothing was decided without her authorization. Concerning her personal work, she had become an experienced chemist and was devoted in particular to the preparation of various radioactive sources. Thus, at the time, her laboratory had the most intense sources of various radioactive substances in the world. A number of researchers began their work in her laboratory and some of them, such as Solomon Rosenblum and Frédéric Joliot, became renowned as scientists.

In 1932 Marie Curie was obliged to take a rest after a fall, and in June 1934 she had to leave her laboratory and was hospitalized. Doctors believed that she suffered from tuberculosis and sent her to a sanatorium in the Alps. There, leukemia was detected and Marie Curie died on July 4. For succeeding generations, Marie Curie was an eminent figure of science and a model personality. Starting from nothing, she had become a successful scientist, owing to her intelligence, her energy, and her tenacity.

See also: RADIOACTIVITY; RADIOACTIVITY, ARTIFICIAL; RADIOACTIVITY, DISCOVERY OF

Bibliography

CURIE, M. *Pierre Curie and Autobiographical Notes* (Macmillan, New York, 1923).
PFLAUM, R. *Grand Obsession: Madame Curie and Her World* (Doubleday, New York, 1989).
QUINN, S. *Marie Curie* (Simon & Schuster, New York, 1995).
REID, R. *Marie Curie* (Dutton, New York, 1974).

JULES SIX

CURIE TEMPERATURE

The common definition of the Curie temperature is that it is the temperature at which spontaneous magnetization disappears in a ferromagnetic material when the temperature is raised. Below the Curie temperature the individual atomic or molecular magnetic moments are coupled, allowing long-range order. This phase is called ferromagnetism. As the temperature is raised through the Curie temperature, the individual magnetic moments become independent from one another, destroying the long-range order. This is the phase transition from ferromagnetism to paramagnetism. Because it is a phase transition, the Curie temperature is also called the Curie point and, in reference to phase transitions in fluid systems, is often called the critical point. Included in the materials that show a Curie temperature are systems of mixed magnetic moments, called ferrimagnets, which show a nonzero magnetic moment in the ordered phase. Some common examples of the Curie temperature for some ferromagnetic metals in units of degrees are Fe (770°C), Co (1,130°C), Ni (358°C), and Gd (16°C). A similar definition of a critical temperature exists for antiferromagnetic materials and is usually called the Neel temperature.

An important alternate description of the Curie temperature arises from considering a system in the paramagnetic phase, that is, at temperatures above the Curie temperature. In this description it is useful to use the Weiss mean-field approach in which each magnetic moment experiences the average field of all the neighboring moments so that the field at the site of each moment is considered to be proportional to the total magnetization. Adding the definition of dc susceptibility, $\chi = M/B$, where M is the magnetization and B is the applied field, results in the Curie–Weiss law: $\chi = C/(T - \theta)$, where C is called the Curie constant and θ is called the Weiss constant. In general, the critical temperature is defined when χ diverges as the temperature is lowered. Inverting the expression for the Curie–Weiss law says that the critical temperature in the mean-field approximation is that temperature at which $1/\chi$ goes to zero, that is, when $T = \theta$. When $\theta = 0$ the Curie–Weiss law reduces to the Curie law, $\chi = C/T$, appropriate for pure paramagnets. For ferromagnets, $1/\chi$ goes to zero at a positive value of T. If the mean-field approximation were correct, the Curie temperature would occur at $T = \theta$. Normally, however, as temperature is lowered and before θ is reached, a departure from the Curie–Weiss law becomes apparent. In this temperature region the energy of the short-range interaction between the magnetic moments begins to overcome the temperature so that fluctuations to long-range order begin to have an effect. This results in a zero intercept of $1/\chi$ at a temperature lower than θ. This lower temperature is the Curie temperature.

To complete the description of the Curie–Weiss law and to introduce the quantum approach, a calculation of the Curie constant may be done in the following way. The energy of interaction of a magnetic moment with the magnetic field is given by $U = -\mu \cdot \mathbf{B}$, where μ is the magnetic moment given by $m_s g \mu_B$, m_s is the magnetic quantum number for spin $\mathbf{S}$, g is the spectroscopic splitting factor, and μ_B is the Bohr magneton. For the simple case of a spin-1/2 particle the magnetization per unit volume M is given by the difference between the populations of the spin-up state and spin-down states. The result is $M = N\mu \tanh(\mu B/kT)$, where N is the number of spins per unit volume. For the more general case of a system with magnetic moments containing both spin ($\mathbf{S}$) and orbital ($\mathbf{L}$) angular momentum quantum numbers, and therefore with multiple levels defined by $\mathbf{J} = \mathbf{L} + \mathbf{S}$, the result is $M = Ng J\mu_B B_J(x)$ where $x = g J\mu_B B/kT$. $B_J(x)$ is called the Brillouin function and is given by

$$BJ(x) = \frac{2J+1}{2J} \operatorname{ctnh}\left[\frac{(2J+1)x}{2J}\right] - \operatorname{ctnh}\left[\frac{x}{2J}\right].$$

For small magnetic fields, $\mu_B B \ll kT$, the susceptibility reduces to $\chi = NJ(J+1)g^2\mu_B^2/3kT$, which when equated to the Curie law, $\chi = C/T$, gives the Curie constant $C = NJ(J+1)g^2\mu_B^2/3k$.

See also: BOHR MAGNETON; CRITICAL POINTS; FERRIMAGNETISM; FERROMAGNETISM; MAGNETIC MOMENT; PARAMAGNETISM; PHASE TRANSITION; TEMPERATURE

Bibliography

JILES, D. *Introduction to Magnetism and Magnetic Materials* (Chapman and Hall, New York, 1989).

KITTEL, C. *Introduction to Solid-State Physics,* 6th ed. (Wiley, New York, 1986).

MARTIN, D. H. *Magnetism in Solids* (Pitman, London, 1967).

JOHN E. DRUMHELLER

CURRENT, ALTERNATING

Electrical current exists when there is a flow of charge through a region, for example, through a wire. Alternating current (ac) is current that reverses its direction periodically, as opposed to the case of direct current (dc), in which the direction of the flow is constant. While alternating current can take many forms, the most commonly encountered form is similar to that supplied by local electric companies—the standard 110 V potential difference and 60 Hz (cycles per second) frequency. In some countries the frequency is 50 Hz and in some the potential difference is higher. The frequency of ac is constant in many applications and is the number of times per second that the current goes through one complete cycle, including one maximum in each direction. In some applications the frequency varies, such as in frequency modulated (fm) radio antennae. The form (current versus time) of the signal is often that of a sine curve for convenience of generation and use but any form is possible.

Measurement and Units

Even in the simplest case, that of ordinary ac applications, the constantly changing direction and value of the current and of the potential difference complicate measurement. The current is in each direction half of the time so a standard average over time will be zero. The amount of work that an alternating current can do is the practical basis of measurement and the root-mean-square (rms) average is the appropriate value. For example, this gives the value of a direct current (the so-called effective current) that would produce the same heating effect in a resistor as a given alternating current. The most commonly used unit is the ampere (A), which is the same as one coulomb per second passing through a region.

Production

Rotating a coil in a magnetic field is a common way to produce alternating current. During rotation each side of the coil spends half its time going one direction and half its time going the opposite direction relative to the field. This results in the current being produced in opposite directions during each half-rotation, thus alternating current. The sides of the coil vary the angle at which they pass through the field in a sinusoidal fashion, so the emf (volts) and, therefore, the current produced, are sinusoidal. Electronic devices produce and use many other wave forms. Amplitude modulated (am) radio antennae currents are a common example.

Use

Alternating current is in wide use specifically because of its constantly varying value and direction. Efficient transmission of electrical energy over long distances requires large potential differences (volts) but using electricity at these potential differences is dangerous and inconvenient. The constant changes in alternating current make possible the use of transformers to increase the potential difference before transmission and lower it again before delivery. The carefully maintained frequency of the current is convenient for timing devices such as clocks that count the alternations in the current.

See also: CURRENT, DIRECT; ELECTRICITY; TRANSFORMER

HOWARD G. VOSS

CURRENT, DIRECT

Electrical current exists where there is a flow of charge through a region, for example, through a wire. Direct current (dc) is current that consists of charge flow that is constant in direction, as opposed to alternating current (ac), in which the direction of the flow of charge changes with time.

Measurement and Units

Ammeters use the magnetic field associated with currents to measure them. Placing the meter in series with the circuit element of interest ensures that the current in the meter is identical to the current in the circuit element. A shunt within the meter and in parallel with the meter movement may carry part of the current so that a known fraction of the current exists in the meter movement. This protects the meter movement and increases the range of current values measurable by the meter.

The ampere (A), a part of the SI system of units, is the most common unit for current. An ampere is the flow of one coulomb of charge per second, say through a wire.

Production

Various electric cells, including wet cells (such as the common automobile storage battery of several wet cells), dry cells, fuel cells, and nickel cadmium cells, are common sources of reasonably constant and unidirectional electromotive force (emf). By means of chemical processes, these cells produce an emf (measured in volts) and a potential difference (also measured in volts) between their terminals or contacts. This potential difference causes charges to move through a conductor in the direction indicated by the polarity $(+, -)$ orientation of the cell and the sign of the charge of the moving charged particles. It is common to arbitrarily take the direction that a positive charge would move in a circuit as the direction of the current even though, in the typical circuits with metal conductors, it is the negative electron that is doing the moving in the direction opposite to the arbitrarily chosen current direction. Direct current generators produce a constant-polarity electromotive force by rotating a coil of conducting wire in a magnetic field. Special slip rings and brushes connect the coil to the external circuit so that the flowing charge always exits the rotating coil into the same part of the external circuit, resulting in direct current in the circuit.

Use

Many electronic and electrical devices require direct current. Where cells or batteries of cells are not convenient or practical, rectifiers can change alternating current, such as that commonly supplied in buildings, into direct current. Direct current is not often appropriate for long-distance delivery of electrical energy because transmission at the potential differences convenient and safe for the end user results in large energy losses along the lines. Alternating current is more appropriate in these cases because transformers can efficiently raise the potential difference for efficient transmission and lower it again before delivery to the customer.

See also: CURRENT, ALTERNATING; ELECTRICITY; ELECTROMOTIVE FORCE

HOWARD G. VOSS

CURRENT, DISPLACEMENT

In 1864 James Clerk Maxwell wrote down the equations, now known by his name, that describe everything we know about the behavior of classical electricity and magnetism. He formulated these equations in terms of electric and magnetic fields. A field is simply a vector defined at each point of space, and we can easily visualize it by drawing lines that follow the direction of this vector everywhere. If an electrically charged particle with electric charge q moves with velocity **v** through a region in which there is an electric field **E** and a magnetic field **B,** it will feel a force, **F**, where

$$\mathbf{F} = q(\mathbf{E} + \mathbf{v} \times \mathbf{B}).$$

The first term on the right-hand side of this equation is the force exerted on the particle by the electric field, parallel to that field, while the second term is a vector cross product, and the resultant magnetic force is proportional to $|\mathbf{v}| \, |\mathbf{B}|$ and is perpendicular to both vectors. We can take the above equation, known as the "Lorentz force law," as a definition of what it is we mean by electric and magnetic fields. Maxwell's equations then tell us how these fields are produced. They summarize everything that was known at that time, and contain an additional mechanism invented by Maxwell and known as the "displacement current." The first of these equations is

$$\oint_{S} \mathbf{E} \cdot d\mathbf{A} = \frac{q}{\varepsilon_0}.$$

The term on the left is the integral of the electric field over any closed surface (S), and it essentially counts the net number of field lines that exit that surface (the number leaving minus the number entering). The term on the right is the total charge q enclosed by that surface divided by a constant, $\varepsilon_0 = 8.854 \times 10^{-12}$ C²/N·m² where electric charge is measured in units of coulombs (SI units). The physical content of this equation is that an electric field is produced by charges, and that the lines of this field begin (end) only on positive (negative) electric charges. It also tells us that the field produced by a point charge is proportional to $1/r^2$, where r is the

314

distance from the charge, and, therefore, that the electrostatic force between two charges is also proportional to $1/r^2$, where r is the distance between them. This is known as "Coulomb's law." A vector field with lines that begin and end at sources and sinks is known as an irrotational field.

The second equation is

$$\oint_C \mathbf{E} \cdot d\mathbf{l} = -\frac{d}{dt} \int_S \mathbf{B} \cdot d\mathbf{A}.$$

The term on the left is the integral of the electric field over any closed curve (C), and the term on the right is the negative rate of change of the flux of magnetic field through any open surface (S) that is bounded by that curve. The flux is simply the number of field lines that pierce the surface. The implication of this equation, also known as "Faraday's law," is that a changing magnetic field also can act as the source of an electric field, but it is an electric field with a very special property; its line integral around a closed path does not vanish. The resulting field lines do not begin or end on any charges; they are closed on themselves. Such a field is known as a rotational or solenoidal field. Its mathematical properties are distinctly different from those of an irrotational field, and no charges are necessary to produce a rotational vector field.

The third equation is

$$\int_S \mathbf{B} \cdot d\mathbf{A} = 0,$$

which says that the integral of the magnetic field over any closed surface (S) vanishes. This should be compared with the first of Maxwell's equations above, which relates the irrotational part of the electric field to the charges that produce it. To the best of our current understanding, there are no magnetic "charges" or "magnetic monopoles," although some modern theories of elementary particles predict their existence, and experimental searches are ongoing. The magnetic field is purely rotational, and its field lines all close on themselves. Up until the time of Maxwell's work, it was thought that the sources of magnetic fields were exclusively electric currents; that is, moving charges, and the equation which related the fields to the currents that produced them was

$$\oint_C \mathbf{B} \cdot d\mathbf{l} = \mu_0 I,$$

where I is the total current flowing through any open surface bounded by the closed curve C, and $\mu_0 = 4\pi \times 10^{-7}$ N·s²/C² (in SI units). Maxwell realized that this equation, known as "Ampère's law," was inconsistent as it stands. The easiest way to see this is to consider two parallel conducting metal plates separated by an insulating gap. Such a device is called a parallel plate capacitor. These plates can be connected to a battery by an external wire, which will allow the flow of charge from the battery onto the plates, leaving them with equal and opposite charges, and a resulting irrotational electric field between the plates, with field lines that begin on the positively charged plate and end on the negatively charged one. Imagine a circle surrounding the wire, with its plane perpendicular to the wire. While the capacitor is being charged, a current flows through the wire, and, hence, through the plane of the circle; Ampère's law predicts that there will be a magnetic field surrounding the wire, and that the magnetic field lines will be closed circles concentric with the wire. But one also can imagine another (curved) surface bounded by the same circle which is not pierced by the wire, but rather extends from its circular perimeter into the gap between the capacitor plates. Since there are no charges or currents in the gap, Ampère's law would then predict no magnetic field at the position of the circular loop. Maxwell realized that one has to add an additional term to Ampère's law:

$$\oint_C \mathbf{B} \cdot d\mathbf{l} = \mu_0(I + I_d),$$

where

$$I_d \equiv \varepsilon_0 \frac{d}{dt} \int_S \mathbf{E} \cdot d\mathbf{A}$$

is called the "displacement current." It will be nonvanishing wherever the electric field is changing with time. But this is exactly what is happening to the electric field inside the gap while the plates are being charged, and the displacement current term inside that gap will yield the same magnetic field predicted by Ampère's law applied to a surface, such as the plane of the circle, outside the gap. Although the field inside the gap, and the amount of charge that flows onto the capacitor plates, will depend on the nature of the insulating material in the gap, it is important to note that there will be a changing electric field between the capacitor plates even if the gap is a vacuum. For Maxwell, the vacuum consisted of a tenuous substance that he called "the ether," and he envisioned the displacement current as a flow of real charges through this ether. We now understand that such an ether does not exist, and the displacement current is not really an ordinary electric current composed of moving charges. With the addition of the displacement current term, a new symmetry appears in the theory; not only can a changing magnetic field give rise to a rotational electric field, but a changing electric field can give rise to a magnetic field in much the same way. In the above example, the changing electric field is itself an irrotational field, produced by the charges on the plates. That is all it is necessary to postulate in order to resolve the inconsistency in the equations.

Maxwell, however, took his development one step further, and argued that a displacement current would exist, and be given by the above formula, in the presence of any changing electric field, be it rotational or irrotational. Recall that a rotational electric field is produced by a changing magnetic field, and does not require any charges or currents as sources. If such a field is also changing in time, then, according to Maxwell's interpretation, it will produce a new magnetic field. We can then ask if there are solutions of Maxwell's equations which involve only time-varying magnetic and electric fields, in the complete absence of charges and currents. Such solutions are easy to find; the resulting **E** and **B** fields propagate through space with a speed

$$v = \frac{1}{\sqrt{\varepsilon_0\mu_0}} = 3 \times 10^8 \text{ m/s},$$

the speed of light. We now term such a configuration of fields "electromagnetic radiation," and visible light is only one possible form of such radiation. Maxwell's introduction of the displacement current successfully unified the apparently disparate phenomena of light (and, of course, all of its humanly invisible forms such as radio waves and x rays) with

electricity and magnetism. It must be ranked as one of the highest intellectual achievements of the nineteenth century.

See also: AMPÈRE'S LAW; ELECTROMAGNETISM; FIELD, ELECTRIC; FIELD, MAGNETIC; FIELD LINES; MAXWELL'S EQUATIONS; VECTOR

Bibliography

MAXWELL, J. C. "A Dynamical Theory of the Electromagnetic Field" in *The Scientific Papers of James Clerk Maxwell,* edited by W. D. Niven (Dover, New York, 1952).

MAXWELL, J. C. "On a Method of Making a Direct Comparison of Electrostatic with Electromagnetic Force, with a Note on the Electromagnetic Theory of Light" in *The Scientific Papers of James Clerk Maxwell,* edited by W. D. Niven (Dover, New York, 1952).

HAROLD S. ZAPOLSKY

CURRENT, EDDY

When a conductor passes through a changing (or nonuniform) magnetic field, currents are induced in it that can be large if the paths the currents follow have low resistance. The magnetic and heating effects of these currents can be both useful and troublesome.

The direction of the induced currents is such as to oppose the motion or change in the magnetic field. This is predicted by Lenz's law. Eddy currents are most pronounced in metallic conductors, although any electrical conductor will support eddy currents, sometimes called Foucault currents after their discoverer. In the 1850s, Jean-Bernard-Léon Foucault first noticed currents induced in a copper disk moving in a strong magnetic field. (This is the same Foucault who used a pendulum to prove that the earth rotates on an imaginary axis roughly through the poles.)

In the spirit of Foucault, we allow a copper disk to roll down an inclined plane and pass through the poles of a strong magnet. The disk will be seen to slow down on entering the field, speed up a little, and then slow down on exiting. Once free of the field it speeds up to its terminal velocity as dictated by friction and the length of the plane. The magnetic field changes most rapidly as the disk enters and leaves the pole pieces. The size of the induced currents are proportional to the rate of change of the magnetic field. The direction of the induced currents is such to oppose the motion of the disk. When the disk is fully between the pole pieces, the magnetic field, passing through the disk, is nearly constant for a moment so the induced current drops, the opposition to the motion of the disk decreases, and the disk accelerates. The eddy currents create their own magnetic fields that act to oppose the motion, repelling the poles as the disk approaches and attracting the poles as the disk tries to pass through. If we cut radial slits into the disk and repeat the experiment, we find that the braking action is less pronounced. We have cut across the paths of the eddy currents reducing their circulation to segments of the disk that offer a higher resistance to the current flow. The induced currents' ability to oppose the motion has been reduced significantly.

One can imagine a small conducting sphere held suspended over a coil in which a large alternating current is flowing. The eddy currents induced in the sphere repel the strong magnetic field of the coil. Electromagnetic forces can overpower gravity in such cases.

The braking effect of eddy currents is useful to slow down rapidly rotating conductors, rotary saw blades in industry, and brakes for trains. The technique is very powerful, because the higher the speed the greater the braking force. Also, no friction is involved since the eddy currents are induced inside the conductor so there are no touching parts that cause wear. A side effect is large heating due to $I^2 R$-type losses in the conductor. These types of brakes must be cooled. The same effect is used to damp out oscillations in light beam galvanometers.

Problems

The magnetic field produced by the eddy currents has a negative effect when they are produced in a ferromagnetic material to be magnetized in a coil. The induced magnetic field acts in opposition to the magnetizing field of the coil, so the resulting magnetic field of the ferromagnet is not quite as strong as it might have been without the intervention of eddy currents.

Eddy currents also represent large heat losses in the core of transformers used to step-up or step-down voltages. Large varying magnetic fields in the primary coil induce varying fields in the secondary coil. Both coils are usually wrapped around a ferromagnetic core to increase the magnetic field. The eddy currents induced in the core can be very large and would represent a serious loss of energy in the conversion from one voltage to another if not controlled. The core, instead of being made from one piece of iron, is typically made from thin sheets of metal separated by insulating layers of paper or thin varnish. This restricts the eddy currents to the thin sheets where there is high resistance to the flow of induced current, which cuts the heat loss. The core laminations have the same effect as the radial slits in the copper disk mentioned above. Bundles of wires and compacted powder also act as transformer cores and produce the same effect.

Applications

Induction or eddy current heating can be used for cooking or for metallurgical heating. This is especially useful when direct heat cannot be applied, for example, to a sample in a vacuum chamber. Polycrystalline materials can be purified by eddy current heating by passing them through a narrow strip of high-frequency coil. The impurities are carried away in the molten metal that passes along under the coil as the crystal moves through, leaving a purified crystal behind. This is known as zone melting or zone refining.

Eddy current brakes are used for industrial saw blades, train brakes, and other braking mechanisms involving rapidly rotating metal objects.

An eddy current tachometer is widely used in car speedometers. A magnet rotates with the main shaft (car axle) and the eddy currents induced in it are proportional to the rate of angular rotation of the shaft. Thus, by measuring the induced current, one can calculate the speed of the car.

Metal detectors, such as those used in airports, are based on eddy current detection. In Geo-surveying, large alternating currents between 500 and 5,000 Hz are sent into the ground to induce eddy currents in metallic deposits. A voltage is then induced in a receiving coil that is out of phase with the primary coil sending the signal. The phase difference between the two voltages can be used to locate ore deposits underground.

See also: CONDUCTOR; ELECTROMAGNETIC INDUCTION; ELECTROMOTIVE FORCE; FERROMAGNETISM; FIELD, MAGNETIC; LENZ'S LAW; TRANSFORMER

Bibliography

HALLIDAY, D.; RESNICK, R.; and KRANE, K. S. *Physics*, 4th ed. (Wiley, New York, 1992).

SMYTHE, W. R. *Static and Dynamic Electricity*, 3rd ed. (Hemisphere, New York, 1989).

HEIDI FEARN

CYCLOTRON

One of the major physical problems of the 1920s and 1930s was production of protons of high energy so that the composition of the nucleus could be investigated. Particles are given energy as they move through a region of potential difference (a particle of charge $+e$ increases its energy in a potential difference of 1.0 V by 1.0 eV, 1.6×10^{-19} J, of energy). One could try to get particles of high energy by achieving a very high potential difference through which the particle could be sent, but this was a very difficult enterprise with then-extant technology above about 250 kV. Attempts were carried out by Robert Van de Graaff and by Merle Tuve that finally in the mid-1930s did result in such high voltages.

In early 1929 Ernest O. Lawrence of the University of California, Berkeley, chanced on an article by Norwegian engineer Ralf Wideröe in the journal *Archiv für Electrotechnic,* based on an idea of the Swedish physicist Gustav Ising, for accelerating particles successively through the same potential difference. By shielding the particle from the field through part of the cycle of oscillation, this method allows a relatively small oscillating potential difference to accelerate the same particle again and again.

Lawrence immediately saw the application of the magnetic force relation in the Lorentz force to this process, and magnetic resonance acceleration, in a machine later dubbed the cyclotron, was born. The idea of using magnetic fields and circular orbits had been discovered independently by many researchers. However, according to Lawrence's graduate student M. Stanley Livingston, Lawrence was the first and only one to have enough confidence to try out the idea.

Cyclotron resonance is based on the fact that a particle with charge p, mass m, and velocity $\mathbf{v}$ moving in a uniform magnetic field $\mathbf{B}$ experiences a Lorentz force $\mathbf{F} = q(\mathbf{v} \times \mathbf{B})$, which is perpendicular to $\mathbf{v}$ and results in circular motion. If the particle moves in a circle in a region of the magnetic field $\mathbf{B}$ perpendicular to $\mathbf{v},$ it is subjected to the centripetal force whose magnitude is given by $F = qvB$. However, particles traveling in a circle are subject to an acceleration $\omega = v^2/r$, requiring a net force $F_{\text{to center}} = mv^2/r$, where r is the arc radius. This force must be the magnetic force, so $qvB = mv^2/r = mv\omega$. Thus, one finds a rotation angular frequency $\omega = qB/m$, which is independent of the radius.

With this in mind, one has only to supply a source of ions at the center of the device and an oscillating electric field. The particles are shielded (using two shields called dees) from the field except for a small opening between the dees. When particles pass into the opening, the external field accelerates them. Every time a particle appears at the opening, its energy is increased. Inside the dees, the particles are forced into a circular trajectory by the magnetic field. The electric field frequency ($f = \omega/2\pi = qB/2\pi m$) is set to produce electric kicks on the particles every time they are in the opening. In this way, the small amplitude oscillating field can produce a particle with large kinetic energy.

In March 1931, in the second cyclotron ever made (an 11-in. cyclotron), Lawrence and Livingston produced a very low current of 80-keV protons from a 2-kV field source. Characteristically, Lawrence said at the time that 1 MeV protons were just a matter of continuing development of the cyclotron. He did achieve this goal just a few years later.

There were two key developments that allowed Lawrence and Livingston to produce high-energy protons; both had to do with focusing. In the original design, there was a grid at the entrance to each dee. The beam current was very low. It was found that, after the grids were removed, there was a net electric focusing of the beam into the central plane of the cyclotron caused by the field lines between the dees. This electric focusing worked only when the protons were at low energy; high-energy protons went through the gap too quickly. The current was still too low, but next they found that metal shims inserted toward the center of the magnet caused the magnetic field to be weaker at the edges than the center; this also caused a net force toward the plane of the beam near the outer edge of the cyclotron, meaning that magnetic focusing had occurred.

Lawrence, a successful fundraiser, built many cyclotrons with private and foundation money, including the 37-in. (1935), the 60-in. (1939), and the 184-in. (1946). Cyclotrons came to be very important in the production of both high-energy x rays for cancer therapy and isotopes for radiomedicine. The elements neptunium and plutonium, as well as the isotope carbon-14, were discovered using the 60-in. cyclotron. This long-lived isotope of carbon allowed living processes to be studied completely nondestructively for the first time and led to the ability to date the age of once-living objects (radiocarbon dating). The production of artificial radioactivity was verified in Lawrence's lab within hours of his learning of its discovery in Europe.

Cyclotrons proliferated across the country and around the world. Simple cyclotrons were operated with a continuous ion input, and relativistic limitations prevented energies much higher than 20 MeV from being achieved. As higher energy was sought, the cyclotron became generally less useful as a tool for research. It was the addition of the idea of phase stability to the cyclotron by Lawrence's coworker Edwin McMillan in the postwar years that gave rise to the truly energetic, but very low current, beams found in the more modern synchrotrons (synchronous cyclotrons).

See also: ACCELERATOR; ACCELERATOR, HISTORY OF; CYCLOTRON RESONANCE; LAWRENCE, ERNEST ORLANDO

Bibliography

BRYANT, P. J. , and JOHNSON, K. *The Principles of Circular Accelerators and Storage Rings* (Cambridge University Press, New York, 1993).

KOLLATH, R. *Particle Accelerators* (Pitman and Sons, London, 1967).

LIVINGOOD, J. J. *Principles of Cyclic Particle Accelerators* (D. Van Nostrand, Princeton, NJ, 1961).

WIDERÖE, R. *Arch. Electrotech.* **21**, 387 (1928).

GORDON J. AUBRECHT II

CYCLOTRON RESONANCE

Cyclotron resonance deals with the interaction of electromagnetic radiation and charged particles moving in a magnetic field. Charged particles in a magnetic field follow either circular or helical trajectories about the field lines with a characteristic frequency, ω_c, called the cyclotron frequency. When electromagnetic radiation with this frequency is present, resonance between the radiation and circulating charges can occur leading to effects such as reflection or absorption of the radiation. This is called cyclotron resonance and is important in problems of solid-state, plasma, and atmospheric physics.

Particles with charge q, mass m, and velocity **v** moving in a uniform magnetic field **B** experience a Lorentz force:

$$\mathbf{F} = q\mathbf{v} \times \mathbf{B},$$

which is perpendicular to both **v** and **B**. The component of **v** along **B** remains unchanged whereas that perpendicular to **B**, denoted $v_\perp$, gives rise to a centripetal force with magnitude

$$F = m\frac{v_\perp^2}{r} = qv_\perp B,$$

and circular motion of radius r in planes transverse to **B**. The frequency of this circular motion is

$$\omega_c = 2\pi\frac{v_\perp}{2\pi r} = \frac{qB}{m}.$$

In general, charged particles follow helical trajectories; however, when **v**⊥**B**, pure uniform circular motion results. Note that ω_c depends on B and not on r or **v**. Particles with higher transverse speeds therefore will have orbits with larger radii. If **B** is nonuniform, then particles in helical motion about the field lines will increase or decrease their orbital radii inversely with B.

Similar motions occur when electrons or holes within semiconductors and metals, and charged particles within a plasma are exposed to magnetic fields. Interesting effects arise when electromagnetic radiation, of frequency ω, interacts with these systems. When $\omega = \omega_c$, resonant absorption of the radiation by electrons (or holes) in a semiconductor occurs. In solids, however, collisions between charge carriers and defects, impurities, or phonons can destroy this resonance. These collisions occur with a characteristic relaxation time, τ, and frequencies for which $\omega_c \tau \gg 1$ must be used. Very pure semiconductors at low temperatures yield the best results if microwave frequencies are used, which require magnetic fields from 0.1 to a few tesla.

In practice, radiation of a known frequency, ω, is applied to a sample oriented carefully with respect to a uniform magnetic field. This field is then tuned for the strongest absorption of the radiation using microwave spectroscopy techniques. At resonance, $\omega = \omega_c = eB/m^*$, where m^* is called the electron (or hole) effective mass. By measuring the absorption or resonance peak and the field strength, m^* for electrons or holes at the fermi surface can be determined from $m^* = eB/\omega$. Effective masses in metals can also be measured using cyclotron resonance although it is more difficult in practice due to high metallic reflectivities at microwave frequencies.

Cyclotron resonance is also important in plasma physics. An important example is the reflection of radio waves by Earth's ionosphere. Free electrons in the ionosphere orbit Earth's magnetic field ($\sim$0.3 G)

with $\omega_c \approx 10^6 - 10^7$ Hz. Radio waves resonating with these frequencies undergo reflection. This provides for both a useful technique for probing the height and density of the ionosphere and radio transmissions over surprisingly long distances.

See also: CENTRIPETAL FORCE; FIELD, MAGNETIC; IONOSPHERE; LORENTZ FORCE; PLASMA PHYSICS; RESONANCE

Bibliography

ASHCROFT, N. W., and MERMIN, N. D. *Solid-State Physics* (Saunders, Philadelphia 1976).

BLAKEMORE, J. S. *Solid-State Physics,* 2nd ed. (Saunders, Philadelphia, 1974).

BURNS, G. *Solid-State Physics* (Academic Press, San Diego, 1990).

KITTEL, C. *Introduction to Solid-State Physics,* 6th ed. (Wiley, New York, 1986).

William F. Oliver III

D

DAMPING

Damped oscillations describe how the dynamics of a system, particularly amplitude, will respond to a given input. Modeling the dynamics of a system with differential equations can provide critical clues to how a system will react as well as what physical properties need to be changed to improve performance.

There are many physical phenomena that exhibit damped oscillations. Examples include a plucked guitar string, the motion of a playground swing, and the change in voltage of an electronic filter. The dynamics or change with respect to time of these systems are described mathematically by a second-order differential equation such as the following one:

$$a\frac{d^2f(t)}{dt^2} + b\frac{df(t)}{dt} + cf(t) = 0,$$

where $f(t)$ is a function of the dependent variable over time, and the parameters a, b, and c are constants derived from the physics of the situation.

To gain insight into the dynamics of these systems and the differential equation describing those dynamics, a playground swing will be used as an example. Figure 1 shows a child ready to swing. The parent releases the swing at a distance x_0 at time t_0. Figure 2 is a plot of the swing's position with respect to time. This graph is a description of the dynamics of the system. The swing initially overshoots the resting position. It then proceeds to oscillate about the resting position until it comes to a stop. The forces acting on the swing (e.g., gravity, friction, the normal force of the rope) all contribute to the second-order differential equation describing its motion.

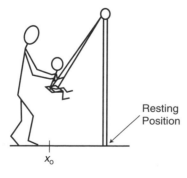

Figure 1

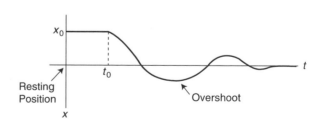

Figure 2

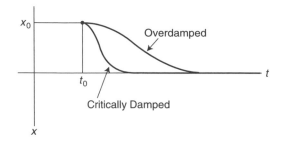

Figure 3

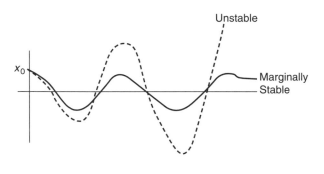

Figure 4

Bibliography

FEYNMAN, R. P.; LEIGHTON, R. B.; SANDS, M. *The Feynman Lectures on Physics,* Vol. 1 (Addison-Wesley, Reading, MA, 1963).

KREYSZIG, E. *Advanced Engineering Mathematics,* 4th ed. (Wiley, New York, 1979).

G. L. ZIMMERMAN

The swing in this example is called an underdamped system and is characterized when the parameters of the system satisfy $b < 2\sqrt{ac}$. The damping is due to frictional forces.

If the frictional forces are great enough, a point would be reached where the system's parameters are $b = 2\sqrt{ac}$. This is called a critically damped system; the swing, when released, would return to the resting position without any overshoot. If $b > 2\sqrt{ac}$, the system is said to be overdamped; when released, it would return to its resting position, but more time is needed to do it. Both of these situations are shown in Fig. 3. In the swing example, this type of damping might occur if the hinges of the swing were really rusty or if the whole swing was immersed in liquid.

There are situations when energy is being added to the system that can equal or surpass the damping forces. When this happens, the system parameters are such that $ac < 0$. In this case, the dynamics of the system can become either marginally stable or unstable, as shown in Fig. 4. An unstable system can have profound consequences on the survival of the whole system.

See also: FRICTION; OSCILLATION; OSCILLATOR, ANHARMONIC; OSCILLATOR, FORCED; OSCILLATOR, HARMONIC

DARK MATTER

Dark matter is material that cannot be seen. Until very recently, astronomers had no way of knowing anything about dark matter because their view of the cosmos was restricted to objects that emitted light. With the advent of more sophisticated instruments and techniques, many astronomers now believe not only that dark matter exists, but that it dominates our universe. Even more remarkable, there is a fair amount of evidence that the dark matter is not the neutrons and protons and electrons of which we and everything around us are composed. Rather, most of the material in the universe appears to be some new form of matter.

Where is the dark matter? It appears to be ubiquitous. Even a small region encompassing our solar system and a hundred other stars contains as much dark matter as it does visible stars. Dark matter makes up close to ninety percent of a typical galaxy's mass. On larger scales, dark matter is the primary component in clusters of hundreds of galaxies. Finally, on the largest scales, one can argue that, besides the mass in clumped objects such as galaxies and clusters, there is a smooth component of mass permeating the universe. This argument implies that the vast regions of space in the universe that appear dark are not empty but rather contain more mass than all the galaxies and clusters that light up our night sky.

Evidence for Dark Matter

Astronomers gain information about dark matter by measuring the mass of a system gravitationally. Such a measurement "weighs" all objects in a galaxy, for example, without worrying about the nature of the objects. A bright star is counted as much as a faint star if they both have the same mass. Similarly, dark matter—matter that does not emit light at all—weighs in as forcefully as a star with similar mass. This gravitational measurement can then be compared with an optical measurement, where the system is weighed by adding up the visible objects in it. If these two measurements disagree, then dark matter must be contributing to the gravitational mass but not to the optical mass.

The most compelling evidence for dark matter comes from applying this comparison to typical galaxies. A galaxy's mass can be measured gravitationally by recording the velocities of its stars. Kepler's laws imply that the mass M inside the orbit of a given star can be determined by measuring the distance of the star from the center of the galaxy R and the star's velocity V. Specifically,

$$M = \frac{RV^2}{G},$$

where G is Newton's constant, which is known experimentally. If the velocities of stars are very large, then the mass in the galaxy must be very large; indeed, it is the large gravitational force due to this mass that causes the stars to move so rapidly.

Kepler's law is so powerful because velocities can be accurately measured by astronomers. To measure velocities, astronomers measure the Doppler effect on photons emitted from a given star. If the star is traveling away from Earth, then we observe longer wavelengths than if the star were at rest. Conversely, light emitted from a star traveling toward Earth has shorter wavelengths than if the star were at rest. This wavelength shift (dubbed a redshift and a blueshift, respectively) can be used to determine the star's velocity with respect to Earth. Thus, using Kepler's law, one can accurately measure the total mass within a given star's orbit around the center of the galaxy.

The masses of most galaxies measured in this fashion are very surprising. As one measures further and further from the center of the galaxy, the mass M continues to increase. This feature persists even as one measures beyond the visible edge of the galaxy. The inescapable conclusion is that, even in the outer regions of a galaxy, where there are no stars shining, there is still a great deal of dark matter contributing to the total mass. A typical galaxy therefore extends much further than do its visible stars. The mass estimates obtained suggest that a typical

dark matter "halo" contains roughly ninety percent of the total galactic mass.

Dark matter also exists inside clusters of hundreds or thousands of galaxies. Using techniques described above, astronomers have shown that the mass of all the stars in the Coma cluster is much smaller than the total mass of the cluster. There is quite a bit of hot gas in such clusters—this is established by observing the x rays emitted by the hot gas—but even this gas falls short of accounting for all of the mass in the cluster. So we conclude that dark matter dominates not only individual galaxies but also huge clusters of galaxies.

Even on the largest scales in our observable universe, one may argue that dark matter prevails. One can calculate the average energy density (or mass density) in the universe and compare this with the critical density above which the universe is closed. This ratio is known as Ω. The total mass in clusters of galaxies can be estimated and compared with the critical density. This estimate yields a value of $\Omega \simeq 0.2$. Such a low value is disturbing because the most natural value of Ω is one. A universe starting at early times with Ω close to one quickly finds Ω dropping to zero. A value of $\Omega = 0.2$ therefore seems very unlikely to many scientists. Sticking with the aesthetically appealing $\Omega = 1$ leads one to the conclusion that, even apart from the dark matter in galaxies and clusters, there is a smooth component of dark matter permeating the universe. This smooth component contains roughly four times as much mass as does the sum of all galaxies and clusters.

Cosmologists who study the formation of structure in the universe have advanced yet another argument for dark matter. This argument is posed within the framework of the big bang model of the universe. Within this framework, when the universe was very young, it was very smooth and very dense. Since it was smooth, the temperature everywhere was almost uniform. Since it was dense, photons interacted very frequently with electrons and therefore shared the same temperature. Any place where the photon temperature was slightly larger than average also contained slightly more electrons (and protons) than average. Regions with more protons and electrons than average accreted more and more matter due to gravity until they ultimately grew into the large structures (such as clusters and galaxies) we see today. Cosmologists can calculate how large the initial inhomogeneity had to be to have grown into a structure as large as a galaxy today. They find that at the time when the photons and electrons stopped interacting, there had to have been roughly one extra proton for every thousand. This translates into a photon temperature hotter than average by one part in a thousand. Today we observe photons that traveled to us from that epoch at which they last interacted with electrons. The Cosmic Background Explorer (COBE) satellite and other experiments *do* observe regions that are slightly hotter than average. However, they are hotter by only one part in one hundred thousand. This is one hundred times smaller than expected. Once again, the only way out of this conundrum seems to be to invoke dark matter. The dark matter could have been much more inhomogeneous than the electrons and protons at the time of last scattering. After the electrons stopped scattering, they quickly fell into the relatively large inhomogeneities already present in the dark matter.

What Is the Dark Matter?

The dark matter in the universe is almost certainly electrically neutral. If not, the charged particles would quickly lose energy by colliding with each other. In so doing, they would rapidly fall into the central region of a galaxy: the extended halo would not exist.

A natural candidate for the dark matter is a population of low-mass stars. Low-mass stars would not shine brightly enough to be observed, but if there were enough of them, they could contribute significantly to the total mass of a galaxy. In fact, using the technique of gravitational lensing, astronomers in 1993 detected several low-mass objects (about ten times less massive than our sun) in our galactic halo.

However, as more data from these lensing experiments has become available, it now appears quite likely that there are not enough of these low-mass objects (dubbed MACHOs, massive compact halo objects) in our halo to account for all of its mass. Indeed, cosmologists in 1984 showed that the dark matter is probably not low-mass stars consisting of ordinary electrons, protons, and neutrons. They based their argument on the successes of big bang nucleosynthesis (BBN). The abundance of light elements (hydrogen, deuterium, helium, and lithium) in the universe is predicted by BBN to be a function of the number of electrons and protons in the universe. For the predictions to agree with observations, it is necessary that the total energy density of these ordinary particles be much less than the critical density. Thus, if BBN is correct and if the observations of light elements are accurate (and this situation has not changed much since the early 1970s, when the argument was first advanced), then the density of ordinary particles cannot even be close to the critical density. In fact, ordinary matter cannot even have a density large enough to account for cluster masses.

Therefore, many scientists have come to the conclusion that the dark matter consists of elementary particles never before observed in a laboratory. There are many candidates for such particles; a successful theory contains a completely new set of particles and forces, beyond the standard model of particles and interactions. Particle physicists have invented many models that contain the basic requirement for a dark matter candidate: a massive, electrically neutral, stable particle. Examples of theories with the required candidate are massive neutrino models, supersymmetry, technicolor, and axion models. It is important to note that all of these theories were developed for other reasons: dark matter was not the motivating factor. Of all the candidates proposed, only the neutrino is known to exist. For the neutrino to be the dark matter, its mass must be roughly 50 eV. There are three types of neutrinos. At present, laboratory experiments have shown that the electron neutrino cannot have a mass this large. The other two types—muon and tau neutrinos—could be the dark matter. Cosmologists have studied this possibility carefully and now consider it very unlikely. A 50-eV neutrino would be traveling very rapidly (it is called hot dark matter) so it would easily escape from overdense regions. Thus, it would be very difficult for structure to form in a universe dominated by massive neutrinos. Similarly, neutrinos would be traveling too fast to be trapped in small galaxies. There would have to be another dark matter component in these galaxies.

Dark Matter Detection

Detecting matter that is known to be dark is extremely challenging. Nonetheless, the promise of at once discovering a new form of matter while at the same time discovering the dominant material in the universe has spurred a number of groups to set up experiments searching for dark matter. Some of the methods involve brute force techniques: the MACHO searches described above have filtered through literally millions of stars in order to obtain just a handful of MACHO candidates. Other experiments involve delicate transfers of energy from a dark matter particle whizzing by Earth to a large detector, wherein a single nucleus would recoil. There are still other experiments, such as those searching for axions, that invoke profound physics: an axion would resonate in an electromagnetic cavity, giving up its energy to a photon. There are also extremely clever and indirect experiments such as those which look for the products from the annihilation of a pair of dark matter particles in the Sun. Since the identity of the dark matter is unknown, none of these experiments can be guaranteed success. So far, only the MACHO experiments have reported any detections. The other experiments have placed limits on the mass of the dark matter candidate under investigation.

See also: BIG BANG THEORY; COSMIC BACKGROUND EXPLORER SATELLITE; COSMOLOGY; COSMOLOGY, INFLATIONARY; DARK MATTER, BARYONIC; DARK MATTER, COLD; DARK MATTER, HOT; KEPLER'S LAWS; NEUTRINO; UNIVERSE, EXPANSION OF; UNIVERSE, EXPANSION OF, DISCOVERY OF

Bibliography

KOLB, E. W., and TURNER, M. S. *The Early Universe* (Addison-Wesley, Reading, MA, 1990).
RUBIN, V. "Dark Matter in Spiral Galaxies." *Sci. Am.* **248** (6) 96–108 (1983).
TREMAINE, S. "The Dynamical Evidence for Dark Matter." *Phys. Today* **2,** 28–36 (1992).

SCOTT DODELSON

DARK MATTER, BARYONIC

Analysis of the internal motions of galaxies and of clusters of galaxies has revealed the presence of substantial amounts of dark matter in the universe. This nonluminous matter amounts to perhaps 90 percent of the total mass in these systems and of the universe as a whole. Observations indicate that a typical spiral galaxy like our own Milky Way is embedded in a halo of dark matter that extends well beyond the visible disk of stars.

By definition, dark matter does not emit electromagnetic radiation: its existence is inferred solely from its gravitational effects on luminous objects. Since the 1970s, astronomers have gathered abundant evidence about the quantity and spatial distribution of dark matter in the universe, but its composition is as yet completely unknown.

Astrophysicists distinguish between two broad categories of dark matter candidates: baryonic, composed primarily of protons and neutrons, or more fundamentally of quarks; and nonbaryonic, some hypothetical elementary particle that interacts only very weakly with ordinary matter.

One candidate for baryonic dark matter in galaxy halos are brown dwarfs or large planets with mass less than about one-tenth of the Sun's mass. Such substellar objects do not achieve sufficiently high internal temperatures to ignite the nuclear fusion of hydrogen into helium—the power source that causes most stars to shine—so they are intrinsically very faint. Other baryonic dark matter possibilities include old, massive stars that have used up their nuclear fuel and collapsed to form white dwarfs, neutron stars, or black holes. These faint stellar and substellar remnants have been collectively dubbed MACHOs (for massive astrophysical compact halo objects).

Independent evidence for baryonic dark matter comes from cosmology. Astronomers conventionally measure cosmic density in terms of the critical density the universe would have if it marginally continues expanding in the indefinite future against the contracting pull of gravity. Primordial nucleosynthesis, the generation of helium, deuterium, and lithium in the first 3 minutes of the big bang, suggests that the cosmic density of baryonic matter is between 1 percent and 10 percent of the critical density. (If the baryon density were outside this range, the predictions of primordial nucleosynthesis would not agree with observations of the abundances of these light elements.) Visible galaxies contribute less than about seven-tenths of 1 percent of the critical density, suggesting that at least some of the dark matter is baryonic. In fact, the dark matter inferred in galaxy halos contributes roughly 2 to 5 percent of the critical density and could thus be made entirely of baryons. Beyond the scale of galaxies, however, a greater density of dark matter is indicated, of order 20 percent of critical or more; it would likely be nonbaryonic in nature. On the largest scales, if the universe has critical density (as predicted in the inflation hypothesis), then the predictions of nucleosynthesis argue strongly for nonbaryonic matter.

Whether the dark matter is predominantly baryonic or of more exotic nonbaryonic composition (or a combination of the two) will ultimately be decided by experiment. MACHOs in the Milky Way amplify the brightness of more distant stars when they pass near the stellar line of sight, a phenomenon known as gravitational microlensing. In 1993, three experimental groups carefully monitoring the brightness of several million distant stars announced the first discoveries of microlensing events. The preliminary indication from the number of microlensing events observed since then is that MACHOs contribute about 30 percent of the dark matter in the halo of our galaxy.

See also: ASTROPHYSICS; BARYON NUMBER; COSMOLOGY; DARK MATTER; DARK MATTER, COLD; DARK MATTER, HOT; GALAXIES AND GALACTIC STRUCTURE; MILKY WAY; NUCLEOSYNTHESIS

Bibliography

LYNDEN-BELL, D., and GILMORE, G., eds. *Baryonic Dark Matter* (Kluwer, Dordrecht, 1990).

JOSHUA FRIEMAN

DARK MATTER, COLD

The dynamics of galaxies and of clusters of galaxies indicate that upwards of 90 percent of the mass of the universe is "dark." While astronomers have gathered abundant evidence about the quantity and spatial distribution of the dark matter, its composition is as yet completely unknown.

Astrophysicists distinguish between two broad categories of dark matter candidates: baryonic, composed primarily of protons and neutrons; and nonbaryonic, some hypothetical elementary particle that interacts only weakly with ordinary matter. Nonbaryonic dark matter candidates have been further subdivided into two classes: hot and cold dark matter. Hot dark matter particles were relativistic, traveling at nearly the speed of light at the time when galaxies are thought to have begun forming, about 10,000 years after the big bang. Cold dark matter

particles would instead have been nonrelativistic, moving at much lower speeds, at that time.

Hot and cold dark matter are distinguished by their different implications for the formation of galaxies and larger structures. It is thought that galaxies form by the amplification of small primordial fluctuations in the density of the expanding universe: Regions where the density was slightly higher than average expanded more slowly, eventually collapsing to form galaxies and clusters. Since cold dark matter particles were initially moving slowly, they could begin clumping gravitationally quite early. In contrast, for hot dark matter, galaxy-sized fluctuations are impeded from collapsing until recently in time and must form from the fragmentation of larger objects. With cold dark matter, the sequence is reversed: smaller objects collapse first, subsequently merging together to form larger systems. Cosmologists have also recently found that a combination of 20 percent hot and 80 percent cold dark matter can perhaps provide an empirically acceptable scenario for galaxy formation.

There are at least two theoretically well-motivated candidates for cold dark matter. Theories of supersymmetry hypothesize that every particle in the standard model of particles and interactions has a supersymmetric partner. In particular, every known boson has a fermionic partner; the lightest of these partners is a prime candidate for cold dark matter. The expected mass of this supersymmetric particle is in the range from about 10 GeV to several hundred giga-electron-volts (compared with the proton mass of just below 1 GeV). Moreover, for supersymmetric particle masses in this range, the present mass density of cold dark matter would be comparable to the critical density the universe would have if it marginally continues expanding in the indefinite future against the contracting pull of gravity. (If the density of the universe is larger than critical, the universe will eventually recollapse to a singularity; if smaller, it will expand forever.) A number of experiments are currently searching for cold dark matter: When a supersymmetric particle from the halo of our galaxy strikes a nucleus in an underground detector on Earth, it deposits a small but measurable amount of energy.

The second cold dark matter candidate currently under active study is the axion, a very light Nambu–Goldstone boson predicted in theoretical extensions of the model of strong interactions. If the axion rest mass is about $1/100,000$ eV, then its cosmic density is also predicted to be close to the critical density. Experiments to search for cosmic axions are currently underway. One experiment is deploying a tunable microwave cavity in a strong magnetic field and search for the expected conversion of axions to electromagnetic waves (photons) in the presence of the strong field.

See also: BIG BANG THEORY; BOSON, NAMBU–GOLDSTONE; DARK MATTER; DARK MATTER, BARYONIC; DARK MATTER, HOT; SUPERSYMMETRY

Bibliography

RIORDAN, M., and SCHRAMM, D. N. *The Shadows of Creation: Dark Matter and the Structure of the Universe.* (W. H. Freeman, New York, 1991).

JOSHUA FRIEMAN

DARK MATTER, HOT

Astrophysicists distinguish between two broad categories of dark matter candidates: baryonic and nonbaryonic. Nonbaryonic dark matter candidates have been further subdivided into two classes: hot and cold dark matter. Hot dark matter particles were relativistic, traveling at nearly the speed of light at the time when galaxies are thought to have begun forming, about 10,000 years after the big bang.

Hot and cold dark matter have quite different implications for the formation of galaxies and larger structures. It is thought that galaxies form by the amplification of small primordial fluctuations in the density of the expanding universe; regions where the density was slightly higher than average expanded more slowly, eventually collapsing to form galaxies and clusters. Since hot dark matter particles were initially moving so rapidly, they streamed out of small overdense regions, effectively erasing the fluctuations. As a result, the first objects to collapse would have been massive clusters of galaxies, which would have subsequently fragmented into individual galaxies. This process would have happened too late in cosmic history to reproduce the observed population of galaxies, and hot dark matter has therefore fallen out of favor among astrophysicists. However, hot dark matter may be compatible with more complex models of galaxy formation that rely on cosmic strings or other defects to act continuously as seeds for new fluctuations. Cosmologists have also recently found that a combination of 20 percent hot and 80 percent cold dark matter can perhaps provide an empirically acceptable scenario for galaxy formation.

The prototypical hot dark matter candidate is a massive neutrino. Astronomers conventionally measure cosmic density in terms of the critical density the universe would have if it marginally continues expanding in the indefinite future against the contracting pull of gravity. If the density of the universe is larger than critical, the universe will eventually recollapse to a singularity; if smaller, it will expand forever. The big bang cosmology predicts that the mass density of neutrinos is about equal to the critical density if the neutrino rest mass-energy is about 30 eV (about 5×10^{-32} g), less than 1/10,000 the mass of the electron.

In the standard model of particles and interactions, the three neutrino species (electron, muon, and tau) are exactly massless. However, a number of hypothetical extensions of the standard model predict that neutrinos have mass. Experiments studying the radioactive decay of tritium, a heavy isotope of hydrogen, have shown that the electron neutrino mass must be less than 8 eV, but the other two neutrino types can be more than heavy enough to account for the critical density. Future neutrino oscillation experiments, which use neutrino beams generated by high-energy particle accelerators to study the possible transformations between different neutrino types, will place stronger constraints on neutrino masses. Additional information on the neutrino masses will come from the duration of the bursts of neutrinos emitted from the next supernova explosion in our galaxy, which should be seen by large underground detectors now being built.

See also: BIG BANG THEORY; COSMIC STRING; DARK MATTER; DARK MATTER, BARYONIC; DARK MATTER, COLD; NEUTRINO; NEUTRINO, HISTORY OF

Bibliography

RIORDAN, M., and SCHRAMM, D. N. *The Shadows of Creation: Dark Matter and the Structure of the Universe* (W. H. Freeman, New York, 1991).

JOSHUA FRIEMAN

DATA ANALYSIS

It is a common misconception that science is an objective enterprise whose task is to let experimental data "speak for themselves." Were this true, scientific publications would consist of little more than tables of recorded measurements. Instead, a principal task of science is the interpretation of data in a way that allows scientists to generalize what they have learned from a particular set of data so that those lessons have implications for other data. In particular, scientists seek knowledge that will help them make predictions about data not yet observed. The search for such knowledge is an inherently subjective enterprise. Its success depends on scientists' ingenuity in using a specific set of data to create general hypotheses that can be used to make predictions about other data. It also depends on their ability to use old and new data, inevitably imperfect, to test and compare rival hypotheses.

To accomplish these tasks, scientists have devised—and continue to devise—a wide array of tools for manipulating data to aid their intuitions in creating and testing hypotheses. The use of these tools is called data analysis. These tools fall roughly into two categories, depending on whether they are used to guide the creation of new hypotheses or to test existing hypotheses. Exploratory data analysis refers to the use of tools designed to help scientists identify unanticipated features or patterns in data in order to guide the creation of new hypotheses. Statistical data analysis refers to the use of the mathematical theory of probability to assess the compatibility of hypotheses with data, taking into account experimental error.

Exploratory data analysis involves the interactive, iterative use of informal tools and techniques; there is no "theory" of exploratory analysis. Exploratory techniques typically combine numerical manipulation of data with visual display of the results, guided by very general considerations of the kinds of patterns that might be present. Underlying many exploratory methods is the notion that data can often be modeled as a combination of two or more parts, with one part usually of greater interest than the other(s). The analyst usually has some amount of knowledge about the uninteresting components and uses this knowledge to create procedures that, hopefully, can isolate and magnify the interesting component, about which little may be known before consideration of the data.

One of the most common goals of exploratory methods is the removal of any distortions or artefacts introduced into data by imperfections in the experiment that produced them. Although of limited intrinsic interest, these imperfections can mask or mimic interesting features in data. Thus identifying, quantifying, and removing them are among the most important of the data analyst's tasks. Three examples of common imperfections will serve to illustrate the methods used and the kind of information required to perform such analyses.

Many detectors produce nonzero data even when not exposed to a signal source. For example, photomultipliers used to detect very dim light sources in particle physics and astrophysics produce a small current signal even when not exposed to light, due to thermal agitation of electrons in the device. This "dark current" must be taken into account in analysis of photomultiplier data. In this case, and more generally, the analyst can model data obtained in the presence of a signal as the sum of an uninteresting background component and the signal of interest. Careful measurement of the detector-induced background can enable the analyst to reliably subtract its contribution from the data, thereby isolating the signal.

Alternatively, many detectors have a sensitivity that varies with some property of the signal. For example, imperfections in the optical system and detector of an astronomical camera often lead to light from different directions being detected with differing sensitivity or efficiency. In this case, it is appropriate to model the data as the product of the signal and an efficiency function that depends on position in the image. By carefully measuring images of artificially produced sources that are uniformly illuminated (called flat fields), an astronomer can measure the efficiency and factor it out of the data.

Finally, many detectors produced "blurred" renditions of the signal. An astronomical camera again provides a good example; the blurring may be due to imperfect optics or to atmospheric turbulence (such as that leading to the "twinkling" of stars). But even nonimaging detectors produce similar "blurring": a scintillator used to detect gamma rays or charged particles can produce signals of various measured energies when a monoenergetic beam is incident upon it, thereby "blurring" the energy of the detected quanta. Mathematically, the data is said to be the convolution of the signal with a point-spread function that describes the blurring. Both for cameras and scintillators, the first step in accounting for blurring is careful measurement of the induced blurring (by observing a known point source with the camera or a known monoenergetic source with the scintillator). A variety of mathematical inverse methods exist that can take this information and use it to at least partially remove the blurring from the data.

Once imperfections are removed, data often require further analysis into component parts to isolate potentially interesting features. One of the most common and general models for data considers each of a sequence of measured values to be the sum of a constant or slowly varying signal component and a randomly varying noise component. Usually the signal component is of most interest, so the analyst seeks to isolate it from the noise component, and perhaps examine whether it is constant or whether it varies across the measurements. To do this, some knowledge about the noise is required. For example, previous experience with a piece of equipment often reveals that sources of noise tend to artificially increase a measured value as frequently as they tend to decrease it. This suggests that averaging a group of measurements may decrease the noise contribution, as the equally frequent increments and decrements may tend to cancel each other. By averaging a subset of the measurements and comparing the result to the averages of other subsets, the analyst might detect variation in the signal that would otherwise be masked by the noise. Patterns in the variation may then suggest one or more hypotheses for the source of the variation or for a mathematical description of the signal, hypotheses that could then be further analyzed with statistical methods.

The kinds of signal variations the analyst anticipates will dictate details of the averaging procedure. For example, one analyst wishing to determine if the signal is steadily increasing or decreasing might compare the average of a group of measurements from the beginning of the data set to an average of a group at the end. Another analyst, searching for periodic variation, may use averages of data points separated by a trial period in order to isolate such variation. Whether an analyst finds a particular type of variation depends not only on the presence of the variation in the data, but also on the type of analysis chosen. For example, a significant periodic variation masked by noise could well be overlooked if the data is analyzed only with tools designed to identify steady trends.

In many cases the signal itself may consist of several components. These may be combined in a sum, product, convolution, or some more complicated manner. The same methods used to correct for instrumental distortions may be suitable for trying to isolate the various components comprising the signal. In correcting for instrumental effects, these methods all relied on the availability of precise measurements of the distortion introduced by the instrument alone. To apply them to the analysis of a multicomponent signal, the analyst must similarly obtain information about one or more of the signal components.

In the most common situation the signal can be considered to be the sum of an uninteresting background component and an interesting source component. Particle detectors intended for operation in the presence of an accelerator beam or a radioactive source often produce a signal when the beam is off or the source removed, due to detection of cosmic rays or their by-products. X-ray and gamma-ray telescopes searching for radiation from an astrophysical object also detect radiation from other sources in the direction of the object, including many distant, unresolved sources that produce a diffuse background of radiation from all directions. Background signals like these may be large, masking the signal of interest. Even when they are small in amplitude, their variation in time or direction might distort the appearance of the signal. One can often obtain the information needed to account for such backgrounds by taking measurements in the absence of a source. For example, a particle detector might be operated in the absence of a source in order to provide measurements of the rate of detection of background cosmic rays; or a telescope can be pointed away from a possible source in order to provide a measurement of the diffuse background.

In other cases, one might have little precise information about the background apart from the knowledge or presumption that it varies on different scales than does the signal. Most commonly, one presumes the background to vary slowly, on the largest scales, with the signal of interest giving rise to the more rapidly varying details in the data. An analyst might devise exploratory tools that seek to identify and remove smoothly varying background components from the data in an effort to isolate a small, potentially significant source component. A wide variety of such tools exist. Some tools mathematically "blur" or smooth the data to smear out the detailed component. The smoothed data is then used as an estimate of the background and subtracted from the original data, hopefully leaving only the detailed source component for further analysis. Other tools attempt to "filter" the data—mathematically separate components that vary on different scales—through the use of Fourier transforms, wavelets, or other specialized mathematical operations. Then the rapidly varying parts of the data can be isolated for further study. Finally, one might find some smoothly varying mathematical function that can be adjusted to fit the data, and examine the residuals resulting from subtracting the best-fit function to see if any patterns or features reveal the presence and nature of a source component.

The tools just mentioned—background subtraction, efficiency correction, inversion, averaging, smoothing, filtering, fitting—all involve numerical manipulation of data. Since the primary function of exploratory tools is to aid the analyst in creating new hypotheses, a crucial part of exploratory data analysis is the visual display of data in a manner that helps the intuition identify unanticipated features. To this end, scientists have developed many techniques for displaying data, ranging from simple graphs and histograms to three dimensional color video representations. These techniques can make subtle features that would evade detection if presented numerically readily visible to the eye.

Once the analyst has specified hypotheses that might account for the data (found by applying exploratory methods to the data, by analysis of other data, or as a result of theoretical arguments), statistical analysis can be used to assess the viability of the hypotheses. In many respects, statistical data analysis takes the opposite view of exploratory analysis: rather than trying to separate the data into components in order to isolate potentially interesting parts of the data, statistical methods start with a hypothesized signal, and combine it with knowledge of the experiment to try to predict as much of the data as possible, incorporating any known backgrounds or distortions.

Two approaches are available for statistical analysis: frequentist and Bayesian. The most common approach is the frequentist one. Frequentist methods require the analyst to specify a procedure that selects from among the specified hypotheses one that best explains the data. Presuming that the chosen hypothesis is correct, the analyst can then simulate many hypothetical data sets that could be produced by the chosen hypothesis (by introducing simulated sources of experimental error), and apply the selection procedure to each of the simulated data sets. In some cases, the procedure will correctly identify the

hypothesis, but in others it will not, due to experimental error. The analysis will determine how well the chosen procedure behaves in the long run; this provides a measure of how confident the analyst can be in the results found by applying the procedure to the one actual data set.

In this frequentist approach, there is no formal theory specifying what procedure to use to select the best hypothesis. Often some simple modification of the exploratory methods used to study the data will be used, usually by minimizing some measure of misfit between data modified by exploratory methods and error-free data predicted by the hypotheses. Misfit is most often quantified by calculating residuals. Suppose there are N values of data available, denoted by d_i, with $i = 1$ to N. If we denote the error-free data predicted by some hypothesized model for the signal and the experiment by m_i, then the N quantities $(d_i - m_i)$ are called residuals—the part of the data unaccounted for by the hypothesis. Detailed knowledge of the experiment will usually also provide some estimate of the experimental error present in each measurement, usually in the form of a root-mean-square error, denoted σ_i. One of the most common measures of misfit is the sum of the residuals divided by their average size, and then squared (so all nonzero residuals contribute a positive amount of misfit to the sum):

$$S^2 = \sum_{i=1}^{N} \frac{(d_i - m_i)^2}{\sigma_i^2}.$$

Since on the average we expect $(d_i - m_i)^2$ to approximately equal σ_i^2, for a successful hypothesis the value of S^2 should not be too far from N. By using probability theory (either analytically or through simulation, as described above), an analyst can find more formal constraints on the size of S^2 and use them to test hypotheses.

In the Bayesian approach to statistical data analysis, specification of the hypoteses and the sources of experimental error automatically identifies the functions of the data that are relevant for assessing the hypotheses. These may be quite different from the functions or procedures used in exploratory analyses or standard frequentist analyses. For example, the best estimate of a signal contaminated by background may not be simply the difference between the data and a best estimate of the background, particularly when the signal is required to be nonnegative (as are many physical quantities measured in experiments, such as light intensity, mass, or particle production rates). When the signal is weak, simple subtraction can easily result in negative signal estimates because it is easy for experimental error to cause slight overestimation of the background. Bayesian methods produce more complicated procedures in such cases that guarantee a nonnegative result. Similarly, for some sources of experimental error, averaging the results of several measurements of a quantity does not necessarily reduce the noise contribution, even when the noise is known to lead to increments and decrements with equal frequency. Bayesian methods automatically determine whether an average or some more complicated combination of the data will best reduce the noise contribution, provided the analyst has accurately described the sources of experimental error. This power comes at the cost of greater complexity and less generality, which may partly explain why frequentist methods remain the favored approach. Nevertheless, Bayesian methods offer the important lesson that the exploratory tools that best aid the intuition in creating hypotheses may not be optimal for assessing those hypotheses once they have been created.

Exploratory and statistical methods strongly interact. Statistical analysis is often used to confirm hypotheses suggested by an exploratory analysis; and an exploratory analysis of the residuals used in a statistical analysis may detect patterns suggesting alternative, superior hypotheses. Finally, both methods share a similar subjectivity: their successful application depends crucially on the analyst's creativity and awareness of the assumptions explicitly or implicitly built into the methods used. An exploratory method can lead one to erroneously conclude that features are present if distortions are not carefully eliminated, or it can miss an important feature if the analyst has not at least partially anticipated the characteristics of the feature. Similarly, a statistical method can lead one to accept or reject a hypothesis incorrectly if it is based on inaccurate probabilities for experimental errors, or if the analyst has not been careful in specifying the hypotheses from which the procedure must choose. There is no mathematical substitute for good judgment and an open mind.

See also: DIMENSIONAL ANALYSIS; ERROR, EXPERIMENTAL; ERROR, RANDOM; ERROR, SYSTEMATIC; ERROR AND FRAUD; EXPERIMENTAL PHYSICS

Bibliography

EADIE, W. T.; DRIJARD, D.; JAMES, F. E.; ROOS, M.; and SADOULET, B. *Statistical Methods in Experimental Physics* (North-Holland, Amsterdam, 1971).

HARTWIG, F., and DEARING, B. E. *Exploratory Data Analysis: Quantitative Applications in the Social Sciences,* Vol. 16 (Sage Publications, Beverly Hills, CA, 1979).

LOREDO, T. J. "The Promise of Bayesian Inference for Astrophysics" in *Statistical Challenges in Modern Astronomy,* edited by E. D. Feigelson and G. J. Babu (Springer-Verlag, New York, 1992).

TAYLOR, J. R. *An Introduction to Error Analysis: The Study of Uncertainties in Physical Measurements* (University Science Books, Mill Valley, CA, 1982).

TUKEY, J. W. *Exploratory Data Analysis* (Addison-Wesley, Reading, MA, 1977).

THOMAS J. LOREDO

DAVISSON–GERMER EXPERIMENT

"The investigation reported in this paper was begun as the result of an accident which occurred in this laboratory in April 1925." This celebrated line opens the article by Clinton Joseph Davisson and Lester Halbert Germer describing their experiment in *The Physical Review,* December 1927, in the "Diffraction of Electrons by a Crystal of Nickel." Davisson and Germer's discovery of low energy electron diffraction was one of the irrefutable proofs that electrons possess a wave-like nature.

There were two developments in the 1920s that led to the demonstration of electron diffraction: one technological and one theoretical. Davisson was in a unique position to take advantage of both. The technology that would prove necessary to observe electron diffraction originated from research begun in 1917 when Davisson started work at the Western Electric Company (now the Bell Telephone Laboratories). During World War I he was responsible for the development of thermionic high vacuum tubes for military applications such as communication. In 1919, after the end of World War I, he was allowed the freedom to pursue basic research. At that time he was joined by Lester Germer, and a year later by Charles Kunsman.

The object of Davisson's initial research was the triode vacuum tube, invented by Lee DeForest in

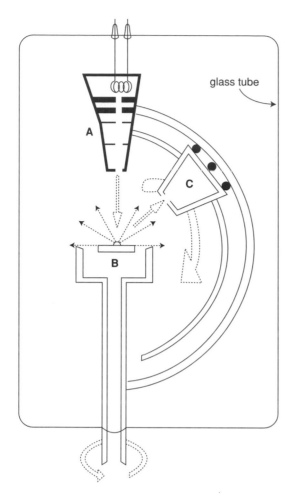

Figure 1 Experimental arrangement for vacuum tube research.

1910. Between 1919 and 1927, this study evolved into the experimental arrangement depicted in Fig. 1. There is a source of monoenergetic electrons A (the *primary* beam), the metallic target B, and a collecter of electrons at point C (the *secondary* beam). The entire arrangement is sealed into a glass tube and brought to a high vacuum.

In 1920, a curious effect was noticed that took hold of Davisson's attention. Davisson and Kunsman found that, among the mostly low energy electrons collected at point C, there were some secondary electrons collected at the point C that possessed the same energy as the primary electrons. Davisson interpreted it to be primary electrons that were scattered through a very large angle. Subsequent experiments focused only on these elastically scattered electrons.

The famous accident occurred when a liquid air container that was being used to improve the vacuum inside the glass tube exploded, the glass tube ruptured, and the surface of the target was oxidized. The target (nickel) had been specially prepared and highly polished before being placed into the tube. At the time of the accident Kunsman had left the laboratory and Germer had revived the electron scattering studies. Rather than creating a replacement for the broken tube, an attempt was made to reseal the tube and to bring the target sample to a high temperature over several days in order to clean it. When the electron scattering experiments were continued, it was found that the scattering distribution had "completely changed." Davisson and Germer, in an attempt to understand this change, cut open the tube and discovered that a secondary effect of the heating was to form a few large crystals from what was previously several small crystals within the target. A schematic of their results is shown in Fig. 2. The atoms are represented by circles, and the lines are superimposed to highlight the possible

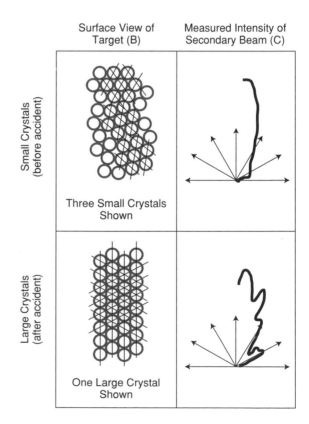

Figure 2 Schematic of the "accident" results.

crystal patterns. The measured intensity is a function of the latitude angle of the detector C.

As a result of the accident, Davisson and Germer realized that the crystal pattern of the target was a significant factor in determining the scattering distribution. Subsequent experiments designed to understand this effect using a single crystal of nickel continued through 1926.

During this time, Germer performed lengthy calculations in order to interpret the experiments. Davisson and Germer assumed, however, that the electron was behaving as a particle. This had been demonstrated by Joseph J. Thomson in 1897. In 1924, Louis de Broglie conjectured that, in analogy with the wave-particle duality of light, the electron may possess a wave-like nature. The wavelength would be determined by $\lambda = h/p$ where p is the electron momentum and h is Planck's constant. In 1925, a graduate student in Germany, Walter Elsasser, wrote a note suggesting that the Davisson–Kunsman experiments on electron scattering provided qualitative evidence that this wave-like property existed. Although, in 1927, Davisson would observe that Elsasser's suggestion was likely incorrect, this note may have provided Davisson with the insight he needed to interpret his and Germer's more recent unpublished results. In August 1926, Davisson was vacationing in England and attended a lecture by Max Born, "Physical Aspects of Quantum Mechanics," given at a meeting of the British Association for the Advancement of Science. In this talk, Born cited Elsasser's conjecture about the Davisson–Kunsman experiments and cited Erwin Schrödinger's recent work on a wave equation for the electron. Davisson had taken with him to England some of the experimental results from the scattering experiments with the single crystal. According to Davisson's recollections, Davisson "showed them to Born, to [Douglas] Hartree and probably to [Patrick Maynard Stuart] Blackett; Born called in another Continental physicist . . . and there was much discussion of them." By the time Davisson returned to the experiment, his attitude toward what he had been observing had changed drastically, and he began a program of search for the scattering maxima that would result from an electron diffraction pattern. This search proved successful in January 1927, and a thorough study was published in *The Physical Review* in December 1927.

There were others in the academic community attempting to do similar experiments at the same time. However, the high vacuum required to observe low energy electron diffraction impeded these research attempts and provided Davisson with a technological advantage. At Aberdeen University, George Paget Thomson, the son of Joseph J. Thomson, began a study of high energy electron diffraction by observing the diffraction pattern obtained from a beam of electrons piercing a thin film.

In 1937, Davisson would share a Nobel Prize in physics for the experimental confirmation of the wave-like nature of the electron with George Paget Thomson.

See also: BORN, MAX; BROGLIE, LOUIS-VICTOR-PIERRE-RAYMOND DE; DIFFRACTION, ELECTRON; SCATTERING; SCHRÖDINGER EQUATION; THOMSON, JOSEPH JOHN

Bibliography

BORN, M. "Physical Aspects of Quantum Mechanics." *Nature* 119, 354–357 (1927).

DAVISSON, C., and GERMER, L. H. "Diffraction of Electrons by a Crystal of Nickel." *Phys. Rev.* 30, 705–740 (1927).

DAVISSON, C., and KUNSMAN, C. H. "The Scattering of Low-Speed Electrons by Platinum and Magnesium." *Phys. Rev.* 22, 242–258 (1923).

GEHRENBECK, R. K. "C. J. Davisson, L. H. Germer, and the Discovery of Electron Diffraction." Ph.D. thesis, University of Michigan, Ann Arbor (1974).

JAMES J. BOYLE

DE BROGLIE, LOUIS-VICTOR-PIERRE-RAYMOND

See BROGLIE, LOUIS-VICTOR-PIERRE-RAYMOND DE

DECAY

See EXPONENTIAL GROWTH AND DECAY

DECAY, ALPHA

The alpha (α) particle is the nucleus of an ^{4}He atom. Alpha decay occurs when a parent nucleus with atomic mass number A and nuclear charge number Z spontaneously emits an alpha particle leaving a residual (daughter) nucleus:

$$^AZ \rightarrow {}^{(A-4)}(Z-2) + {}^4\text{He}.$$

This form of radioactivity was first observed in 1896 by Antoine Henri Becquerel from his study of uranium salts, and named in 1899 by Ernest Rutherford after his observations that there were two different kinds of radioactivity: one that was easily absorbed (alpha), and one that was more penetrating (beta). In 1909, Ernest Rutherford and Thomas Royds established that alpha particles were the nuclei of helium atoms and have atomic number $A = 4$ and nuclear charge number $Z = 2$. The alpha decay of a given parent nucleus often leads to daughter nuclei that are themselves alpha or beta radioactive, thus giving rise to a disintegration series. By 1935 the detailed decay schemes for three naturally occurring series that started with ^{238}U ($Z = 92$), ^{235}U, and ^{232}Th ($Z = 90$) had been discovered. The alpha particles observed for these naturally occurring decays have energies in the range of 5 to 10 MeV, and were used as a source of projectiles for nuclear reaction experiments until the use of particle accelerators took over in the 1940s. In 1932 James Chadwick discovered the neutron in a reaction that turned out to be ^{9}Be $+ {}^4$He $\rightarrow {}^{12}$C $+ n$.

Spontaneous alpha decay is allowed when the Q value for the decay is positive. The energy of the emitted alpha particle is given by $E_\alpha = QM_d/(M_d + M_\alpha)$, where M_d and M_α are the masses of the daughter nucleus and alpha particle, respectively. The alpha-decay Q value becomes positive above $Z \approx 50$ and, generally, for nuclei that are proton-rich compared to the most stable. (An exceptional case is that of ^{8}Be for which the Q value is positive for the decay into two alpha particles.) Alpha decay is more important than other light element emissions because of the relatively large binding energy of the alpha particle together with its small Z value. The alpha decay of nuclear ground states competes with beta decay and fission decay. The alpha decay of nuclear excited states competes, in addition, with gamma decay. Human-made heavy elements are often most easily identified by their characteristic alpha decay series. For example, the element with $Z = 110$ discovered in 1994 at GSI (*Gesellshaft für Swerionenforshung*) in Darmstadt, Germany, was identified from the detection of an alpha particle with an energy of 11.13 MeV and a half-life of about 400 μs in coincidence with the alpha particle from the decay of the $Z = 108$ daughter, whose properties had been studied in previous experiments.

The basic theory for alpha decay was developed by George Gamow and others in 1930. One postulates an alpha particle moving in the potential well of an attractive strong interaction. The potential energy diagram for the ^{238}U decay is shown in Fig. 1. The radius of the strong interaction potential R_t is determined by the distance between the centers when the surfaces of the alpha particle and the daughter nucleus touch. This is given approximately by $R_t = R_d + R_\alpha$, where $R_d = 1.2A^{1/3}$ fm is the radius of the daughter nucleus and $R_\alpha = 2.15$ fm for the alpha particle. For illustrative purposes the magnitude of the potential inside R_t has been set to zero. The dashed line shows the Q value energy of 4.27 MeV. Beyond the distance R_t the interaction between the alpha particle and the daughter nucleus is determined by the repulsive Coulomb potential $V(r) = 2Z_d e^2/r$, where Z_d is the nuclear charge number of the daughter nucleus. The semiclassical picture developed by Gamow envisions an alpha particle moving back and forth classically inside the radius R_t and hitting the potential barrier at R_t with a decay rate

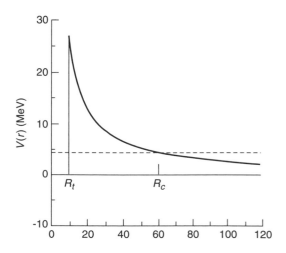

Figure 1 Potential energy diagram for ^{238}U decay.

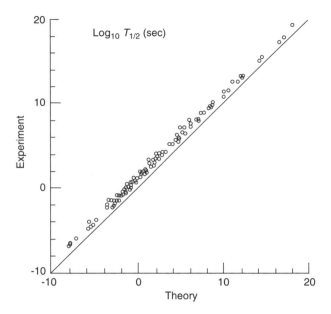

Figure 2 Comparison of the alpha decays of 119 nuclei with the Gamow estimate.

$$W_c = \sqrt{Q/(2\mu R_t^2)},$$

where μ is the reduced mass $M_d M_\alpha/(M_d + M_\alpha)$. When the alpha particle hits the potential barrier it has a probability T in the theory of quantum mechanics to tunnel through the barrier given by

$$T = \exp\{-2\int_{R_t}^{R_c} \sqrt{2\mu[V(r) - Q]/\hbar^2}\, dr\}.$$

The radius $R_c = 2Z_d e^2/Q$ at which $V(r) = Q$ is referred to as the classical turning radius since classically an alpha particle approaching the daughter nucleus from a large radius cannot go beyond this point. Integration gives

$$T = \exp\left\{-4Z_d e^2 \sqrt{2\mu/(Q\hbar^2)}\left[\cos^{-1}(x) - x\sqrt{1 - x^2}\right]\right\},$$

where

$$x = \sqrt{R_t/R_c}.$$

The alpha-decay half-life is given by $T_{1/2} = \ln 2/W$, where W is the alpha-decay rate $W = W_c T$. The experimental half-lifes for the alpha decays of 119

nuclei with ground-state angular momenta of $J = 0$ are compared to the Gamow estimate in Fig. 2. One observes quite good agreement with the Gamow estimate over 25 orders of magnitude in the half-life with a systematic trend for the experimental half-lifes to be about 100 times longer than predicted. This hindrance can be interpreted as a 1 percent probability that the actual many-body wave function of the parent nucleus has an overlap with the wave function representing the alpha particle plus daughter nucleus. A simple empirical formulae that reproduces these experimental half-lifes to a good approximation is

$$\log_{10} T_{1/2} = \frac{-51.37 + 9.54 Z_d^{0.6}}{\sqrt{Q}}.$$

The Gamow model can be extended to the decay of nonzero spin nuclei by adding a centrifugal barrier to the potential $V(r)$. The model can be refined by considering a more realistic shape for the interior and surface region of the potential. A more quantitative understanding of alpha decay relies on using many-body wave functions for all three particles involved in the decay that incorporate the individual proton and neutron degrees of freedom. The experimental data as well as the microscopic calculations indicate deviations from the Gamow model that are related to change in shape and change in shell-structure between nuclei.

See also: DECAY, BETA; DECAY, GAMMA; DECAY, NUCLEAR; NUCLEAR BINDING ENERGY

Bibliography

ROMER, A. *The Restless Atom: The Awakening of Nuclear Physics* (Dover, New York, 1982).

WILLIAMS, W. S. C. *Nuclear and Particle Physics* (Clarendon Press, Oxford, Eng., 1991).

B. ALEX BROWN

DECAY, BETA

Beta decay is a form of radioactivity that was first observed in 1896 by Antoine-Henri Becquerel from his study of uranium salts. It was named in 1899 by Ernest Rutherford after his observations that there were two different kinds of radioactivity—one that was easily absorbed (alpha) and one that was more penetrating (beta). Becquerel (1899) and Walter Kaufmann (1902) identified the beta radiation with the electron by its deflection in electric and magnetic fields. The basic theory for beta decay was constructed by Enrico Fermi in 1934 after Wolfgang Pauli's suggestion in the same year about the existence of the neutrino. Beta decay is due to the weak interaction. The weak interaction is one of the four known interactions in physics—the others being the strong, the electromagnetic, and the gravitational. The weak interaction is so-named because it is the weakest of these, being about 10^{14} times weaker than the strong interaction. An example of beta decay is that of the neutron into a proton, an electron, and an electron antineutrino:

$$n \rightarrow p + e^- + \bar{v}_e.$$

Nuclei are composed of protons and neutrons bound together by the strong interaction. In the beta decay of nuclei, a given initial nuclear state $^{A_i}Z_i$ (A is the atomic mass number and Z is the atomic number) is converted into the ground state or an excited state of the final nucleus $^{A_f}Z_f$, where $Z_f = Z_i \pm 1$. The transition rate for nuclear beta decay is determined by the Q value or energy release and the structure of the initial and final nuclear states. The beta-decay Q value is given in terms of the nuclear masses by

$$Q_\beta = M(^{A_i}Z_i) c^2 - M(^{A_f}Z_f) c^2 - \delta_e,$$

where δ_e accounts for the electron mass. Beta minus, β^-, decay ($\delta_e = m_e c^2 = 0.511$ MeV) involves the emission of an electron and electron antineutrino:

$$^AZ \rightarrow {}^A(Z + 1) + e^- + \bar{v}_e.$$

Beta plus, β^+, decay ($\delta_e = m_e c^2 = 0.511$ MeV) involves emission of a positron and electron neutrino:

$$^AZ \rightarrow {}^A(Z - 1) + e^+ + v_e.$$

Another form of beta decay that competes with β^+ decay is electron capture in which one of the atomic electrons is captured by the nucleus and an electron-neutrino is emitted:

$$e^- + {}^AZ \rightarrow {}^A(Z - 1) + v_e.$$

In this case, the energy release depends also on the initial energy of the captured electron. Beta decay also occurs between the states of hadrons, between the states of mesons, and between leptons, and is responsible for the instability of particles containing any of the more massive quarks or leptons.

The energy released in beta decay is shared between the recoiling nucleus, the electron, and the neutrino. Since the nucleus is heavy compared to the electron and neutrino, most of the energy is shared between the electron and the neutrino with a probability distribution for each that can be accurately calculated. Usually only the electron or positron is detected, and it has a range of kinetic energies ranging from zero up to Q_β (the end-point energy), assuming that the mass of the neutrino is zero. If the neutrino has a mass, the end-point energy of the electron would be reduced. The end-point energy of the tritium beta decay has been used to set a limit of about $m_v < 9$ eV/c^2 for the mass of the electron antineutrino.

Nuclear states are characterized by their angular momentum $\vec{J}$. Beta decay between them can take place only if the emitted electron and neutrino carry away an amount of angular momentum $\vec{\ell}$ such that $\vec{J}_f = \vec{J}_i + \vec{\ell}$, which means that $|J_f - J_i| \le \ell \le J_f + J_i$ where $J = |\vec{J}|$. The nuclear angular momenta are quantized to be either integer (for even A) or half-integer (for odd A) values in units of $\hbar$, and ℓ must be an integer times $\hbar$. Those beta decays with the fastest rate have $\ell = 0$ and are referred to as "allowed" transitions. The dependence upon the energy release can usually be calculated to a precision of about 0.1 percent, and beta decay thus provides a precise test of the strength of the weak interaction as well as of the internal structure of particles and nuclei. In the limit when Z is small and Q is large, the transition rate for "allowed" beta transitions is proportional to Q_β^5.

In 1956 T. D. Lee and C. N. Yang suggested that beta decay should violate the principle of parity conservation, and they proposed an experiment to test this idea. In 1957 parity nonconservation was confirmed by experiments carried out by C. S. Wu, E. Ambler, R. W. Hayward, D. D. Hoppes, and R. P. Hudson on the beta decay of ^{60}Co.

The modern theory of beta decay is based upon the standard model, which unifies the weak and electromagnetic interactions. The standard model of beta decay involves the $W^{\pm}$ bosons at an intermediate stage of the decay process. The most elementary of these processes involved in nuclear beta decays are

$$d \to u + W^- \to u + e^- + \bar{v}_e,$$

and

$$u \to d + W^+ \to d + e^+ + v_e, \text{ (only in nuclei)},$$

where u and d are the "up" and "down" quarks, respectively. These transformations are examples of a larger class of transformations that involve all quarks and leptons. Each step in the elementary decay process is proportional to the weak-interaction coupling constant g. The standard model relates the value of g to the mass of the W boson and value of the electric charge e. Also in the standard model, beta decay is unified with a larger class of weak interaction processes that involve the Z boson as an intermediate particle.

Nuclear double-beta decay takes place in situations where a nucleus is energetically stable to single-beta decay but unstable to the simultaneous emission of two electrons (or two positrons). For example, for the nuclei with atomic mass number $A = 100$, $^{100}\text{Mo} \to {}^{100}\text{Ru}$ double-beta decay may occur. There are two types of double-beta decay: the standard $(2e, 2v)$ type in which two neutrinos are emitted:

$$^A Z \to {}^A(Z + 2) + 2e^- + 2\bar{v}_e,$$

or

$$^A Z \to {}^A(Z - 2) + 2e^+ + 2v_e,$$

and the $(2e, 0v)$ type in which no neutrinos are emitted:

$$^A Z \to {}^A(Z + 2) + 2e^-,$$

or

$$^A Z \to {}^A(Z - 2) + 2e^+.$$

The $(2e, 2v)$ double-beta decay mode has been observed in recent experiments. The $(2e, 0v)$ double-beta decay mode is in principle easier to observe experimentally since the total energy of the two electrons (or positrons) must add up to the Q value and is not spread out as in the $(2e, 2v)$ decay. The $(2e, 0v)$ decay mode is being searched for experimentally, but it has not yet been observed. The $(2e, 0v)$ decay mode is possible only if the neutrinos have a finite mass, and the nonobservation of this decay allows one to set an upper limit on the neutrino mass of about $1 \text{ eV}/\text{c}^2$.

See also: DECAY, ALPHA; DECAY, GAMMA; ELECTROMAGNETIC RADIATION; INTERACTION, ELECTROMAGNETIC; INTERACTION, WEAK; NEUTRINO; PARITY

Bibliography

CLOSE, F.; MARTEN, M.; and SUTTON, C. *The Particle Explosion* (Oxford University Press, Oxford, Eng., 1987).

LEDERMAN, L., and SCHRAMM, D. M. *From Quarks to the Cosmos: Tools of Discovery* (W. H. Freeman, New York, 1995).

WILLIAMS, W. S. C. *Nuclear and Particle Physics* (Clarendon Press, Oxford, Eng., 1991).

B. ALEX BROWN

DECAY, GAMMA

The gamma ray is a high-energy form of electromagnetic radiation. In 1896 Antoine-Henri Becquerel found radiations emitted from uranium salts. The radiations, which were observed to be bent by magnetic fields, were named alpha (α) and beta (β) by Ernest Rutherford in 1899. Rutherford observed that one type of radiation was easily absorbed (alpha) and one was more penetrating (beta). In 1900, Paul Villard identified a third form of penetrating radiation, which could not be bent by a magnetic field, and it was described as gamma (γ) radiation. Researchers soon realized that gamma radiation was a high-energy form of the electromagnetic radiation described by Maxwell's equations. Today "gamma" radiation generally refers to the high-energy region of the electromagnetic-radiation spectrum associated with the decay of particles and nuclei. Gamma-ray energies range from a few kilo-electron-volts (where their energy range overlaps with those of x rays that are emitted in the decay of atoms and molecules) up to 10^2 TeV (10^{14} eV) for those found in cosmic rays. The gamma ray, like the x ray, consists of photons that have no mass and no charge.

For an electromagnetic transition from an initial nuclear state i (where the nucleus is at rest) to nuclear state f, the momentum of the nucleus in state f (after the transition) and the emitted gamma ray are equal and opposite. The nucleus recoils with a kinetic energy $T_f = (\Delta E)^2/(2m_f c^2)$ and the gamma ray has an energy $E_\gamma = \Delta E - T_f$, where $\Delta E = E_i - E_f$ (the transition energy) is the rest-mass energy difference between initial and final nuclear states. T_f is much smaller than ΔE and thus to a good approximation $E_\gamma = \Delta E$. The energy and angular frequency ω of the photon are related by $E_\gamma = \hbar\omega$. The wavelength is $\lambda = hc/E_\gamma = 1{,}237$ MeV-fm$/E_\gamma$.

Nuclear states are characterized by their angular momentum $\vec{J}$. The electromagnetic transition between them can take place only if the emitted gamma ray carries away an amount of angular momentum $\vec{\ell}$ such that $\vec{J}_f = \vec{J}_i + \vec{\ell}$, which means that $|J_i - J_f| \leq \ell \leq J_i + J_f$ where $J = |\vec{J}|$. The nuclear angular momenta are quantized to be either integer (for even atomic mass numbers) or half-integer (for odd atomic mass numbers) values in units of $\hbar$, and ℓ must be an integer times $\hbar$. Transitions with $\ell = 0$ are forbidden, and hence gamma transitions with $J_i = 0 \rightarrow J_f = 0$ are not allowed. A specific ℓ value determines the multipolarity of the gamma radiation; $\ell = 1$ is called dipole, $\ell = 2$ is called quadrupole, and so on. In addition, when states can be labeled with a definite parity $\pi_i = \pm 1$ and $\pi_f = \pm 1$, the transitions between them are restricted to the "electric" type of radiation when $\pi_i \pi_f (-1)^\ell$ is even and the "magnetic" type of radiation when $\pi_i \pi_f (-1)^\ell$ is odd.

The gamma transition rate is determined by the transition energy ΔE, the multipolarity, and a factor that depends upon the details of the internal nuclear structure. For example, the power for electric-dipole radiation (E1) from classical electromagnetism is given by $P = \omega^4 e^2 d^2/(3c^3)$, where d is the average distance between the positive and negative charge, $\omega = 2\pi f$, f is the frequency of the vibration in the distance between the positive and negative charge, and c is the speed of light. The E1 transition

rate $W(\text{E1})$ (the number of gamma rays per second) is the power divided by the energy per gamma ray ($E_\nu = \hbar\omega$): $W = P/E_\gamma = \omega^3 e^2 d^2/(3\hbar c^3) = 2.9 \times 10^{15} E_\gamma^3 d^2$ MeV^{-3} fm^{-2}. The quantity d^2 depends upon the internal structure of the nuclear states. An estimate for the lifetime associated with electric dipole radiation can be obtained by taking a typical nuclear transition energy of $E_\gamma = 1$ MeV and a typical nuclear size scale of $d^2 = 1$ fm^2, which gives a mean lifetime of $\tau = 1/W \approx 0.3 \times 10^{-15}$ s.

The lowest allowed multipolarity in the decay rate dominates over the next higher one (when more than one is allowed) by several orders of magnitude. The most common types of transitions are electric dipole (E1), magnetic dipole (M1), and electric quadrupole (E2). Electromagnetic transition rates provide one of the most unambiguous tests for models of nuclear structure. The strong interaction conserves parity, and to the extent that the protons and neutrons are held together in the nucleus by the strong interaction, their states can be labeled by a definite parity. Since the weak interaction does not conserve parity, the weak interaction between protons and neutrons leads to nuclear states that have a slightly mixed parity. The electromagnetic decay between nuclear states that have a mixed parity gives rise to "mixed" transitions (such as E1 plus M1), which produce circularly polarized gamma rays (gamma rays in which the electric field vector rotates around the axis of propagation). Recent observations of circular polarized gamma rays have provided a test of the weak interaction between protons and neutrons in the nucleus.

See also: DECAY, ALPHA; DECAY, BETA; DECAY, NUCLEAR; ELECTRIC MOMENT; ELECTROMAGNETIC RADIATION; INTERACTION, ELECTROMAGNETIC; INTERACTION, STRONG; INTERACTION, WEAK; MAGNETIC MOMENT; MAXWELL'S EQUATIONS; NUCLEAR STRUCTURE; PARITY; PHOTON; POLARITY

Bibliography

CLOSE, F.; MARTEN, M.; and SUTTON, C. *The Particle Explosion* (Oxford University Press, Oxford, Eng., 1987).

LEDERMAN, L., and SCHRAMM, D. M. *From Quarks to the Cosmos: Tools of Discovery* (W. H. Freeman, New York, 1995).

WILLIAMS, W. S. C. *Nuclear and Particle Physics* (Clarendon Press, Oxford, Eng., 1991).

B. ALEX BROWN

DECAY, NUCLEAR

Atomic nuclei are composed of nucleons (protons and neutrons) held together by strong interaction. The processes by which a nucleus decays into lighter mass products are governed by elementary interactions and their conservation laws. The rate of decay is determined by the kinetic energy released in the decay (the Q-value) together with the strong and electromagnetic interactions between the nucleons as well as the interaction of the nucleus with the weak and electromagnetic fields. For a sample of N radioactive nuclei, the rate of decay at any given time is proportional to the number of nuclei at that time:

$$\frac{dN(t)}{dt} = -WN(t),$$

where W is the decay-rate constant. The number of radioactive nuclei as a function of time is thus given by

$$N(t) = N_o e^{-Wt} = N_o e^{-t/\tau},$$

where τ is the mean lifetime and N_o is the number of nuclei at time $t = 0$. The half-life $T_{1/2}$ is the time at which $N(t) = N_o/2$ and is thus related to τ by $T_{1/2} = \ln 2\tau = 0.693\tau$. Lifetimes can be directly measured down to about 10^{-15} s (1 fs). Lifetimes that are less than this are usually observed and reported in terms of the "width," $\Gamma = \hbar/\tau = 0.658$ eV fs$/\tau$, which has units of energy. When several decay modes compete, the decay constant is given by the sum, $W = \sum_i W_i$, and the decay is characterized by a total lifetime, $\tau = 1/W$, together with the branching ratio for the individual modes, $B_i = W_i/W$. Nuclear decay occurs by electromagnetic, weak, and strong interaction processes.

Electromagnetic decay proceeds by the emission of a gamma ray (γ) from an excited state of a given nucleus, $^A Z^*$ (A is the atomic mass number and Z is the atomic number), to a lower energy excited state or ground state of the same nucleus, $^A Z$:

$$^A Z^* \rightarrow {}^A Z + \gamma,$$

where $A = N + Z$ is the atomic mass number, Z is the number of protons (often labeled by the chemical element symbol for a given Z), and N is the number of neutrons.

The weak interaction is responsible for beta decay, which proceeds by the emission of a neutrino together with an electron or positron:

$$^A Z \rightarrow {}^A(Z - 1) + e^+ + v_e,$$

or

$$^A Z \rightarrow {}^A(Z + 1) + e^- + \bar{v}_e,$$

where v_e is the electron neutrino and $\bar{v}_e$ is the electron antineutrino. A third form of nuclear beta decay, referred to as "electron capture," takes place when one of the atomic electrons is captured by the nucleus and an electron neutrino is emitted:

$$e^- + {}^A Z \rightarrow {}^A(Z - 1) + v_e.$$

In these weak decays, a given initial nuclear state $^A Z$ (usually, but not always, the ground state) is converted into an excited state or the ground state of the nucleus $^A(Z \pm 1)$. Near the region of maximum nuclear stability the lowest energy state for a given A value can sometimes be reached only by double-beta decay:

$$^A Z \rightarrow {}^A(Z - 2) + 2e^+ + 2v_e,$$

$$^A Z \rightarrow {}^A(Z + 2) + 2e^- + 2\bar{v}_e,$$

and

$$2e^- + {}^A Z \rightarrow {}^A(Z - 2) + 2v_e.$$

The double-beta decay mode has recently been experimentally observed.

Most nuclei that are heavier than ^{58}Fe (the nucleus with the maximum binding energy per nucleon) can energetically decay by the strong interaction into the two or more lighter products:

$$^A Z \rightarrow \sum_f {}^{A_f} Z_f,$$

where the atomic mass $A = \sum_i A_i$ and charge $Z = \sum_i Z_i$ are conserved. Common examples of these types of decay include alpha decay, proton decay, fission, and fission accompanied by neutrons. The decay rate for these decays is determined by the probability with which the nucleus can be decomposed into the two or more fragments together with the probability with which the fragments are able to tunnel through the Coulomb barrier. More exotic forms of these decays include light cluster emissions, such as $^{223}\text{Ra} \rightarrow ^{209}\text{Bi} + ^{14}\text{C}$.

See also: ATOMIC NUMBER; DECAY, ALPHA; DECAY, BETA; DECAY, GAMMA; INTERACTION, ELECTROMAGNETIC; INTERACTION, STRONG; INTERACTION, WEAK; NEUTRINO; NEUTRON; NUCLEAR BINDING ENERGY; NUCLEON; PROTON

Bibliography

LEDERMAN, L., and SCHRAMM, D. M. *From Quarks to the Cosmos: Tools of Discovery* (W. H. Freeman, New York, 1995).

WILLIAMS, W. S. C. *Nuclear and Particle Physics* (Clarendon Press, Oxford, Eng., 1991).

B. ALEX BROWN

DEGENERACY

Degeneracy, for example in an atom or nucleus, is the occurrence of two or more states having the same energy. All combinations of degenerate states in any proportion have the same energy and are thus degenerate with the others. The degree of degeneracy (or sometimes, just the degeneracy) of an energy level is the largest number of linearly independent states that can be found at that energy. The states of motion of a free particle at a fixed speed are infinitely degenerate in quantum mechanics because there are infinitely many possible directions of travel.

Degeneracy is related to symmetry operations that leave the energy of a system unchanged. The symmetries in a physical system can be inferred from the observed degeneracies, and conversely each symmetry causes a characteristic pattern of degeneracy. For example, if a single particle is subject only to a central force, such as electrostatic attraction by a point charge, then its energy levels exhibit a pattern of degeneracy that follows from the rotational symmetry of a sphere. The states then fall into classes labeled by angular momentum quantum numbers $l = 0, 1, 2, \ldots$ and $m_l = -l, \ldots, +l$. Each energy level labeled by l has a degeneracy $2l + 1$, and rotations of the coordinate system transform the $2l + 1$ states into combinations of one another.

A modification of the physical environment that makes it less symmetric, such as application of an external field, may lift the degeneracy, that is, cause a separation of the level into two or more different energy levels, which may themselves also be degenerate but to a lower degree. The new pattern of degeneracies that results is characteristic of the overall resulting symmetry. Levels that remain degenerate are prone to being mixed by weak influences because no transfer of energy is required to mix them.

The accidental degeneracy of the Coulomb potential, which gives each energy level of the nonrelativistic hydrogen atom a degeneracy n^2 (rather than $2l + 1$), results from a rotational symmetry in an abstract four-dimensional space (*not* the spacetime of relativity theory). The isotropic harmonic oscillator in three dimensions and the continuous spectrum of the Coulomb potential (the scattering states) also exhibit higher symmetries. When energy levels depend on a continuous parameter, such as a field or position, they may cross (coincide) for certain parameter values. This degeneracy at the crossing indicates that the levels respond differently under the transformations of some underlying, perhaps hidden, symmetry.

Degeneracies may be associated with continuous or discrete symmetry operations. Continuous symmetries include rotation, translation, or gauge transformations, while discrete symmetries include inversion (parity), reflection, charge conjugation, exchange of identical particles or shifts of position within a periodic lattice.

The masses of elementary particles (protons, neutrons, electrons, etc.) are associated with discrete energy levels of a microscopic quantum-mechanical world by Albert Einstein's equation $E = mc^2$. The near equality of m_n and m_p implies an approximate symmetry of the strong nuclear force. A hierarchy of symmetries can be used as a unifying concept to classify particles (baryons, mesons, leptons, quarks) whose actual mass energies differ greatly but arguably would coincide, that is, be degenerate, except for perturbations of lower symmetry that are present in nature.

See also: ENERGY LEVELS; JAHN–TELLER EFFECT; PARTICLE MASS; PAULI'S EXCLUSION PRINCIPLE; QUANTUM MECHANICS; SOLITON; SYMMETRY

Bibliography

BOHM, D. *Quantum Theory* (Dover, New York, 1989).
MERZBACHER, E. *Quantum Mechanics* (Wiley, New York, 1967).

K. B. MACADAM

DEGREE OF FREEDOM

The number of degrees of freedom of a mechanical system is the minimum number of independent coordinates which must be specified to determine completely the mechanical configuration of the system. The simplest system with only one degree of freedom would be a point mass traveling along a line or curve, such as a bead sliding along a wire or on a wire loop, or even on a helical spring. But a slider on an air track also acts like a point mass and only has one degree of freedom. In all these cases the obvious coordinate is distance along the path. Examples where an angular coordinate might be used instead would include a ball rolling in a groove without slipping, a rigid body rotating about a fixed axis, or a simple pendulum swinging on a fixed horizontal axis.

The simplest system with two degrees of freedom would be a point mass on a plane surface, like a nonrotating puck on a hockey rink. The most familiar pair of coordinates here are rectangular Cartesian coordinates (x, y), and polar coordinates (R, θ), but one is free to use any other pair that may be convenient. For the motion of a point mass on some curved surface, such as a sphere or a torus, latitude and longitude make a likely coordinate pair. A plumb bob suspended by a wire so as to swing freely in any horizontal direction, like a Foucault pendulum, is a similar example: The motion of its tip describes a spherical surface. Quite different systems with two degrees of freedom include a pair of beads on a wire, with or without an interactive force between them, or a pair of simple pendula, with or without interaction, or even a double pendulum formed by hanging one simple pendulum from the bottom of another.

To locate a point mass in space requires three coordinates: Therefore it has three degrees of freedom. This can be confusing because point masses in space, in our experience, are usually on a center-of-mass trajectory, which is a one-dimensional path. The distinction which should be kept in mind is that the subject of mechanics deals with all possible trajectories. In general, a system of n particles moving freely in space has $3n$ degrees of freedom. If there are m constraints, the system will have $3n - m$ degrees of freedom.

The position and orientation of an extended rigid body in space involves six degrees of freedom—three for the position of the center of mass, two for the orientation of some axis fixed in the object and passing through the center of mass, and one for the angle of rotation around that axis.

According to the theorem of the equipartition of energy in classical statistical mechanics, the average kinetic energy in each mechanical degree of freedom of a statistical system, at equilibrium, should be $(1/2) k_B T$, where k_B is Boltzmann's constant ($k_B = 1.38 \times 10^{-23}$ J/K) and T is the absolute temperature in degrees Kelvin.

However, this theorem is subject to an important quantum-mechanical constraint: The energies of all mechanical degrees of freedom are actually quantized, and if the energy level spacing of a particular degree of freedom is large compared to the average thermal kinetic energy $(1/2) k_B T$, that degree of freedom will not be thermally excited but will be "frozen out." Thus the average kinetic energy of a molecule of a monatomic gas is $(3/2) k_B T$, but the average kinetic energy of a molecule of a diatomic gas is not the expected $(6/2) k_B T$ but is usually only about $(5/2) k_B T$ because in most diatomic molecules at room temperature the vibrational motion between the two atoms is not excited. Similarly all the electrons in atoms and molecules could be expected to contribute many additional degrees freedom, but they are not much excited thermally until the gas is hot enough to ionize and form a plasma.

These considerations apply even to crystalline solids such as metals. A crystal containing N atoms has $3N$ degrees of freedom, six of which represent the position and orientation of the crystal, the remaining $(3N - 6)$ representing various internal vibrational degrees of freedom. If $3N$ is very much greater than 6, then we expect the crystal to have $(3/2)Nk_BT$ of thermal kinetic energy when the vibrations are fully excited and, because they *are* vibrations, an equal $(3/2)Nk_BT$ of average potential energy, for a total of $3Nk_BT$. This is found to be true of many metals (copper, silver, tin, etc.) at room temperature and is known as the law of Dulong and Petit. Because these vibrations are mostly not excited at temperatures near absolute zero, the heat capacity of the crystal must rise from zero to $3Nk_B$ as the crystal warms up. This phenomenon was first explained qualitatively by Albert Einstein, then quantitatively by Peter Debye, who showed that the frequency spectrum of the bulk elastic vibrations of the solid crystal give rise to a T^3 dependence of the heat capacity near absolute zero.

See also: CENTER OF MASS; COORDINATE SYSTEM, CARTESIAN; EQUIPARTITION THEOREM; FRAME OF REFERENCE; HEAT CAPACITY; PENDULUM, FOUCAULT

Bibliography

GOLDSTEIN, H. *Classical Mechanics* (Addison-Wesley, Reading, MA, 1980).

REIF, F. *Berkeley Physics Course*, Vol. 5: *Statistical Physics* (McGraw-Hill, New York, 1964).

SOMMERFELD, A. *Lectures on Theoretical Physics*, Vol. 5: *Thermodynamics and Statistical Physics* (Academic Press, New York, 1956).

PAUL C. MANGELSDORF JR.

DE HAAS–VAN ALPHEN EFFECT

See HAAS–VAN ALPHEN EFFECT, DE

DENSITY

The ratio of mass to volume of any substance is called its density. Density is one of the fundamental properties of matter and is a measure of how compactly (or "densely") packed are its constituent particles. The concept applies to matter in solid, liquid, gaseous, and plasma states and is used for both pure substances and mixtures, whether homogeneous or not. The density of all substances depends on temperature. Density also depends on pressure, especially for gases, which can be easily compressed. In the case of compounds and mixtures, one may need to specify density range since local density will depend on how uniformly the substance has been mixed. The unit of density in the SI system is kg/m^3.

If an object of mass m has a uniform density ρ and a volume V, its weight W is given by $W = \rho Vg = mg$, where g is the magnitude of the gravitational acceleration on the surface of the earth. For inhomogeneous materials, density ρ is replaced with the average density, ρ_{av}. A related concept is the specific gravity of matter, which is a normalized density based on a standard material under specified conditions. That standard material is usually water at 4°C for solids and liquids, and air at normal conditions (at a temperature of 20°C and atmospheric pressure) for gases.

The range of densities of the elemental substances at normal conditions can be glimpsed from the following examples: hydrogen (H), atomic number $Z = 1$, density of H_2 gas $\rho = 0.08375$ kg/m^3; carbon (C), $Z = 6$, $\rho = 2.25 \times 10^3$ kg/m^3; nitrogen (N), $Z = 7$, N_2 gas $\rho = 1.1649$ kg/m^3; platinum (Pt), $Z = 78$, $\rho = 21.45 \times 10^3$ kg/m^3; lead (Pb), $Z = 82$, $\rho = 11.36 \times 10^3$ kg/m^3. The elements with greater atomic number Z do not always have greater densities. Gases have much lower densities than solids, typically by three or four orders of magnitude. In solids, atoms are arranged in a crystal lattice holding them closely together, whereas gaseous molecules are unattached to each other and occupy more space. Liquids, with weak bonds between the molecules, usually have densities lower than solids but higher than gases.

For most substances at constant pressures, density decreases when the temperature increases. As the temperature rises, the average kinetic energy of the molecules rises proportionally, causing them to spread further apart and thus increasing the volume. Among solids, thermal expansion is especially pronounced in metals, necessitating gaps between joints in bridges and other structures. Not all substances behave that way, however. Water is a common substance with an uncommon dependence of density on temperature, $\rho(T)$. Water density reaches maximum $\rho = 1.0 \times 10^3$ kg/m^3 at about 4°C, and decreases monotonicly (but nonlinearly!) as the temperature is increased or decreased from this value. This is the reason why water at the bottom of a lake does not freeze. Crystals that shrink isotropically when heated from 200°C to 800°C have recently been devised from a compound of oxygen, vanadium, and phosphorus; their density rises with temperature in this range.

Nonuniform Density

Unless it is well-stirred, a mixture of two materials with different densities cannot be characterized by a single density value. It is better described by a local and an average density, or by a range of densities. Local density is a function $\rho = \rho(\mathbf{x})$ giving the distribution of mass within some object of volume V ($\mathbf{x}$ denotes vector position). By integrating $\rho(\mathbf{x})$ over the volume one obtains the mass m contained within. The average density is now given by $\rho_{av} = m/V$. In a thoroughly stirred mixture, ρ is approximated well by the constant ρ_{av}.

Mixtures of gases tend to stir through the natural convection and diffusion. In solid and liquid mixtures, density variations are sometimes quite regular. If a container with stones of varying sizes, for example, is vigorously shaken for a sufficiently long time, the larger stones will move to the top and the smallest will settle at the bottom. This phenomenon is analogous to stratified liquids.

Stratified liquids are mixtures of two or more different and immiscible liquids and are characterized by layers of different densities arranged, due to gravity, in the order of increasing density from top to bottom. In some solutions, the density varies continuously with depth, reaching a maximum at the bottom of the vessel. An example of this is a salt lake, such as the Dead Sea, in which the salinity increases with depth.

Measuring Density

For solids, a standard method for finding density is by measuring the mass of an object (by weighing it) and separately obtaining its volume. For irregularly shaped solids, volume can be measured using a suitable pycnometer, which is a calibrated vessel filled with a liquid in which the object in question is immersed. A container of stratified liquid provides a way to measure densities of small objects, since they will float in the fluid at the depth corresponding to their density. Another approach is to mix two misci-ble liquids of known density in such proportion that the object floats suspended within the mixture. The density of the solid is then the same as that of the mixture.

Archimedes' principle is useful in measuring fluid densities. A common device based on this principle is a hydrometer, which is essentially a buoy with a calibrated stem protruding above the surface of a liquid. A sample of fluid can be weighed and its volume determined using a pycnometer. For accurate measurements, the mass has to be corrected for buoyancy in the air. Another technique, called the Schilling effusion method, consists of measuring the time it takes a gas sample of known volume to escape from a container through an opening of known aperture. Based on Bernoulli's principle, the density is proportional to the square of the escape time.

See also: ARCHIMEDES' PRINCIPLE; BERNOULLI'S PRINCIPLE; BUOYANT FORCE; HYDROMETER; MASS; SPECIFIC GRAVITY; WEIGHT

Bibliography

LIDE, D. R., ed. *Handbook of Chemistry and Physics*, 76th ed. (CRC Press, Boca Raton, FL, 1995).

TUMA, J. J. *Handbook of Physical Calculations* (McGraw-Hill, New York, 1983).

ANDRZEJ HERCZYŃSKI

DENSITY OF STATES

See STATES, DENSITY OF

DETERMINISM

A venerable tradition holds that the world is deterministic just in case every event has a cause. Such a view of determinism presupposes an account of causality, and the two concepts are certainly linked. But whereas the cause-effect relationship describes a property of particular states, determinism in physics can be understood as a property of theories, and can be discussed without commitment to any particular account of causality.

A central tenet of determinism is due to Pierre-Simon Laplace, who, influenced by the Newtonian synthesis, depicted a deterministic universe as one in which knowledge of the positions of all physical bodies and the forces acting on them at any instant would be sufficient to predict all future and past positions. Laplace's vision can be adapted to a discussion of determinism in all of physics by treating theories as collections of models. A model of a Newtonian particle theory might consist of a set of particle trajectories defined on a four-dimensional amalgam of three-dimensional Euclidean space and one-dimensional time (Newtonian spacetime). Newtonian gravitational theory can be thought of as encompassing the collection of such models associated with Newton's gravitational force law. A model of Newtonian heat flow might consist of a continuous vector "heat field" defined on the same background. Two models in any theory agree at a particular time if their physical object specifications—discrete or continuous fields of values—are the same on an instantaneous space. And if for any pair of models that do so agree, they agree for every other space, the theory is Laplacian deterministic. This systematic framework for characterizing theories and the criterion for determinism are applicable across all of physics, from Newtonian theories to special relativity, general relativity, and quantum mechanics.

Newtonian physics seems to offer nonproblematic determinism. The outcomes of mechanical experiments are readily predictable by Newtonian mechanics. However, arbitrarily fast causal interactions are integral components of Newtonian theories, for example, the particle theories associated with Newton's law of gravity. Infinite velocities permit effects that instantaneously rush in from and rush out to infinity. With such effects, the specification of physical object fields at a particular time is, in general, not sufficient for unique future and past evolution. Boundary conditions can be imposed on the equations of motion associated with the physical object fields to eliminate such effects. But in imposing such boundary conditions one must be careful not to impose determinism simply by fiat. Much discussion of determinism in Newtonian theories is associated with such problems of the "stability" of equations of motion and the character of boundary conditions.

Equations of motion that are generically unstable are the subject of chaos theory. Chaotic Newtonian systems, those that describe fluid flow, for example, are so unstable that the tiniest variation in specification of their instantaneous states can lead to wildly divergent predictions of future and past states. Since physical properties can only be specified with finite precision, such systems, while in principle deterministic, are in practice indeterministic. Statistical mechanics is concerned with how the microscopic determinism of some Newtonian systems—for example, the time-symmetric, Newtonian motions of the individual molecules of a gas—can give rise to the macroscopic irreversibility of the behavior of the system as a whole—the evolution of the gas to an equilibrium state.

In special relativity the limiting signal velocity of light is built into the spacetime structure. This precludes effects that rush to and from infinity and difficulties associated with boundary conditions, so most special relativistic theories are Laplacian deterministic. Were faster-than-light particles to exist, however, such tachyons under certain conditions could be used to send signals into the past, creating paradoxes for determinism.

In general relativity the dynamics of physical fields is integrated into spacetime structure, eliminating the distinction between background geometry and physical fields of Newtonian physics and special relativity. This introduces problems of defining the equality of two models at a time, for some general relativistic models do not contain a global instantaneous space. As with boundary conditions in Newtonian theories, one must be careful in dismissing such models as "physically unrealistic" not simply to impose determinism by fiat. Many physically interesting models do have global instantaneous spaces, and are uniquely specified by their instantaneous spacetime structure. For such models Laplacian determinism can be said to hold in general relativity.

A problem for determinism in general relativity is associated with singularities, points of spacetime for which the geometry breaks down. Predictions of dynamical behavior to the future of a naked singularity are impossible, but singularities do not threaten determinism if they are hidden within black holes. Whether general relativity guarantees the formation of black holes in all physically realistic cases of gravitational collapse is an open question.

If experimental Newtonian mechanics seems strongly to support determinism, experimental quantum mechanics seems equally strongly to support indeterminism. Outcomes of experiments on individual quantum systems are unpredictable. But the discussion of determinism in quantum mechanics is hindered by a lack of consensus on what constitutes a quantum mechanical model. There is consensus on the mathematical characterization of quantum mechanical states and on their unique free evolutions: but such states are defined and evolve in an abstract space that may not be connected to the physical space and time presupposed by Laplacian determinism.

Connecting quantum states to physical spatial and temporal properties is part of the quantum theory of measurement. In one account of quantum measurement, the full quantum state discontinuously collapses into a state associated with a particular value of the measured property. This account is radically indeterministic. Bell's theorem places severe constraints on any other account of measurement that would connect unmeasured to measured properties of quantum systems deterministically.

See also: BELL'S THEOREM; BLACK HOLE; CAUSALITY; CHAOS; GRAVITATIONAL FORCE LAW; LAPLACE, PIERRE-SIMON; NEWTONIAN MECHANICS; NEWTONIAN SYNTHESIS; NEWTON'S LAWS; MODELS and THEORIES; QUANTUM MECHANICS; QUANTUM THEORY OF MEASUREMENT; RELATIVITY, GENERAL THEORY OF; RELATIVITY, SPECIAL THEORY OF; SPACETIME; STATISTICAL MECHANICS; TACHYON

Bibliography

ALBERT, D. *Quantum Mechanics and Experience* (Harvard University Press, Cambridge, MA, 1992).

EARMAN, J. *A Primer on Determinism* (Reidel, Dordrecht, 1986).

KELLERT, S. *In the Wake of Chaos* (University of Chicago Press, Chicago, 1993).

SKLAR, L. *Physics and Chance: Philosophical Issues in the Foundations of Statistical Mechanics* (Cambridge University Press, Cambridge, Eng., 1993).

THORNE, K. *Black Holes and Time Warps: Einstein's Outrageous Legacy* (W. W. Norton, New York, 1994).

E. ROGER JONES

DEUTERON

"Deuteron" is the name given to the nucleus of deuterium (or heavy hydrogen). It is a stable nonradioactive isotope of hydrogen comprised of a bound proton and neutron. It is represented by the symbol ^{2}H or D.

Deuterium was first discovered by the American chemist Harold Urey and his associates in 1932. It was the first isotope to be separated in pure form.

Physical Properties

The atomic weight of deuterium is 2.01363. It has one unit of angular momentum. The fact that the binding energy of the deuteron is positive and small (2.22452 MeV) has important implications for nuclear force. It has a magnetic moment of 0.85741 nuclear magnetons and a quadrupole moment of 0.00278 $e^2 \cdot$ barns. These quantities also give useful information about the nuclear force.

Importance in Nuclear Physics

The properties of the deuteron provide important clues as to the nature of the nuclear force. The nuclear force is quite complex. In order to unravel this complexity much work has been devoted to studying the simplest nuclear interactions, that is, those involving only two nucleons. Of the possible two-nucleon combinations, proton-proton, neutron-neutron, and proton-neutron, only the proton-neutron system occurs in a bound state. This implies that the nuclear force between the proton and neutron is stronger than the force between either two protons or two neutrons. This is a consequence of the spin dependence of the nuclear force.

In a deuteron, the neutron and proton can line up with their spins parallel (referred to as a triplet state). Their two spins sum to one unit of angular momentum and they have no orbital angular momentum. This combination is referred to as a triplet S state, denoted $^3S_1$, where the 3 denotes the parallel spin alignment, the 1 denotes the total angular momentum, and S denotes that there is no orbital angular momentum. Two protons or two neutrons, on the other hand, must have their spins aligned antiparallel in what is called a singlet state. This is to satisfy the Pauli exclusion principle, which states that two identical particles (e.g., two neutrons or two protons) cannot occupy the same quantum orbital unless their spins are antiparallel. The fact that the deuteron is bound shows that the parallel spin configuration leads to a stronger nucleon-nucleon force than the antiparallel configuration. Hence nuclear forces depend upon the relative spin orientation of the nucleons.

The properties of the deuteron also give information about the complexity of the nuclear force. If indeed the force between a proton and neutron only depended upon the distance between the two nucleons and their relative spins, then it can be shown that the deuteron would necessarily be in the lowest energy $^3S_1$ state. Such a state, however, is spherically symmetric and therefore should have no quadrupole moment. The fact that the deuteron is observed to have a small but nonzero quadrupole moment means that, in addition to the $^3S_1$ configuration, there is a small probability that the deuteron will be in a triplet state with *two* units of angular momentum, denoted $^3D_2$.

The deuteron wave function can be written

$$\psi_d = a\psi(^3S_1) + b\psi(^3D_2),$$

where a and b are the probabilities of observing the deuteron in a $^3S_1$ or $^3D_2$ state, respectively, such that $a^2 + b^2 = 1$. The observed quadrupole moment would imply that $b \approx 0.07$.

Such a configuration involving states with different amounts of angular momentum cannot arise from a spherically symmetric potential (like the Coulomb or Gravitational potential). Such a configuration could, however, arise from what is called a tensor force, that is, an interaction that depends upon both the relationship between the directions of the spin vectors of the two nucleons *and* the direction of the line joining them. An example of such a force is the interaction between two bar magnets, although the nuclear tensor force implied by the deuteron quadrupole moment is much stronger than the magnetic interaction between the nucleons. This is an additional nuclear interaction that is required by the observed properties of the deuteron, as well as by results from nucleon scattering experiments.

Importance in Cosmology

The cosmic abundance of deuterium has gained particular importance as a constraint on the matter density and early history of the universe. Since the binding energy of the deuteron is so low, it is impossible for stars to produce deuterium. This is contrary to most of the elements that are produced during various stages of thermonuclear burning in stars. It seems most likely that deuterium can only be destroyed in stars at the high temperatures and densities of thermonuclear burning. On the other hand, there is an epoch in the early universe (about 200 seconds into the big bang) when temperatures are high enough for thermonuclear burning to occur, but the densities are low enough that some of the deuterium produced can survive. The most likely origin for deuterium, therefore, is probably from the first few moments of cosmic expansion in the big bang.

The abundance of deuterium produced during the big bang depends sensitively upon the density of baryonic matter (normal nucleons and nuclei) during the big bang. More specifically, it depends upon the ratio of the density of baryons to the density of photons. If there are too many nucleons present the deuterium is burned into heavier nuclei (mostly helium). Hence, the observed cosmic abundance of deuterium has been used to place stringent limits on the density of matter in the universe.

The presently adopted lower limit to the primordial deuterium to hydrogen ratio is $D/H > 1.6 \times 10^{-5}$. Among other things, this limit is based upon observations of the deuterium abundance in the interstellar medium by the Hubble Space Telescope. This represents a lower limit, since the original primordial value would have been diminished as some of the deuterium was destroyed in stars over the lifetime of the Galaxy. This conservative lower limit implies that the ratio of baryons to photons in the universe is now less than 9×10^{-10}. This number is less than the mass density inferred from the motions of galaxies, which implies that at least some of the matter associated with galaxies is nonbaryonic dark matter. Although a number of variations of the big bang model have been proposed to increase the baryonic content of the universe allowed by the deuterium abundance, there is general agreement that some amount of nonbaryonic dark matter is still required.

See also: BARYON NUMBER; BIG BANG THEORY; DARK MATTER, BARYONIC; ELEMENTS; ELEMENTS, ABUNDANCE OF; HYDROGEN BOND; ISOTOPES; MAGNETIC MOMENT; NEUTRON; NEUTRON, DISCOVERY OF; NUCLEAR FORCE; NUCLEAR MOMENT; NUCLEAR PHYSICS; NUCLEON; NUCLEOSYNTHESIS; NUCLEUS; NUCLEUS, ISOMERIC; PAULI, WOLFGANG; PAULI'S EXCLUSION PRINCIPLE; PROTON; QUANTUM MECHANICS; QUANTUM NUMBER; SPECTROSCOPY; SPIN

Bibliography

COPI, C. J.; SCHRAMM, D. N.; and TURNER, M. S. "Big Bang Nucleosynthesis and the Baryon Density of the Universe." *Science* **267**, 192 (1995).

DE-SHALIT, A., and FESHBACH, H. *Theoretical Nuclear Physics* (Wiley, New York, 1974).

FESHBACH, H. *Theoretical Nuclear Physics: Nuclear Reactions* (Wiley, New York, 1992).

HARVEY, B. G. *Introduction to Nuclear Physics and Chemistry,* 2nd ed. (Prentice Hall, Englewood Cliffs, NJ, 1969).

HEYDE, K. L. G. *Basic Ideas and Concepts in Nuclear Physics: An Introductory Approach* (Institute of Physics, Philadelphia, 1994).

MALANEY, R. A., and MATHEWS, G. J. "Probing the Early Universe: A Review of Primordial Nucleosynthesis Beyond the Standard Big Bang." *Phys. Rep.* **229**, 145 (1993).

PRESTON, M. A., and BHADURI, R. K. *Structure of the Nucleus* (Addison-Wesley, Reading, MA, 1975).

GRANT J. MATHEWS

DIAMAGNETISM

Diamagnetism is a form of magnetic response characterized by a negative magnetic susceptibility. If M is the magnetization of a material in an applied field H, then the magnetic susceptibility is defined as

$$\chi = \frac{M}{H}.$$

For diamagnets $\chi < 0$, whereas $\chi > 0$ for paramagnets. Susceptibilities are given units normalized to the unit volume, mass, or mole. While paramagnetism results from the alignment of moments due to electron spin in the direction of the field, diamagnetism follows from orbital motion of the electrons in the solid.

The physical basis of diamagnetism is as follows. Electrons in solids can be regarded as either being in atomic orbits or, in the case of conductors, as nearly free. The electrons can be thought of as carrying currents which are influenced by a changing magnetic field. Lenz's law of electromagnetism states that when the magnetic flux through a circuit changes, a current is induced in the circuit in such a direction as to oppose the change that caused it.

In general, materials can have both diamagnetic and paramagnetic contributions to the susceptibility. For example, in the case of a solid with both atomic core electrons that give rise to a diamagnetic term χ_d, and conduction electrons leading to a paramagnetic term χ_p, the total susceptibility is given by

$$\chi = \chi_d + \chi_p.$$

This means that the sign of χ is determined by the magnitudes of the two contributions. Although χ_p often is larger than χ_d, there are cases in which the reverse is true. Magnitudes of χ for some common elements are Bi, -1.31×10^{-6}; Au, -2.74×10^{-6}; Ge, -0.56×10^{-6}; Al, 1.65×10^{-6}; Pt, 21.0×10^{-6}; Mn, 66.1×10^{-6}. In order to achieve a situation where $\chi_p = 0$, all the electrons must be paired and the orbital moments must be essentially zero or cancel each other. Most molecules with an even number of electrons (an exception is O_2) satisfy this condition and thus are diamagnetic.

A simple treatment of the diamagnetism of atoms and ions follows from Larmor's theorem, which says that the frequency of precession of electrons in orbits around a nucleus is

$$\omega = \frac{eH}{2mc} \text{ (Gaussian units)}.$$

The application of the field causes an equivalent current given by

$$I = (-Ze)\left(\frac{1}{2\pi} \cdot \frac{eH}{2mc}\right),$$

where Z is the number of electrons orbiting the nucleus. If $<\rho^2>$ is the mean square of the radius of the current loop, then the induced magnetic moment is

$$\mu = -\frac{Ze^2H}{4mc}<\rho^2>.$$

For a spherically symmetric charge distribution the diamagnetic contribution to the susceptibility is thus

$$\chi_{\text{core}} = \frac{-NZe^2}{6mc^2} <r^2>,$$

where N is the number of atoms per unit volume and $<r^2>$ is the mean square distance of the electrons from the nucleus. A quantum mechanical calculation gives precisely the same result as the above classical one.

There are a number of other special cases in which diamagnetic effects have been studied and understood. For example, Lev Landau considered quantum mechanically the diamagnetism of free (conduction) electrons, which vanishes classically. He found that the magnitude of this effect was given by

$$\chi_{\text{free}} = -\frac{2}{3}\mu_B^2 N(E_F),$$

where μ_B is the Bohr magneton, and $N(E_F)$ is the density of states at the Fermi level. Another effect is caused by the quantization of the conduction electrons into Landau levels that expand and move through the Fermi surface in wave-vector space as H is increased. This causes a fluctuation in energy of the system as a function of H, which causes the susceptibility to oscillate periodically in H^{-1}. This effect, called the de Haas–van Alphen effect, has been used to determine the geometry of the Fermi surfaces of most elemental metals and many metallic compounds.

In a superconductor the flux lines of the field are essentially completely excluded (Meissner effect). Thus, the superconductor is a "perfect" diamagnet for which $\chi \approx -1/4\pi$.

See also: ATOMIC NUMBER; BOHR MAGNETON; CONDUCTOR; ELECTROMAGNETISM; ELECTRON, CONDUCTION; FIELD, MAGNETIC; GAUSS'S LAW; LANDAU, LEV DAVIDOVICH; LENZ'S LAW; MAGNETIC FLUX; MAGNETIC MATERIAL; MAGNETIC MOMENT; MAGNETIC SUSCEPTIBILITY; MAGNETIZATION; PARAMAGNETISM; SPIN; SPIN, ELECTRON; SUPERCONDUCTIVITY

Bibliography

KITTEL, C. *Introduction to Solid-State Physics,* 7th ed. (Wiley, New York, 1996).

PATTERSON, J. D. *Introduction to the Theory of Solid-State Physics* (Addison-Wesley, Reading, MA, 1971).

DAVID J. SELLMYER

DICHROISM

In many crystals, the transmission and absorption of light depends upon the polarization of the incident beam. The asymmetry in adsorption is referred to as dichroism. Dichroism was first observed by Jean Baptiste Biot in 1815 in crystals of the mineral tourmaline. In 1852 William Herapeth discovered that the small crystals of iodoquinine sulfate (herapathite) were strongly dichroic. Crystals as thin as 1/500th of an inch thick produced total extinction of one polarization. Larger crystals of herapathite were difficult to grow, and it took about eighty years after Herapeth's discovery before the first sheet of Polaroid, a substance used commonly in sun glasses, was produced by Edwin Land, who used strain to align crystals of herapathite in a supporting matrix. Modern Polaroid sheets are formed from polymers, such as polyvinyl alcohol (PVA), that exhibit strong dichroism and are easier to work with than the brittle crystals of herapathite. These are all examples of linear dichroism, which is the asymmetric absorption of linearly polarized light. Crystals that exhibit dichroism appear highly anisotropic at the atomic level. For example, crystals of PVA have atoms within the polymer chains held together by strong covalent bonding, while the bonding between the chains is formed by the weaker van der Waals interactions.

Light can also be polarized circularly into left or right polarizations. Since its discovery in the nineteenth century, circular dichroism has been frequently used in organic chemistry. Chiral molecules (molecules that differ from their mirror image) exhibit absorption characteristics of the chemical groups present. Circular dichroism is also often observed in magnetic solids, where the chiral symmetry is a consequence of the direction of magnetization. Measured absorption spectra, when interpreted using simple selection rules, provide estimates of both spin and orbital magnetic moments.

Many liquids exhibit linear dichroism when an electric or magnetic field is applied to align anisotropic molecules. Electric dichroism has proven to be a useful technique for studying large biological macromolecules such as DNA, RNA, and certain viruses. Measurement of the dichroism in an applied electric field provides information about both permanent and induced electric dipole moments. The method has been used extensively in studies of charge distributions, orientation of molecular units, internal rigidity, and molecular length as a function of temperature, pH, and ionic strength.

Flow dichroism has been observed for liquids between a rotating and fixed cylinder. Orientation of anisotropic molecular units resulting from the imposed rotational motion is responsible for this form of dichroism, which depends on temperature, optical properties of the isotropic liquid, and the optical properties of the molecules forming the liquid.

See also: BIREFRINGENCE; DICHROISM, CIRCULAR; POLARIZED LIGHT

Bibliography

CRABBE, P. *ORD and CD in Chemistry and Biochemistry, An Introduction* (Academic Press, London, 1972).

SHURCLIFF, W. A. *Polarized Light* (Harvard University Press, Cambridge, MA, 1966).

SWINDELL, W. *Polarized Light* (Dowden, Hutchinson, and Ross, Inc., Stroudsburg, PA, 1975).

J. M. MACLAREN

DICHROISM, CIRCULAR

Circular dichroism is exhibited by a material when it absorbs left- and right-circularly polarized light unequally. It was experimentally discovered by Aime Cotton in 1895 and is sometimes called the Cotton effect. Cotton found that optically active compounds of tartaric acid containing copper or chromium showed this unequal absorption behavior in the region where absorption occurs. This phenomenon of circular dichroism was soon understood to arise from a similar mechanism as optical rotation, whose colorful effects in quartz were reported first by Francois Arago and a year later interpreted by his colleague Jean Baptiste Biot in 1812 in experiments with turpentine vapor.

The extent of the circular dichroism activity of a material is measured as a difference in the material's molar absorption coefficients for left- and right-circularly polarized light. Both optical rotation and circular dichroism have their roots in the chirality (handedness) of the material. Such chiral or handed or enantiomorphous or optically active materials were called dissymmetric materials by Louis Pasteur

in 1848 in his famous experiments with racemic acid, an optically inactive mixture of sodium ammonium tartrate (a salt formed by tartaric acid produced during fermentation of wine). On crystallization, the racemic acid formed into two (mirror image) dissymmetric crystals, which exhibited opposite optical activity. (Following the tradition of this great discovery the term "racemic mixture" denotes equal mixture of a dissymmetric pair of molecules.)

To see the difference between these two effects (i.e., circular dichroism versus optical rotation), consider the experimental setup in Fig. 1. Here 100 percent linearly polarized light is separately incident through an optically active material. Linearly polarized is a coherent sum of equal amplitudes of left- and right-circularly polarized light. The optical rota-

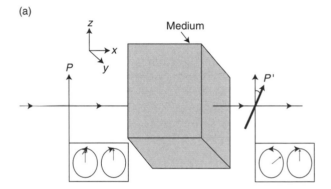

(a)

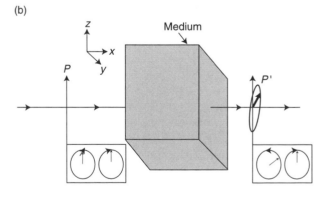

(b)

Figure 1 Diagrams showing the difference between (a) optical rotation and (b) circular dichroism. The boxes to the right of the diagram show the superposed circular light components that make up the polarized light to the left. Arrows indicate the E vectors. The incident linear polarized light has its P (polarization) vector along the z axis while the propagation is along the x axis.

tion of the linearly polarized light in the material is a consequence of the spatial variation of the induced electromagnetic dipole in the material. This results in a difference in refractive indexes (and hence the speed of light in the material) for right- and left-circularly polarized light. The spatial retardation of one circular component relative to the other results in a temporal phase difference between the two circular polarization components. This manifests itself as a progressive spatial rotation of the plane of the polarization of the linearly polarized light. Hence, in the case of purely optical rotation, 100 percent linearly polarized light stays 100 percent linearly polarized light. In contrast, if the material causes circular dichroism, it preferentially absorbs one of the circularly polarized components that make up the linearly polarized light. This results in a nonzero circular component when the light exits out of the circularly dichroic material, that is, elliptically polarized light results. In this case, 100 percent linearly polarized light incident on the material results in less than 100 percent linearly polarized light exiting the material. The transformation of polarization of the light by optical rotation and circular dichroism is generally handled using the well-known Jones matrix, named after C. Clark Jones.

Both optical rotation and circular dichroism reverse sign when the chirality (handedness) of the medium is reversed. Both are zero when the medium has no chirality. However, circular dichroism is localized to the structure of particular light absorbing groups in the molecules. Circular dichroism proved an effective probe in early stereochemistry because it is simpler to interpret than optical rotation. Its application to stereochemistry was made by H. G. Kuhn in 1958. The quantitative understanding of optical rotation relies on a (complicated) knowledge of the dipole polarizabilities of the many bonds in the molecule.

In modern times, availability of molecular data (bond lengths and knowledge of the symmetry of molecules) has enabled numerical calculations of circular dichroic spectra, a sample of which is shown in Fig. 2. Further applications of optical activity can be obtained by inducing handedness in molecules and atoms by holding them in a constant magnetic field and directing linearly polarized light along the magnetic field axis. The molecules are caused to orient themselves along the field lines and a handed sample results. The handedness can be reversed by

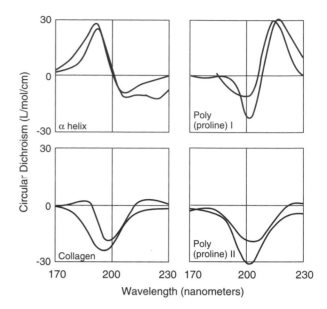

Figure 2 A sample of calculated circular dichroisms for several polypeptides. The repeating group that causes the dichroism here is –NH–CHR–CO–, where R is an organic functional group.

flipping the magnetic field. The resulting optical rotation of light is called the magnetic optical rotatory dispersion, and the resulting circular dichroism is called magnetic circular dichroism. This magneto-optical activity is also called the "Faraday effect" after Michael Faraday, who discovered this in lead glass around 1845.

A similar use of optical activity is to be found in the Kerr effect (discovered by John Kerr in 1875), where an electric field is used to orient the molecules (e.g., nitrobenzene). Here optical rotation is used to produce a fast optical switch when the nitrobenzene is placed between a pair of crossed linear polarizers and an electric field is applied across the optical axis. However, some circular dichroism is also induced along with this optical rotation.

Following the discovery of the Faraday effect, Louis Pasteur attempted to induce optical activity in achiral crystals by growing them in the presence of magnetic fields. The results obtained in 1884 were negative. Conjecture on repeating Pasteur's experiments in a chiral combination of electric fields and magnetic fields to synthesize optically active materials from achiral samples has been refuted, at least for the case of uniform static fields, by applications of Eugene Wigner's group theoretical general principle. This principle states that a complete experiment subject to time reversal and space inversion (parity invariance) should give a transformed experiment, which is possible in practice.

Optical activity in the chemical industry has been found to be significant in that enantiomers display strikingly different chemical properties. One enantiomer of the drug Thalidomide is a sedative, whereas the other produces tragic birth defects. This effect on the human body is due to the chirality of key molecules in the human body (e.g., DNA). Today the drug industry pays careful attention to the separation of enantiomers. Circular dichroism plays a major role in such monitoring.

Furthermore, modern science has revealed that mirror symmetry is often absent in nature. However, the universe is dissymmetric at all levels from subatomic to macroscopic. To understand how the universe may have acquired a handedness, circular dichroism has been put to powerful use in the investigation of parity violations in atoms and molecules. The electroweak theory by Sheldon L. Glashow, Steven Weinberg, and Abdus Salam in the late 1960s linked the weak interaction (which was discovered experimentally to be parity violating by Tsung-Dao Lee and Chen Ning Yang in 1956) with the electron interaction via the W and Z forces (so-called electroweak theory). The Z^0 particle (neutral boson) is about 100 times more massive than the proton, with a range of interaction of approximately 10^{-7} times typical atomic dimensions. The painstaking experiments that were used to seek out this (small) parity violation in atoms were done at several places using heavy atoms (since the parity violation term depends on the overlap of the electron charge cloud with the nucleus and scales up approximately as Z^3, where Z is the nuclear charge). For example, at the University of Boulder, Colorado, cesium vapor was pumped to the $7S_{1/2}$ state by left- and then right-circularly polarized light. The energy diagram is shown in Fig. 3. The fluorescence monitored was that of the $7S \rightarrow 6P$ transition at 1.36 and 1.47 um. Crossed electric and magnetic fields were used to control the absorption rate. The parity violation experiment took many years to accomplish and was enhanced by controlling the $7S \rightarrow 6P$ transition using a modulated Stark field. The experimental precision to detect parity violation was in excellent agreement with theory.

Similarly in chiral molecules, calculations indicate that the weak neutral current (parity violation)

(a)

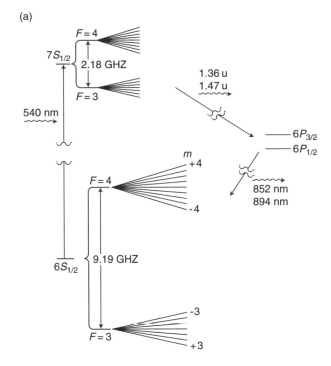

(b)

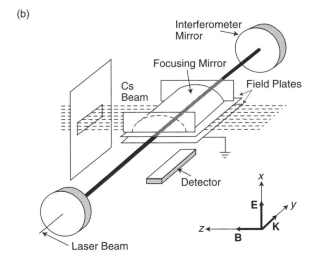

Figure 3 (a) Energy level scheme for the University of Boulder parity violation experiment. The weak-field Zeeman splitting enables an interference experiment to be made between the $6S_{F=4} \rightarrow 7S_{F'=3}$ and the $6S_{F=3} \rightarrow 7S_{F'=4}$ hyperfine transitions; (b) Experimental setup. Some of the salient features are the fluorescence detector and the crossed electric E (for Stark mixing) and magnetic B fields (for weak-field Zeeman splitting of the energy levels), which enhance the dichroic signal by mixing the parity nonconservation interaction term with the Stark term.

interaction gives rise to an energy shift of typically 5×10^{-19} eV, that is, of the order of 1×10^{-19} as compared to molecular bonding energies. Although this is a very small effect, it nevertheless indicates a handedness in nature, which could set a preference for the fabrication of one enantiomer over the other. In the long term, this would provide for a strong bias for one handedness over the other. Experiments employing circular dichroism will continue to be useful to probe the fundamental physics and chemistry of the handedness of our universe.

See also: DICHROISM; FARADAY EFFECT; INTERACTION, WEAK; KERR EFFECT; PARITY; POLARIZED LIGHT; POLARIZED LIGHT, CIRCULARLY; STARK EFFECT; ZEEMAN EFFECT

Bibliography

APPLEQUIST, J. "Optical Activity: Biots Bequest." *Am. J. Phys.* **75**, 59–67 (1987).

HEGSTROM, R. A., and KONDEPUDI, D. K. "The Handedness of the Universe." *Sci. Am.* **262** (1), 108–115 (1990).

LAKHTAKIA, A., and THOMPSON, B. J. *Selected Papers on Natural Optical Activity: SPIE Milestone Series* (SPIE Optical Engineering Press, Washington, DC, 1990).

MASON, S. F. "Origins of Biomolecular Handedness." *Nature* **311**, 19–23 (1984).

MURTADHA A. KHAKOO

DIELECTRIC CONSTANT

When a material body is placed in an electric field, electric charge rearrangement takes place in one form or another. If the material is a metal, its conduction electrons move freely and the field inside immediately vanishes. In the case of nonmetallic materials, charges are constrained from moving and the response of the material to the external field is a charge displacement within each constituent unit, such as a molecule. This effect was first observed by Michael Faraday when he inserted a plate of "shell-lac" between a charged parallel plate capacitor. He found that the effective capacitance of the capacitor increased by a factor κ, which is called the dielectric constant (or specific inductive capacity as Faraday called it). Substances that have this property are called dielectrics. The increased capacitance C' can then be expressed in terms of the original capacitance C as $C' = \kappa C$. Since the dielectric constant is a ratio, it is dimensionless (i.e., it is a pure number and has no units). Expressions in electrostatics where ε_0 appears for the vacuum case (or approximately for air) can be rewritten in the presence of dielectrics by $\varepsilon_0 \kappa$. For instance, the energy density u of a capacitor filled with a dielectric of dielectric constant κ in an electric field E is given by $u = (1/2) \varepsilon_0 \kappa E^2$.

The behavior of dielectrics in an electric field can be understood from a microscopic standpoint. If the dielectric has permanent electric dipole, such as water, then these dipoles tend to align themselves in the direction of the external field when placed between the charged plates of a parallel plate capacitor isolated from a charging battery after charging. Even without permanent dipole moments, dipole moments are always induced within the dielectrics when the field is turned on. The combined or separate effect of dipole alignment and/or creation (induction) is to shift the center of positive charge in the body and that of negative charge relative to each other. In other words, the dielectric exhibits polarization. The dielectric block itself remains electrically neutral but the charges appear on the surfaces facing the capacitor plates in such a way that the electric field inside the dielectric opposes the external field. The net result is the reduction of the electric field within the dielectric and reduces the potential difference between the charged isolated capacitor plates. Since the capacitance is defined as $C = Q/V$, where Q is (the absolute value of) the real charge on a single capacitor plate, an increase in capacitance results if nothing is done to change Q on each plate and the potential difference between the capacitor plates V is reduced.

There have been considerable efforts to find or develop materials of large and small dielectric constants for numerous applications. To reduce the size and increase the capacitance of capacitors, unusually large dielectric constants are desired in industry, including the computer industry. Ceramics with a large dielectric constant, such as barium titanate and strontium titanate, are used to reduce the mass and volume of capacitors. On the other hand, engineers are constantly in search of thin films of small dielectric constant to reduce the power loss in a computer chip as its interconnections become thinner.

See also: CAPACITANCE; CAPACITOR; DIELECTRIC PROPERTIES; DIPOLE MOMENT; ELECTRIC MOMENT; ELECTRON, CONDUCTION; POLARIZATION

Bibliography

SERWAY, R. A. *Physics for Scientists and Engineers,* 3rd ed. (Saunders, Philadelphia, 1990).

CARL T. TOMIZUKA

DIELECTRIC PROPERTIES

In electromagnetic theory, a dielectric is, loosely, an electric insulator; more precisely it is a material that exhibits polarization and (ideally) does not conduct electricity. Polarization is the macroscopic manifestation of the presence of electric dipole moments at the molecular level. An electric dipole moment $\mathbf{p}$ can be thought of as a pair of equal and opposite charges (of magnitude q, say) separated by a small distance $\mathbf{s}$, $\mathbf{p} = q\mathbf{s}$, as shown in Fig. 1. Macroscopically, polarization $\mathbf{P}$ can be written

$$\mathbf{P} = n\mathbf{p}, \tag{1}$$

where n is the average number of dipole moments per unit volume.

In analogy with magnetic materials, dielectrics can be classified as either paraelectric or ferroelectric. Paraelectric materials are polarized by an applied electric field. For moderate fields, the polarization of many materials is directly proportional to the field; for linear, homogeneous, isotropic dielectrics, $\mathbf{P} =$

Figure 1 A schematic for an electric dipole moment, where $\mathbf{p} = q\mathbf{s}.$

$\chi\varepsilon_0\mathbf{E,}$ where χ is the susceptibility constant and ε_0 is the permittivity of free space, a universal constant. For sufficiently strong fields, dielectric breakdown occurs and electricity flows; the magnitude of the electric field at the point of breakdown is called the dielectric strength E_S. (Typical values of E_S are presented in Table 1.)

Ferroelectric materials exhibit permanent polarization (much like a magnet exhibits permanent magnetization). They possess a "Curie" temperature below which they spontaneously polarize and above which they behave as paraelectrics. Ferroelectric materials include such substance as Rochelle salt and barium titanate, $BaTiO_3$. Ferroelectrics will not be discussed further here.

Dielectric absorption refers to the persistence of polarization following a change in the applied electric field. One can think of "internal friction" retarding the relaxation. Times associated with dielectric absorption can vary from 10^{-10} seconds to 10^3 years, depending on the material. Materials with very long relaxation times are called electrets; they resemble ferroelectric materials in holding their polarization after the applied electric field is removed, but differ from them in other ways.

Macroscopic Behavior

To understand the bulk behavior of dielectrics, it is important to understand capacitors. A capacitor is a device for storing electric charge and energy; it consists of two conductors, often parallel metal plates, separated by empty space or a dielectric. When a potential difference V is applied across the plates (e.g., by a battery), a charge Q, proportional to V, is transferred from one plate to the other, leaving one plate charged $-Q$ and the other $+Q$. Capacitance is defined as the ratio of the magnitude of Q to the magnitude of V,

Table 1 Dielectric Constant and Strength

Dielectric	Constant κ	Strength $E_s(kV/mm)$
Air	1	3
Glass	5–10	15
Mica	3–6	10–100
Plexiglas	3.4	40

$$C = Q/V. \qquad (2)$$

Michael Faraday discovered experimentally that capacitance is increased by filling the space between the plates with dielectric. The ratio of the capacitance with the dielectric in place to that with a vacuum between the plates, C_0, is defined as the dielectric constant κ:

$$\kappa = C/C_0. \qquad (3)$$

The dielectric constant is always greater than or equal to one and varies considerably with the material; for linear, isotropic materials (as we shall assume from here on) κ depends primarily on the temperature and the frequency of the applied potential difference, if time dependent.

Consider a simple parallel-plate capacitor of plate area A and separation distance d. With a vacuum between the plates, its capacitance $C_0 = \varepsilon_0 A/d$. If a charge of magnitude Q_0 is placed on the plates, a potential difference $V_0 = Q_0/C_0$ develops across the plates, by Eq. (2), and a uniform electric field $E_0 = Q_0/A\varepsilon_0 = V_0/d$ exists between the plates. If a dielectric of constant κ is inserted so as to fill the space between the plates, without disturbing the charge, the capacitance rises to κC_0, by Eq. (3); thus the potential difference drops to $V = Q_0/\kappa C_0$ and the electric field $E = V/d = Q_0/\kappa A\varepsilon_0$. The dielectric has, in effect, reduced the charge from Q_0 to Q_0/κ by providing an induced charge $Q' = Q_0 - Q_0/\kappa$ (of opposite sign to that on the adjacent plates) on the surfaces of the dielectric, as illustrated in Fig. 2.

In the parallel-plate capacitor considered above, P like E is uniform; the polarization can be thought of as owing to one large dipole of magnitude $Q'd$ occupying a volume Ad. Thus $P = Q'd/Ad = Q'/A$, the induced surface charge density. Again from above, $Q' = (1 - 1/\kappa)Q_0 = (1 - 1/\kappa)\kappa\varepsilon_0 AE$; thus $\chi = \kappa - 1$ and

$$\mathbf{P} = (\kappa - 1)\varepsilon_0\mathbf{E}. \qquad (4)$$

Molecular Theory

The molecules (or atoms) that make up a dielectric can be described as either polar or nonpolar. In both types, the local electric field $\mathbf{E}'$ (here, the local electric field $\mathbf{E}'$ in the vicinity of an atom is different from the average electric field $\mathbf{E}$ in the dielectric) forces the positive nuclear charges one way and the negative electronic cloud the other, producing an electric dipole moment $\mathbf{p}$. By definition, $\mathbf{p} = q\mathbf{s}$, where q is the positive charge and $\mathbf{s}$ is the displacement from the center of the negative charge to the center of positive charge.

For this "distortional" polarization

$$\mathbf{p} = \alpha\mathbf{E}', \qquad (5)$$

where α is the distortional polarizability; it is essentially constant.

Polar molecules possess permanent dipole moments within their molecular structures. For example, in hydrogen chloride, HCl, the H^+ ion is displaced about 0.13 nm from the Cl^- ion producing a permanent dipole moment $p_0 = 2.1 \times 10^{-29}$ C·m. For polar molecules the local electric field $\mathbf{E}'$ produces a torque tending to align the dipole with the field, as illustrated in Fig. 3. Thermal agitation prevents the alignment from being perfect, however; on average, for moderate electric fields, the component p of $\mathbf{p_0}$ in the direction of $\mathbf{E}'$ is approximately $p = (p_0E'/3k_BT)p_0$, where k_B is Boltzmann's constant and T is the absolute temperature. Thus, the average dipole moment p is proportional to E' and parallel to it. We can write

$$\mathbf{p} = (p_0^2/3k_BT)\mathbf{E}', \qquad (6)$$

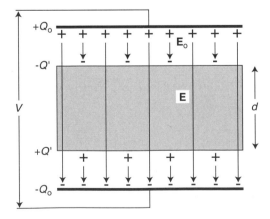

Figure 2 An "exploded" view of a parallel-plate capacitor showing the distribution of surface charges and electric fields in a parallel-plate capacitor. Space between the plates and the dielectric is shown only for the sake of clarity.

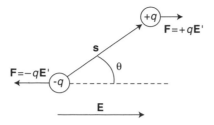

Figure 3 A dipole $\mathbf{p} = q\mathbf{s}$ at an angle θ to an electric field $\mathbf{E}$. A torque of magnitude $pE \sin \theta$ tends to align the dipole with the field.

where $p_0^2/3k_BT$ is the orientational polarizability. A polar molecule can exhibit both types of polarization, sometimes with comparable magnitudes.

The local electric field $\mathbf{E}'$ is different from the average electric field $\mathbf{E}$ in the dielectric because of the influence of surrounding dipoles; although beyond the scope of this article, it can be shown that $\mathbf{E}' = (\kappa + 2)\mathbf{E}/3$. From Eqs. (1), (4), (5), and (6), we can finally relate the dielectric constant to the sum of the polarizabilities:

$$\frac{\kappa - 1}{\kappa + 2} = \frac{n}{3\varepsilon_0}\left(\alpha + \frac{p_0^2}{3k_BT}\right), \qquad (7)$$

which is called the Debye equation.

Dynamic Properties

The frequency dependence of a dielectric depends on the elasticity of the molecular dipole moments. For a linear medium, the force to produce a dipole (i.e., to displace the electrons, since the relatively massive nuclei remain essentially fixed) is proportional to the displacement of the electrons; consequently, each electron oscillates with some natural frequency ω_0, which, for transparent dielectrics can be in the ultraviolet range. In response to a sinusoidally varying electric field (e.g., an electromagnetic wave of frequency ω) the amplitude of oscillation, thus the polarizability, is essentially that of a driven harmonic oscillator. Equation (7) is replaced by

$$\frac{\eta^2 - 1}{\eta^2 + 2} = \frac{N}{3\varepsilon_0}\left(\frac{e^2/m}{|\omega_0^2 - \omega^2|}\right), \qquad (8)$$

which is called the Lorenz–Lorentz equation. Because this equation is of particular importance in optics, we have introduced the index of refraction η in place of the dielectric constant κ; we know that $\kappa = \eta^2$. Also, e is the charge and m the mass of the electron, and N is the number of electrons per unit volume. It is assumed that all the electrons oscillate with the same natural frequency.

See also: CAPACITANCE; CAPACITOR; DIELECTRIC CONSTANT; DIPOLE MOMENT; ELECTRET; ELECTROMAGNETISM; FIELD, ELECTRIC; PARAMAGNETISM; POLARIZABILITY; POLARIZATION

Bibliography

GRIFFITHS, D. J. *Introduction to Electrodynamics*, 2nd ed. (Prentice Hall, Englewood Cliffs, NJ, 1981).

SERWAY, R. A. *Physics for Scientists and Engineers*, 3rd ed. (Saunders, Philadelphia, PA, 1990).

WANGSNESS, R. K. *Electromagnetic Fields* (Wiley, New York, 1979).

F. R. YEATTS

DIFFRACTION

The distortion of a wave front as it passes by an object is called diffraction. Diffraction causes shadows to be indistinct. If one looks closely at the area where a shadow begins, one notices that the transition from light to dark is not sharp, but consists of a series of closely spaced lines. The object that causes the shadow is disrupting the light, and the closely spaced lines cannot be made to disappear. The light is being diffracted. If the light had traveled in a perfectly straight line, there would have been a sharp demarcation between light and shadow. Light, however, does not travel in a straight line as it passes an object; instead, part of it "bends around" the object. Any wave-like phenomenon (e.g., light waves, waves on the ocean, sound waves) can be diffracted. The effect is most noticeable when the distance between crests on the wave, the wavelength, is comparable to the dimensions of the object causing the diffraction. This phenomenon is easily seen with water waves. As the waves pass a large solid object, part of the wave can be seen behind the object. Sound waves, even though they are longitudinal, also undergo diffraction, allowing the listener to hear around corners. This bending can be understood better by considering Huygens's principle.

Huygens's method constructs a wave front out of many smaller, spherical waves (secondary wavelets) with the envelope of the wavelets forming the larger wave front. At the point where diffraction occurs, each element of the original wave front sends out a wavelet. Huygens's principle states that the superposition of all the wavelets reaching a point constitutes the optical field at that point. In other words, the intensity at a point is just the summation of the various wave amplitudes reaching that point. When the incident wave is diffracted, wavelets originate at different points along the object, and their phases measured at a distant point will be different. These wavelets may interfere constructively or destructively, resulting in a variation in intensity along the image plane. Furthermore, since wavelets are also sent out at points near the surface of the object, part of the wave can travel behind the object; there is no distinct division of the wave as it passes by, as one might expect if a wave behaved as a ray.

In general, diffraction can be divided into two types. The first type, Fraunhofer diffraction, considers the case in which the incident wave fronts consist of parallel planes before they are diffracted, and, once they are diffracted, are still parallel. Because this often happens when the wave source is far away from the diffracting object, this is known as the far-field case. The second type, Fresnel diffraction, occurs when the incident wave fronts are spherical. Because this usually happens when a point source lies near the diffracting object, this is known as the near-field case. For the discussion which follows, light waves will be used as examples, but the same results occur for any wave.

As an example, let us consider the diffraction of a plane wave of light as it passes through a slit. At the slit, we construct a wave front consisting of many smaller waves. Each of these wavelets is then allowed to propagate (see Fig. 1). As can be seen, the individual wavelets will interfere with each other as they move from the slit, causing a variation in the light intensity on a screen parallel to the slit. The angular position of bright areas is given by

$$\sin(\theta) = \frac{n\lambda}{w} \qquad (n = 0, 1, 2, \ldots),$$

where λ is the wavelength and w is the slit width. When $\lambda \ll w$, $\sin(\theta) \approx \theta$, and

$$\theta = \frac{n\lambda}{w}.$$

For waves passing through a circular aperture, the mathematics is a little more complicated, but the results have the same simple form as the slit. The Fraunhofer diffraction pattern consists of a series of concentric rings, with the angle formed by the radius of the first dark ring given by

$$\sin(\theta) = 1.22\frac{\lambda}{2R}.$$

By multiplying this result by the distance to the image plane D, we obtain the radius of the Airy disk, which defines the diffraction-limited resolution of an instrument with circular optics:

$$R_{\text{Airy}} = 1.22\frac{\lambda D}{2R}.$$

From Fresnel diffraction comes much more interesting results. As stated earlier, Fresnel diffraction occurs when either the source or the observation point, or both, are so near the diffracting object that the wave fronts describe spheres. The mathematical description for this case is much more complicated

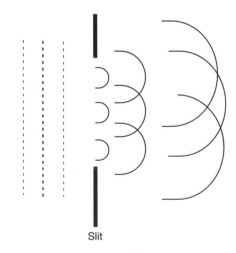

Figure 1 Huygens's construction of a plane wave diffracted by a slit. Only three wavelets are shown.

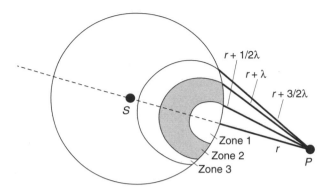

Figure 2 Fresnel zones from a source at S and observed at P.

than for Fraunhofer diffraction, and only the results will be given here.

In the Fresnel construction, as shown in Fig. 2, the wave front is divided into concentric, circular rings such that the distance from a ring on the surface of a sphere generated by source S to a point P differs from the distance from an adjacent ring by one-half of a wavelength. These rings are called the Fresnel zones. Waves from adjoining zones interfere destructively with each other, while waves from zones separated by one zone will interfere constructively. Hence, the amplitude at P will be given by the alternating series

$$A = A_0 - A_1 + A_2 - A_3 + A_4 - A_5 + \ldots .$$

This expression can be cast in a different form:

$$A = \tfrac{1}{2}A_0 + (\tfrac{1}{2}A_0 - A_1 + \tfrac{1}{2}A_2) + \ldots .$$

The amplitude from each successive zone decreases because the wave front is spherical, but the decrease is gradual. Hence, the series may be approximated as $A = \tfrac{1}{2}A_0$ for an infinite source plane (i.e., a source for which no diffraction occurs).

When part of the wave front is obstructed, only some of the zones contribute, and the situation becomes much more interesting. For example, suppose a screen with a circular aperture is positioned so that light strikes it at normal incidence. If we locate the observation point in line with the center of the aperture and in such a way that only the first Fresnel zone passes through the aperture, the amplitude at point P will be twice the value of the case

where no screen is present. That is, since the illumination is proportional to the square of the amplitude, the light is four times brighter than if the screen were not there! By either moving the point along the axis, or changing the size of the aperture, we can change the number of Fresnel zones reaching P. If the number of zones is odd, the net result of the interference among the zones will be constructive, and brightness prevails. If the number is even, P will be in darkness.

Another surprising result occurs when a disk or sphere obstructs the light in a Fresnel diffraction. In this case, the first n zones are obstructed. At points slightly beyond the object along the center line, there will always be a bright spot. This result was so surprising that it is called the Poisson spot, after the famous French mathematician who argued that this bright spot should not exist.

The consequences of diffraction can be significant. The variation in intensity caused by diffraction limits the resolution of instruments used to observe the waves. For example, optical telescopes can be diffraction limited because the entrance apertures of telescopes diffract the light. Consequently, the image of an object is spread over a larger area, the Airy disk, than if no diffraction were present. The minimum size of a laser spot that can be projected onto the Moon from Earth is given by the Airy disk. If two objects are closer together than their diffraction patterns, the light from each will blend into the other, and the objects cannot be resolved. It is interesting to note that the rods and cones in the retina of the eye are separated by the optimum distance given by the Airy disk.

At the other extreme, diffraction also limits the extent to which we can examine objects at very close range. In the making of integrated circuits, diffraction restricts the amount by which the photolithography process can miniaturize components. Diffraction also limits the magnification achieved by optical microscopes. As shown above, the amount of diffraction is directly proportional to the wavelength, leading to the use of very short wavelength light to extend the magnification range of microscopes. As will be discussed later, electrons have an even shorter wavelength. Consequently, electron microscopes are capable of extreme magnification. Scientists have recently succeeded in eliminating diffraction in a new type of microscope that takes advantage of the fact that diffraction requires a full wavelength. They simply intercept the light with their optics before the light can complete one wavelength.

Diffraction allows the measurement of many properties of waves. If one creates a series of finely spaced, narrow slits, a diffraction grating is created. A diffraction grating has the property that the angles formed by successive bright areas, measured from the plane of incidence, are given by the famous Bragg relation

$$\sin(\theta) = \frac{n\lambda}{d} \quad (n = 0, 1, 2,...),$$

where d is the distance between the slits, λ is the wavelength, and n is the order of the diffraction. The same effect can be achieved by cutting a series of closely spaced grooves on a metallic surface, creating a reflective diffraction grating. Hence, by knowing the dimensions of the slits composing the grating, one can determine the wavelength of the light striking the grating. Conversely, if light of a known wavelength strikes an object that has a regular array of diffracting points, the dimensions of these points can be determined. Laser holograms are now commonly used to produce gratings.

Diffraction gratings are widely used in analytical chemistry and optical spectroscopy for precisely these reasons. The diffraction grating, as noted above, can separate different wavelengths of light, much like a prism. An important distinction, however, is that a prism absorbs some of the light, reducing its effectiveness. Gratings absorb very little light and are much more useful when careful measurements are to be made. In analytical chemistry, it is often important to know the wavelengths at which chemical compounds either emit or absorb light. Diffraction gratings provide the dispersive element in almost all spectrometers, including high resolution dye lasers. Diffraction gratings also find use as monochromators in which, following separation, one color is retained.

Another type of diffraction occurs if the wave is scattered. In this case, the object causing the diffraction is itself distorted by the passing wave and so modifies the wave. Usually this happens when light of extremely short wavelength interacts with matter. The treatment of this situation is very involved. Suffice it to say that each atom can be considered to be a source of a scattered wave, and that like atoms scatter light in a like manner.

If these atoms are arranged in a uniform array, such as in a crystal, they form a three-dimensional diffraction grating. The distance between the atoms can be found by using the Bragg equation, which

rests on the assumption that the atoms can be considered to form parallel planes of infinite extent. In this case, the angular dependence for the position of diffracted bright spots is again given by the Bragg relationship

$$\sin(\theta) = \frac{n\lambda}{2d} \quad (n = 0, 1, 2,...),$$

where d in this case is the separation of the planes. The pattern of bright spots is called the Laue pattern. By measuring the diffracted angles, and by knowing the incident wavelength, one can measure the distance between the planes. In the case of x-ray crystallography, by rotating either the crystal or the source of x rays, one can measure all of the distances between pairs of atoms.

Diffraction provided convincing proof of the wave-like nature of moving objects, and partially validated quantum mechanics. According to quantum mechanics, all moving particles have associated with them a wavelength given by the de Broglie equation $\lambda = h/p$, where h is Planck's constant and p is the object's momentum. Because of their wave-like nature, moving objects obey all the rules of wave mechanics, including diffraction and interference. Diffraction of electrons by atoms was first observed by Clinton Joseph Davisson and Lester Halbert Germer in the 1920s when they shot an electron beam at a crystal and observed a diffraction pattern similar to that caused by x rays.

Today, diffraction of subatomic particles is a valuable tool in analyzing materials. For example, low-energy electrons are used to analyze the structure of areas near the surface of materials by observing the diffraction patterns created. Analysis of the diffraction patterns caused by neutrons can reveal the deep structure of materials.

See also: DAVISSON–GERMER EXPERIMENT; DIFFRACTION, ELECTRON; DIFFRACTION, FRAUNHOFER; DIFFRACTION, FRESNEL; GRATING, DIFFRACTION; INTERFERENCE

Bibliography

HALLIDAY, D.; RESNICK, R.; and WALKER, J. *Fundamentals of Physics* (Wiley, New York, 1993).

KLEIN, M. V. *Optics,* 2nd ed. (Wiley, New York, 1986).

JOHN E. MATHIS
ROBERT N. COMPTON

DIFFRACTION, ELECTRON

In 1924 Louis de Broglie suggested that matter may exhibit both wave and particle properties. The wavelength for matter waves is called the de Broglie wavelength and is given by

$$\lambda = \frac{h}{p} \qquad (1)$$

where h is Plank's constant (6.63×10^{-34} J·s), and p is the particle momentum (mass times velocity). This relation is consistent with the definition of wavelength for light since

$$\lambda = \frac{h}{p} = \frac{hc}{E} = \frac{hc}{hf} = \frac{c}{f} \qquad (2)$$

where f is the frequency, E is the energy, and c is the speed of light.

Classical physics views electrons as particles with a given mass and charge, and the interaction of the electron with electric and magnetic fields can be explained in terms of the motion of a particle. Early experiments using cathode-ray tubes, which provide a beam of electrons, showed that small objects (objects about 1 cm in size) placed in the tube cast sharp shadows on phosphor screens. This experiment is clearly consistent with the classical picture of the electron as a particle. The de Broglie wavelength of a 1,000-eV electron, that is, an electron accelerated by a voltage of 1,000 V, is 4×10^{-11} m, and thus, because this is much smaller than the size of the object, diffraction effects are too small to be noticeable.

Soon after de Broglie suggested matter should exhibit wave properties, Walter Elsasser suggested that the diffraction of electrons by a crystal surface should be observable. It was already known that x rays were diffracted by the regular arrangement of atoms in a crystal in exactly the same manner as light is diffracted by a diffraction grating. Further, interatomic spacings, which are typically about 10^{-10} m, are comparable to the electron's de Broglie wavelength, and thus diffraction effects should be evident. Clinton J. Davisson and Lester H. Germer demonstrated the diffraction of electrons from the surface of a single crystal of nickel. In this experiment, a monoenergetic electron beam was acceler-

ated with a known voltage, allowing a determination of the electron velocity and momentum to be made. The intensity of the elastically scattered electrons was measured as a function of angle. Diffraction peaks were observed at the Bragg scattering angles, from which the wavelength of the electron could be estimated. The wavelength extracted from the position of the diffraction peaks was exactly that predicted by de Broglie's relationship.

These low-energy electron diffraction experiments were difficult to perform for a number of reasons. To produce a diffraction pattern, clean surfaces and a high vacuum are required. Control of the accelerating voltage, and hence the electron wavelength, is also essential. Another complication is that the diffracted beams are weak and thus difficult to measure accurately. And finally, there are theoretical difficulties associated with strong multiple scattering of the electron. Unlike x-ray diffraction, where the diffracted beam scatters only once from the atoms in the solid, at energies around 100 eV, the electrons scatter off many atoms before leaving as a diffracted beam. As a result it was easier, experimentally, to view diffraction of high energy electrons transmitted through thin slices of material. In this case, multiple scattering is less important, leading to a simpler analysis of the diffracted intensities. Also, lower vacuums can be tolerated because the electrons are scattered less at high energies. The transmission electron microscope was developed to study images formed by the transmission of 100 to 1,000 keV electrons. Structural determinations of small crystals can be accomplished in the electron microscope. High-energy electron diffraction has also successfully been used to study molecular structure.

By about 1960, it had become possible both to produce an ultrahigh vacuum (leading to surfaces that were stable over longer periods of time) and to prepare single crystal surfaces of many materials. In addition, post-acceleration grids allowed weak diffracted beams to be accelerated sufficiently to fluoresce efficiently, increasing sensitivity. There was a resurgence of interest in low-energy electron diffraction (LEED), such as that originally performed by Davisson and Germer. It was found that electrons accelerated to a few hundred electron volts did not penetrate deeply into the solid; thus the observed diffraction pattern consisted of electrons scattered by atoms at or close to the surface. The accelerating voltage could be accurately adjusted, altering the electron's wavelength in accordance with the de Broglie relation. Experiments recorded the intensity of the diffraction peaks as a function of accelerating voltage. This information, combined with advances in theoretical simulation of the diffraction intensities, led to detailed structural determinations of surfaces. Several hundred surface structures have been examined with LEED, including clean metal, semiconductor and oxide surfaces, along with surfaces with co-adsorbed atoms. Usually, the positions of atoms at metal surfaces are close to those of an ideal truncation of the bulk. Semiconductor and oxide surfaces often reconstruct with the surface atomic positions that can be very different from those of the truncated bulk. These reconstructions have important consequences for semiconductor devices. Further, a detailed knowledge of adsorption sites has helped understand surface chemistry in, for example, heterogeneous catalysis.

Reflection high energy electron diffraction (RHEED) is often used to assess growth of artificial layered materials. Semiconductor lasers, for example, can be made by growing alternating layers of GaAs and GaAlAs. By adjusting the relative thicknesses of the constituent layers, the electronic and optical properties can be tailored. Many of the properties depend upon the quality of the interfaces between the layers. A flat surface produces a characteristic diffraction pattern, which disappears during the growth of the next layer because of the random arrangements of atoms in the growing layer. If the growth mode is "layer by layer," which is often the case of semiconductor systems, then, once sufficient atoms have been deposited to cover completely the surface and form the next layer, a sharp RHEED pattern will be seen. RHEED oscillations associated with completion and growth of layers are monitored to assess the quality of the superlattice.

See also: BROGLIE WAVELENGTH, DE; DAVISSON-GERMER EXPERIMENT; DIFFRACTION; ELECTRON; ELECTRON MICROSCOPE; SEMICONDUCTOR; WAVE-PARTICLE DUALITY; WAVE-PARTICLE DUALITY, HISTORY OF

Bibliography

DAVISSON, C. J., and GERMER, L. H. "The Scattering of Electrons by a Single Crystal of Nickel." *Nature* **119,** 558–560 (1927).

DAVISSON, C. J., and GERMER, L. H. "Diffraction of Electrons by a Crystal of Nickel." *Phys. Rev.* **30,** 705–740 (1927).

PENDRY, J. B. *Low-Energy Electron Diffraction* (Academic Press, London, 1974).

J. M. MACLAREN

DIFFRACTION, FRAUNHOFER

The bending of a wave front as it passes through an aperture or around the edge of an obstacle is a general property of wave motion and can be observed, for example, with light, sound, and matter waves. When a wave is bent in such a manner, part of the wave front will have its amplitude and phase altered by the presence of the obstacle. The disturbance, found by propagating the wave front, will exhibit interference resulting in regions with increased and decreased intensity. Common examples of this diffraction phenomenon include the bending of water waves around harbor walls and the pattern of light and dark fringes seen in Young's double-slit experiment. Fraunhofer diffraction is the interference pattern observed when the wave front at both the obstacle and observation point can be treated as a plane wave. In practice this can be achieved either by placing the source and observation point far from the obstacle or by placing the source at the focal point of one lens and the obstacle at the focal point of a second lens.

Consider the aperture shown in Fig. 1. In order to calculate the resulting diffraction intensity, each point $p(x, y, z)$ in the aperture is considered to be a source of spherical waves of angular frequency ω ($f = \omega/2\pi$) and wave vector k ($\lambda = 2\pi/k$). The wave amplitude (dE) at the observation point $P(X, Y, Z)$ from one of these spherical wave sources is given by

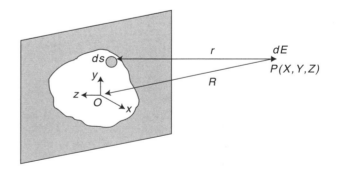

Figure 1 The wave amplitude dE at a point $P(X, Y, Z)$ resulting from a spherical wave source at $p(x, y, z)$, area dS, in the aperture. The distance between the source in the aperture and the point P is r, and the distance between the origin of the aperture and the point P is R. The aperture is located in the yz plane.

$$dE = \frac{\epsilon_0}{r}e^{i(\omega t - kr)}dS, \qquad (1)$$

where ϵ_0 is the incident plane wave amplitude. The resultant amplitude is found by adding up all the waves leaving the aperture. The summation over all points in the aperture becomes an integration. In the case of Fraunhofer diffraction, where the wave front can be treated as planar, the following approximation can be made:

$$r = R \cdot \frac{1 - (Yy + Zz)}{R^2}. \qquad (2)$$

Thus,

$$E = \frac{\epsilon_0 e^{i(\omega t - kR)}}{R} \iint_{\text{aperture}} e^{ik(Yy + Zz)/R}dydz. \qquad (3)$$

The wave amplitude, given by Eq. 3, is a two-dimensional Fourier transform of the aperture. Consider the specific case of a rectangular aperture with width a and height b. The diffraction amplitude is given by

$$E = \frac{\epsilon_0 e^{i(\omega t - kR)}}{R} \int_{-\frac{b}{2}}^{\frac{b}{2}} dy e^{ikYy/R} \int_{-\frac{a}{2}}^{\frac{a}{2}} dz e^{ikZz/R} \qquad (4)$$

This leads to an intensity at point P of

$$I = |E|^2 = I(0)\left(\frac{\sin\alpha}{\alpha}\right)^2\left(\frac{\sin\beta}{\beta}\right)^2 \qquad (5)$$

where $\alpha = kaZ/2R$, and $\beta = kbY/2R$.

Fraunhofer diffraction is important in many aspects of optics. For example, the resolving power of the human eye is determined by diffraction. Light leaving a distant object will be diffracted since the pupil is a circular aperture. Two nearby objects can only be resolved by the eye if the central diffraction peaks do not overlap. The same principle applies to optical disk drives, where the storage density is determined by the smallest bit size that can be resolved. Optical filtering and image enhancement involve modifying the Fraunhofer diffraction pattern before reconstructing the image. In this way details in blurred photographs can be enhanced.

See also: DIFFRACTION; DIFFRACTION, FRESNEL; DOUBLE-SLIT EXPERIMENT; GRATING, DIFFRACTION; IMAGE, OPTICAL; INTERFERENCE; OPTICS; RESOLVING POWER; WAVE MOTION

Bibliography

HECHT, E., and ZAJAC, A. *Optics*, 2nd ed. (Addison-Wesley, Reading, MA, 1987).

J. M. MACLAREN

DIFFRACTION, FRESNEL

Fresnel diffraction is observed at close proximity to an aperture or obstacle, in contrast to Fraunhofer diffraction, which is seen at greater distances. In both cases, the diffraction pattern can be explained from Huygens's principle, which treats each point of a wave front as a source of wavelets spreading out in all directions. It is the interference of these wavelets from the same wave front that gives the resultant diffraction pattern. In the Fraunhofer case, the wave fronts of the wavelets are essentially parallel, whereas in the Fresnel case the wave fronts are spherical and therefore more complicated.

To be specific, consider light of wavelength λ passing through a circular aperture as seen at a point P opposite the center of the aperture, such that the difference between the distance of P from the center of the aperture and its distance from the edge of the aperture is equal to λ. The situation is sketched in Fig. 1. The wavelets emanating from the aperture can constructively or destructively interfere due to phase differences caused by the varying distances traveled by wavelets to P. Also, the distances traveled by the wavelets will affect their intensity, or amplitude. Both the phase and intensity effects can be taken into account in the analysis if one imagines the aperture region to be split up into (Fresnel) zones. The wavelets from one end of a zone travel a path to P that differs by $\lambda/2$ with those coming from the opposite end. For each zone in the top half of the aperture region there is a corresponding similar zone in the bottom half. The number of zones depends on the distance to P relative to the size of the opening.

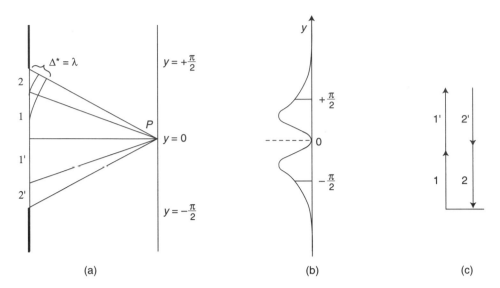

Figure 1 The light-intensity pattern (b) for the special case (a), where the opening in the barrier corresponds to exactly four Fresnel zones, two above and two below the midpoint. (c) The addition (superposition) of the individual disturbances from the various zones in order to obtain the net disturbance.

In the example under consideration, there are two zones in each half of the aperture region. The wavelets reaching P from zone 2 (or 2′) travel a path that differs by $\lambda/2$ from those coming from zone 1 (or 1′). Also, the intensities of the wavelets are not too different here because the path difference is small. Consequently, essentially complete destructive interference results at P. If the circular aperture is replaced by a disk or sphere of the same diameter, then instead of a dark spot occurring at P, a bright spot will be seen in the middle of the shadow region cast by the obstacle. In either situation, if the observation point P is moved farther away, or equivalently, if the size of the aperture or obstacle is reduced, zonal analysis will show that the diffraction pattern will alternate between one having a bight spot at the center and one having a dark spot at the center.

Augustin Fresnel proposed this theory in 1815, the first mathematical theory of diffraction and interference based on the wave nature of light. His theory also explained the double-slit interference first reported by Thomas Young in 1801. But Siméon Poisson tried to discredit the theory by pointing out that it predicted the "absurd" result of the bright spot at the center of the shadow region cast by a circular obstacle. When Fresnel and Do-

minique Arago experimentally demonstrated the existence of the "Poisson spot" in 1818, the wave theory of light quickly won universal acceptance.

See also: Diffraction; Diffraction, Fraunhofer; Fresnel, Augustin-Jean; Interference; Light, Wave Theory of

Bibliography

Benson, H. *University Physics* (Wiley, New York, 1991).

Carr, H. Y., and Weidner, R. T. *Physics from the Ground Up* (McGraw-Hill, New York, 1971).

Sears, F. W. *Optics*, 3rd ed. (Addison-Wesley, Reading, MA, 1949).

Tipler, P. A. *Physics for Scientists and Engineers* (Worth, New York, 1991).

Sam J. Cipolla

DIFFRACTION GRATING

See Grating, Diffraction

DIFFUSION

Diffusion is the random migration of molecules or large particles. In most cases of interest, the "driving force" for this migration is thermal energy. Fundamental dynamical processes in nature, such as ionic transport through cell membranes, molecular transport in liquids, hydrogen diffusion in metals, ionic diffusion in glassy material, transport of electrons and excitons in condensed phases, atomic and molecular diffusion on surfaces, surface growth, motion of microorganisms, and dynamics of populations, as well as many other examples, have all been described in terms of diffusion.

The motion of a diffusing species is, in most cases, characterized by a mean-squared displacement that, apart from very short times, increases linearly with time:

$$\langle r^2(t) \rangle = 2dDt, \tag{1}$$

where d is the spatial dimension and D is the diffusion constant. More specifically, the diffusive motion is characterized by the probability $P(r,t)$ that the species is in a unit volume around r at time t when starting at $t = 0$ from the origin. For noninteracting species, $P(r,t)$ satisfies the diffusion equation

$$\frac{\partial P(r,t)}{\partial t} = D\nabla^2 P(r,t) \tag{2}$$

and is a Gaussian,

$$P(r,t) = (2\pi Dt)^{-d/2} \exp(-dr^2/4Dt). \tag{3}$$

Equations (1) and (3) provide almost everything one needs to know about classical diffusion, where interactions among the particles and with the substrate can be neglected.

A useful model for diffusion is the random walk model, where a species advances in one unit of time to a randomly chosen site nearby.

For charged particles, the diffusion constant is related to the conductivity σ of the medium by the Nernst–Einstein relation

$$\sigma \cong (ne^2/kT)D, \tag{4}$$

where n is the density of particles with charge e, k is the Boltzmann constant, and T is the absolute temperature.

There is a growing body of evidence that classical diffusion does not hold generally. Rather, anomalous laws of diffusion exist, both slower and faster relative to classical diffusion:

$$\langle r^2(t) \rangle \sim t^\alpha, \qquad \alpha \neq 1. \tag{5}$$

Cases with $\alpha < 1$, known as the dispersive transport regime, have been the topic of many theoretical and experimental studies and have been attributed to random walks on random self-similar structures or to temporal disorder, both having underlying scale-invariant properties. Prominent examples for dispersive transport are diffusion in disordered materials (amorphous semiconductors, polymer networks, gels, and ionic glasses) and defect diffusion in relaxation. For random walks on fractals, one finds that

$$\alpha = 2/d_w, \qquad d_w > 2, \tag{6}$$

where $d_w = 2d_f/d_s$ is the fractal dimension of a random walk, d_f is the fractal dimension, and d_s is the spectral dimension. In addition, Eq. (3) is no longer valid but substituted by a stretched Gaussian.

The enhanced diffusion regime, $\alpha > 1$ in Eq. (5), has also been studied extensively. Probably the most known example of enhanced diffusion is Lewis F. Richardson's observation of turbulent diffusion, where $\alpha \approx 3$. Other examples include diffusion in elongated micelles, transport of high-density excitons, recent suggestions on the analysis of director fluctuations in liquid crystals in the nematic phase, tracer diffusion in underground water, and the analysis of DNA nucleotide sequences.

To model enhanced diffusion, Paul Lévy generalized the random walk model to what is now called the Lévy flight. While in random walks the step length is bounded, a Lévy flight takes steps of size $\{x_i\}$ chosen from the probability density $p(x)$

$$p(x) \sim |x|^{-1-\gamma}, \qquad \text{for } x \gg 1. \tag{7}$$

The moments $\langle |x|^\delta \rangle$ are finite for $\delta < \gamma$ and are infinite for $\delta \geq \gamma$. Modifications to spacetime coupled behavior, called Lévy walks, have been useful in understanding a broad range of enhanced dynamical properties in fluid mechanics, interfacial diffusion, and polymers.

See also: DISPERSION; FLUID DYNAMICS; FRACTAL; LIQUID CRYSTAL; POLYMER

Bibliography

AVNIR, D., ed. *The Fractal Approach to Heterogeneous Chemistry* (Wiley, New York, 1989).

BARABÁSI, A.-L., and STANLEY, H. E. *Fractal Concepts in Surface Growth* (Cambridge University Press, Cambridge, Eng., 1995).

BERG, H. C. *Random Walks in Biology,* 2nd ed. (Princeton University Press, Princeton, NJ, 1996).

BUNDE, A., and HAVLIN, S., eds. *Fractals and Disordered Systems,* 2nd ed. (Springer-Verlag, Berlin, 1996).

KLAFTER, J.; SHLESINGER, M. F.; and ZUMOFEN, G. "Beyond Brownian Motion." *Phys. Today* **49** (2), 33–39 (1996).

WEISS, G. H. *Aspects and Applications of the Random Walk* (North-Holland, Amsterdam, 1994).

ZSCHOKKE, I., ed. *Optical Spectroscopy of Glasses* (Redel, Dordrecht, 1986).

ARMIN BUNDE

SHLOMO HAVLIN

JOSEPH KLAFTER

H. EUGENE STANLEY

DILATION

See TIME DILATION

DIMENSIONAL ANALYSIS

Dimensional analysis is an important tool used to check the correctness of equations relating physical quantities and to make educated guesses, under favorable circumstances, on the combinations of variables that determine a specific aspect of a system's behavior. In this context the *dimensions* of a physical variable are specified by their dependence on fun-

damental quantities, usually taken to be time (T), length (L), and mass (M). When electromagnetic phenomena need to be considered, the electrical current (I) (charge per unit time) is often taken to be a fourth fundamental quantity. For example, it is said that the dimensions of speed are the dimension of distance (or length) divided by the dimension of time. In more formal terms, it is written that [speed] $= [v] = L/T = L\,T^{-1}$. The square brackets are read as "the dimensions of," so $[v]$ means "the dimensions of speed." In a similar fashion it is seen that the dimensions of acceleration are LT^{-2} and those of energy (e.g., kinetic energy) are ML^2T^{-2}.

The distinction between the dimensions of a variable and the *units* used to express that variable should be carefully noted. The units for speed are meters/second in the SI system of units or feet/second in the British system. On the other hand, the dimensions are independent of the particular system of units used. In the examples that follow, the analysis can be carried out either with dimensions or with a particular set of units.

A valuable benefit of dimensional analysis is that it can be used to check the correctness of a calculation. Suppose an individual claims to have calculated the maximum vertical height achieved by a projectile to be $h = v_0t - gt^3$, where v_0 is the initial upward velocity, g is the usual acceleration due to gravity (approximately 9.8 m/s^2 near the surface of the earth), and t is the time of flight. However, the individual expects that an algebraic mistake has been made. Comparing the dimensions of the terms in the equation can tell something about the correctness of the equation because all the terms in an equation must be dimensionally consistent. For example, the height h has dimensions of length (L). Thus, the terms on the right-hand side of the equation must also have dimensions of length. The first term has dimensions of [speed] $\cdot [t] = (L/T)\,T = L,$ which is fine. However, the second term has dimensions of $[g] \cdot [t]^3 = (L/T^2)\,T^3 = L/T,$ which is not the correct dimension. Hence, it can be said with certainty that the individual's result is not correct. Note that the dimensions may be treated just like algebraic quantities to find the overall dimensions of a combination of variables. It should be kept in mind, however, that while dimensional analysis can indicate correctness based on dimensional consistency, it cannot determine errors involving nondimensional factors. For example, suppose that the individual returns with an improved calculation of the projectile's trajectory height: $h = v_0t - gt^2$. This result is dimensionally consistent, but a factor of one-half is missing from the second term on the right; dimensional analysis cannot detect this missing factor.

Dimensional analysis can be used to find combinations of variables that determine other properties of a system. As an example, consider a simple pendulum consisting of a mass M suspended at the end of a massless string of length l. Suppose it is believed that the time t_0 it takes the pendulum to undergo one oscillation depends only on l and the acceleration due to gravity g, but the important combination of g and l is unknown. The correct combination can be determined by considering the relevant dimensions. The time t_0 has a dimension of time T. Since $[g] = L/T^2$ and $[l] = L$, the appropriate combination must be $\sqrt{l/g}$. The actual result, for small angle oscillations, is $t_0 = 2\pi\sqrt{l/g}$. Dimensional analysis cannot provide the factor of 2π, but it will give the correct dependence on l and g.

Two special cases deserve additional attention. The first involves "pseudo-units" such as radians for angle measure and cycles for frequency measure, as in cycles per second. The quantities are called pseudo-units because they represent dimensionless quantities. For example, the radian measure of an angle can be defined as the ratio of the arc length s to the radius r of a segment of a circular arc subtending this angle. Thus, the dimensions of the angle measure $\theta = s/r$ cancel (L/L). Similarly, a "cycle" is just a dimensionless counting unit, recording the number of events such as the number of swings of a pendulum. For dimensional analysis purposes, units of radians or cycles must be ignored.

The second special case involves electromagnetic quantities such as electric current, electric field, and magnetic field. As mentioned previously, the fundamental dimension is most often taken to be current (I), although other choices are possible. Electric charge then has dimensions of current multiplied by time, IT. The dimensions of the electric field $\mathbf{E}$ are the dimensions of $[\text{force}]/[\text{charge}] = MLT^{-2}Q^{-1} = MLT^{-3}I^{-1}$ from the definition of electric field as electric force per unit charge. The magnetic field $\mathbf{B}$ is defined in many ways, but a common method uses the so-called Lorentz force on a point charge moving with velocity $\mathbf{v}$: $\mathbf{F} = q\mathbf{v} \times \mathbf{B}$ (in SI units). Thus, the magnetic field must have dimensions of $[\text{force}]/([\text{charge}] \times [\text{velocity}]) = MLT^{-2}/(QLT^{-1}) = MQ^{-1}T^{-1} = MI^{-1}T^{-2}$.

See also: ELECTROMAGNETISM; LORENTZ FORCE; PENDULUM; SI UNITS

Bibliography

BRIDGMAN, P. W. *Dimensional Analysis* (Yale University Press, New Haven, CT, 1922).

BRODY, B. "Dimensional Analysis in Calculus." *Phys. Teach.* **32**, 367 (1994).

JACKSON, J. D. *Classical Electrodynamics,* 2nd ed. (Wiley, New York, 1975).

REMILLARD, W. J. "Applying Dimensional Analysis." *Am. J. Phys.* **51** (2), 137–140 (1983).

ROBERT C. HILBORN

DIODE

A diode is an electric circuit element with two leads, which allows current to flow through it in one direction, but not the other. An ideal diode can be thought of as a resistor whose resistance is zero in the forward direction, and infinite in the backward direction. Diodes are typically used to rectify alternating current by blocking off half of the cycle.

Real diodes (cat-whiskers and crystals in primitive radios, vacuum tubes, or modern solid-state devices) are not quite so simple: They do not conduct perfectly in the forward direction (at least, if the driving voltage is too small), and they allow some current to flow in the backward direction. The behavior of a real diode is represented by its characteristic curve—the graph of current I versus voltage V. Figure 1 shows the characteristic curve for a typical

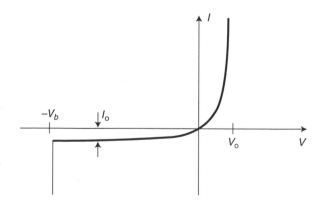

Figure 1 Characteristic curve for a typical semiconductor diode.

semiconductor diode. For voltages above V_b the graph fits the theoretical formula

$$I(V) = I_0(e^{qV/kT} - 1),$$

where $q = 1.60 \times 10^{-19}$ C is the magnitude of the electron charge, $k = 1.38 \times 10^{-23}$ J/K is Boltzmann's constant, T is the absolute temperature, and I_0 (the backward current) depends on the material and the temperature (it is about 0.01 μA for silicon at room temperature). In the forward direction the graph begins to climb very steeply at voltage V_0 (about 0.6 V for silicon and 0.2 V for germanium, at room temperature); above this it behaves as an essentially perfect diode.

In the backward direction the current is extremely small as long as the magnitude of the driving voltage is less than the breakdown value V_b. At this point the diode conducts with essentially zero resistance in the backward direction. For ordinary diodes the breakdown voltage is quite large (75 V or more); Zener diodes are specially designed to have quite low breakdown voltages (12 V or less is typical). Zener diodes are used to regulate a fluctuating signal, for whenever substantial current is flowing in the backward direction the voltage across the diode is precisely V_b.

When an electron passes through a diode in the forward direction, it gives up an energy qV_0, which is emitted as a photon of frequency $f = qV_0/h$ (where $h = 6.63 \times 10^{-34}$ J·s is Planck's constant). For ordinary diodes this frequency is in the infrared, but for light-emitting diodes (LEDs) it is in the visible region, and the LED lights up when the current flows.

A photovoltaic cell is a diode with the special property that incident light pushes the characteristic curve downward. In the fourth quadrant, the current flows against the voltage (just as it does in a battery), and the device acts as a source of electromotive force.

A semiconductor diode consists of n-type (electron rich) material in contact with p-type (electron deficient) material, forming a "p-n junction." Free electrons on the n-side encounter a potential energy hill, equal in height to the difference in the Fermi energies (ΔE). A forward voltage $V_0 = \Delta E/q$ enables them to make it over the hill, and current flows (the negatively charged electrons go from n to p, so the current itself flows from p to n). Similar considerations, leading to the same result, apply to the positively charged holes on the p-side.

See also: ELECTRICAL RESISTANCE; RESISTOR; SEMICONDUCTOR; SOLAR CELL

Bibliography

BROPHY, J. J. *Basic Electronics for Scientists,* 5th ed. (McGraw-Hill, New York, 1990).

SIMPSON, R. E. *Introductory Electronics for Scientists and Engineers,* 2nd ed. (Allyn & Bacon, Boston, 1987).

DAVID GRIFFITHS

DIPOLE MOMENT

There are two types of dipole moments: the electric dipole moment and the magnetic dipole moment. The electric dipole moment of a charge distribution is defined by

$$\mathbf{p} = \int \mathbf{x}' \rho(\mathbf{x}') \, dV$$

where $\mathbf{p}$ is the electric dipole moment, $\mathbf{x}'$ is the vector pointing from the origin to volume element dV, and $\rho(\mathbf{x}')$ is the volume charge density (see Fig. 1).

The electric dipole moment for a collection of point charges is given by

$$\mathbf{p} = \sum_{i=1}^{n} q_i x_i',$$

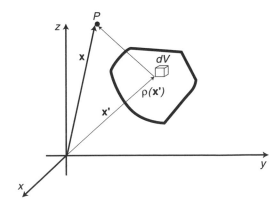

Figure 1 The electric dipole moment of a charge distribution.

where q_i is the charge value of the ith point charge. The physical dipole is a system of two equal and opposite charges separated by a distance d. The electric dipole moment of the physical dipole is

$$\mathbf{p} = q\mathbf{d}$$

where $\mathbf{d}$ is the vector pointing from the negative charge to the positive one.

The dipole contribution to the electric potential at the position $\mathbf{x}$ is given by

$$V_{\text{Dipole}} = \frac{p \cdot \mathbf{x}}{4\pi\varepsilon_0 |\mathbf{x}|^2},$$

where ε_0 is the permittivity of free space. The electric field of a dipole in terms of the electric dipole moment is

$$E_{\text{Dipole}} = \frac{p}{4\pi\varepsilon_0 |\mathbf{x}|^3} [3(\mathbf{p} \cdot \mathbf{x})\hat{\mathbf{x}} - \mathbf{p}],$$

where $\hat{\mathbf{x}}$ is the unit vector in the $\mathbf{x}$ direction.

A dipole placed in an external electric field experiences a torque given by the vector cross product of the electric dipole moment and the electric field, that is, $\boldsymbol{\tau} = \mathbf{p} \times \mathbf{E}$.

An example that illustrates the importance of the electric dipole moment is the atomic description of dielectrics. The electric field inside a capacitor is given by

$$|\mathbf{E}| = \frac{|\mathbf{E}_0|}{\kappa},$$

where $\mathbf{E}_0$ is the field in the capacitor if it had no dielectric filler and κ is a dimensionless constant that is characteristic of the dielectric material. This reduction in the field $\mathbf{E}$ occurs if the dielectric filler is composed of tiny dipoles that tend to line up and oppose the applied field. A material that exhibits this behavior is polarized. A material is polarized if, at the atomic level, the positive and negative charges are slightly separated. Molecules that exhibit permanent electric dipole moments are called polar molecules. An external electric field can induce a charge separation (and hence an electric dipole moment) in nonpolar molecules, and hence induce di-

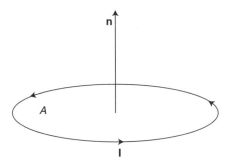

Figure 2 The magnetic dipole moment of a current loop.

electric behavior in otherwise unpolarized material. Capacitors are filled with dielectrics to enhance their performance.

The magnetic dipole moment of a current loop is given by

$$\boldsymbol{\mu} = \tfrac{1}{2}I\oint(\mathbf{x}' \times d\mathbf{l}),$$

where I is the current in the loop, $\mathbf{x}'$ is the vector from the origin to the current element, and $d\mathbf{l}$ is the infinitesimal length element of the loop (see Fig. 2). For planar (or "flat") current loops the above expression simplifies to

$$\boldsymbol{\mu} = IA\hat{\mathbf{n}},$$

where A is the area of the loop and $\hat{\mathbf{n}}$ is the unit vector perpendicular to the surface that represents the area A. The unit vector $\hat{\mathbf{n}}$ is determined from the right-hand rule, with the fingers wrapping in the direction of the current and the thumb pointing in the direction of $\hat{\mathbf{n}}$.

The magnetic field due to a magnetic dipole is given by

$$\mathbf{B} = \frac{\mu_0}{4\pi|\mathbf{x}|^3} [3(\boldsymbol{\mu} \cdot \mathbf{x})\hat{\mathbf{x}} - \boldsymbol{\mu}],$$

where μ_0 is the permeability of free space, which is not to be confused with the magnetic dipole moment $\boldsymbol{\mu}$. Note the similarity between the equation for the magnetic field in terms of the magnetic dipole moment and that of the electric field in terms of the electric dipole moment.

It should be noted that dipole moments are vectors and all the rules thereof apply.

See also: CAPACITOR; DIELECTRIC CONSTANT; DIELECTRIC PROPERTIES; ELECTRIC MOMENT; MAGNETIC MOMENT; POLARIZATION; VECTOR

Bibliography

JACKSON, J. D. *Classical Electrodynamics,* 2nd ed. (Wiley, New York, 1975).
GRIFFITHS, D. J. *Introduction to Electrodynamics,* 2nd ed. (Prentice Hall, Englewood Cliffs, NJ, 1981).
SERWAY, R. A. *Physics for Scientists and Engineers,* 3rd ed. (Saunders, Philadelphia, 1990).

SHANE STADLER

DIRAC, PAUL ADRIEN MAURICE

b. Bristol, England, August 8, 1902; *d.* Miami, Florida, October 20, 1984; *quantum theory, quantum electrodynamics, relativity, cosmology.*

Dirac was born into a Bristol middle-class family. His mother, Florence Hannah Holten, was British, and his father, Charles Adrien Ladislas Dirac, was an emigré from French Switzerland. The father, a school teacher, was a domineering person who disliked social contacts and brought Paul up in an atmosphere of rigid discipline in which there was no room for social and cultural life. The result was that Paul's early years were unhappy, and he developed into a taciturn person for whom mathematical physics became a substitute for a fuller life. In school, Dirac showed a talent for mathematics and physics, but he did not appreciate the humanistic subjects. From 1918 to 1921 he studied electrical engineering at Bristol University, but in the postwar depression he was unable to find a job. After two years of study in the mathematics department at Bristol University, Dirac entered Cambridge University, where his supervisor was Ralph Fowler, a specialist in statistical physics and atomic theory. Dirac remained in Cambridge for most of his life. When he completed his Ph.D. thesis in 1926, he was already an expert in the new quantum mechanics and recognized as a genius in theoretical physics. Dirac

became a Fellow of St. John's College in 1927, and three years later he was elected a Fellow of the Royal Society. In 1932 he was appointed Lucasian Professor at Cambridge, a position he held until his retirement in 1969. He was awarded the Nobel Prize in 1933 for his "discovery of new fertile forms of the theory of atoms and for its applications." After 1969, Dirac went to Florida, where he stayed as a member of the physics department of Florida State University for the rest of his life.

Shortly after Werner Heisenberg had invented quantum mechanics, Dirac developed his own, original version in the fall of 1925. With this version, known as q-number algebra, he developed a more general formalism of quantum mechanics that included both Heisenberg's theory and Erwin Schrödinger's wave mechanics of 1926. In 1927 he extended his theory to cover also atoms interacting with electromagnetic fields and showed how to quantize such fields. This work marked the foundation of quantum electrodynamics, a subject to which Dirac contributed throughout his life. The original quantum mechanics of Heisenberg and Schrödinger was nonrelativistic, that is, only valid for small velocities. The problem of finding a relativistic quantum equation for the electron was solved by Dirac in 1928 in one of the most important works of twentieth-century theoretical physics. Dirac found an equation—known as the Dirac equation—that differed from the ordinary Schrödinger equation, was relativistically invariant, explained electron spin, and predicted the correct fine structure in the spectrum of hydrogen. By analyzing the mathematical properties of the new equation, Dirac *predicted* the existence of a new particle, the *antielectron* (and also suggested that antiprotons would exist). He pictured this hypothetical particle as a "hole" in the "sea" filled up with unobservable negative-energy electrons that followed from his equation. According to Dirac, the antielectron would have the same mass as an ordinary electron, but with the opposite charge, and might be created, together with an electron, from high-energy electromagnetic radiation. (This process, $\gamma \rightarrow e^- + e^+$, is known as pair creation). In 1932 Dirac's *prediction* was confirmed when Carl D. Anderson detected a positive electron in the cosmic radiation, and later experiments proved that this "positron" was identical with Dirac's *antielectron.* The corresponding prediction of a negatively charged proton was verified in 1955 when the first antiproton was discovered. In 1931, Dirac also concluded that magnetic monopoles—elementary particles with an isolated magnetic charge—might exist according to the laws of quantum mechanics, and he calculated some of their properties. However, although magnetic monopoles are widely believed to exist, and many experiments have been made in order to detect them, they are still hypothetical objects.

Dirac was a rationalist who was strongly attracted by the mathematical structure of physical theories, which, he argued, should be as beautiful as possible. Although he never explained what he meant, exactly, by "mathematical beauty," he believed that there is a preestablished harmony between the fundamental laws of nature and the theoretical formulations that can be expressed in mathematically beautiful ways. This "principle of mathematical beauty" was an important inspiration for most of Dirac's work after about 1935. Together with numerological reasoning it led him in 1937 and 1938 to suggest a new cosmological theory in which the gravitational constant G varies inversely with time (i.e., that $G \sim t^{-1}$). As another unorthodox feature, Dirac concluded that the number of particles in the universe would increase in time by some kind of continuous creation of matter (thus violating energy conservation). In the 1970s Dirac returned to cosmology and further developed his varying-G theory. However, the theory was met with skepticism by most astronomers, and experiments in the 1980s showed that the variation in G predicted by Dirac does not exist. In general, most of Dirac's revolutionary work was made between 1925 and 1933, and few of his subsequent contributions had lasting value.

Dirac was an ivory tower physicist who concentrated his intellectual efforts on theoretical physics and had little interest in matters outside physics. Much of his pioneering and highly original work was indebted to an aesthetic view of science and an intuitive philosophy of beautiful mathematics and its manifestations in physics, but Dirac was neither interested in philosophy nor other branches of the humanities. He was a worshiper of mathematical logic and had little appreciation for emotions and what most people would call the human aspects of life. Characteristically, even his view of religion was rationalistic. In 1963, he wrote that "God is a mathematician of a very high order, and He used very advance mathematics in constructing the universe."

See also: POSITRON; PROTON; QUANTUM ELECTRODYNAMICS; QUANTUM MECHANICS, CREATION OF

Bibliography

CORBY HOVIS, R., and KRAGH, H. "P. A. M. Dirac and the Beauty of Physics." *Sci. Am.* **268** (5), 104–109 (1993).

DIRAC, P. A. M. *The Principles of Quantum Mechanics* (Cambridge University Press, Cambridge, MA, 1930).

DIRAC, P. A. M. "The Proton." *Nature* **126**, 605–606 (1930).

DIRAC, P. A. M. "The Evolution of the Physicist's Picture of Nature." *Sci. Am.* **208** (5), 45–53 (1963).

DIRAC, P. A. M. *Directions in Physics* (Wiley, New York, 1978).

KRAGH, H. *Dirac: A Scientific Biography* (Cambridge University Press, Cambridge, Eng., 1990).

KURSUNOGLU, B. N., and WIGNER, E. P., eds. *Paul Adrien Maurice Dirac: Reminiscences About a Great Physicist* (Cambridge University Press, Cambridge, Eng., 1987).

TAYLOR, J. G., ed. *Tributes to Paul Dirac* (Adam Hilger, Bristol, Eng., 1987).

HELGE KRAGH

DIRECT CURRENT

See CURRENT, DIRECT

DISINTEGRATION, ARTIFICIAL

When a radioactive element such as uranium or radium disintegrates by throwing off an alpha or beta particle, it is transmuted or transformed into a different element with different chemical properties. That was recognized in 1903 by Ernest Rutherford and Frederick Soddy at McGill University in Montreal. Hence, in 1919, when Rutherford, then in Manchester, England, announced that he could produce hydrogen by bombarding nitrogen atoms with alpha particles, this was also seen as a disintegration, which was called "artificial" to distinguish it from the natural disintegrations of radioactivity.

That discovery was an unexpected outcome of Rutherford's use of alpha particles to explore the interior of atoms between 1911 and 1913. He had pictured the atom as essentially empty, with a tiny, positively charged nucleus at its center from which a positively charged alpha particle would sheer away. This had been verified by his Manchester associates, Hans Geiger and Ernest Marsden, who studied the scattering pattern of the emerging alphas, observing them by the scintillations, or sparks of light, they produced on a fluorescent screen.

Those particles were scattered by the fixed atoms in thin foils of various metals. Rutherford's next step was to try hydrogen to get a closer encounter between the alpha particle and the target nucleus. Here, he expected the light hydrogen atoms to shoot off from the collisions, outrunning and outdistancing the alpha particles that struck them. In fact, when Marsden began the work, he could see two sets of scintillations and could distinguish them easily by the difference in their ranges.

Marsden's next discovery was puzzling. The long-range scintillations also appeared when there was no hydrogen in his apparatus. Then World War I intervened; Marsden left Manchester, and Rutherford could pick at the problem only in the intervals of his war work. By the spring of 1919 he had convinced himself, first, that these long-range particles were actually hydrogen, and next, that they appeared only when nitrogen was present. He suggested then that the nitrogen nucleus (of mass 14) might be built of three helium nuclei (each of mass 4) and two hydrogen nuclei (of mass 1), one of which would be released in the collision.

That summer, Rutherford moved to Cambridge, England, bringing along James Chadwick, and together they worked to extend the discovery. With increasingly ingenious experiments, they obtained long-range H particles from five other elements, then short-range particles from seven more. These were not always driven forward by the impact but came off in all directions, some with more energy than they could have gained from the collision.

Apparently, the alpha particle was captured by the target nucleus, which then expelled a hydrogen nucleus, or proton, as Rutherford had proposed to call it. This was confirmed in 1925 by P. M. S. Blackett's cloud chamber photographs of the tracks in nitrogen of some 400,000 alpha particles, 8 of which forked to show a heavy track for the incoming alpha particle, a light track for the outgoing proton, and a dense track for the recoiling atom. There was no trace of the alpha particle beyond. Clearly, it had been drawn into the nitrogen nucleus, raising its charge from 7 to 9 and its atomic weight from 14 to 18, to make an underweight nucleus of fluorine. That had then disintegrated by expelling the proton, to leave the recoiling atom a slightly heavy, stable isotope of oxygen.

Evidently, the repulsion between a nucleus and an alpha particle would change to attraction at close quarters. New scattering experiments by E. S. Bieler with light elements suggested as much although Rutherford and Chadwick found no trace of it in the scattering by uranium and gold. Here was another puzzle. Uranium was naturally radioactive. Suppose that its emitted alpha particles gained energy only from the repulsion of the nucleus behind them. Then it was easy to calculate how far away from it the attraction finally ceased and the repulsion began. Unfortunately, that was farther out than the point of deepest penetration at which the higher-energy, scattered alpha particles were still strongly repelled.

This puzzling contradiction caught George Gamow's eye in June 1928, soon after his arrival from Leningrad at Max Born's institute in Göttingen. He needed a topic for his summer visit, and he saw at once how Schrödinger's new quantum mechanics might deal with radioactive disintegration.

Quantum mechanics used energy relations to build wave functions, mathematical expressions that told where an atomic particle was likely to be found. Gamow's first task was to imagine an energy barrier around the uranium nucleus that would hold its alpha particles from escaping, yet repel any that happened to get outside. His next task was to piece together a three-part wave function for his three energy regions: inside the nucleus, within the body of the barrier, and outside it.

By July 27 he had a wave function that gave an alpha particle some small chance of being outside the nucleus, as the slow disintegration of uranium required. It also specified the energy of its escape. There was a known link between the lives of the radioactive elements and the energies of their alpha particles, which had been established from the experimental data by Hans Geiger and J. M. Nuttall. This assigned an alpha particle an energy directly proportional to the logarithm of the decay constant of its emitting element. Gamow's three-piece wave function for uranium delivered precisely that relation. A second paper, written jointly with Fritz Houtermans, another summer visitor, extended the theory to the other radioactive elements. In a third paper, Gamow developed the quantum mechanics of the alpha particle captures in Rutherford's *H*-particle experiments.

Meanwhile, Ronald W. Gurney and Edward U. Condon at Princeton University were also tackling the quantum mechanics of the uranium nucleus. They sent off a preliminary sketch on July 29, a full treatment in November, and although they differed from Gamow in details, got the same results for uranium's disintegration.

This was highly satisfactory in showing—for the first time—that quantum mechanics could be applied to the nucleus. It described nothing of how the alpha particle penetrated its barrier, but, strictly speaking, this was none of its business. Not content to be left in ignorance, we have turned to metaphors that we are careful not to believe. Gamow, Gurney, and Condon wrote of "leaking" through the barrier; nowadays we speak of "tunneling."

In September 1928, Gamow traveled to Copenhagen to spend a year at Niels Bohr's institute, and in January 1929 Bohr sent him to Cambridge to present his quantum mechanics at first-hand. It was a timely visit. Rutherford was eager for other high-speed ions to extend the disintegration experiments, and E. T. S. Walton was busy with electrical schemes for producing them. When Gamow's calculations showed at what moderate energies protons could enter nuclei, J. D. Cockroft was assigned to join Walton in building an accelerator for them. By 1932 they had it operating, and their protons were entering lithium nuclei to split them into pairs of alpha particles. Nuclear physics was freed from its dependance on radium and the accelerator age was beginning.

See also: ATOM, RUTHERFORD–BOHR; DECAY, ALPHA; DECAY, BETA; DECAY, NUCLEAR; RADIOACTIVITY, ARTIFICIAL; RUTHERFORD, ERNEST; SCATTERING, RUTHERFORD

Bibliography

ANDRADE, E. N. DA C. *Rutherford and the Nature of the Atom* (Doubleday, Garden City, NY, 1964).

GAMOW, G. *Thirty Years that Shook Physics: The Story of the Quantum Theory* (Doubleday, Garden City, NY, 1964).

ALFRED ROMER

DISORDER

See ORDER AND DISORDER

DISPERSION

Dispersion is the bending or spreading of a beam, resulting in the spatial separation of its constituent parts. For example, in color dispersion a beam of "white" light separates into a spectrum of colors. What humans perceive as color is actually a property of light called wavelength (λ), which is the longitudinal distance between two identical points (e.g., peaks or troughs) on a traveling wave of electromagnetic radiation. The visible portion of the light spectrum spans the range of wavelengths from violet ($\lambda = 240$ nm) to red ($\lambda = 640$ nm). When "white" light enters a prism, each color (λ) bends or refracts to a different angle, the light rays diverge, and then all colors bend back along parallel paths as the rays exit the prism.

Rainbows form when sunlight refracts through water droplets in the sky. Each spherical water droplet acts like a prism to disperse rings of colors back toward the Sun. Only within one or two arcs in the sky are the angles between Sun, water droplet, and a human observer just right to direct each color back into the observer's eye. Alas, we can never find the "pot of gold" at the end of the rainbow, because as we move, so does the rainbow.

The bending or refraction of light rays occurs because the effective velocity of propagation of light v varies according to λ and the index of refraction of the gas or solid medium through which it travels. The simplest definition of index of refraction is the ratio of these velocities: $n = c/v$, where c is the velocity of light in a vacuum. For glass, $n \cong 1.6$; for water (rain drops), $n \cong 1.3$. Even air has an index of refraction greater than 1.0. Usually, when light travels through a dispersive medium, v is approximately inversely proportional to λ^3. However, this is only true for normal dispersion. If the dispersive medium is strongly absorbing for some band of wavelengths (e.g., a prism filled with a colored dye) then abnormal dispersion occurs for light with wavelengths near the absorption band. Since laser beams are nearly monochromatic (one single wavelength), a prism bends, but does not disperse, laser light. However, a laser beam spreads when traveling through air because individual photons (particles of light in the beam) refract at random angles. That is, dispersion in air causes the diameter of a laser beam to increase as a function of distance traveled.

Velocity dispersion for large objects is easier to understand. Consider automobiles traveling around a curve. When the driver turns the steering wheel, the frictional force between the tire and the surface of the roadway steers the car around the curve. Friction exerts a transverse force that causes a rotation of the velocity vector **v** describing the motion of the automobile. However, for speeds (magnitudes of velocity) above a critical value v_0, the inertial force (tendency for the car to move straight ahead) overcomes the transverse frictional force and the car begins to skid. Cars traveling faster than v_0 disperse as a function of speed. If the roadway is tilted or banked such that a component of **v** points down into the road surface, then v_0 increases and drivers can steer safely around the curve. An extreme example of this principle is an oval automobile racetrack where the 180° turns bank very steeply to allow high speeds with minimum dispersion.

Dispersion is very important in the design and operation of accelerators used in various fields of physics research. These machines use a combination of electric and magnet fields to accelerate and focus beams of charged particles (e.g., electrons, protons, heavy ions). Typically, an accelerator complex starts with an ion source, which produces very heterogeneous beams with many atomic and molecular ion components. Dispersion can be used to filter the beam into a single component with the same charge (q), mass (m), and a narrow range of velocities (**v**). First, the beam from the ion source passes through a region of transverse electric (**E**) and/or magnetic (**B**) fields. Each beam component experiences a different transverse force given by $\mathbf{F} = q\mathbf{E} + q/c(\mathbf{v} \times \mathbf{B})$. These forces, which act perpendicular to the direction of ion motion, disperse the individual beam components along different trajectories. Then, beam-defining apertures (slits with narrow openings) pass only the desired component for injection into the accelerator.

Dispersion is also an important beam-loss mechanism, particularly in circular heavy-ion machines, such as the Relativistic Heavy-Ion Collider (RHIC) now under construction at Brookhaven National Laboratory. RHIC dipole magnets will steer two counter-rotating beams of bare (all electrons removed, $q = 79$) gold nuclei along similar paths in opposite directions. The physics program will study what happens when the ions undergo head-on collisions, but ions that do not collide are recycled millions of times. Ions that just miss each other can produce an electron-positron pair, and the electron can emerge from the interaction attached to one of the gold nuclei, thereby changing q from 79 to 78. The smaller trans-

verse force then causes that ion to spiral out of the main beam orbit and be lost. Such beam losses will be significant at RHIC due to the large probability for electron capture from pair production.

See also: ACCELERATOR; COLOR; ELECTROMAGNETIC RADIATION; ELECTROMAGNETIC SPECTRUM; FIELD, ELECTRIC; FIELD, MAGNETIC; ION; LASER; LIGHT; RAINBOW; REFRACTION; REFRACTION, INDEX OF; WAVE MOTION

Bibliography

EBERT, H., ed. *Physics Pocketbook* (Interscience, New York, 1967).

FEYNMAN, R. P.; LEIGHTON, R. B.; and SANDS, M. *The Feynman Lectures on Physics,* Vol. 1 (Addison-Wesley Reading, MA, 1963).

BRANT M. JOHNSON

DISPLACEMENT

Webster's New World Dictionary defines "displace" as "to move from its usual or proper place." In the context of stationary fluids (hydrostatics) and moving fluids (hydrodynamics), the term "displacement" has several interpretations.

In hydrostatics, fluid displacement refers to the amount of fluid moved aside when a solid body is partially or wholly submerged in it. In the presence of gravity, if the density of the solid body is larger than the density of the fluid, it will sink to the bottom of the fluid container, and the volume of fluid that is displaced will be equal to the volume of the body. If the density of the body is less than the density of the fluid, it will float on the surface, with only part of its volume submerged. In this case, the weight of the fluid that is displaced is equal to the weight of the body. These are Archimedes' well-known rules of buoyancy. Ships are frequently characterized by their displacement, which is the weight of the water they displace, measured in long tons (2,040 lb). A somewhat related usage occurs in characterizing internal combustion engines by their displacement. In this case, the term refers to the total volume of gas (usually measured in liters) that is displaced from the engine cylinders in a single cycle of the engine.

In hydrodynamics, the state of a moving fluid is often characterized by specifying its density, pressure, and velocity at each point in space, as a function of time. Since velocity has both a magnitude and a direction, the fluid motion is described by a velocity field $\mathbf{v}(\mathbf{r},t)$, where $\mathbf{v}$ is the (vector) velocity of the fluid at position $\mathbf{r}$ (measured relative to a fixed origin) at time t. The quantity $\mathbf{v}(\mathbf{r},t)\,\delta t$ is sometimes referred to as the Eulerian displacement of the fluid that was located at the point $\mathbf{r}$ at time t during the infinitesimal time δt; $\mathbf{v}(\mathbf{r},t)$ is called the Eulerian velocity field after Leonhard Euler, the Swiss mathematician who first wrote down the equations describing fluid flow. It is frequently useful to reformulate these equations in terms of Lagrangian coordinates. (Joseph Louis Lagrange was a French mathematician who greatly simplified the study of the motion of systems of particles under the action of known forces.) If we move along with the fluid at any point, we can imagine dividing it up into infinitesimal volume elements called parcels or (sometimes) particles. At some instant of time t_0 we specify the location of each fluid parcel by a position vector $\mathbf{r}_0$. Then the position of each parcel at some later time t is specified by its Lagrangian displacement $\mathbf{d}(\mathbf{r}_0, t)$, and the corresponding Lagrangian velocity field is the time rate of change of the Lagrangian displacement. This quantity specifies the vector velocity of each parcel of fluid (labeled by $\mathbf{r}_0$) at any time and differs from the Eulerian velocity field, which specifies the velocity of the different fluid parcels that are located at any fixed point $\mathbf{r}$ at different times.

See also: ARCHIMEDES' PRINCIPLE; BUOYANT FORCE; FLUID DYNAMICS; HYDRODYNAMICS

Bibliography

LAMB, H. *Hydrodynamics,* 6th ed. (Dover, New York, 1945).
TRITTON, D. J. *Physical Fluid Dynamics,* 2nd ed. (Clarendon Press, Oxford, Eng., 1988).

HAROLD S. ZAPOLSKY

DISPLACEMENT CURRENT

See CURRENT, DISPLACEMENT

DISTRIBUTION

See BOLTZMANN DISTRIBUTION; DISTRIBUTION FUNCTION; MAXWELL SPEED DISTRIBUTION

DISTRIBUTION FUNCTION

A distribution tells you how a quantity is shared among the individual members of a given set.

A simple example is the distribution of grades on an examination. Suppose thirty-nine students take an exam, on which scores could be anything from 0 to 100 percent. A histogram shows the number of students $N(n)$ who get the score n percent.

In this example, shown in Fig. 1, one person got a low score of 9 percent, one got 39 percent, and so on. The histogram shows the distribution of scores in this particular exam, or how scores are "shared out" among the students. The average score is 67.5 percent and the standard deviation, σ, is 14.5 percent.

Binomial Distribution

There is another kind of distribution that is often useful: one that shows how likely a particular number—a score, for example—is to occur. It is a graph of the probability of the number n occurring, $P(n)$, versus n. Consider the following example.

If you toss an unloaded die you get 1, 2, 3, 4, 5, or 6. If the die is truly unloaded the probability of getting any one of these numbers is the same as for any other number. If you throw the die just once you will get a 1, or 2, or one of the other possibilities, at random.

The probability of throwing one particular number—4, for example—is clearly 1/6, since the die can land in any one of six ways, with equal probability. Call this probability $p = 1/6$. If you throw a 4, call it a "success." The probability of throwing any number other than a 4 is clearly 5/6. Call this probability $q = 5/6$. Suppose you throw the die N times, where $N = 30$. How many times will you throw a 4? The most likely answer, $\tilde{n}$, is also the average number, $\bar{n}$, for this kind of distribution:

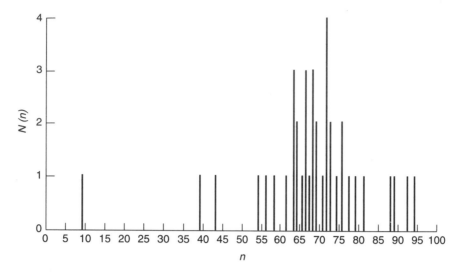

Figure 1 Number of students $N(n)$ scoring n percent.

$$\tilde{n} = \overline{n} = pN = \frac{30}{6} = 5.$$

But it is easy to believe that the die might land showing 4 a few times more or less than this. If the die is really not loaded it is unlikely that you will never throw a 4, not even once in thirty trials, but it is not totally impossible. In fact, if you throw the die thirty times, the possible number of successes could be 0, 1, 2, $\cdots$, 30. The binomial distribution tells the probability of obtaining n successes in N trials. It is given by

$$P(n) = \left(\frac{N!}{n!\,(N-n)!} \right) p^n\, q^{N-n},$$

where p is the probability of a success (e.g., throwing a 4) and q is the probability of "failure" (throwing any other number) in any single throw of the die.

The distribution function for the die experiment is shown in Fig. 2 for thirty trials. For example, the probability of seeing $n = 7$ fours in thirty trials is

$$P(7) = \left(\frac{30!}{7!\,(30-7)!} \right)\left(\frac{1}{6} \right)^7 \left(\frac{5}{6} \right)^{30-7} = 0.11.$$

Clearly the sum

$$\sum_{n=0}^{30} P(n) = 1,$$

since if you throw the die thirty times, you must see a 4 some number of times: never, once, twice, or other possibilities up to thirty.

Gaussian Disstribution

The Gaussian distribution is an approximation to the binomial distribution when N is very large, or when n is allowed to vary continuously. The average value of n is $\overline{n} = Np$ as before. The probability of seeing n successes in N trials is

$$P(n) = \left(\frac{1}{\sqrt{2\pi\sigma^2}} \right) \exp -\left(\frac{(n - \overline{n})^2}{2\sigma^2} \right),$$

where $\sigma^2 = Npq$.

The distribution is "normalized" to make

$$\int_0^N P(n)\ dn = 1,$$

showing that the probability of getting some value of n, between 0 and N, must add up to 1.

Figure 3 shows a typical Gaussian distribution for $N = 50$, $p = 0.4$, and $q = 0.6$ with $\overline{n} = Np = 20$ and $\sigma = 5$.

Gaussian distributions are very important in physics. Statistical mechanics is the name of the statistical theory that predicts the values of thermal properties (such as energy, pressure, or entropy) of

374

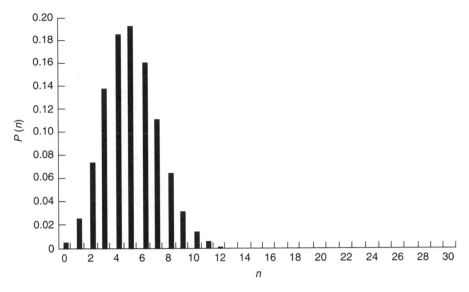

Figure 2 Binomial distribution.

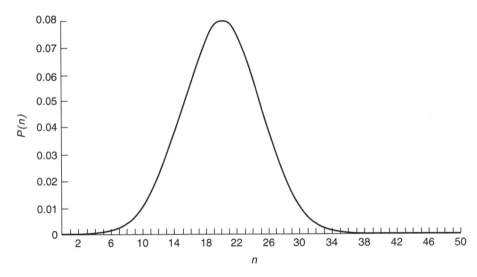

Figure 3 Gaussian distribution.

a system in thermal equilibrium. The premise of statistical mechanics is that in a system containing a large number of particles, the probability of the system being in any one of the states possible is represented by a Gaussian distribution, sharply peaked about the most probable, or average value. You are then overwhelmingly likely to see the system at, or close to, the average value when it is in thermal equilibrium. This theory allows calculation of the thermal quantities; for example, it predicts that an ideal *gas* obeys the equation of state $pV = nRT$

(p is pressure, V volume, n the number of moles present, R the gas constant, and T the temperature in kelvins).

Maxwell–Boltzmann Speed Distribution

In 1852 James Clerk Maxwell found a way to predict the speed distribution of molecules in an ideal gas. This has many applications, including understanding planetary atmospheres in the solar system.

Let M be the mass of a mole of gas at temperature T. The probability that a molecule has speed v is given by

$$P(v) = 4\pi \left(\frac{M}{2\pi RT}\right)^{3/2} v^2 \exp - \left(\frac{Mv^2}{2RT}\right).$$

The quantity $P(v)\,dv$ is the fraction of molecules whose speeds lie in the range v to $v + dv$.

Dirac Delta Distribution

This distribution is useful in many branches of physics, especially quantum mechanics. It is defined by

$$\delta(x) = 0$$

if $x \neq 0$, and

$$\int \delta(x)\,dx = 1,$$

where the domain of integration includes $x = 0$.

Equivalently, if $f(x)$ is an arbitrary function, continuous at $x = 0$,

$$\int f(x)\,\delta(x)\,dx = f(0).$$

The distribution $\delta(x)$ is clearly singular. Qualitatively, it may be visualized as 0 everywhere except at the origin, where it is a large spike. It is used when describing orthonormality relations between wave functions in quantum mechanics.

See also: BOLTZMANN DISTRIBUTION; IDEAL GAS LAW; MAXWELL–BOLTZMANN STATISTICS; MAXWELL SPEED DISTRIBUTION; PROBABILITY; QUANTUM MECHANICS; STATISTICAL MECHANICS

Bibliography

DAS, A., and MELISSINOS, A. *Quantum Mechanics* (Gordon and Breach, New York, 1986).
REIF, F. *Statistical Physics* (McGraw-Hill, New York, 1967).

ROBERT H. DICKERSON

GAYLE COOK

DONOR

The room-temperature electrical conductivity and other electrical properties of semiconductors are profoundly affected by the presence of very small amounts of certain impurities or crystal imperfections. The deliberate addition of small amounts of impurities that increase the electron concentration of the semiconductor is called donor doping, and the impurities are called donors.

The simplest and most important type of donor is substitutional; that is, the impurity atom occupies a position in the crystal lattice that in a perfect crystal would be occupied by one of the atoms of the semiconductor itself. Semiconductors from group IV of the periodic table (Si or Ge) are normally donor doped using the group V impurities P, As, or Sb.

A surprisingly accurate picture of donor doping can be obtained using a simple adaptation of the Bohr model of the hydrogen atom. Consider, for example, phosphorus-doped silicon. The substitutional P atom has one more valence electron and one more proton than the Si atom it replaces. Imagine these two charged particles as a hydrogen atom immersed in a medium of positive and negative charges that constitutes the rest of the crystal. The Bohr model then applies with two modifications: We allow for the polarizability of the medium by means of a dielectric constant ϵ (11.7 for Si), and for the effect of the periodicity of the crystal by assigning an effective mass m^* to the electron ($m^* \simeq 0.2\ m_e$ for Si). The ionization energy for the hydrogen-like atom then becomes

$$E_d = \frac{m^* e^4}{8(\epsilon \epsilon_0 h)^2} \simeq 3 \times 10^{-2}\ \text{eV}.$$

Here ϵ_0 is the permittivity of free space, h is Planck's constant, and e the elementary charge; the numerical value given is that for silicon. The experimental value is 4.5×10^{-2} eV.

The energy E is comparable with the mean room-temperature thermal energy $k_B T = 2.6 \times 10^{-2}$ eV, and therefore a large proportion of the donor electrons are ionized and free to carry electric current through the crystal. In contrast, the ionization energy of the intrinsic electrons (those bound to Si atoms) is 0.55 eV. Although the concentration of intrinsic electrons is much greater than that of donor electrons, far fewer of them are ionized, and the crystal can readily be doped so that the donor electrons dominate the electrical properties of the crystal.

It is useful to picture the electrons in a semiconductor on an energy-level diagram. The ionized donor electrons lie in the conduction band, and the donor states (the unionized ground states) are located a small distance E_d below the conduction band edge, in the forbidden gap of the semiconductor.

In real semiconductor crystals, small concentrations of acceptor impurities are inevitably present. A sufficient concentration of donor impurities serves to fill the acceptor states, so that the crystal exhibits the properties of a purely donor-doped semiconductor. A semiconductor that satisfies these conditions is called n type.

At low temperatures ($k_B T \ll E$) most donor electrons are not ionized and will not be present in the conduction band. Nevertheless, conduction can take place if donor atoms are sufficiently close together to allow significant overlap of their wave functions. In this case, electron transfer from impurity to impurity can take place, a process called impurity conduction or hopping conduction.

See also: ACCEPTOR; DOPING; ELECTRICAL CONDUCTIVITY; ENERGY LEVELS; IONIZATION; SEMICONDUCTOR

Bibliography

ASHCROFT, N. W., and MERMIN, N. D. *Solid-State Physics* (Saunders, Orlando, FL, 1976).

KITTEL, C. *Introduction to Solid-State Physics,* 7th ed. (Wiley, New York, 1996).

LAWRENCE S. LERNER

DOPING

Doping is the intentional, controlled introduction of specific types and amounts of impurity atoms into a host semiconductor crystal. It is the means by which the electrical properties of the semiconductor are controlled. The resistivity of the material can be manipulated over many orders of magnitude, from highly insulating to highly conducting. Electronic devices are made by combining different types and amounts of doping in various geometries. There are many techniques for growing semiconductor crystals and doping them. In some cases the dopant atoms are incorporated as the crystal is grown, in others they are introduced afterwards.

In an "intrinsic" semiconductor, with no dopants or other impurity atoms, the number of free electrons available to conduct current is low (conductivity, the inverse of resistivity, is proportional to the density of mobile charges). This is because nearly every electron in the solid is bound to an atom, and therefore immobile. The inner shell, or "core" electrons, of each atom is tightly bound in atomic orbitals. The outer shell, or "valence" electrons, is shared between two neighboring atoms to form a bond that holds the atoms together. For example, a silicon (Si) atom has fourteen electrons, arranged in shells around the nucleus. In a Si crystal, which has the diamond-type crystal structure, the ten innermost core electrons of each atom are in atomic orbitals that are virtually unaffected by the presence of nearby Si atoms. The four outermost valence electrons are shared with the four neighboring Si atoms, each of which also shares an electron. The two shared electrons between each atom form a strong covalent ("equally shared") bond. At room temperature, a small number of valence electrons have sufficient thermal energy to escape their covalent bonds, becoming free to move about the crystal and conduct current. However, the vast majority of electrons remain bound, so they cannot contribute to the conductivity.

Doping generally increases the number of mobile charges in the semiconductor. A dopant atom directly substitutes for one of the original atoms in the crystal. The dopant has a different number of valence electrons, usually one more or one less than the atom it replaces. For example, phosphorous (P) has five valence electrons, one more than Si. Since only four electrons are shared with the neighboring atoms in a Si crystal, the fifth electron is an extra, which does not fit into any of the bonds. It is still attracted to the P atom by the Coulomb force between the negatively charged electron and the positively charged atomic nucleus. However, the positive charge of the nucleus is screened by the negative charge of the surrounding core and valence electrons, and is further screened by the polarizability of the nearby atoms (i.e., the extra electron finds itself in a medium with a dielectric constant of the host), so the Coulomb force holding the extra electron is much weaker than the intrinsic covalent bonds. At room temperature it has sufficient thermal energy to escape this weak bond and move freely through the crystal. The result is a positively ionized dopant atom, and a free electron that can contribute to current conduction. Each dopant atom contributes one free electron, so the total number of free electrons is determined by the concentration of dopant atoms. This type of dopant, with an extra valence electron, is called a "donor" because it donates a freely moving electron to the crystal. Elements from the same column in the periodic table of elements have the same number of valence electrons, so the column V elements arsenic (As) and antimony (Sb) are also donors in the Si crystal.

Now consider a boron atom replacing a Si atom within the Si crystal. Boron has three valence electrons, one less than Si. Since four electrons are required to complete the covalent bonds with its four neighbors in the crystal, one of the bonds contains only one electron. The absence of a second electron in this incomplete bond is called a "hole." With the thermal energy available at room temperature, a valence electron from a neighboring Si–Si bond can easily jump into the incomplete bond, thus exchanging the locations of the complete and incomplete bonds. In this way the hole can move successively from one bond site to another, and freely move through the crystal. The hole is often thought of as a particle. Since a hole is the absence of an electron in a solid that is, on average, charge neutral, it carries a positive charge equal in magnitude to the charge of an electron. This type of dopant is called an acceptor, because it acquires (accepts) an electron from the crystal to fill its incomplete bond, thus introducing a positively charged hole that can move freely and contribute to current conduction. Other acceptor dopants for Si include gallium, indium, and aluminum, from column III of the periodic table of elements. Material that is doped with an acceptor impurity is called p type (short for positive type, because the carriers introduced by doping are positive holes); material doped with a donor impurity is called n type (short for negative type).

The density of free electrons and holes in an intrinsic semiconductor is usually very low at room temperature. Pure Si has about 5×10^{22} atoms per cubic centimeter (cm^{-3}), but only about 10^{10} cm^{-3} free electrons at room temperature (and the same density of free holes, because when an electron escapes from a valence bond a hole is also created). Therefore a small addition of dopant atoms can have a large effect on conductivity. For example, adding one donor atom per *billion* Si atoms (a donor density of 5×10^{13} cm^{-3}) increases the free electron density (and conductivity) by a factor of over 1,000. Similarly, doping with an acceptor increases the number of free holes, which also increases the conductivity. Typical doping densities fall in the range of about 10^{16} to 10^{18} cm^{-3}. The amount of doping is an important quantity to measure and control. The Hall effect measurement is routinely used to determine both the type (electron or hole) and density of free charges resulting from doping.

Electronic devices are made by combining n-type and p-type material as well as undoped material. One of the most basic configurations is the p-n junction, which is formed by doping one region p-type and an adjacent region n-type. There are various ways of doing this. For instance, n-type material can be grown first, then the top part of this material can be converted to p-type material by ion implantation or some other technique, resulting in a p-type layer on top of an n-type material. A single p-n junction with electrical connections to each side forms a diode. A bipolar transistor contains two regions of one type, separated by a thin region of the opposite type (either n-p-n or p-n-p).

See also: ACCEPTOR; ATOM; CHARGE, ELECTRONIC; CONDUCTION; CONDUCTOR; COULOMB'S LAW; COVALENT BOND; CRYSTAL; CRYSTAL STRUCTURE; DONOR; ELECTRICAL CONDUCTIVITY; ELECTRICAL RESISTANCE; ELECTRICAL RESISTIVITY; ELECTRON, CONDUCTION; ELECTRONICS; HOLES IN SOLIDS; INSULATOR; SEMICONDUCTOR; TRANSISTOR

Bibliography

PIERRET, R. F. *Semiconductor Fundamentals*, 2nd ed. (Addison-Wesley, Reading, MA, 1988).

STREETMAN, B. G. *Solid-State Electronic Devices*, 4th ed. (Prentice Hall, Englewood Cliffs, NJ, 1995).

PAUL G. SNYDER

DOPPLER EFFECT

The Doppler effect is the change in the observed frequency of waves caused by a relative motion between the source of the wave and the observer. This change of frequency occurs for all waves, including water waves (mechanical effect), sound or pressure waves (acoustical effect), and light (optical Doppler effect).

The classical example of the acoustical Doppler effect is the variation of sound from the whistle of a moving train. Today, a more commonly encountered manifestation is the change of pitch of noise made by a passing car or truck on a highway. The pitch is perceived as higher when the car is approaching and as lower when the car speeds away. Bats depend on the Doppler effect to echolocate insects for their meal, as well as each other, by emitting high-pitched sound waves usually not audible to humans. Doppler velocimetry is a method of reconstructing a velocity profile in a fluid flow by measuring the shift of a signal emitted and received by transducer crystals. A remarkable application of this method in medicine is diagnosing blood-flow irregularities *in vivo* through an ultrasonic probe inserted via a catheter into an artery. Laser doppler anemometry is another technique for obtaining velocity in a fluid; a beam of laser light is scattered by particles moving with the fluid, so the frequency shift can then be determined using interferometry. Police car speed detectors exploit the Doppler effect in electromagnetic waves in several microwave frequency bands. The newest police detectors use light emitted by lasers. The optical Doppler effect is important in astronomy, allowing scientists to deduce the speed with which a galaxy is receding from the earth based on the shift of its light toward lower frequencies (the redshift). By observing spectra of various galaxies, one can measure the rate at which the universe is expanding, which is expressed through the Hubble constant. The Doppler shift also can be used to determine rates of rotation for celestial bodies such as binary pulsars.

The phenomenon was first described and analyzed by Johann Christian Doppler, an Austrian mathematician and physicist who worked in Vienna and Prague. In his 1842 article "On Color Light of Double Stars and Some Other Stars in the Heavens," Doppler includes the correct formula for the motion of the source or observer along the line between them. The application of the principle to the titular double stars was flawed, however. The first experimental verification of the acoustical Doppler effect was carried out in Utrecht in 1845 by Buys Ballot, who used a moving, open train car carrying a group of trumpeters pulled by a locomotive. The first laboratory confirmation of the optical effect is due to A. Bélopolsky, who devised in 1900

an ingeneous apparatus with multiple rotating mirrors.

Mechanical and Acoustical Waves

One can readily observe the Doppler effect for waves in water. If the water is stationary, a mechanical disturbance confined to a point, such as a penny dropped into the water container, produces a circular wave emanating from the point. A series of coins dropped at the same point will result in a series of concentric circular waves (Fig. 1a). When the source is moving, however, for example when the coins are dropped from a hand moving across the container, the resulting circular waves are no longer concentric (Fig. 1b). Figure 1b shows that the wavelength varies depending on the direction.

The same pattern of Fig. 1b will be created by a stationary source and a uniformly moving medium (e.g., when the coins are dropped from a fixed location in the middle of a wide river). Hence, for a stationary observer, the wave pattern depends on the relative motion of the source and the medium rather than on each independently.

More generally, for mechanical and acoustical (pressure) waves, the observed frequency depends on the motion of the source (emitter) and the observer (listener) relative to the medium. For a source and an observer that are stationary relative to the medium, the frequency f of the wave depends solely on the source, while the wave speed u is a property of the medium. The wavelength in this case depends on both the source and the medium and is given by $\lambda = u/f$, or $\lambda = uT$, where $T = 1/f$ is the period.

If the source of frequency f_s is moving with the constant speed u_s relative to the medium, and the observer is stationary, the wave moving in the same direction as the source is shortened and its wave-

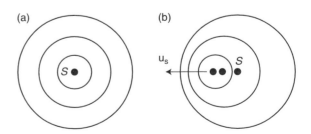

Figure 1   (a) Stationary source; (b) moving source.

(a)

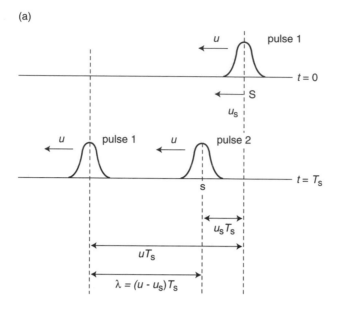

(b)

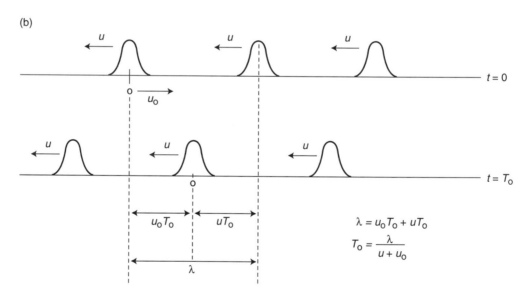

Figure 2 (a) Moving source, s; (b) moving observer, o.

length is given by $\lambda = (u - u_s)/f_s$ (Fig. 2a). The wave crests are crowding in this direction. If the observer is moving through the medium with the speed u_o toward the source, the successive wave fronts will be passing him at a faster rate, namely with the frequency $f_o = (u + u_o)/\lambda$, as shown in Fig. 2b. Combining these two effects gives

$$f_o = \frac{f_s(u + u_o)}{u - u_s}. \qquad (1)$$

In Eq. (1), u_o is taken to be positive when the observer is moving toward the source, and u_s is taken to be positive when the source is moving toward the observer. It follows that the observed frequency is larger than the source frequency ($f_o > f_s$) when $u_o + u_s > 0$; smaller than the source frequency ($f_o < f_s$) when $u_o + u_s < 0$; and equal to the source frequency ($f_o = f_s$) when $u_o = u_s = 0$ or $u_o = -u_s$, that is, when neither the observer nor the source are moving relative to the medium, or they are moving with the same speed and in the same direction.

Equation (1) is valid only when the source and the object are moving along a straight line between them. If this is not the case, but both the source and the observer are moving along straight lines (relative to the medium), and the distance between the source and the observer is large compared with the distance either moves during one (source) period T_s, the Doppler shift will depend on the components of the source and observer velocities along the direction from one to the other, rather than on u_s and u_o as in Eq. (1). Still further modifications of Eq. (1) are needed in the general case of motion along arbitrary paths with variable speeds.

Electromagnetic Waves and Light

The Doppler shift for light is a fundamentally different effect from its acoustical counterpart and is characteristic of the relativistic nature of light. According to Albert Einstein's theory of relativity, electromagnetic waves, such as light, propagate with a constant velocity c (approximately 3×10^8 m/s) as observed in any inertial frame of reference. Hence, there is no preferred frame of reference or material medium, and the Doppler effect for light depends only on the relative velocity between the source and the observer.

Derivation of the expression for the Doppler shift is based on the relativistic effect of time dilation. If the source and the observer are moving along a straight line between them, and u is their relative speed (taken to be positive for approach), then one obtains

$$f_o = \frac{f_s(1 + u/c)^{1/2}}{(1 - u/c)^{1/2}}. \tag{2}$$

When u/c is small (e.g., in the case of radar waves reflecting off a car speeding along a straight stretch of a highway), Eq. (2) can be approximated with good results by

$$f_o \approx f_s(1 + u/c). \tag{3}$$

Equation (3) can be used to calculate the speeds with which stars are moving toward or away from the earth, using the shifts of their light. Since, in the limit of u/c approaching 0, relativistic formulae should reduce to their classical counterparts, Eq. (3)

should be compatible with Eq. (1). This can be verified by replacing u with c in Eq. (1) and developing a series expansion for the resulting expression in terms of u/c, where $u = u_o + u_s$. The first term of the expansion series is the right-hand side of Eq. (3).

As in the case of mechanical waves, correction must be introduced into Eq. (2) if the source and the observer are not moving along a straight line between them. If θ_o is the angle, measured in the frame of the observer, between a straight line passing from the observer to the source and the vector difference of their velocities, then

$$f_o = \frac{f_s(1 - u^2/c^2)^{1/2}}{1 - (u/c) \cos \theta_o}. \tag{4}$$

For $\theta_o = 0$, Eq. (4) is equivalent to Eq. (2). In the general case, Eq. (4) can be understood to be composed of two factors. The observed frequency is proportional to $f_s(1 - u^2/c^2)^{1/2}$, which is a purely relativistic factor independent of the direction of motion. This is called the transverse Doppler effect. At the same time, f_o is inversely proportional to $[1 - (u/c) \cos \theta_o]$, which depends on the direction of the relative velocity and can be derived based on a classical argument, such as that used for the acoustical effect. This second factor represents the radial Doppler effect.

If $\theta_o = 90°$, Eq. (4) reduces to a relativistic formula for the transverse effect:

$$f_o = f_s(1 - u^2/c^2)^{1/2}. \tag{5}$$

Equation (5) is essentially the time dilation formula as given in Lorentz transformations. Based on the predictions of classical mechanics, there should be no frequency shift in this case since the radial component of the relative velocity is zero. By measuring the transverse Doppler effect, one can therefore verify the predictions of the special theory of relativity. Such an experiment was performed by Herbert Ives and G. R. Stilwell in 1938 using a beam of hydrogen ions and measuring the wavelength of the emitted $H_β$ spectral line. The experiment confirmed precisely the predicted relativistic behavior.

See also: ELECTROMAGNETIC WAVE; LORENTZ TRANSFORMATION; REDSHIFT; RELATIVITY, SPECIAL THEORY OF; SOUND; UNIVERSE, EXPANSION OF; WAVELENGTH; WAVE SPEED

Bibliography

BELOPOLSKY, A. "On an Apparatus for the Laboratory Demonstration of the Doppler–Fizeau Principle." *Astrophys. J.* **13,** 15 (1901).

DRAIN, L. E. *The Laser Doppler Technique* (Wiley, New York, 1980).

GILLIPSIE, C., ed. *Dictionary of Scientific Biography,* Vol. IV (Scribner, New York, 1971).

HALLIDAY, D., and RESNICK, R. *Fundamentals of Physics,* 4th ed. (Wiley, New York, 1992).

IVES, H. E., and STILWELL, G. R. "An Experimental Study of the Rate of a Moving Atomic Clock." *J. Opt. Soc. Am.* **31,** 369 (1941).

RADIN, S., and FOLK, R. *Physics for Scientists and Engineers* (Prentice Hall, Englewood Cliffs, NJ, 1982).

ANDRZEJ HERCZYŃSKI

DOUBLE-SLIT EXPERIMENT

Two sets of light waves can be made to cross each other with neither producing any modification in the amplitude, frequency, polarization, or phase of the other. In the region of crossing, the resultant electromagnetic field at any point is given by the sum of those due to each wave separately. This is known as the principle of superposition and was first clearly stated by the English physician, architect, Egyptologist, and physicist Thomas Young in 1802. The general phenomenon whereby two or more waves overlap to produce a resultant wave is called interference. The resultant intensity in the overlap region can be either greater or less than the separate intensities, correspondingly referred to as constructive and destructive interference. However, in order for interference effects in the overlap region to be stationary and thereby observable, it is necessary for the two light sources to be coherent, that is, to have precisely and persistently the same frequency and phase difference.

In a conventional optical light source, energy is pumped into an atomic sample by collisions, and the light is produced by spontaneous emission from individual atoms within a coherence time of approximately 10^{-9} to 10^{-8} s. This leads to sudden changes of phase on this time scale and precludes persistent coherence between separate sources of this type. This "granular" nature of radiation was demon-strated by Albert Einstein to be required by the atomic properties of matter and a necessary replacement for the earlier continuum interpretation of James Clerk Maxwell. This limitation in coherence time is in contrast to modern laser devices, where energy is pumped into ensembles of atoms that can then be triggered into coherent stimulated emission by the presence of other photons of the proper frequency. Electrical charges can also be driven in phase at lower frequencies in a radio transmitter, and coherent sources of radio frequency radiation can be obtained by connecting two spatially separated antennas to the same oscillator. Although separate, independent, coherent sources of visible light did not exist until the development of modern lasers, in 1801 Thomas Young observed interference with visible light by an ingenious method that effectively created spatially separated sources from a single wave front.

Young's experiment can be understood in terms of a principle advanced in 1678 by the Dutch physicist Christiaan Huygens, who proposed that each point on a wave front can be regarded as a new source of waves. This leads to a geometrical construction that predicts the position of a given wave front at any time in the future as the surface of tangency of the common envelope of these secondary wavelets. Young allowed sunlight to fall on a pinhole punched in a screen, thus creating a point source of light by diffraction. The emerging light was allowed to spread and to fall upon a second screen through which two pinholes with a small separation had been punched. These two spherical waves also spread and overlapped on the downstream side of the second screen. The overlapping region was not a simple area of intensified light but was observed to exhibit a striped pattern alternating light and darkness, analogous to the phenomenon of sound beats that Young had also studied.

Young's experiment was especially convincing because he was able to use the separation between the holes d and the angle θ, at which the mth constructive interference fringe occurred, to deduce the wavelength of the light λ, using $d \sin\theta = m\lambda$ (Fig. 1). Young obtained a value for the average wavelength of sunlight that is quite close to the presently accepted value. The method can also be inverted to measure an unknown value for d given a known value for λ. Subsequent to Young's measurement it has been found convenient to replace the pinholes with narrow slits and to use a monochromatic light source. This yields two dimensional cylindrical wave

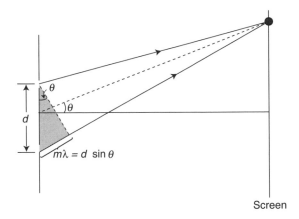

Figure 1 Set-up for the double-slit experiment.

fronts that produce light and dark lines at the positions of constructive and destructive interference. The transmission can also be enhanced by using a large number of equally spaced slits to form a diffraction grating. Similar interference patterns can be readily observed in water waves, for example, in the region of overlap of the ripples that spread from the impacts of two adjacent raindrops striking the surface of a body of water.

Young's work culminated more than a century of controversy as to whether light consisted of particles or waves. Objections were initially raised that the pattern might be an artifact caused by the edges of the slits, but these were resolved when the work was confirmed by Augustin Fresnel and others using alternative devices such as the Fresnel biprism, the Fresnel double-mirror, and Lloyd's mirror. Young's work so dramatically demonstrated the wave properties of light that this view was accepted for more than 100 years, until the advent of the modern quantum theory. Although Young's work seemed to refute Newton's corpuscular theory of light, Newton had partially foreseen this modern view (often described as wave–particle duality), having proposed that light is a stream of material particles capable of

setting up vibrations in the quintessential ether. It is now known that light (and all other forms of electromagnetic radiation) consists of photons that propagate in a wavelike fashion with a single characteristic speed, but which behave like streams of particlelike concentrations of energy when interacting with matter in the processes of emission and absorption. Thus light is neither a "wave" nor a "particle" but can exhibit either of these macroscopic attributes under appropriately limiting conditions.

See also: COHERENCE; EINSTEIN, ALBERT; FRESNEL, AUGUSTIN-JEAN; HUYGENS, CHRISTIAAN; INTERFERENCE; LASER; LIGHT; LIGHT, ELECTROMAGNETIC THEORY OF; LIGHT, WAVE THEORY OF; MAXWELL, JAMES CLERK; NEWTON, ISAAC; PHOTON; POLARIZATION; POLARIZED LIGHT; WAVE–PARTICLE DUALITY; YOUNG, THOMAS

Bibliography

HECHT, E. *Physics* (Brooks/Cole, Pacific Grove, CA, 1994).
JENKINS, F. A., and WHITE, H. E. *Fundamentals of Optics,* 4th ed. (McGraw-Hill, New York, 1976).

LORENZO J. CURTIS

DRIFT SPEED

See ELECTRON, DRIFT SPEED OF

DUALITY

See WAVE–PARTICLE DUALITY

383